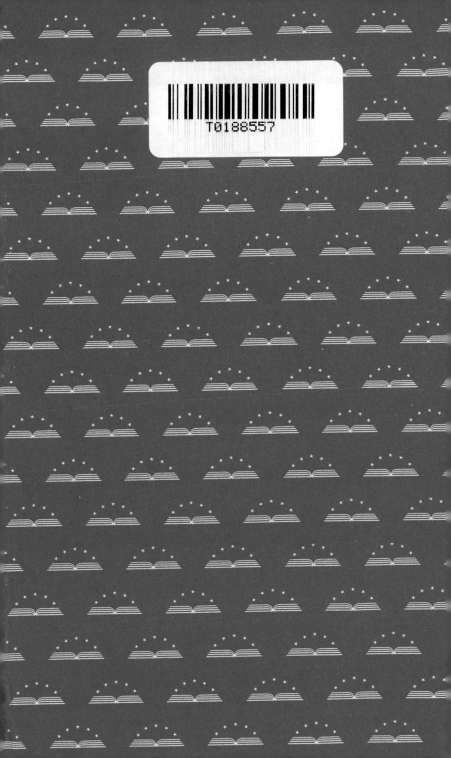

Library of America, a nonprofit organization,
champions our nation's cultural heritage
by publishing America's greatest writing in
authoritative new editions and providing resources
for readers to explore this rich, living legacy.

WENDELL BERRY

WENDELL BERRY

ESSAYS 1969–1990

INCLUDING
The Unsettling of America

AND SELECTIONS FROM
The Long-Legged House
The Hidden Wound
A Continuous Harmony
Recollected Essays
The Gift of Good Land
Standing by Words
Home Economics
What Are People For?

Jack Shoemaker, *editor*

THE LIBRARY OF AMERICA

WENDELL BERRY: ESSAYS 1969–1990
Volume compilation, notes, and chronology copyright © 2019 by
Literary Classics of the United States, Inc., New York, N.Y.
All rights reserved.
No part of this book may be reproduced in any manner whatsoever without
the permission of the publisher, except in the case of brief
quotations embodied in critical articles and reviews.

Published in the United States by Library of America.
Visit our website at www.loa.org.

All texts reprinted by arrangement with Counterpoint Press.

This paper exceeds the requirements of
ANSI/NISO Z39.48–1992 (Permanence of Paper).

Distributed to the trade in the United States
by Penguin Random House Inc.
and in Canada by Penguin Random House Canada Ltd.

Library of Congress Control Number: 2018952818
ISBN 978–1–59853–606–5

Second Printing
The Library of America—316

Manufactured in the United States of America

Wendell Berry: Essays 1969–1990
is kept in print with a gift from

WALTER E. ROBB
with the deepest appreciation and gratitude
for the words, life, and inspiration of Wendell Berry

to the Guardians of American Letters Fund,
established by Library of America
to ensure that every volume in the series
will be permanently available.

Wendell Berry: Essays 1969–1990
is published with support from

THE GOULD FAMILY FOUNDATION

WALTER E. ROBB

To Wes Jackson and David Kline

and to the memory of Gene Logsdon and Maury Telleen

Contents

The Rise

WE PUT THE canoe in about six miles up the Kentucky River from my house. There, at the mouth of Drennon Creek, is a little colony of summer camps. We knew we could get down to the water there with some ease. And it proved easier than we expected. The river was up maybe twenty feet, and we found a path slating down the grassy slope in front of one of the cabins. It went right into the water, as perfect for launching the canoe and getting in as if it had been worn there by canoeists.

To me that, more than anything else, is the excitement of a rise: the unexpectedness, always, of the change it makes. What was difficult becomes easy. What was easy becomes difficult. By water, what was distant becomes near. By land, what was near becomes distant. At the water line, when a rise is on, the world is changing. There is an irresistible sense of adventure in the difference. Once the river is out of its banks, a vertical few inches of rise may widen the surface by many feet over the bottomland. A sizable lagoon will appear in the middle of a cornfield. A drain in a pasture will become a canal. Stands of beech and oak will take on the look of a cypress swamp. There is something Venetian about it. There is a strange excitement in going in a boat where one would ordinarily go on foot—or where, ordinarily, birds would be flying. And so the first excitement of our trip was that little path; where it might go in a time of low water was unimaginable. Now it went down to the river.

Because of the offset in the shore at the creek mouth, there was a large eddy turning in the river where we put in, and we began our drift downstream by drifting upstream. We went up inside the row of shore trees, whose tops now waved in the current, until we found an opening among the branches, and then turned out along the channel. The current took us. We were still settling ourselves as if in preparation, but our starting place was already diminishing behind us.

There is something ominously like life in that. One would always like to settle oneself, get braced, say "Now I am going to begin"—and then begin. But as the necessary quiet seems about to descend, a hand is felt at one's back, shoving. And that is the way with the river when a current is running: once the connection with the shore is broken, the journey has begun.

We were, of course, already at work with the paddles. But we were ahead of ourselves. I think that no matter how deliberately one moved from the shore into the sudden fluid violence of a river on the rise, there would be bound to be several uneasy minutes of transition. It is another world, which means that one's senses and reflexes must begin to live another kind of life. Sounds and movements that from the standpoint of the shore might have come to seem even familiar now make a new urgent demand on the attention. There is everything to get used to, from a wholly new perspective. And from the outset one has the currents to deal with.

It is easy to think, before one has ever tried it, that nothing could be easier than to drift down the river in a canoe on a strong current. That is because when one thinks of a river one is apt to think of *one* thing—a great singular flowing that one puts one's boat into and lets go. But it is not like that at all, not after the water is up and the current swift. It is not one current, but a braiding together of several, some going at different speeds, some even in different directions. Of course, one *could* just let go, let the boat be taken into the continuous mat of drift—leaves, logs, whole trees, cornstalks, cans, bottles, and such—in the channel, and turn and twist in the eddies there. But one does not have to do that long in order to sense the helplessness of a light canoe when it is sideways to the current. It is out of control then, and endangered. Stuck in the mat of drift, it can't be maneuvered. It would turn over easily; one senses that by a sort of ache in the nerves, the way bad footing is sensed. And so we stayed busy, keeping the canoe between the line of half-submerged shore trees and the line of drift that marked the channel. We weren't trying to hurry—the currents were carrying us as fast as we wanted to go—but it took considerable labor just to keep straight. It was like riding a spirited horse not fully bridle-wise: We kept our direction *by intention*; there could be no dependence on habit or inertia; when our minds wandered the river took over and turned us according to inclinations of its own. It bore us like a consciousness, acutely wakeful, filling perfectly the lapses in our own.

But we did grow used to it, and accepted our being on it as one of the probabilities, and began to take the mechanics of it for granted. The necessary sixth sense had come to us, and we began to notice more than we had to.

There is an exhilaration in being *accustomed* to a boat on dangerous water. It is as though into one's consciousness of the dark violence of the depths at one's feet there rises the idea of the boat, the buoyancy of it. It is always with a sort of triumph that the boat is realized—that it goes *on top of the water*, between breathing and drowning. It is an ancient-feeling triumph; it must have been one of the first ecstasies. The analogy of riding a spirited horse is fairly satisfactory; it is mastery over something resistant—a buoyancy that is not natural and inert like that of a log, but desired and vital and to one's credit. Once the boat has fully entered the consciousness it becomes an intimate extension of the self; one feels as competently amphibious as a duck, whose feet are paddles. And once we felt accustomed and secure in the boat, the day and the river began to come clear to us.

It was a gray, cold Sunday in the middle of December. In the woods on the north slopes above us we could see the black trunks and branches just faintly traced with snow, which gave them a silvery, delicate look—the look of impossibly fine handwork that nature sometimes has. And they looked cold. The wind was coming straight up the river into our faces. But we were dressed warmly, and the wind didn't matter much, at least not yet. The force that mattered, that surrounded us, and inundated us with its sounds, and pulled at or shook or carried everything around us, was the river.

To one standing on the bank, floodwater will seem to be flowing at a terrific rate. People who are not used to it will commonly believe it is going three or four times as fast as it really is. It is so all of a piece, and so continuous. To one drifting along in a boat this exaggerated impression of speed does not occur; one is going the same speed as the river then and is not fooled. In the Kentucky when the water is high a current of four or five miles an hour is about usual, I would say, and there are times in a canoe that make that seem plenty fast.

What the canoeist gets, instead of an impression of the river's speed, is an impression of its power. Or, more exactly, an impression of the *voluminousness* of its power. The sense of the volume alone has come to me when, swimming in the summer-time, I have submerged mouth and nose so that the plane of the water spread away from the lower eyelid; the awareness of its bigness that comes then is almost intolerable; one feels how falsely assuring it is to look down on the river, as we usually do. The sense of the power of it came to me one day in my boyhood when I attempted to swim ashore in a swift current, pulling an overturned rowboat. To check the downstream course of the boat I tried grabbing hold of the partly submerged willows along the shore with my free hand, and was repeatedly pulled under as the willows bent, and then torn loose. My arms stretched between the boat and the willow branch might have been sewing threads for all the holding they could do against that current. It was the first time I realized that there could be circumstances in which my life would count for nothing, absolutely nothing—and I have never needed to learn that again.

Sitting in a canoe, riding the back of the flooding river as it flows down into a bend, and turns, the currents racing and crashing among the trees along the outside shore, and flows on, one senses the volume and the power all together. The sophistications of our age do not mitigate the impression. To some degree it remains unimaginable, as is suggested by the memory's recurrent failure to hold on to it. It can never be remembered as wild as it is, and so each new experience of it bears some of the shock of surprise. It would take the mind of a god to watch it as it changes and not be surprised.

These long views that one gets coming down it show it to move majestically. It is stately. It has something of the stylized grandeur and awesomeness of royalty in a Sophoclean tragedy. But as one watches, there emanates from it, too, an insinuation of darkness, implacability, horror. And the nearer look tends to confirm this. Contained and borne in the singular large movement are hundreds of smaller ones: eddies and whirlpools, turnings this way and that, cross-currents rushing out from the shores into the channel. One must simplify it in order to speak of it. One probably simplifies it in some way in order to look at it.

There is something deeply horrifying about it, roused. Not, I think, because it is inhuman, alien to us; some of us at least must feel a kinship with it, or we would not loiter around it for pleasure. The horror must come from our sense that, so long as it remains what it is, it is not subject. To say that it is indifferent would be wrong. That would imply a malevolence, as if it could be aware of us if only it wanted to. It is more remote from our concerns than indifference. It is serenely and silently not subject—to us or to anything else except the other natural forces that are also beyond our control. And it is apt to stand for and represent to us all in nature and in the universe that is not subject. That is its horror. We can make use of it. We can ride on its back in boats. But it won't stop to let us get on and off. It is not a passenger train. And if we make a mistake, or risk ourselves too far to it, why then it will suffer a little wrinkle on its surface, and go on as before.

That horror is never fully revealed, but only sensed piecemeal in events, all different, all shaking, yet all together falling short of the full revelation. The next will be as unexpected as the last.

A man I knew in my boyhood capsized his motorboat several miles upriver from here. It was winter. The river was high and swift. It was already nightfall. The river carried him a long way before he drowned. Farmers sitting in their houses in the bottoms heard his cries passing down in the darkness, and failed to know what to make of them. It is hard to imagine what they could have done if they had known.

I can't believe that anyone who has heard that story will ever forget it. Over the years it has been as immediate to me as if I had seen it all—almost as if I had *known* it all: the capsized man aching and then numb in the cold water, clinging to some drift log in the channel, and calling, seeing the house lights appear far off across the bottoms and dwindle behind him, the awful power of the flood and his hopelessness in it finally dawning on him—it is amazingly real; it is happening to him. And the families in their lighted warm kitchens, eating supper maybe, when the tiny desperate outcry comes to them out of the darkness, and they look up at the window, and then at each other.

"Shh! Listen! What was that?"

"By God, it sounded like somebody hollering out there on the river."

"But it *can't* be."

But it makes them uneasy. Whether or not there *is* somebody out there, the possibility that there *may* be reminds them of their lot; they never know what may be going by them in the darkness. And they think of the river, so dark and cold.

The history of these marginal places is in part the history of drownings—of fishermen, swimmers, men fallen from boats. And there is the talk, the memory, the inescapable *feeling* of dragging for the bodies—that terrible fishing for dead men lost deep in the currents, carried downstream sometimes for miles.

Common to river mentality, too, are the imaginings: step-offs, undertows, divers tangled in sunken treetops, fishermen hooked on their own lines.

And yet it fascinates. Sometimes it draws the most fearful to it. Men must test themselves against it. Its mystery must be forever tampered with. There is a story told here of a strong big boy who tried unsuccessfully to cross the river by walking on the bottom, carrying an iron kettle over his head for a diving bell. And another story tells of a young man who, instead of walking under it, thought he would walk *on* it, with the help of a gallon jug tied to each foot. The miracle failing, of course, the jugs held his feet up, and his head under, until somebody obliged him by pulling him out. His pride, like Icarus', was transformed neatly into his fall—the work of a river god surely, *hybris* being as dangerous in Henry County as anywhere else.

To sense fully the power and the mystery of it, the eye must be close to it, near to level with the surface. I think that is the revelation of George Caleb Bingham's painting of trappers on the Missouri. The painter's eye, there, is very near the water, and so he sees the river as the trappers see it from their dugout —all the space coming down to that vast level. One feels the force, the aliveness, of the water under the boat, close under the feet of the men. And there they are, isolated in the midst of it, with their box of cargo and their pet fox—men and boat and box and animal all so strangely and poignantly coherent on the wild plain of the water, a sort of island.

But impressive as the sights may be, the river's wildness is most awesomely announced to the ear. Along the channel, the area of the most concentrated and the freest energy, there is silence. It is at the shore line, where obstructions are, that the currents find their voices. The water divides around the trunks of the trees, and sucks and slurs as it closes together again.

Trunks and branches are ridden down to the surface, or suddenly caught by the rising water, and the current pours over them in a waterfall. And the weaker trees throb and vibrate in the flow, their naked branches clashing and rattling. It is a storm of sound, changing as the shores change, increasing and diminishing, but never ceasing. And between these two storm lines of commotion there is that silence of the middle, as though the quiet of the deep flowing rises into the air. Once it is recognized, listened to, that silence has the force of a voice.

After we had come down a mile or two we passed the house of a fisherman. His children were standing on top of the bank, high at that place, waiting for him to come in off the river. And on down we met the fisherman himself, working his way home among the nets he had placed in the quieter water inside the rows of shore trees. We spoke and passed, and were soon out of sight of each other. But seeing him there changed the aspect of the river for us, as meeting an Arab on a camel might change the aspect of the desert. Problematic and strange as it seemed to us, here was a man who made a daily thing of it, and went to it as another man would go to an office. That race of violent water, which would hang flowing among the treetops only three or four days, had become familiar country to him, and he sunk his nets in it with more assurance than men sink wells in the earth. And so the flood bore a pattern of his making, and he went his set way on it.

And he was not the only creature who had made an unexpected familiarity with the risen water. Where the drift had matted in the shore eddies, or caught against trees in the current, the cardinals and chickadees and titmice foraged as confidently as on dry land. The rise was an opportunity for them, turning up edibles they would have found with more difficulty otherwise. The cardinals were more irresistibly brilliant than ever, kindling in the black-wet drift in the cold wind. The sight of them warmed us.

The Kentucky is a river of steep high banks, nearly everywhere thickly grown with willows and water maples and elms and sycamores. Boating on it in the summer, one is enclosed in a river-world, moving as though deep inside the country. One sees only the river, the high walls of foliage along the banks, the hilltops that rise over the trees first on one side and then the other. And that is one of the delights of this river. But one of the delights of being out on a winter rise is in seeing the country, and in seeing it from a vantage point that one does not usually see it from. The rise, that Sunday, had lifted us to the bank tops and higher, and through the naked trees we could look out across the bottoms. It was maybe like boating on a canal in Holland, though we had never done that. We could see the picked cornfields, their blanched yellow seeming even on that cloudy day to give off a light. We could see the winter grain spiking green over the summer's tobacco patches, the thickly wooded hollows and slews, the backs of houses and farm buildings usually seen only by the people who live there.

Once, before the man-made floods of modern times, and before the automobile, all the river country turned toward the river. In those days our trip would probably have had more witnesses than it did. We might have been waved to from house windows, and from barn doors. But now the country has turned toward the roads, and we had what has come to be the back view of it. We went by mostly in secret. Only one of the fine old river houses is left on this side of the river in the six miles of our trip, and it is abandoned and weathering out; the floods have been in it too many times in the last thirty-five years, and it is too hard to get back to from the road. We went by its blank windows as the last settlers going west passed the hollow eyes of the skulls of their predecessors' oxen.

The living houses are all out along the edges of the valley floor, where the roads are. And now that all the crops had been gathered out of the bottoms, men's attention had mostly turned away. The land along the river had taken on a wildness that in the summer it would not have. We saw a pair of red-tailed hawks circling low and unafraid, more surprised to see us than we were to see them.

Where the river was over the banks a stretch of comparatively quiet water lay between the trees on the bank top and the new shore line. After a while, weary of the currents, we turned into one of these. As we made our way past the treetops and approached the shore we flushed a bobwhite out of a brush pile near the water and saw it fly off downstream. It seemed strange to see only one. But we didn't have to wait long for an explanation, for presently we saw the dogs, and then the hunters coming over the horizon with their guns. We knew where their bird had gone, but we didn't wait to tell them.

These men come out from the cities now that the hunting season is open. They walk in these foreign places, unknown to them for most of the year, looking for something to kill. They wear and carry many dollars' worth of equipment, and go to a great deal of trouble, in order to kill some small creatures that they would never trouble to know alive, and that means little to them once they have killed it. If those we saw had killed the bobwhite they would no doubt have felt all their expense and effort justified, and would have thought themselves more manly than before. It reminds one of the extraordinary trouble and expense governments go to in order to kill men—and consider it justified or not, according to the "kill ratio." The diggers among our artifacts will find us to have been honorable lovers of death, having been willing to pay exorbitantly for it. How much better, we thought, to have come upon the *life* of the bird as we did, moving peaceably among the lives of the country that showed themselves to us because we were peaceable, than to have tramped fixedly, half oblivious, for miles in order to come at its death.

We left the hunters behind and went down past a green grainfield where cattle were grazing and drinking at the waterside. They were not disturbed that the river had come up over part of their pasture, no more troubled by the height of today's shore line than they were by the height of yesterday's. To them, no matter how high it was, so long as the ground was higher it was as ordinary as a summer pond. Surely the creatures of the fifth day of Creation accepted those of the sixth with equanimity, as though they had always been there. Eternity is always present in the animal mind; only men deal in beginnings and ends. It is probably lucky for man that he was created last. He would have got too excited and upset over all the change.

Two mallards flew up ahead of us and turned downriver into the wind. They had been feeding in the flooded corn rows, reminding us what a godsend the high water must be for ducks. The valley is suddenly full of little coves and havens, so that they can scatter out and feed safer and more hidden, and more abundantly too, than they usually can, never having to leave the river for such delicacies as the shattered corn left by the pickers. A picked cornfield under a few inches of water must be the duck Utopia—Utopia being, I assume, more often achieved by ducks than by men.

If one imagines the shore line exactly enough as the division between water and land, and imagines it rising—it comes up too slowly for the eye usually, so one *must* imagine it—there is a sort of magic about it. As it moves upward it makes a vast change, far more than the eye sees. It makes a new geography, altering the boundaries of worlds. Above it, it widens the freehold of the birds; below it, that of the fish. The land creatures are driven back and higher up. It is a line between boating and walking, gill and lung, standing still and flowing. Along it, suddenly and continuously, all that will float is picked up and carried away: leaves, logs, seeds, little straws, bits of dead grass.

And also empty cans and bottles and all sorts of buoyant trash left behind by fishermen and hunters and picnickers, or dumped over creek banks by householders who sometimes drive miles to do it. We passed behind a house built on one of the higher banks whose backyard was simply an avalanche of kitchen trash going down to the river. Those people, for all I know, may be champion homebodies, but their garbage is well-traveled, having departed for the Gulf of Mexico on every winter rise for years.

It is illuminating and suitably humbling to a man to recognize the great power of the river. But after he has recognized its power he is next called upon to recognize its limits. It can neither swallow up nor carry off all the trash that people convenience themselves by dumping into it. It can't carry off harmlessly all the sewage and pesticides and industrial contaminants that we are putting into it now, much less all that we will be capable of putting into it in a few years. We haven't accepted—we can't really believe—that the most characteristic product of our age of scientific miracles is junk, but that is so. And we still think and behave as though we face an unspoiled continent, with thousands of acres of living space for every man. We still sing "America the Beautiful" as though we had not created in it, by strenuous effort, at great expense, and with dauntless self-praise, an unprecedented ugliness.

The last couple of miles of our trip we could hear off in the bottoms alongside us the cries of pileated woodpeckers, and we welcomed the news of them. These belong to the big trees and the big woods, and more than any other birds along this river they speak out of our past. Their voices are loud and wild, the cries building strongly and then trailing off arrhythmically and hesitantly as though reluctant to end. Though they never seemed very near, we could hear them clearly over the commotion of the water. There were probably only a pair or two of them, but their voices kept coming to us a long time, creating beyond the present wildness of the river, muddy from the ruin of mountainsides and farmlands, the intimation of another wildness that will not overflow again in *our* history.

The wind had finally made its way into our clothes, and our feet and hands and faces were beginning to stiffen a little with the cold. And so when home came back in sight we thought it wasn't too soon. We began to slant across the currents toward the shore. The river didn't stop to let us off. We ran the bow out onto the path that goes up to my house, and the current rippled on past the stern as though it were no more than the end of a stranded log. We were out of it, wobbling stiff-legged along the midrib on our way to the high ground.

With the uproar of the water still in our ears, we had as we entered the house the sense of having been utterly outside the lives we live as usual. My warm living room was a place we seemed to have been away from a long way. It needed getting used to.

The Long-Legged House

SOMETIME IN THE TWENTIES, nobody knows exactly when any more, Curran Mathews built a cabin of two small rooms near Port Royal between the Kentucky River and the road, on a narrow strip of land belonging to my mother's father. Curran Mathews was my grandmother's bachelor brother. He died of cancer when I was a child and I have come to know him better since his death than I did while he was alive. At the time he died I knew him mostly as a sort of wanderer visiting in the family households, an inspired tinkerer with broken gadgetry and furniture, a man of small disciplines and solutions without either a home or a profession, and a teller of wonderful bed-time stories.

At that time the family still owned the building in Port Royal in which James Mathews, my great-grandfather, an immigrant from Ireland, had made shoes and run an undertaking business. There was a little partitioned-off space at the back of this building that Curran used as a workshop. He was good with tools, and a thorough workman—though I don't believe he ever made any money in that way, or ever intended to. I loved that little shop, and the smells of it, and spent some happy hours there. It smelled of varnish, and was filled with tools and objects of mysterious use. It was tucked away there, as if secretly, out of the way of the main coming and going of the town, looking out across the back lots into a pasture. And everything about it partook of the excitement I felt in this man, home from his wanderings, with his precise ways and peculiarly concentrated silences. I sat in the open doorway there one afternoon, a rich plot of sunlight on the floor around me, Curran quietly at work in the room at my back; I looked up at the ridge beyond the town, the open, still sunlit country of the summer afternoon, and felt a happiness I will never forget.

As for the bedtime stories, they were inexhaustible and always lasted a fine long while past bedtime. He was a reader of adventure stories, and a deeply restless man who spent a large part of his life in distant places. And his stories fed on his reading and on his travels. They could keep going night after night

14

for weeks, and then he would go off someplace. By the time he got back he would have "forgot" the story, and we would have to tell it to him up to where he left off, and then after grunting a little with effort he would produce the next installment. Some of these were straight out of books—I discovered later that he had told us, almost page by page, the first of the *Tarzan* stories; and some were adapted from Zane Grey. But others are harder to account for. I remember that the best one was a sort of Swiftian yarn about a boy who was kidnapped by the fairies, and escaped, and was captured and then adopted by a tribe of Indians, and escaped, and so on and on through a summer in which bedtime was, for a change, looked forward to.

He had a family side, and another side. The family side I knew, I suppose, well enough. The other side I still pick up the hearsay of in conversations here and there. The other side is wild, extravagant, funny in the telling, and for me always more than a little troubling and sad. There was something he needed that he never found. That other side of his life he lived alone. Though some of his doings are legendary among people who remember him, I believe he kept quiet about them himself, and I think I will do the same.

He built the cabin on the river because he thought it would improve his health to live more in the open. His health had never been good. He had permanently injured his leg in a buggy accident in his youth—he disguised his limp so well that he was taken into the army during World War I. And there seems reason to think that the disease that finally killed him had begun years back in his life.

But health must have been only the ostensible reason; the best reason for the cabin he built here must be that it was in his nature to have a house in the woods and to return now and again to live in it. For there was something deep about him, something quiet-loving and solitary and kin to the river and the woods. My mother remembers summer evenings in her girlhood when they would drive his old car five or six miles out in the country to a railroad section pond, where as it grew dark they would sit and listen for a certain bullfrog to croak—and when the frog croaked at last they would drive back to Port Royal. He did that, of course, partly for the entertainment of his young niece; but partly, too, he was entertaining himself.

Given the same niece and the same need for entertainment, another man would not have thought of that. It must have been meaningful to him—a sort of ritual observance. Hard as it may be to know the needs and feelings of a man who said little about himself, dead now more than twenty years, that jaunting to hear the bullfrog does suggest how far his life escaped the categories. It was a life a man could hardly have carried in his wallet, or joined comfortably to the usual organizations.

And so his building of the cabin on the riverbank had a certain logic. The state road had not yet been built up the river on our side—that would have to wait for the Depression and the W.P.A. There was only a sort of rocked wagon-track going up through the bottoms, winding around the heads of slew-hollows, fording the creeks; farm fences crossed it, and one could not travel it without opening and closing gates. Only a few hundred yards down the river from the site of the cabin was Lane's Landing, the port of Port Royal, with its coal tipple and general store, but even so Curran Mathews' wooded stretch of riverbank was far more remote and isolated then than it is now. The voices of the frogs were here—there must have been summer nights when he sat awake a long time, listening to them. And there were other wild voices, too, that I know he heard, because I have heard them here myself. Through a notch in the far bank a little stream poured into the river, a clear moist sound, a part of the constancy of the place. And up in the thickets on the hillside there would be a cowbell, pastoral music, *tinkle-tink tink tink tinkle*, a sort of rhythmless rhythm late into the night sometimes, and nearby, when all else would be still. It was a place where a man, staying by himself, could become deeply quiet. It would have been a quiet that grew deeper and wider as the days passed, and would have come to include many things, both familiar and unexpected. If the stories about him are correct, and they doubtless are, Curran Mathews was not always alone here, and not always quiet. But there were times when he was both. He is said to have owned an old fiddle that he played at these times.

He did most of the building himself. And I imagine that those were days of high excitement. Except for the floors and the one partition, the lumber came out of an old log house that was built up on the hilltop by my grandfather's great-grandfather,

Ben Perry. It surely must have been one of the first houses built in this part of the country. In my childhood the main part of it was still standing, and was known then as the Aunt Molly Perry House, after the last member of the family to live in it. It was in bad shape when I knew it, and hogs were being fed in it. Now there is nothing left but the stones of the chimneys piled up, and the well. The lot where it stood is still known, though, as the Old House Lot, and it is a lovely place, looking out over the woods into the river valley. Some of the old locusts that stood in the yard are still there, and in the spring the little white starry flowers of old-time dooryards still bloom in the grass. It is a place I like, and like to go to and sit down. I am not oppressed by it as I would be by an ancient and venerated family seat, full of old records and traditions and memories. I figure a ponderous amount of my historical inheritance took place there—it is one of the main routes on my way here—but none of it was written down, and most of it has been forgotten, and the house itself is gone. Last summer for a while I pastured a mare and colt up there in the Old House Lot, and where my forebears had sat down to meals three times a day for generations, the horses grazed and thought nothing of it, it being only daylight and grass to them. I sat sometimes on the piled stones of the old chimneys, watching them graze, and it seemed to me that my line had issued out of the ground there like a spring, as regardless of itself, in the historical sense anyhow, and as little able to memorialize itself, as water—and had trickled off into oblivion, as I said to myself that I would too, leaving the hill there looking into the valley, and the horses grazing in the sunlight. And I was more grateful for the silence of their departure than I would be for the lineage of a king. Not knowing who had planted the flowers there, I did not have to weep over them, or grow reminiscent, but could enjoy them as they were. I had them free as wildflowers.

But the house—or, anyhow, part of it—had a very pleasing resurrection. I think of Curran Mathews coming there one summer morning, full of the excitement of his vision of a cabin on the riverbank, to begin work. I can see him carefully prizing off the boards of the kitchen L—broad poplar and walnut boards cut to irregular thicknesses in some crude sawmill; and fine handmade tongue-and-groove paneling out of, I think,

—some memorable, some forgotten. Those visits put the place deeply into my mind. It was a place I often thought about. I located a lot of my imaginings in it. Very early, I think, I began to be bound to the place in a relation so rich and profound as to seem almost mystical, as though I knew it before birth and was born for it. It remained so attractive to me, for one reason, because I had no bad associations with it. It was the family's wilderness place, and lay beyond the claims and disciplines and obligations that motivated my grownups. From the first I must have associated it with freedom. And I associated it with Curran, who must have associated it with freedom. It always had for me something of the charming strangeness of that man's life, as though he had carried it with him in his travels out in the West—as indeed I know he had, in his mind. Like Curran's life, the Camp seemed open to experiences not comprehended in the regularities of the other grown people. That is only to suggest the intensity and the nature of the bond; such feelings, coming from so far back in childhood, lie deeper than the reasons that are thought of afterward. But given the bond, it was no doubt inevitable that I would sooner or later turn to the place on my own, without the company of the grownups.

The first such visit I remember was during the flood of March, 1945. With my brother and a friend, I slipped away one bright Saturday and hitchhiked to Port Royal. We walked down to the Camp from there. It is one of the clearest days I have in my memory. The water was up over the floor of the Camp, and we spent a good while there, straddling out along the porch railings over the water—a dangerous business, of course, but we were in danger the rest of that day. There is something about a flooded house that is endlessly fascinating. Mystery again. One world being supplanted by another. But to a boy on a bright spring day it is without horror. I remember well the fascination I felt that day: the river pouring by, bright in the sun, laced with the complex shadows of the still naked trees, its currents passing the porch posts in little whirlpools the size of half dollars. The world suddenly looked profoundly alive and full of new interest. I had a transfiguring sense of adventure. The world lay before me, and Saturday lay before the world.

We left the Camp after a while and walked on up the river to

where several johnboats were tied, and helped ourselves to the
biggest one, a white one, built long and wide and heavy, as a
boat must be when it is used for net fishing. We unfastened its
mooring chain and pushed off. For oars we had an oak board
about four feet long and an old broom worn off up to the
bindings, but that seemed the paltriest of facts. We were free
on the big water. Once we broke with the shore and disappeared into the bushes along the backwater, we were farther
beyond the reach of our parents than we had ever been, and
we knew it—and if our parents had known it, what a dark day it
would have been for them! But they didn't. They were twelve
miles away in New Castle, their minds on other things, assuming maybe that we had just gone to somebody's house to play
basketball. And we had the world to ourselves—it would be
years before we would realize how far we were from knowing
what to do with it. That clumsy boat and our paddles seemed
as miraculous as if we were cave men paddling the first dugout.

We paddled up to the Cane Run bridge and found the water over the road. Another wonder! And since it was unthinkable to refuse such an opportunity, we boated over the top
of the road and entered the creek valley. We worked our way
up through the treetops for some way, and broke out into a
beautiful lake of the backwater, the sun on it and all around it
the flooded woods. It was deeply quiet—nobody around anywhere, and nobody likely to be. The world expanded yet again.
It would have taken an airplane to find us. From being mere
boys—nine, ten, and eleven years old, I believe—we were enlarged beyond our dreams.

We went up the long backwater and into the woods. The
woods is still there and I often go there these days to walk or
to sit and look. It is a large stand of trees, and there is some big
timber in it. I know it well now, but then I had only been there
a few times to gather ferns with the rest of the family, and it was
still new to me. The woods stands on a series of narrow low
ridges divided by deep hollows. On the day of our adventure
the backwater was in all these hollows, and we could go far into
the woods without ever setting foot on the ground. That was
another wondrous opportunity, and the hollows were numerous and long, allowing for endless variation and elaboration.
The first wildflowers were blooming among the dead leaves

above the water line, and I remember how green and suggestive the pads of moss looked in the warm water light, that time
of the year. I gathered some and wrapped it in my handkerchief
and put it in the boat, as if to take that place and the day with
me when I left. We found a groundhog basking on a point of
land, and got out and gave chase, thinking we had him cor-
nered, only to see him take to the water and swim away. It
never had occurred to me that groundhogs could swim, and I
felt like a naturalist because of that discovery.

But it had to end, and it ended hard. When we started back
the wind had risen, and drove strong against our boat. Our
crude paddles lost their gleeful aura of makeshift, and turned
against us. We moved like the hands of a clock, and only by
the greatest effort. I don't think it occurred to any of us to
worry about getting out, or even about being late, but being
thwarted that way, pitted so against the wind and the water,
had terror in it. Before we got safely back it had become a
considerable ordeal—which, as soon as we *were* safely back,
began to look like the best adventure of all. We tied the boat
and walked up to the road, feeling like Conquerors, weary as
Conquerors.

And there my grandparents were, waiting anxiously for us.
We had been spied upon, it turned out. We had been seen
taking the boat. Phone calls had been made. And my grandparents had driven down—in great fear, I'm sure—to see what was
to be done. When we finally came off the river they were ready
to give us up for drowned. They told us that. And suddenly we
felt a good deal relieved to know we weren't.

That was the second bad flood to hit the Camp. These were
modern floods, and man-made. Too many of the mountain
slopes along the headwater streams had been denuded by
thoughtless lumbering and thoughtless farming. Too little humus remained in the soil to hold the rains. After this second
flood more of the old houses built near the river in earlier times
would be abandoned. And in that spring of 1945 the war was
about to end. Before long the country, as never before, would
be full of people with money to spend. Men's demands upon
nature were about to begin an amazing increase that would
continue until now. The era when Curran Mathews conceived
and built the Camp was coming to an end.

And two springs later Curran lay dying in the town of Carrollton at the mouth of the river. He had gone to spend his last
days in the home of two of my aunts. From his bedroom in
Carrollton the windows looked out across back yards and gardens toward the bluff over the valley of the Ohio. He couldn't
see the river, but as always in a river town it would have been
present to his mind, part of his sense of the place, the condition of everything else. The valley of the river that had been
home to him all his life, and where his cabin stood in its woods
patch on the bank, lay across town, behind him. He knew he
wouldn't see it again.

That spring he watched the lengthening of the buds and
the leafing out of a young beech near his window. And after
the leaves had come he continued to watch it—the lights and
moods and movements of it. He spoke often to my aunts of its
beauty. I learned of this only a year or so ago. It has made his
illness and death real to me, as they never were before. It has
become one of the most vivid links in my kinship with him.
How joyous the young tree must have been to him, who could
have had so little joy then in himself. That he watched it and
did not turn away from it must mean that he found joy in it,
affirmed the life in it, even though his own life was painfully
going out.

In those years of the war and of Curran's illness the Camp
fell into disuse and neglect. The screens rotted; leaves piled up
on the porch, and the boards decayed under them; the weeds
and bushes grew up around it; window glasses were broken
out by rock-throwers; drunks and lovers slipped in and out
through the windows. And then one evening in early summer
—it must have been the year after Curran's death, when I was
nearly fourteen—I came down by myself just before dark, and
cooked a supper on the riverbank below the house, and spent
the night in a sleeping bag on the porch. I don't remember
much about that night, and don't remember at all what reason
I had for coming. But that night began a conscious relation
between me and the Camp, and it has been in my mind and
figured in my plans ever since.

I must have come, that first time, intending to stay for a
while, though I can't remember how long. The next morning
as I was finishing breakfast a car pulled in off the road and

an old man got out, saying he had come to fish. Did I mind?
Maybe I felt flattered that my permission had been asked. No, I
didn't mind, I said. In fact, I'd fish with him. He was a friendly
white-haired old man, very fat, smiling, full of talk. Would he
like to fish from a boat? I asked him. I'd go get one. He said
that would be fine, and I went down along the bank to the
landing and got one—again, I think, by the forthright method
of "borrowing"—and brought it back. It was a good big boat,
but I had managed to find only a broken oar to paddle with.
He got in with his equipment and we set off. We fished the
whole day together. Not much was said. I don't remember that
we ever introduced ourselves. But it was a friendly day. I was
slow in moving us between fishing spots with my broken oar
and I remember how he sat in the middle seat and hunched us
along like a kid in a toy car. It helped a great deal. We fished
until the late afternoon and caught—or, rather, *he* caught—a
sizable string of little perch and catfish. When he left he gave
me all the catch. He fished, he said, for the love of it. I don't
yet know who he was. I never saw him again. In the evening,
after I had taken the boat home and had started cleaning the
fish on the porch, my friend Pete stopped by with his father.
They had been working on their farm up the river that day,
and were on their way home. Pete liked the look of things and
decided to stay on with me. And that began a partnership that
lasted a long time. The next morning we began scraping out
the dried mud that the last high water had deposited on the
floors—and suddenly the place entered our imagination. We
quit being campers and became settlers. Twenty or so years
before, the same metamorphosis must have taken place in the
mind of Curran Mathews. A dozen or two strokes, flaking up
the sediment from the floor—and we had got an idea, and
been transformed by it: we'd clean her out and fix her up.

And that is what we did. We gave it a cleaning, and then
scrounged enough furnishings and food from our families
to set up in free and independent bachelorhood on the river.
Pete's family had a well-stocked frozen-food locker that we
acquired a deep respect for. The idea was to be well enough
supplied that we wouldn't need to go back home for a while.
That way, if somebody wanted to impose work or church on us
they would have to come get us. That was another realization

the Camp had suddenly lit up for us: We were against civilization, and wanted as little to do with it as possible. On the river we came aware of a most inviting silence: the absence of somebody telling us what to do and what not to do. We swam and fished and ate and slept. We could leave our clothes in the cabin and run naked down the bank and into the water. We could buy cigars and lie around and smoke. We could go out on the porch in the bright damp shadowy mornings and pee over the railing. We made a ritual of dispensing with nonessentials; one of us would wash the dishes with his shirttail, and the other would dry them with his. We were proud of that, and thought it had style.

For several years we came to the Camp every chance we got, and made good memories for ourselves. We will never live days like those again, and will never live happier ones. I suppose it was growing up that put an end to them—college, jobs, marriages and families of our own. We began to have a stake in civilization, and could no longer just turn our backs on it and go off. Back in our free days we used to tell people—a good deal to their delight, for we were both sons of respected lawyers—that we intended to be bootleggers. I think the foreknowledge of our fate lay in that. For to contrive long to be as free and careless as we were in those days we would have had to become bootleggers, or something of the kind. Ambition and responsibility would take us a different way.

During those years when I would come down to the Camp with Pete, I would also fairly often come alone. Pete, who was older than I was, would be having a date. Or he would have to stay at home and work at a time when I could get free. And at times I came alone because I wanted to. I was melancholic and rebellious, and these moods would often send me off to the river.

Those times were quiet and lonely, troubled often by the vague uneasiness and dissatisfaction of growing up. The years I spent between childhood and manhood seem as strange to me now as they did then. Clumsy in body and mind, I knew no place I could go to and feel certain I ought to be there. I had no very good understanding of what I was rebelling against: I was going mostly by my feelings, and so I was rarely calm. And I didn't know with enough certainty what I wanted to be purposeful about getting it. The Camp offered no escape from these troubles, but it did allow them the dignity of solitude. And there were days there, as in no other place, when as if by accident, beyond any reason I might have had, I was deeply at peace, and happy. And those days that gave me peace suggested to me the possibility of a greater, more substantial peace —a decent, open, generous relation between a man's life and the world—that I have never achieved; but it must have begun to be then, and it has come more and more consciously to be, the hope and the ruling idea of my life.

I remember one afternoon when I tied my boat to a snag in the middle of the river outside the mouth of Cane Run, and began fishing. It was a warm sunny afternoon, quiet all around, and I had the river to myself. There is some nearly mystical charm about a boat that I have always been keenly aware of, and tied there that afternoon in midstream, a sort of island, it made me intensely alive to the charm of it. It seemed so intact and dry in its boatness, and I so coherent and satisfied in my humanness. I fished and was happy for some time, until I became *conscious* of what a fine thing I was doing. It came to me that this was one of the grand possibilities of my life. And suddenly I became deeply uneasy, even distressed. What I had been at ease with, in fact and without thinking, had become, as a possibility, too large. I hadn't the thoughts for it. I hadn't the background for it. My cultural inheritance had prepared me to exert myself, work, move, "get someplace." To be idle, simply to live there in the sunlight in the middle of the river, was something I was not prepared to do deliberately. I tried to stay on, forcing myself to do what I now *thought* I ought to be doing, but the spell was broken. That I had nothing to do but what I wanted to do, and what I was in fact doing, had become utterly impotent as an idea. I had to leave. I would have to live to twice that age before I could do consciously what I wanted so much then to do. And even now I can do it only occasionally.

I read a good deal during those stays I made at the river. I read *Walden* here then—though I can't remember whether it was for the first or second time. And I was beginning to read poetry with some awareness that it interested me and was important to me—an awareness I had not yet come by in any classroom. I had a paperback anthology of English and American poems, and I would lie in bed at night and read it by the light of a kerosene lamp. One night I lay there late and read for the first time and was deeply stirred by Gray's "Elegy Written in a Country Churchyard." It seemed to me to be a fine thing, and I thought I understood it and knew why it was fine. That was a revealing experience. While I read a dog was howling somewhere down the river, and an owl was calling over in the Owen County bottom, and my father's old nickel-plated Smith and Wesson revolver lay in the bedding under my head. I loved this place and had begun to understand it a little—but I didn't love it and understand it well enough yet to be able to trust it in the dark. That revolver belonged on top of the bookcase at home, but nobody ever thought of it, and I could borrow it and bring it back without fear that it would be missed.

The Camp was not the only place that was important to me. But during the winters of those years, starting when I was fourteen, when I was away at school, it lay in my mind with the other places that were important, part of what I dreamed of coming home to. During the years of high school, particularly, I believe that the thought of it was indispensable to me. The school I attended was a military school. There military correctness and regularity were always the aim—thwarted constantly, to be sure, by the natural high spirits of the students and by the natural mediocrity of most of the teachers—but when thwarted always exacting vengeance on somebody. Sympathy and intelligence were in everything replaced by rules, and by a long ago outworn—hence, threatened and fanatical—moral dogmatism. The highest aim of the school was to produce a perfectly obedient, militarist, puritanical moron who could play football. That aim, or course, inspired a regime that was wonderfully vindictive against anything that threatened to be exceptional. And having a lively and independent mind, I became a natural enemy of the regime. Take a simpleton and give him power and confront him with intelligence—and you have a tyrant. I was once struck by one of the teachers for using a dictionary (not an authorized textbook) during a study hall, and another time was openly chastised for reading a story by Balzac entitled "A Passion in the Desert," the only passion authorized by the regime being a passionate servility. The "discipline" exercised by the student officers was often equally stupid, and often more violent. I waged four years there in sustained rebellion against everything the place stood for, paying the cost both necessarily and willingly. I was not, during all those years, well equipped for such a struggle—though I was a conscious student of resistance, and got pretty good at it toward the end. I had, maybe because of the prolonged awkwardness of my adolescence, an enormous craving for personal dignity—and in the military school dignity simply was not possible for one who was not an athlete and who could not regard mechanical obedience as the summit of virtue. I don't think I could have survived that struggle intact if I hadn't had a history that taught me that there was dignity of another kind, and more desirable. I had known from the beginning a few men who accepted and required of themselves as men with a great simplicity of pride, who could be lonely in their virtues and excellences if they had to be, and who could move in their lives without either crawling or marching—and the thought of those men was before me. But also I had lived days of my own, perhaps mainly at the Camp, when my life had seemed to come to me naturally, with an ease and rightness, as life must come to the kingfishers on the river. I knew that these were my best days, and knew they had not come to me on the orders of anybody or because of anybody's opinion of them or because somebody had allowed me to have them. I knew they were my holy days, my sabbaths, and they had come to me freely because I was free. Knowing that, I knew that men were most admirable singly, and that standing in lines under the command of other men is the least becoming thing they do. Though I stood in line many times each day, my privilege was that my mind was *not* in line. It didn't *have* to be. It had better places to go. Now that I think of it, I had, by that time, a superbly furnished mind. It was not yet much furnished with books, for I had not read many books, or many good ones, then; it was furnished with the knowledge of a few good men and good places and good days that had come to me in my life. I was not servile because I knew what it was *not* to be servile. It is bound to be in some sense true that a man is not a slave if his mind is free. It is the man who can think of no alternative to his enslavement who is truly a slave. Though I found the life at the military school to be often painful and interminably hateful, I was coherent and steadfast in my rebellion against it because I knew, I *must* have known, that I was the creature of another place, and that my life was already given to another way.

What I remember best from those years are the days in early summer when I would first come down to the Camp to clean the place up. By the time school would be out the weeds there on the damp riverbank would have already grown waist-high. I would come down with ax and scythe, preparatory to some stay I intended to make, and drive back the wilderness. I would mow the wild grass and horseweeds and nettles and elderberries, and chop down the tree sprouts, and trim obstructive branches off the nearby trees. It would be hard to describe the satisfaction this opening up would give me. I would sit down now and again to rest and dry the sweat a little, and look at what I had done. I would meditate on the difference I had made, and my mind would be full of delight. It was some instinctive love of wildness that would always bring me back here, but it was by the instincts of a farmer that I established myself. The Camp itself was not imaginable until the weeds were cut around it; until that was done I could hardly bring myself to enter it. A house is not simply a building, it is also an enactment. That is the first law of domesticity, even the most meager. The mere fact must somehow be turned into meaning. Necessity must be made a little ceremonious. To ever arrive at what one would call home even for a few days, a decent, thoughtful approach must be made, a clarity, an opening.

Only after I had mowed and trimmed around it, so that it stood clear of its surroundings, or clearly within them, could I turn to the house. The winter's leaves would have to be swept off the porch, the doors opened, the shutters opened, daylight filling the rooms for the first time in months, the sashes drawn back in their slots, the walls and floors swept. And then, finished, having earned the right to be there again, I would go out on the porch and sit down. Tired and sweaty, the dust of my sweeping still flavoring the air, I would have a wonderful sense of order and freedom. The old recognitions would come back, the familiar sights and sounds slowly returning to their places in my mind. It would seem inexpressibly fine to be living, a joy to breathe.

During the last of my college years and the year I spent as a graduate student, the Camp went through another period of neglect. There were distractions. I had begun to be preoccupied with the girl I was going to marry. In a blundering, half-aware fashion I was becoming a writer. And, as I think of it now, school itself was a distraction. Although I have become, among other things, a teacher, I am skeptical of education. It seems to me a most doubtful process, and I think the good of it is taken too much for granted. It is a matter that is overtheorized and overvalued and always approached with too much confidence. It is, as we skeptics are always discovering to our delight, no substitute for experience or life or virtue or devotion. As it is handed out by the schools, it is only theoretically useful, like a randomly mixed handful of seeds carried in one's pocket. When one carries them back to one's own place in the world and plants them, some will prove unfit for the climate or the ground, some are sterile, some are not seeds at all but little clods and bits of gravel. Surprisingly few of them come to anything. There is an incredible waste and clumsiness in most efforts to prepare the young. For me, as a student and as a teacher, there has always been a pressing anxiety between the classroom and the world: how can you get from one to the other except by a blind jump? School is not so pleasant or valuable an experience as it is made out to be in the theorizing and reminiscing of elders. In a sense, it is not an experience at all, but a hiatus in experience.

My student career was over in the spring of 1957, and I was glad enough to be done with it. My wife and I were married in May of that year. In the fall I was to take my first teaching job. We decided to stay through the summer at the Camp. For me, that was a happy return. For Tanya, who was hardly a country girl, it was a new kind of place, confronting her with hardships she could not have expected. We were starting a long way from the all-electric marriage that the average modern American girl supposedly takes for granted. If Tanya had been the average modern American girl, she would probably have returned me to bachelorhood within a week—but then, of course, she would have had no interest in such a life, or in such a marriage, in the first place. As it was, she came as a stranger into the country where I had spent my life, and made me feel more free and comfortable in it than I had ever felt before. That seems to me the most graceful generosity that I know.

For weeks before the wedding I spent every spare minute at the Camp, getting it ready to live in. I mowed around it, and cleaned it out, and patched the roof. I replaced the broken windowpanes, and put on new screens, and white-washed the walls, and scrounged furniture out of various family attics and back rooms. As a special wedding gift to Tanya I built a new privy—which never aspired so high as to have a door, but did sport a real toilet seat.

All this, I think, was more meaningful and proper than I knew at the time. To a greater extent than is now common, or even possible for most men, I had by my own doing prepared the house I was to bring my wife to, and in preparing the house I prepared myself. This was the place that was more my own than any other in the world. In it, I had made of loneliness a good thing. I had lived days and days of solitary happiness there. And now I changed it, to make it the place of my marriage. A complex love went into those preparations —for Tanya, and for the place too. Working through those bright May days, the foliage fresh and full around me, the river running swift and high after rain, was an act of realization: as I worked getting ready for the time when Tanya would come to live there with me, I understood more and more what the possible meanings were. If it had gone differently—if it had followed, say, the prescription of caution: first "enough" money, and then the "right" sort of house in the "right" sort of place —I think I would have been a poorer husband. And my life, I know, would have been poorer. It wasn't, to be sure, a permanent place that I had prepared; we were going to be there only for the one summer. It was, maybe one ought to say, no more than a ritual. But it was a meaningful and useful ritual.

That is more than I am able to say of our wedding. Our wedding reminds me a little of the Kentucky Derby; the main event, which lasted only a couple of minutes, required days of frantic prologue. During this commotion one must discontinue one's own life and attempt to emulate everybody. Nobody can be still until convinced that this marriage will be like everybody else's. The men insinuate. The women gloat. The church is resurrected and permitted to interfere. As at funerals, the principals cannot be decently let alone, but must be over-hauled, upgraded, and messed with until they are not recognizable. During the week up to and including the wedding I am sure I was at the center of more absurdity than I hope ever to be at the center of again. One of my unforgettable moments, as they are called, was a quaint session with the minister who having met me for the first time that moment, and being scarcely better acquainted with Tanya, undertook to instruct us in the marital intimacies. A display of preacherish cant and presumption unusual even for a preacher, and all carried out with a slogging joyless dutifulness. As we were leaving he handed us a book on marriage. In a country less abject before "expert advice," the effrontery of it would be incredible. Well, be damned to him and his book, too. We thrive in spite of him, and in defiance of some of his rules. We are, I like to think, his Waterloo—though I know that, like most of his kind, he had come to his Waterloo before that, and survived by his inability to recognize it.

In any sense that is meaningful, our wedding was made in our marriage. It did not begin until the ceremony was over. It began, it seems to me, the next morning when we went together to the Camp for the first time since I started work on it. I hesitate to try to represent here the pleasure Tanya may have felt on this first arrival at our house, or the pride that I felt—those feelings were innocent enough, and probably had no more foundation than innocence needs. The point is that, for us, these feelings were substantiated by the Camp; they had its atmosphere and flavor, and partook of its history. That morning when Tanya first came to it as my wife, its long involvement in my life was transformed, given a richness and significance it had not had before. It had come to a suddenly illuminating promise. A new life had been added to it, as a new life had been added to my life. The ramshackle old house and my renewal of it particularized a good deal more for us than we could have realized then. We began there. It began that morning to have as profound a significance in our marriage as it had already had in my life.

We carried in what we had brought in the way of baggage, and then went out and bought the kitchen utensils and groceries we needed. In that way we began our marriage. And in that way—which may be only to say the same thing—we out-distanced what the sterile formalities of the wedding had expected of us or prescribed to us. We escaped the dead hands of the conventions and the institutions into a life that was distinctly our own. Our marriage became then, and has remained, the center of our life. And it is peculiarly true that the Camp is at the center of our marriage, both as actuality and as symbol. The memories of that first summer are strong and clear, and they stay in our minds. After we left the Camp at the summer's end, we continued to think of it and to talk about it and to make plans for it. We lived in other places in Kentucky, in California, in Europe, in New York—and now we have come back to live in sight and in calling distance of the place where we began. It is our source and our emblem, and it keeps its hold.

It would be a mistake to imply that two lives can unite and make a life between them without discord and pain. Marriage is a perilous and fearful effort, it seems to me. There can't be enough knowledge at the beginning. It must endure the blundering of ignorance. It is both the cause and the effect of what happens to it. It creates pain that it is the only cure for. It is the only comfort for its hardships. In a time when divorce is as accepted and conventionalized as marriage, a marriage that lasts must look a little like a miracle. That ours lasts—and in its own right and its own way, not in pathetic and hopeless parody of some "expert" notion—is largely, I believe, owing to the way it began, to the Camp and what it meant and came to mean. In coming there, we avoided either suspending ourselves in some honeymoon resort or sinking ourselves into the stampede for "success." In the life we lived that summer we represented to ourselves what we wanted—and it was *not* the headlong pilgrimage after money and comfort and prestige. We were spared that stress from the beginning. And there at the Camp we had around us the elemental world of water and light and earth and air. We felt the presences of the wild creatures, the river, the trees, the stars. Though we had our troubles, we had them in a true perspective. The universe, as we could see any night, is unimaginably large, and mostly empty, and mostly dark. We knew we needed to be together more than we needed to be apart.

There were physical hardships, or what pass these days for

physical hardships, that scandalized certain interested onlookers. How, they wondered, could I think of bringing a girl like Tanya into a place like that? The question ought to have been: How could a girl like Tanya think of it? They will never know. We had no electricity, no plumbing, no new furniture. Our house would, no doubt, have been completely invisible to the average American bride and groom of that year, and when it rained hard enough the roof leaked. I think our marriage is better for it. By these so-called "hardships"—millions of people put up with much worse as a matter of course and endlessly —we freed our marriage of things. Like Thoreau at Walden, we found out what the essentials are. Our life will never be distorted by the feeling that there are luxuries we cannot do without. We will not have the anxiety of an abject dependence on gadgets and corporations. We are, we taught ourselves by our beginning, the dependents of each other, not of the local electric company.

That summer has no story; it has not simplified itself enough in my memory to have the consistency of a story and maybe it never will; the memories are too numerous and too diverse, and too deeply rooted in my life.

One of the first things I did after we got settled was to put some trotlines in the river—an early outbreak of male behavior that Tanya, I think, found both mystifying and depraved. On a dark rainy night in early June we had stayed up until nearly midnight, making strawberry preserves, and I decided on an impulse to go and raise my lines. Working my way along the line in pitch dark a few minutes later, I pulled out of the middle of the river a catfish that weighed twenty-seven pounds. Tanya had already gone to bed, and had to get up again to hear me congratulate myself in the presence of the captive. And so, indelibly associated with the early days of my marriage is a big catfish. Perhaps it is for the best.

The night of the Fourth of July of that year there came one of the worst storms this part of the country ever knew. For hours the rain spouted down on our tin roof in a wild crashing that did not let up. The Camp had no inner walls or ceilings; it was like trying to sleep inside a drum. The lightning strokes overlapped, so that it would seem to stay light for minutes at a stretch, and the thunder kept up a great knocking at the walls.

After a while it began to seem unbelievable that the rain did not break through the roof. It was an apocalyptic night. The next morning we went out in bright sunshine to find the river risen to the top of its banks. There was a lasting astonishment in looking at it, and a sort of speculative fear; if that storm had reached much farther upstream we would have had to swim out of bed. Upstream, we could see several large trees that Cane Run had torn out by the roots and hurled clear across the river to lodge against the Owen County bank. The marks of that rain are still visible here.

And I remember a quiet night of a full moon when we rowed the boat up into the bend above the mouth of Cane Run, and let the slow current bring us down again. I sat on the rower's seat in the middle of the boat, and Tanya sat facing me in the stern. We stayed quiet, aware of a deep quietness in the country around us, the sky and the water and the Owen County hills all still in the white stare of the moon. The wooded hill above the Camp stood dark over us. As it bore us, the current turned us as though in the slow spiral of a dance.

Summer evenings here on the river have a quietness and a feeling of completion about them that I have never known in any other place, and I have kept in mind the evenings of that summer. The wind dies about sundown, and the surface of the river grows smooth. The reflections of the trees lie inverted and perfect on it. Occasionally a fish will jump, or a kingfisher hurry, skreaking, along the fringe of willows. In the clearing around the house the phoebes and pewees call from their lookout perches, circling out and back in their hunting flights as long as the light lasts. Out over the water the swallows silently pass and return, dipping and looping, climbing and dipping and looping, sometimes skimming the surface to drink or bathe as they fly. The air seems to come alive with the weaving of their paths. As I sat there watching from the porch those evenings, sometimes a profound peacefulness would come to me, as it had at other times, but now it came of an awareness not only of the place, but of my marriage, a completeness I had not felt before. I was there not only because I wanted to be, as always before, but now because Tanya was there too.

But most of all I like to remember the mornings. We would get up early, and I would go out on the river to raise my lines the first thing. There is no light like that on the river on a clear early morning. It is fresh and damp and full of glitters. The intense linear reflections off the wind waves wobble up the tree trunks and under the leaves. It was fishing that paid well, though not always in fish. When I came back to the Camp Tanya would have a big breakfast waiting.

After we ate I would carry a card table out into a corner of the little screened porch, and sit down to write. I would put in the morning there, conscious always as I worked of the life of the river. Fish would jump. A kingfisher would swing out over the water, blue and sudden in the water light, making his harsh ratcheting boast to startle the world. The green herons would pass intently up and down, low to the water, just outside the willows, like busy traffic. Or one would stop to fish from a snag or a low-bending willow, a little nucleus of stillness; sitting at my own work on the other side of the river I would feel an emanation of his intent silence; he was an example to me. And in the trees around the Camp all the smaller birds would be deep in their affairs. It might be that a towboat—the *Kentuckian* or the old *John J. Kelly*, that summer—would come up, pushing two bargeloads of sand to Frankfort. Or there would be somebody fishing from the other bank, or from a boat. And the river itself was as intricately and vigorously alive as anything on it or in it, always shifting its lights and its moods.

That confirmed me in one of my needs. I have never been able to work with any pleasure facing a wall, or in any other way fenced off from things. I need to be in the presence of the world. I need a window or a porch, or even the open outdoors. I have always had a lively sympathy for Thoreau's idea of a hypaethral book, a roofless book. Why should I shut myself up to write? Why not write and live at the same time?

There on the porch of the Camp that summer I wrote the first poetry that I still feel represented by—a long poem rather ostentatiously titled "Diagon," about a river—and did some of the most important reading I have ever done.

In the spring of that year I had read attentively through the poems of Andrew Marvell, and had felt a strong kinship with him. The poem of his that interested me most was not one of the familiar short poems, but the strange, imperfect long one entitled "Upon Appleton House, to my Lord *Fairfax*." This is a complimentary piece, evidently very deliberately undertaken, and in long stretches it is amply boring. But in his description of the countryside Marvell's imagination seems abruptly to break out of the limitations of subject and genre, and he wrote some stunning poetry. It has remained for me one of the most exciting poems I know—not just in spite of its faults, but to a considerable extent because of them. I need to quote extensively from it, both to show the quality of the best work in the poem and to show what I think is remarkable about it. To begin with, all this is most *particularly* observed—as rarely happens in English nature poetry; the scene is of interest in itself, not just as the manifestation of something transcendent or subjective. The grass is grass, and one feels the real rankness and tallness of it. Equally, the killed bird is real; its blood is on the scythe's edge, and we feel the mower's regret of the useless death. What natural things manifest, if observed closely enough, is their nature, and their nature is to change. Marvell's landscape is in constant metamorphosis, and so metaphor is peculiarly necessary to its poetry—it is continuously being carried beyond what it was. The comparative image is not imposed from without by the poet, but is seen by the poet to be implicit in the nature of the thing: it is in the nature of a meadow to be like a sea.

> And now to the Abbyss I pass
> Of that unfathomable Grass,
> Where Men like Grashoppers appear,
> But Grashoppers are Gyants there:
> They, in there squeking Laugh, contemn
> Us as we walk more low then them:
> And, from the Precipices tall
> Of the green spir's, to us do call.

> To see Men through this Meadow Dive,
> We wonder how they rise alive.
> As, under Water, none does know
> Whether he fall through it or go.
> But, as the Marriners that sound,
> And show upon their Lead the Ground,
> They bring up Flow'rs so to be seen,
> And prove they've at the Bottom been.

No Scene that turns with Engines strange
Does oftner then these Meadows change.
For when the Sun the Grass hath vext,
The tawny Mowers enter next;
Who seem like *Israalites* to be,
Walking on foot through a green Sea.
To them the Grassy Deeps divide,
And crowd a Lane to either Side.

With whistling Sithe, and Elbow strong,
These Massacre the Grass along:
While one, unknowing, carves the *Rail*,
Whose yet unfeather'd Quils her fail.
The Edge all bloody from its Breast
He draws, and does his stroke detest . . .

The Mower now commands the Field;
In whose new Traverse seemeth wrougth
A Camp of Battail newly fought:
Where, as the Meads with Hay, the Plain
Lyes quilted ore with Bodies slain . . .

When after this 'tis pil'd in Cocks,
Like a calm Sea it shews the Rocks:
We wondring in the River near
How Boats among them safely steer.

This *Scene* again withdrawing brings
A new and empty Face of things;
A levell'd space, as smooth and plain,
As Clothes for *Lilly* stretched to stain.
The World when first created sure
Was such a Table rase and pure.

For to this naked equal Flat,
Which *Levellers* take Pattern at,
The Villagers in common chase
Their Cattle, which it closer rase . . .

They feed so wide, so slowly move,
As *Constellations* do above.
Then, to conclude these pleasant Acts,
Denton sets ope its *Cataracts*;

And makes the Meadow truly be
(What it but seem'd before) a Sea.

The River in it self is drown'd,
And Isl's th' astonished Cattle round.
Let others tell the *Paradox*,
How Eels now bellow in the Ox;
How Horses at their Tails do kick,
Turn'd as they hang to Leeches quick;
How Boats can over Bridges sail;
And Fishes do the Stables scale.
How *Salmons* trespassing are found;
And Pikes are taken in the Pound.

As I worked that summer I had these lines of Marvell very
fresh in my mind, and they were having a deeper influence
on me than I knew. My problem as a writer, though I didn't
clearly know it yet, was that I had inherited a region that had
as a literary tradition only the corrupt and crippling local col-
orism of the "Kentucky" writers. This was both a mytholo-
gizing chauvinism and a sort of literary imperialism, tirelessly
exploiting the clichés of rural landscape, picking and singing
and drinking and fighting lazy hillbillies, and Bluegrass Col-
onels. That is a blinding and tongue-tying inheritance for a
young writer. And one doesn't even have to read the books
to get it; it is so thoroughly established and accredited that it
is propagated by schoolteachers, politicians, official bulletins,
postcards, and the public at large. Kentucky is a sunny, beautiful land, full of happy country folks, whose very failures are
quaint and delightful and to be found only here. It is surely no
accident that along with this tradition of literary falsification
there has been a tradition, equally well-established, and one
could almost say equally respected, of political and industrial
exploitation that has defaced and destroyed more of the state's
beauty and wealth than there is left. The truth was too hard to
tell; the language of the state's writers was dead in their mouths
and they could not tell it. And when one finally told the clean
truth, as Harry Caudill did in *Night Comes to the Cumberlands*,
how few could hear it! And for the same reason. The people's
ears are stuffed with the dead language of their literature. All

my life as a writer I have had this rag bag of chauvinistic clichés to struggle with.

It is not difficult to see how serviceable and clarifying I found those lines from "Upon Appleton House." They showed me the poet's vision breaking out of its confines into the presence of its subject. I feel yet the exhilaration and release when Marvell turns from his elaborate overextended compliment to the noble family, and takes up a matter that really interested him; the full powers of his imagination and intelligence become suddenly useful to him, and necessary:

> And now to the Abbyss I pass
> Of that unfathomable Grass. . . .

He is talking about a river valley of farms and woodlands such as I had known all my life, and now had before me as I wrote and read through that summer. As a child I had even believed, on what I then considered the best advice, in the metamorphosis of horsehairs that Marvell alludes to in the last stanza of his description of the meadows. I would put hairs from tails and manes into the watering trough at night, confident that by morning they would be turned to snakes. But I was a poor scientist, and in the mornings always forgot to look —and so kept the faith.

With Marvell's work in my mind, I began that summer of my marriage the surprisingly long and difficult labor of *seeing* the country I had been born in and had lived my life in until then. I think that this was peculiarly important and necessary to me; for whereas most American writers—and even most Americans —of my time are displaced persons, I am a placed person. For longer than they remember, both sides of my family have lived within five or six miles of this riverbank where the old Camp stood and where I sit writing now. And so my connection with this place comes not only from the intimate familiarity that began in babyhood, but also from the even more profound and mysterious knowledge that is inherited, handed down in memories and names and gestures and feelings, and in tones and inflections of voice. I never, for reasons that could perhaps be explained, lost affection for this place, as American writers have almost traditionally lost affection for their rural birthplaces. I have loved this country from the beginning, and I believe I was grown before I ever really confronted the possibility that I could live in another place. As a writer, then, I have had this place as my fate. For me, it was never a question of *finding* a subject, but rather of learning what to do with the subject I had had from the beginning and could not escape. Whereas most of this country's young writers seem able to relate to no place at all, or to several, I am related absolutely to one.

And this place I am related to not only shared the state's noxious literary inheritance, but had, itself, never had a writer of any kind. It was, from a writer's point of view, undiscovered country. I have found this to be both an opportunity and a disadvantage. The opportunity is obvious. The disadvantage is that of solitude. Everything is to be done. No beginnings are ready-made. One has no proof that the place can be written about, no confidence that it can produce such a poet as one suspects one might be, and there is a hesitance about local names and places and histories because they are so naked of associations and assigned values—none of which difficulties would bother a poet beginning in Concord, say, or the Lake District. But here I either had to struggle with these problems or not write. I was so intricately dependent on this place that I did not begin in any meaningful sense to be a writer until I began to see the place clearly and for what it was. For me, the two have been the same.

That summer I was only beginning, and my poem "Diagon" came out of the excitement of that first seeing, and the first inklings that there might be viable meanings in what I knew. I was seeing consciously the lights and colors and forms of my own world for the first time:

> The sun sets vision afloat,
> Its hard glare down
> All the reaches of the river,
> Light on the wind waves
> Running to shore. Under the light
> River and hill divide. Two dead
> White trees stand in the water,
> The shimmering river casts
> A net of light around them,
> Their snagged shapes break through.

*

It is a descriptive poem mostly, and I have worried at times because in my work I have been so often preoccupied with description. But I have begun to think of that as necessary. I had to observe closely—be disciplined by the look and shape and feel of things and places—if I wanted to escape the blindness that would have made my work sound like an imitation of some Kentucky politician's imitation of the Romantic poets.

Sustaining and elaborating the effect of Marvell were the poems of William Carlos Williams, whose work I had known before but read extensively and studiously during that summer of 1957. I had two books of his, the *Collected Earlier Poems*, and his newest one, *Journey to Love*. I saw how his poems had grown out of his life in his native city in New Jersey, and his books set me free in my own life and my own place as no other books could have. I'll not forget the delight and hopefulness I felt in reading them. They relieved some of the pressure in the solitude I mentioned earlier. Reading them, I felt I had a predecessor, if not in Kentucky then in New Jersey, who confirmed and contemporized for me the experience of Thoreau in Concord.

Another book that deeply affected me that summer was Kenneth Rexroth's *100 Poems from the Chinese*, which immediately influenced my work and introduced me to Oriental poetry, not to mention the happy reading it made. I still think it is one of the loveliest books I know.

All these—my new marriage, the Camp, the river, the reading and writing—are intimately associated in my mind. It would be impossible to do more than imply the connections. Those were probably the three most important months in my life, as well as the happiest. When the summer was over it was a sharp sorrow to have to go. I remember us loading our borrowed household things into and onto our old Jeep station wagon and driving off up the river road on a brilliant day, the fields in the bottoms all yellow with the fall flowers. And I remember the troubling sense that what we were going to would be more ordinary than what we were leaving behind. And it was.

II

It has been almost exactly a year since I began this history. My work was interrupted by the spring weather, when gardening and other outside concerns took me away from writing. But now it is deep winter again. Yesterday snow fell all day and covered the ground. This morning, though the sun came up clear, the thermometer read four above—a good morning to sit in the Camp in the warmth of the stove and the brisk snow light from the big window over the table. It is a morning for books and notebooks and the inviting blank pages of writing paper.

For people who live in the country there is a charming freedom in such days. One is free of obligations to the ground. There is no outside work that one ought to do, simply because, with the ground frozen deep and covered with snow, no such work is possible. Growth has stopped; there is plenty of hay and grain in the barn; the present has abated its urgencies. And the mind may again turn freely to the past and look back on the way it came. This morning has been bearing down out of the future toward this bit of riverbank forever. And for perhaps as long, in a sense, my life has been approaching from the opposite direction. The approach of a man's life out of the past is history, and the approach of time out of the future is mystery. Their meeting is the present, and it is consciousness, the only time life is alive. The endless wonder of this meeting is what causes the mind, in its inward liberty of a frozen morning, to turn back and question and remember. The world is full of places. Why is it that I am here?

What has interested me in telling the history of the Camp is the possibility of showing how a place and a person can come to belong to each other—or, rather, how a person can come to belong to a place, for places really belong to nobody. There is a startling reversal of our ordinary sense of things in the recognition that we are the belongings of the world, not its owners. The social convention of ownership must be qualified by this stern fact, and by the humility it implies, if we are not to be blinded altogether to where we are. We may deeply affect a place we own for good or ill, but our lives are nevertheless included in its life; it will survive us, bearing the results. Each of us is a part of a succession. I have come here following Curran Mathews. Who was here or what was done before he came, I do not know. I know that he had predecessors. It is certain that at some time the virgin timber that once stood here was cut down, and no doubt somebody then planted corn among the stumps, and so wore out the ground and allowed the trees to return. Before the white men were the Indians, who generation after generation bequeathed the country to their children, whole, as they received it. The history is largely conjecture. The future is mystery altogether: I do not know who will follow me. These realizations are both aesthetic and moral; they clear the eyes and prescribe an obligation.

At the point when my story was interrupted, my life no longer seemed to be bearing toward this place, but away from it. In the early fall of 1957 Tanya and I left the Camp, and through the following year I taught at Georgetown College. In the spring of 1958 our daughter Mary was born, and in the fall we left Kentucky for the West Coast. The Camp was closed and shuttered. Even though it had taken a new and lasting hold on my mind, it had entered another time of neglect. Three years would pass before I would come back to it. Toward the end of that summer of 1958 my friend Ed McClanahan and I made a canoe trip down the river and spent a night in the Camp, sleeping on the floor. For me, that night had the sadness of a parting. I was about to leave the state; the past was concluded, and the future, not yet begun, was hardly imaginable. The Camp was empty, dark, full of finished memories, already falling back into the decay of human things that humans have abandoned. I was glad when the morning came, and we loafed on down along the shady margin of the river, watching the muskrats and the wood ducks.

I did not go back to the Camp again until the May of 1961. After two years on the West Coast, we had spent another year in Kentucky, this time on the farm, and again we had a departure ahead of us; late in the summer we would be going to Europe for a year. In order to prepare myself for this experience I began spending some mornings and rainy days at the Camp. My intention at first was to do some reading that would help me to understand the life of the places I would be seeing in Europe. But as I might have expected it was not Europe that most held my attention on those days, but the Camp and the riverbank and the river. It soon became clear that I was not so much preparing for an important experience as *having* an important one. I had been changed by what had happened to me and what I had learned during the last three years, and I was no sooner back at the Camp, with the familiar trees around me and the river in front of me again, than I began to see it differently and in some ways more clearly than I had before. Through that summer I wrote a sort of journal, keeping account of what I saw.

I first went back to the Camp that year on a rainy Monday, May 8. Heavy rains had begun the Saturday before, and the river was in flood. The water was under the house, within about a foot and a half of the floor. I had come to read, but mostly I sat and looked. How can one read history when the water is rising? The presence of the present had become insistent, undeniable, and I could not look away; the past had grown still, and could be observed at leisure in a less pressing time. The current was driving drift logs against the legs of the cabin beneath my chair. The river flexed and throbbed against the underpinnings like a great muscle, its vibrations too set in my nerves to permit thinking of anything else.

The river had become a lake, but a lake *flowing*, a continuous island of drift going down the channel, moving swiftly and steadily but forever twisting and eddying within itself; and along the edges the water was picking up little sticks and leaves and bits of grass as it rose. The house's perspective on the river had become the same as that of a boat. I kept an uneasy sense of its nearness, knowing that it was coming nearer; in an hour and a quarter it rose five inches. And there was a sort of permanent astonishment at its massiveness and flatness and oblivious implacable movement.

That day a new awareness of the Camp came to me, an awareness that has become a part of my understanding of all houses. It was a boat—a futile, ill-constructed, doomed boat—a boat such as a child might make on a hilltop. It had been built to stand there on the bank according to the rules of building on solid ground. But now the ground was under water, and the water was rising. The house would have the river to contend with. It would be called on to be a boat, as it had been called on before—as in the 1937 flood it had been called on, and had made its short voyage downstream and up the bank until it lodged among the trees. It was a boat by necessity, but not by nature, which is a recipe for failure. It was built with kinder hopes, to fare in a gentler element than water. All houses are not failed boats, but all are the failed, or failing, vehicles of some alien element; of wind, or fire, or time. When I left that night the river was only a foot beneath the floor, and still rising.

It rose two feet into the Camp, cresting sometime Wednesday. It was out of the house again by Thursday afternoon, and I opened the doors to dry the mud. On Sunday I was at work at a table on the screened porch, sitting in a chair where I could have sat in a boat a few days before, looking down into a landscape that still bore everywhere the marks of overflowing. As far as I could see, up and down across the valley, there was the horizon of the flood, a level in the air below which everything was stained in the dull gray-brown of silt, and the tree branches were hung with tatters of drift, as though the flood was still there in ghostly presence. Above that horizon the spring went on uninterrupted, the new clear green of the leaves unfolding. It was as though I sat with my feet in one world and my head in another. And so with the mud still drying on the floors, I resumed my connection with the Camp.

Other creatures had worse luck getting started that spring than I did. On the Monday of the rise I watched a pair of prothonotary warblers hovering and fluttering around their nest hole near the top of a box elder snag. The snag stood on the bank directly below the porch of the Camp, its top about level with the floor. And so that night when the river crept into the Camp it had already filled the warblers' nest. When I came back after the water went down the birds were back. They nested again in the same hole. And then sometime around the middle of June the snag blew over. Since then I believe that no pair has nested near the Camp, though they nest around the slew across the river and I often see them feeding here.

That spring a pair of phoebes nested under the eaves in front of the house just above the door to the screened porch. A dead elm branch reaching over the porch made them a handy place to sit and watch for insects. I would often pause in my writing or reading to watch them fly out, pick an insect out of the air, and return to the branch. Sometimes I could hear their beaks snap when they made a catch.

A pair of starlings was nesting in a woodpecker hole in a maple down the bank in front of my writing table and a pair of crested flycatchers in another hollow maple a few yards upstream. Titmice were in a woodpecker hole in the dead locust near the kitchen door. High up in an elm, in the fork of a branch hanging over the driveway, pewees built their neat cup of a nest and covered the outside with lichens, so that it looked strangely permanent and ancient, like a small rock.

Wood thrushes lived in the thicker woods upriver. They never came near the house while I was there, but their music did, as though their feeling toward me was both timid and generous. One of the unforgettable voices of this place is their exultant fluting rising out of the morning shadows.

On two separate days, while I sat at work, a hummingbird came in through a hole in the rotted porch screen to collect spiderwebs for his nest. He would stand in the air, deliberate as a harvester, and gather the web in his beak with a sort of winding motion.

Later, in July, I would often watch a red-bellied woodpecker who hunted along the tree trunks on this side of the river to feed his young in a hollow snag on the far side. He would work his way slowly up the trunks of the sycamores, turning his head to the side, putting his eye close to the trunk to search under the loose bark scales. And then he would fly to the far side of the river, where his snag jutted up over the top of a big willow. With the binoculars I could see him perch at his hole and feed his nestlings.

For the first time in all my staying at the Camp I had a pair of binoculars, and perhaps more than anything else during that time, they enlarged and intensified my awareness of the place. With their help I began to know the warblers. At a distance these little birds usually look drab, and the species are hardly distinguishable, but the binoculars show them to be beautifully colored and marked, and wonderfully various in their kinds. There is always something deeply enticing and pleasing to me in the sight of them. Perhaps because I was only dimly aware of them for so long, I always see them at first with a certain unexpectedness, and with the sense of gratitude that one feels for any goodness unearned and almost missed. In their secretive worlds of treetop and undergrowth, they seem among the most remote of the wild creatures. They see little of us, and we see even less of them. I think of them as being aloof somehow from common life. Certain of the most beautiful of them, I am sure, have lived and died for generations in some of our woods without being recognized by a human being.

But the binoculars not only give access to knowledge of lives that are usually elusive and distant; they make possible a peculiar imaginative association with those lives. While opening and clarifying the remote, they block out the immediate. Where one is is no longer apparent. It is as though one stood at the window of a darkened room, lifted into a world that cannot be reached except by flying. The treetops are no longer a ceiling, but a spacious airy zone full of perching places and nervously living lights and shadows. One sees not just the bird, but something of how it is to *be* the bird. One's imagination begins to reach and explore into the sense of how it would be to be without barriers, to fly over the river, to perch at the frailest, most outward branchings of the trees.

In those days I began the long difficult realization of the complexity of the life of this place. Until then—at the level of consciousness, at least—I had thoughtlessly accepted the common assumption of my countrymen that the world is merely an inert surface that man lives on and uses. I don't believe that I had yet read anything on the subject of ecology. But I had read Thoreau and Gilbert White and a little of Fabre, and from seeing natural history displays I knew the concept of the habitat group. And that summer, I remember, I began to think of myself as living within rather than upon the life of the place. I began to think of my life as one among many, and one kind among many kinds. I began to see how little of the beauty and the richness of the world is of human origin, and how superficial and crude and destructive—even self-destructive—is man's conception of himself as the owner of the land and the master of nature and the center of the universe. The Camp with its strip of riverbank woods, like all other places of the earth, stood under its own widening column of infinity, in the neighborhood of the stars, lighted a little, with them, within the element of darkness. It was more unknown than known. It was populated by creatures whose ancestors were here long before my ancestors came, and who had been more faithful to it than I had been, and who would live as well the day after my death as the day before.

Seen as belonging there with other native things, my own nativeness began a renewal of meaning. The sense of belonging began to turn around. I saw that if I belonged here, which I felt I did, it was not because anything here belonged to me. A man might own a whole county and be a stranger in it. If I belonged *in* this place it was because I belonged *to* it. And I began to understand that so long as I did not know the place fully, or even adequately, I belonged to it only partially. That summer I began to see, however dimly, that one of my ambitions, perhaps my governing ambition, was to belong fully to this place, to belong as the thrushes and the herons and the muskrats belonged, to be altogether at home here. That is still my ambition. I have made myself willing to be entirely governed by it. But now I have come to see that it proposes an enormous labor. It is a spiritual ambition, like goodness. The wild creatures belong to the place by nature, but as a man I can belong to it only by understanding and by virtue. It is an ambition I cannot hope to succeed in wholly, but I have come to believe that it is the most worthy of all.

Whenever I could during that summer I would come to the Camp—usually on Sundays, and on days when it was too wet for farm work. I remember vividly the fine feeling I would have, starting out after breakfast with a day at the river ahead of me. The roadsides would be deep in the fresh clear blue of chicory flowers that in the early morning sun appeared to give off a light of their own. And then I would go down into the fog that lay deep in the valley, and begin work. Slowly the sun would burn through the fog, and brighten the wet foliage along the riverbanks in a kind of second dawn. When I got tired of reading or writing I would cut weeds around the house, or trim the trees to open up the view of the river. Or I would take a walk into the woods.

And as before, as always, there was the persistent consciousness of the river, the sense of sitting at the edge of a great opening passing through the country from the Appalachians to the Gulf of Mexico. The river is the ruling presence of this place. Here one is always under its influence. The mind, no matter how it concentrates on other things, is never quite free of it, is always tempted and tugged at by the nearness of the water and the clear space over it, ever widening and deepening into the continent. Its life, in the long warm days of summer when the water is low, is as leisurely and self-preoccupied as the life of a street in a country town. Fish and fishermen pass along it, and so do the kingfishers and the herons. Rabbits and squirrels and groundhogs come down in the late afternoon to drink. The birds crisscross between shores from morning to night. The muskrats graze the weed patches or browse the overhanging willows or carry whole stalks of green corn down out of the fields to the water's edge. Molting wood ducks skulk along the banks, hiding under the willows and behind the screens of grapevines. But of all the creatures, except the fish, I think the swallows enjoy the river most. Whole flocks of barn swallows and rough wings will spend hours in the afternoon and evening circling and dipping over the water, feeding, bathing, drinking—and rejoicing, too, as I steadfastly believe, for I cannot imagine that anything could fly as splendidly as the swallows and not enjoy it.

One afternoon as I was sitting at my worktable on the porch a towboat came up with two barges loaded with sand. A man and a boy sat on the edge of the bow of the head barge, their feet hanging over the water. They were absorbed in their talk, remote from observance, the river world wholly surrounding and containing them, like the boatmen in the paintings of George Caleb Bingham. For the moment they belonged to movement and I to stillness; they were bound in kinship with the river, which is always passing, and I in kinship with the trees, which stand still. I watched them out of sight, intensely aware of them and of their unawareness of me. It is an eloquent memory, full of the meaning of this river.

And on those days, as on all the days I have spent here, I was often accompanied by the thought of Curran Mathews. When I am here, I am always near the thought of him, whether I think about him or not. For me, his memory will always be here, as indigenous and congenial as the sycamores. As long as I am here, I think, he will not be entirely gone. Shortly after the May flood of that year of 1961 I was walking in the weeds on the slope above the original site of the Camp, and I came unexpectedly on a patch of flowers that he had planted there thirty years before. They were trillium, lily of the valley, woodland phlox, Jacob's-ladder, and some ferns. Except for the lily

of the valley, these flowers are native to the woods of this area
—though they apparently had not returned to this stretch of
the riverbank since it was cleared and plowed. And so in digging them up and bringing them here, Curran was assisting
amiably in the natural order. They have remained through all
the years because they belong here; it is their nature to live in
such a place as this. But his pleasure in bringing them here is
an addition to them that does not hamper them at all. All their
lives they go free of him. Because the river had carried the
Camp some distance from the old site, leaving the flowers out
of the way of our usual comings and goings, I had never seen
them before. They appeared at my feet like some good news
of Curran, fresh as if he had spoken it to me, tidings of a day
when all was well with him.

Another insistent presence that summer was that of time.
The Camp was rapidly aging and wearing out. It had suffered
too much abandonment, had been forgotten too much, and
the river had flowed into it too many times. Its floors were
warped and tilted. The roof leaked where a falling elm branch
had punched through the tin. Some of the boards of the walls
had begun to rot where the wet weeds leaned against them.
What was purposive in it had begun to be overtaken by the
necessary accidents of time and weather. Decay revealed its
kinship with the earth, and it seemed more than ever to belong to the riverbank. The more the illusion of permanence fell
away from it, the easier it fit into the flux of things, as though
it entered the fellowship of birds' nests and of burrows. But as
a house, it was a failed boat. As a place to sit and work, it was a
flimsy, slowly tilting shelf. As a shelter, it was like a tree.

And the day was coming when I would leave again. And
again I did not know when I would be back. But this time I
did consciously intend to come back. Tanya and I had even
begun to talk of building a house someday there on the riverbank, although the possibility seemed a long way off, and the
plan was more to comfort ourselves with than to act on. But
the plan, because it represented so deep a desire, was vivid to
us and we believed in it. Near the original site of the Camp
were two fine sycamores, and we thought of a house standing
on the slope above them, looking down between them at the
river. As a sort of farewell gesture, and as a pledge, I cut down

some box elders and elms whose branches had begun to grow obstructively into the crowns of the sycamores. And so, before leaving, I made the beginning of a future that I hoped for and dimly foresaw.

On the twelfth of March the following spring, a letter came to us in Florence, Italy, saying that the river was in the Camp again. Far away as I was, the letter made me strangely restless and sad. I could clearly imagine the look of it. The thought of inundation filled me: the river claiming its valley, making it over again and again—the Camp, my land creature, inhabited by water. The thought, at that distance, that what I knew might be changed filled me and held me in abeyance, as the river filled and held the Camp.

After the year in Europe we lived in New York, and I taught at New York University. We had a second child by then, a boy. In the winter of 1963, I accepted a teaching job at the University of Kentucky to begin in the fall of the following year. That suited us. Our hopes and plans had already turned us back toward Kentucky. We had already spent several years living in other places, and after a second year in New York we would be ready to think of settling down at home.

We returned to Kentucky when school was out that year to spend the summer. And encouraged by the prospect that my relation to the place might soon be permanent, I planned to rebuild the Camp. For one reason, I would be needing a place set aside to do my work in. Another reason, and the main one, was that I needed to preserve the Camp as an idea and a possibility here where it had always been. So many of the good days of my life had been lived here that I could not willingly separate myself from it.

At first I considered repairing the Camp as it stood. But as I looked it over it appeared to be too far gone to be worth the effort and money I would have to spend on it. Besides, it was too near the river; as the watershed deteriorated more and more through misuse, the spring rises had begun to come over the floors too often. Much as I still valued it, the old house had become a relic, and there were no more arguments in its favor. And so on the sixth of June, 1963, I began the work of tearing it down and clearing a place farther up the bank to build it back again. That afternoon I took out the partition between the two rooms, and then cut down the elm that stood where the new building was to be.

The next afternoon I cleared the weeds and bushes off the building site, and with that my sadness at parting with the old house began to give way to the idea of the new. I was going to build the new house several feet higher up the slope than the old one, and to place it so that it would look out between the two big sycamores. Unlike a wild place, a human place gone wild can be strangely forbidding and even depressing. But that afternoon's work made me feel at home here again. My plans suddenly took hold of me, and I began to visualize the new house as I needed it to be and as I thought it ought to look. My work had made the place inhabitable, had set my imagination free in it. I began again to belong to it.

During the next several days I worked back and forth between the old and the new, tearing down and preparing to build. The tearing down was slow work, for I wanted to save and reuse as much of the old lumber as I could, and the floods had rusted all the nails tight in their holes. The sediment of the flood of 1937 still lay on the tops of the rafters. And on the sheeting under the tin of the roof I found the wallpaper the boards had worn when they stood in the walls of the old family house up on the top of the hill. I squared the outlines of the new house on the slope—measuring off a single room twelve by sixteen feet, and a porch eight by eight feet—and dug the holes for the posts it would stand on.

By June 18 I was ready to begin building, and that day two carpenters, friends of mine I had hired to help me, came early in the morning, and we began work. By the night of the twenty-fifth the new house was up, the roof and siding were on, and from that day I continued the work by myself. The old Camp provided the roof, the floor, and three walls of the new, as well as the two doors and some windows. This dependence on the old materials determined to a considerable extent the shape of the new house, for we would shorten or lengthen the dimensions as we went along in accordance with the lengths of the old boards. And so the new house was a true descendant of the old, as the old in its time was the descendant of one still older.

That summer I was deep into the writing on a long book called *A Place on Earth*. And as soon as the heavier work on the house was done and I no longer needed the carpenters, I returned to work on the book, writing in the mornings and continuing work on the house in the afternoons. I nailed battens over the cracks between the boards, braced the underpinnings, made screens and screen doors and shutters and steps, painted the roof and the outside walls. At the end of the summer I had a satisfactory nutshell of a house, green-roofed and brown-walled, that seemed to fit well enough into the place. Standing on its long legs, it had a peering, aerial look, as though built under the influence of trees. But it was heron-like, too, and made for wading at the edge of the water.

The most expensive member of the new house was a big window, six feet by four and a half feet, with forty panes. This was the eye of the house, and I put it in the wall facing the river and built a long worktable under it. In addition, three window sashes from the old Camp were set into each of the end walls. And so the house became a container of shifting light, the sunlight entering by the little windows in the east wall in the morning, and by those in the west wall in the afternoon, and the steadier light from the northward-staring big window over the worktable.

When I began tearing down the old Camp a pair of phoebes was nesting as usual under one of the eaves. Four eggs were already in the nest. I took an old shutter and fixed a little shelf and a sort of porch roof on it, and nailed it to a locust tree nearby, and set the nest with its eggs carefully on it. As I feared, the old birds would have nothing to do with it. The nest stayed where I had put it for another year or so, a symbol of the ended possibilities of the old Camp, and then it blew away.

As I wrote through the mornings of the rest of that summer, a green heron would often be fishing opposite me on the far bank. An old willow had leaned down there until it floated on the water, reaching maybe twenty feet out from the bank. The tree was still living, nearly the whole length of it covered with leafy sprouts. It was dead only at the outer end, which bent up a few inches above the surface of the river. And it was there that the heron fished. He stooped a little, leaning a little forward, his eyes stalking the river as it flowed down and passed beneath him. His attention would be wonderfully concentrated for minutes at a stretch, he would stand still as a dead branch

on the trunk of his willow, and then he would itch and have
to scratch among his feathers with his beak. When prey swam
within his reach, he would crouch, tilt, pick it deftly out of
the water, sit back, swallow. Once I saw him plunge headlong
into the water and flounder out with his minnow—as if the
awkward flogging body had been literally yanked off its perch
by the accurate hunger of the beak. When a boat passed he did
not fly, but walked calmly back into the shadows among the
sprouts of the willow and stood still.

Another bird I was much aware of that summer was the sycamore warbler. Nearly every day I would see one feeding in
the tall sycamores in front of the house, or when I failed to
see the bird I would hear its song. This is a bird of the tall
trees, and he lives mostly in their highest branches. He loves
the sycamores. He moves through their crowns, feeding, and
singing his peculiar quaking seven-note song, a voice passing
overhead like the sun. I am sure I had spent many days of my
life with this bird going about his business high over my head,
and I had never been aware of him before. This always amazes
me; it has happened to me over and over; for years I will go in
ignorance of some creature that will later become important to
me, as though we are slowly drifting toward each other in time,
and then it will suddenly become as visible to me as a star. It
is at once almost a habit. After the first sighting I see it often.
I become dependent on it, and am uneasy when I do not see
it. In the years since I first saw it here and heard its song, the
sycamore warbler has come to hold this sort of power over me.
I never hear it approaching through the white branches of its
trees that I don't stop and listen; or I get the binoculars and
watch him as he makes his way from one sycamore to another
along the riverbank, a tiny gray bird with a yellow throat, singing from white branch to white branch, among the leaf lights
and shadows. When I hear his song for the first time in the
spring, I am deeply touched and reassured. It has come to be
the most characteristic voice of this place. He is the Camp's
emblem bird, as the sycamore is its emblem tree.

From the old Camp the new Camp inherited the fate of a
river house. High up as I had built it, I hadn't been able to
move it beyond the reach of the river; there was not room
enough here below the road to do that. Like the old house,

the new was doomed to make its way in the water—to be a failed boat, and survive by luck. The first spring after it was built the river rose more than two feet over the floor. When the water went down, my uncle hosed out the silt and made the place clean again, and when I came back from New York later in the spring it looked the way it had before. But the idea of it had changed. If it had been built in the hope that the river would never rise into it and in the fear that it would, it now lived in the fact that the river had and in the likelihood that it would again. Like all river houses, it had become a stoical house. Sitting in it, I never forget that I am within the reach of an awesome power. It is a truthful house, not indulging the illusion of the permanence of human things. To be here always is not its hope. Long-legged as it is, it is responsive to the natural vibrations. When the dog scratches under the table there is a tremor in the rafters.

Our return from New York in early June of 1964 changed our lives. We were coming back to Kentucky this time with the intention of staying and making a home. Our plans were still unsettled, but our direction was clear. For the first time, we were beginning to have a foreseeable future. From then on, my relation to my native country here might be interrupted occasionally, but it would not be broken. For the summer, we would stay on the farm and I would spend the days working at the Camp, as before.

The previous summer, when I had moved the phoebes' nest from under the eaves of the old Camp before tearing it down, I had said to myself that it would be a good omen if the birds should nest the next year at the new Camp. And they did, building their mossy nest under the roof of the porch. I felt honored by this, as though my work had been received into the natural order. The phoebes had added to its meaning. Later in the summer a pair of Carolina wrens also nested under the porch roof. Instead of one room, I had begun to have a house of apartments where several different kinds of life went on together. And who is to say that one kind is more possible or natural here than another? My writing and the family life of the phoebes go along here together, in a kind of equality.

In that summer of 1964 one of my first jobs was to insulate the Camp, and to wall and ceil it on the inside with six-inch tongue and groove. Once that was done, I resumed my mornings of writing. The afternoons I often spent working around the Camp, or reading, or walking in the woods. As fall approached I had a bottle-gas heater and a two-burner cooking stove installed, which made the Camp ready for cold weather use and for what is known as batching.

We still were unsure what shape our life in Kentucky was going to take, and so we had rented a furnished house in Lexington for the winter. Since I was still at work on my book, the plan was that I would do my teaching at the university on Tuesdays, Wednesdays, and Thursdays, and then drive down to the Camp to write during the other four days. Difficult as this was, it seemed the best way of assuring the quiet and the concentration that I needed for my work. But as a sort of by-product, it also made for the most intense and prolonged experience of the Camp and the river that I had ever known. In many ways that was to be a most critical time for me. Before it was over it would make a deep change in my sense of myself, and in my sense of the country I was born in.

From the beginning of September, when school started, until the first of May, when it was over, I would leave Lexington after supper Thursday night and stay at the Camp through Monday afternoon. Occasionally I would stay for some meeting or another at the university on into Thursday night, and those late drives are the ones I remember best. I would leave the university and drive across Lexington and the suburbs, and then the sixty miles through the country and down along the river to the Camp. As I went, the roads would grow less traveled, the night quieter and lonelier, the darkness broken by fewer lights. I would reach the Camp in the middle of the night, the country quiet and dark all around when I turned off the car engine and the headlights. I would hear an owl calling, or the sound of the small stream on the far bank tumbling into the river. I would go in, light the lamp and the stove, read until the room warmed, and then sleep. The next morning I would be at work on my book at the table under the window. It was always a journey from the sound of public voices to the sound of a private quiet voice rising falteringly out of the roots of my mind, that I listened carefully in the silence to hear. It was a journey from the abstract collective life of the university and the city into the intimate country of my own life. It is only in a country that is well known, full of familiar names and places, full of life that is always changing, that the mind goes free of abstractions, and renews itself in the presence of the creation, that so persistently eludes human comprehension and human law. It is only in the place that one belongs to, intimate and familiar, long watched over, that the details rise up out of the whole and become visible: the hawk stoops into the clearing before one's eyes; the wood drake, aloof and serene in his glorious plumage, swims out of his hiding place.

One clear morning as the fall was coming on I saw a chipmunk sunning on a log, as though filling himself with light and warmth in preparation for his winter sleep. He was wholly preoccupied with the sun, for though I watched him from only a few steps away he did not move. And while he mused or dozed rusty-golden sycamore leaves bigger than he was were falling all around him.

With the approach of winter the country opened. Around the Camp the limits of seeing drew back like the eyelids of a great eye. The foliage that since spring had enclosed it slowly fell away, and the outlook from its windows came to include the neighboring houses. It was as though on every frosty night the distances stole up nearer.

On the last morning of October, waking, I looked out the window and saw a fisherman in a red jacket fishing alone in his boat tied against the far bank. He sat deeply quiet and still, unmoved as a tree by my rising and the other events that went on around him. There was something heron-like in his intent waiting upon what the day might bring him out of the dark. In his quietness and patience he might have been the incarnation of some river god, at home among all things, awake while I had slept.

One bright warm day in November it was so quiet that I could hear the fallen leaves ticking, like a light rain, as they dried and contracted, scraping their points and edges against each other. That day I saw the first sapsuckers, which are here only in the winter.

Another day I woke to see the trees below the house full of birds: chickadees, titmice, juncos, bluebirds, jays. They had found a red screech owl asleep in a hollow in one of the water maples. It was a most noisy event, and it lasted a long time. The bluebirds would hover, fluttering like sparrow hawks, over the owl's hole, looking in. The titmice would perch on the very lip of the hole and scold and then, as if in fear of their own bravery, suddenly startle away. They all flew away and came back again two or three times. Everybody seemed to have a great backlog of invective to hurl down upon the head of the owl, who apparently paid no attention at all—at least when I climbed up and looked in he paid no attention to me.

I believe that the owl soon changed his sleeping place; when I next looked in he was gone. But for days afterward the birds of the neighborhood pretended he was still there, and would stop in passing to enact a sort of ritual of outrage and fright. The titmice seemed especially susceptible to the fascination of that hole. They would lean over it, yell down into it, and then spring away in a spasm of fear. They seemed to be scaring themselves for fun, like children playing around a deserted house. And yet for both birds and children there must be a seriousness in such play; they mimic fear in order to be prepared for it when it is real.

While I was eating breakfast the morning after the birds' discovery of the owl, I heard several times a voice that I knew was strange to this place. I was reading as I ate, and at first I paid little attention. But the voice persisted, and when I put my mind to it I thought it must be that of a goose. I went out with the binoculars, and saw two blue geese, the young of that year, on the water near the far bank. Though the morning was clear and the sun well up, there was a light fog blowing over the river, thickening and thinning out and thickening again, making it difficult to see, but I made out their markings clearly enough. While I watched they waded out on shore at a place where the bank had slipped, preened their feathers, and drank, and after about fifteen minutes flew away.

That afternoon I found them again in the same slip where I had seen them in the morning. I paddled the canoe within twenty feet of them, and then they only flew out onto the water a short distance away. I thought that since these birds nest to the north of Hudson Bay and often fly enormous distances in migration, I might have been the first man these geese had ever seen. Not wanting to call attention to them and so get them shot, I went on across the river and walked into the bottom on the far side. I spent some time there, looking at the wood ducks and some green-winged teal on the slews, and when I returned to the canoe a little before five the geese were gone. But my encounter with them cast a new charm on my sense of the place. They made me realize that the geography of this patch of riverbank takes in much of the geography of the world. It is under the influence of the Arctic, where the winter birds go in summer, and of the tropics, where the summer birds go in winter. It is under the influence of forests and of croplands and of strip mines in the Appalachians, and it feels the pull of the Gulf of Mexico. How many nights have the migrants loosened from their guide stars and descended here to rest or to stay for a season or to die and enter this earth? The geography of this place is airy and starry as well as earthy and watery. It has been arrived at from a thousand other places, some as far away as the poles. I have come here from great distances myself, and am resigned to the knowledge that I cannot go without leaving it better or worse. Here as well as any place I can look out my window and see the world. There are lights that arrive here from deep in the universe. A man can be provincial only by being blind and deaf to his province.

In December the winter cold began. Early in the mornings when it would be clear and cold the drift logs going down the channel would be white with frost, not having moved except as the current moved all night, as firmly embedded in the current, almost, as in the ground. A sight that has always fascinated me, when the river is up and the water swift, is to see the birds walking about, calmly feeding, on the floating logs and the mats of drift as they pass downward, slowly spinning in the currents. The ducks, too, like to feed among the drift in the channel at these times. I have seen mallards drift down, feeding among the uprooted tree trunks and the cornstalks and the rafted brush, and then fly back upstream to drift down again.

A voice I came to love and listen for on the clear cold mornings was that of the Carolina wren. He would be quick and busy, on the move, singing as he went. Unlike the calls of the other birds, whose songs, if they sang at all, would be faltering and halfhearted in the cold, the wren's song would come big and clear, filling the air of the whole neighborhood with energy, as though he could not bear to live except in the atmosphere of his own music.

Toward the end of that December a gray squirrel began building a nest in a hollow sycamore near the house. He seemed unable or unwilling to climb the trunk of the sycamore; in all the time I watched him he never attempted to do so. He always followed the same complicated route to his nest: up a grapevine, through the top of an elm and then, by a long leap, into the sycamore. On his way down this route offered him no difficulties, but the return trip, when he carried a load of sycamore leaves in his mouth, seemed fairly risky. The big leaves gave him a good deal of trouble; he frequently stopped and worked with his forepaws to make the load more compact. But because he persisted in carrying as big loads as he could, his forward vision seemed usually to be blocked altogether. I believe that his very exacting leaps were made blind, by memory, after a bit of nervous calculation beforehand. When he moved with a load of leaves, apparently because of the obstruction of his vision, he was always extremely wary, stopping often to listen and look around.

I remember one night of snow, so cold that the snow squeaked under my feet when I went out. The valley was full of moonlight; the fields were dazzling white, the woods deep black, the shadows of the trees printed heavily on the snow. And it was quiet everywhere. As long as I stood still there was not a sound.

That winter a pair of flickers drilled into the attic and slept there. Sometimes at night, after I went to bed, I could hear them stirring. But contrary to my sense of economy—though, I suppose, in keeping with theirs—they did not make do with a single hole, but bored one in one gable and one in the other. I shared my roof with them until the cold weather ended, and then evicted them. Later I put up a nesting box for them, which was promptly taken over by starlings.

One sunny morning of high water in April while I sat at work, keeping an eye on the window as usual, there were nineteen coot, a pair of wood ducks, and two pairs of blue-winged teal feeding together near the opposite bank. They fed facing upstream, working against the current, now and then allowing themselves to be carried downstream a little way, then working upstream again. After a while the four teal climbed out onto a drift log caught in the bushes near the shore. For some time they sat sunning and preening there. And then the log broke loose and began to drift again. The four birds never moved. They rode it down the river and out of sight. They accepted this accident of the river as much as a matter of fact as if it had been a purpose of theirs. Able both to swim and to fly, they made a felicity of traveling by drift log, as if serendipity were merely a way of life with them.

On the tenth of April, I woke at about six o'clock, and the first sound I heard was the song of the sycamore warbler, returned from the South. With that my thoughts entered spring. I went into the woods and found the bloodroot in bloom. Curran's flowers were coming up on the slope beside the house. In the warm evening I noticed other spring music: the calling of doves, and the slamming of newly hung screen doors.

During that winter I had spent many days and nights of watchfulness and silence here. I had learned the power of silence in a place—silence that is the imitation of absence, that permits one to be present as if absent, so that the life of the place goes its way undisturbed. It proposes an ideal of harmlessness. A man should be in the world as though he were not in it, so that it will be no worse because of his life. His obligation may not be to make "a better world," but the world certainly requires of him that he make it no worse. That, at least, was man's moral circumstance before he began his ruinous attempt to "improve" on the creation; now, perhaps, he is under an obligation to leave it better than he found it, by undoing some of the effects of his meddling and restoring its old initiatives—by making his absence the model of his presence.

But there was not only the power of silence; there was the power of attentiveness, of permanence of interest. By coming back to Kentucky and renewing my devotion to the Camp and the river valley, I had, in a sense, made a marriage with the place. I had established a trust, and within the assurance of the trust the place had begun to reveal its life to me in moments of deep intimacy and beauty. I had to come here unequivocally, accepting the place as my fate and privilege, before I could see it with clarity. I had become worthy to see what I had inherited by being born here. I had been a native; now I was beginning to belong. There is no word—certainly not *native* or *citizen* —to suggest the state I mean, that of belonging willingly and gladly and with some fullness of knowledge to a place. I had ceased to be a native as men usually are, merely by chance and legality, and had begun to be native in the fashion of the birds and animals; I had begun to be born here in mind and spirit as well as in body.

For some months after our return to Kentucky we assumed that we would settle more or less permanently in Lexington, near the university, and perhaps have a place here in the country to come to for the summers. We had thought of building a couple of more rooms onto the Camp for that purpose, and we had thought of buying a piece of woods out of reach of the river and building there. But in November of that first winter the Lane's Landing property, adjoining the Camp on the downriver side, came up for sale, and I was able to buy it. We would, we told ourselves, fix it up a little, use it in the summers, and perhaps settle there permanently some day. The previous owners moved out in February, and I began spending some of my afternoons there, working to get ready for the carpenters who would work there later on. But on the weekends Tanya and the children would often come down from Lexington, and we would walk over the place, and through the rooms of the house, talking and looking and measuring and planning. And soon we began to see possibilities there that we could not resist. Our life began to offer itself to us in a new way, in the terms of that place, and we could not escape it or satisfy it by anything partial or temporary. We made up our minds to live there. By the morning in April when I first heard the sycamore warbler, we had begun a full-scale overhaul of the house, and I had planted two dozen fruit trees. Early in July, with work completed on only three of the rooms, we moved in. After eight years, our lives enlarged in idea and in concern, our marriage enlarged into a family, we had come back to where we had begun. The Camp, always symbolically the center of our life, had fastened us here at last.

It has been several years now since I first consciously undertook to learn about the natural history of this place. My desire to do this grew out of the sense that the human life of the country is only part of its life, and that in spite of the extreme *effects* of modern man's presence on the land, his relation to it is largely superficial. In spite of all his technical prowess, nothing we have built or done has the permanence, or the congeniality with the earth, of the nesting instincts of the birds.

As soon as I felt a necessity to learn about the nonhuman world, I wished to learn about it in a hurry. And then I began to learn perhaps the most important lesson that nature had to teach me: that I could not learn about her in a hurry. The most important learning, that of experience, can be neither summoned nor sought out. The most worthy knowledge cannot be acquired by what is known as study—though that is necessary, and has its use. It comes in its own good time and in its own way to the man who will go where it lives, and wait, and be ready, and watch. Hurry is beside the point, useless, an obstruction. The thing is to be attentively present. To sit and wait is as important as to move. Patience is as valuable as industry. What is to be known is always there. When it reveals itself to you, or when you come upon it, it is by chance. The only condition is your being there and being watchful.

Though it has come slowly and a little at a time, by bits and fragments sometimes weeks apart, I realize after so many years of just being here that my knowledge of the life of this place is rich, my own life part of its richness. And at that I have only made a beginning. Eternal mysteries are here, and temporal ones too. I expect to learn many things before my life is over, and yet to die ignorant. My most inspiring thought is that this place, if I am to live well in it, requires and deserves a lifetime of the most careful attention. And the day that will finally enlighten me, if it ever comes, will come as the successor of many days spent here unenlightened or benighted entirely. "It requires more than a day's devotion," Thoreau says, "to know and to possess the wealth of a day."

At the same time my days here have taught me the futility of living for the future. Men who drudge all their lives in order to retire happily are the victims of a cheap spiritual fashion invented for their enslavement. It is no more possible to live in the future than it is to live in the past. If life is not now, it is never. It is impossible to imagine "how it will be," and to linger over that task is to prepare a disappointment. The tomorrow I hope for may very well be worse than today. There is a great waste and destructiveness in our people's desire to "get some-where." I myself have traveled several thousand miles to arrive at Lane's Landing, five miles from where I was born, and the knowledge I gained by my travels was mainly that I was born into the same world as everybody else.

Days come to me here when I rest in spirit, and am involuntarily glad. I sense the adequacy of the world, and believe that everything I need is here. I do not strain after ambition or heaven. I feel no dependence on tomorrow. I do not long to travel to Italy or Japan, but only across the river or up the hill into the woods.

And somewhere back of all this, in a relation too intricate and profound to trace out, is the life that Curran Mathews lived here before me. Perhaps he, too, experienced holy days here. Perhaps he only sensed their possibility. But if he had not come here and made a firm allegiance with this place, it is likely that I never would have. I am his follower and his heir. "For an inheritance to be really great," René Char says, "the hand of the deceased must not be seen." The Camp is my inheritance from Curran Mathews, and though certain of his meanings continue in it, his hand is not on it. As an inheritance, he touched it only as a good man touches the earth—to cherish and augment it. Where his hand went to the ground one forgotten day the flowers rise up spring after spring.

Now it is getting on toward the end of March. Just as the grass had started to grow and the jonquils were ready to bloom, we had a foot of snow and more cold. Today it is clear and thawing, but the ground is still white. Though the redbird sings his mating song, it is still winter, and my thoughts keep their winter habits. But soon there will come a day when, without expecting to, I will hear the clear seven-note song of the sycamore warbler passing over the Camp roof. Something will close and open in my mind like a page turning. It will be another spring.

A Native Hill

Pull down thy vanity, it is not man
Made courage, or made order, or made grace,
Pull down thy vanity, I say pull down.
Learn of the green world what can be thy place. . . .
                              —Ezra Pound, Canto LXXXI

THE HILL IS NOT a hill in the usual sense. It has no "other side." It is an arm of Kentucky's central upland known as The Bluegrass; one can think of it as a ridge reaching out from that center, progressively cut and divided, made ever narrower by the valleys of the creeks that drain it. The town of Port Royal in Henry County stands on one of the last heights of this upland, the valleys of two creeks, Gullion's Branch and Cane Run, opening on either side of it into the valley of the Kentucky River. My house backs against the hill's foot where it descends from the town to the river. The river, whose waters have carved the hill and so descended from it, lies within a hundred steps of my door.

Within about four miles of Port Royal, on the upland and in the bottoms upriver, all my grandparents and great-grandparents lived and left such memories as their descendants have bothered to keep. Little enough has been remembered. The family's life here goes back to my mother's great-great-grandfather and to my father's great-grandfather, but of those earliest ones there are only a few vague word-of-mouth recollections. The only place antecedent to this place that has any immediacy to any of us is the town of Cashel in County Tipperary, Ireland, which one of my great-grandfathers left as a boy to spend the rest of his life in Port Royal. His name was James Mathews, and he was a shoemaker. So well did he fit his life into this place that he is remembered, even in the family, as having belonged here. The family's only real memories of Cashel are my own, coming from a short visit I made there five years ago.

And so such history as my family has is the history of its life here. All that any of us may know of ourselves is to be known in relation to this place. And since I did most of my growing up here, and have had most of my most meaningful experiences here, the place and the history, for me, have been inseparable, and there is a sense in which my own life is inseparable from the history and the place. It is a complex inheritance, and I have been both enriched and bewildered by it.

I began my life as the old times and the last of the old-time people were dying out. The Depression and World War II delayed the mechanization of the farms here, and one of the first disciplines imposed on me was that of a teamster. Perhaps I first stood in the role of student before my father's father, who, halting a team in front of me, would demand to know which mule had the best head, which the best shoulder or rump, which was the lead mule, were they hitched right. And there came a time when I knew, and took a considerable pride in knowing. Having a boy's usual desire to play at what he sees men working at, I learned to harness and hitch and work a team. I felt distinguished by that, and took the same pride in it that other boys my age took in their knowledge of automobiles. I seem to have been born with an aptitude for a way of life that was doomed, although I did not understand that at the time. Free of any intuition of its doom, I delighted in it, and learned all I could about it.

That knowledge, and the men who gave it to me, influenced me deeply. It entered my imagination, and gave its substance and tone to my mind. It fashioned in me possibilities and limits, desires and frustrations, that I do not expect to live to the end of. And it is strange to think how barely in the nick of time it came to me. If I had been born five years later I would have begun in a different world, and would no doubt have become a different man.

Those five years made a critical difference in my life, and it is a historical difference. One of the results is that in my generation I am something of an anachronism. I am less a child of my time than the people of my age who grew up in the cities, or than the people who grew up here in my own place five years after I did. In my acceptance of twentieth-century realities there has had to be a certain deliberateness, whereas most of my contemporaries had them simply by being born to them.

In my teens, when I was away at school, I could comfort myself by recalling in intricate detail the fields I had worked and played in, and hunted over, and ridden through on horseback —and that were richly associated in my mind with people and with stories. I could recall even the casual locations of certain small rocks. I could recall the look of a hundred different kinds of daylight on all those places, the look of animals grazing over them, the postures and attitudes and movements of the men who worked in them, the quality of the grass and the crops that had grown on them. I had come to be aware of it as one is aware of one's body; it was present to me whether I thought of it or not.

I believe that this has made for a high degree of particularity in my mental processes. When I have thought of the welfare of the earth, the problems of its health and preservation, the care of its life, I have had this place before me, the part representing the whole more vividly and accurately, making clearer and more pressing demands, than any *idea* of the whole. When I have thought of kindness or cruelty, weariness or exuberance, devotion or betrayal, carelessness or care, doggedness or awkwardness or grace, I have had in my mind's eye the men and women of this place, their faces and gestures and movements. It seems to me that because of this I have a more immediate feeling for abstract principles than many of my contemporaries; the values of principles are more vivid to me.

I have pondered a great deal over a conversation I took part in a number of years ago in one of the offices of New York University. I had lived away from Kentucky for several years —in California, in Europe, in New York City. And now I had decided to go back and take a teaching job at the University of Kentucky, giving up the position I then held on the New York University faculty. That day I had been summoned by one of my superiors at the university, whose intention, I had already learned, was to persuade me to stay on in New York "for my own good."

The decision to leave had cost me considerable difficulty and doubt and hard thought—for hadn't I achieved what had become one of the almost traditional goals of American writers? I had reached the greatest city in the nation; I had a good job; I was meeting other writers and talking to them and learning from them; I had reason to hope that I might take a still larger part in the literary life of that place. On the other hand, I knew I had not escaped Kentucky, and had never really wanted to. I was still writing about it, and had recognized that I would probably need to write about it for the rest of my life. Kentucky was my fate—not an altogether pleasant fate, though it had much that was pleasing in it, but one that I could not leave behind simply by going to another place, and that I therefore felt more and more obligated to meet directly and to understand. Perhaps even more important, I still had a deep love for the place I had been born in, and liked the idea of going back to be part of it again. And that, too, I felt obligated to try to understand. Why should I love one place so much more than any other? What could be the meaning or use of such love?

The elder of the faculty began the conversation by alluding to Thomas Wolfe, who once taught at the same institution. "Young man," he said, "don't you know you can't go home again?" And he went on to speak of the advantages, for a young writer, of living in New York among the writers and the editors and the publishers.

The conversation that followed was a persistence of politeness in the face of impossibility. I knew as well as Wolfe that there is a certain *metaphorical* sense in which you can't go home again—that is, the past is lost to the extent that it cannot be lived in again. I knew perfectly well that I could not return home and be a child, or recover the secure pleasures of childhood. But I knew also that as the sentence was spoken to me it bore a self-dramatizing sentimentality that was absurd. Home—the place, the countryside—was still there, still pretty much as I left it, and there was no reason I could not go back to it if I wanted to.

As for the literary world, I had ventured some distance into that, and liked it well enough. I knew that because I was a writer the literary world would always have an importance for me and would always attract my interest. But I never doubted that the world was more important to me than the literary world; and the world would always be most fully and clearly present to me in the place I was fated by birth to know better than any other.

And so I had already chosen according to the most intimate and necessary inclinations of my own life. But what keeps me thinking of that conversation is the feeling that it was a confrontation of two radically different minds, and that it was a confrontation with significant historical overtones.

I do not pretend to know all about the other man's mind, but it was clear that he wished to speak to me as a representative of the literary world—the world he assumed that I aspired to above all others. His argument was based on the belief that once one had attained the metropolis, the literary capital, the worth of one's origins was canceled out; there simply could be nothing *worth* going back to. What lay behind one had ceased to be a part of life, and had become "subject matter." And there was the belief, long honored among American intellectuals and artists and writers, that a place such as I came from could be returned to only at the price of intellectual death; cut off from the cultural springs of the metropolis, the American countryside is Circe and Mammon. Finally, there was the assumption that the life of the metropolis is *the* experience, the *modern* experience, and that the life of the rural towns, the farms, the wilderness places is not only irrelevant to our time, but archaic as well because unknown or unconsidered by the people who really matter—that is, the urban intellectuals.

I was to realize during the next few years how false and destructive and silly those ideas are. But even then I was aware that life outside the literary world was not without honorable precedent: if there was Wolfe, there was also Faulkner; if there was James, there was also Thoreau. But what I had in my mind that made the greatest difference was the knowledge of the few square miles in Kentucky that were mine by inheritance and by birth and by the intimacy the mind makes with the place it awakens in.

I can ask now why it should be so automatically supposed that a man is wrong when he turns to face what he cannot escape, and permits himself to accept what he has desired. And I know that the answer is that in so turning back I turned directly against the current of intellectual fashion, which for so long in America has paralleled the movement of the population from country to city. At times I have found the going difficult. Though at the time of my return I felt that I was acting in faithfulness to my nature and possibilities, and now am certain that I was, there were many months when virtually none of my friends agreed with me. During the first year I was back in Kentucky, though my work apparently gave no evidence of decline, I received letters warning me against the Village Virus and the attitudes of Main Street, counseling me to remain broad-minded and intellectually aware, admonishing that I should be on the lookout for signs of decay in my work and in my mind. It was feared that I would grow paunchy and join the Farm Bureau.

It is of interest to me that, at the start I was not at all immune to this sort of thing. I felt an obligation to listen, and to take these warnings to heart. I kept an eye on myself, as though the promised signs of decay might appear at any moment. Even at the first I never really thought they would, but I continued, as though honor-bound, to expect them even after I had become sure that they wouldn't. Which is to say that, although I had maintained a vital connection with my origins, I had been enough influenced by the cultural fashion to have become compulsively suspicious both of my origins and of myself for being unwilling to divide myself from them.

I have come finally to see a very regrettable irony in what happened. At a time when originality is more emphasized in the arts, maybe, than ever before, I undertook something truly original—I returned to my origins—and it was generally thought by my literary friends that I had worked my ruin. As far as I can tell, this was simply because *my* originality, my faith in my own origins, had not been anticipated or allowed for by the *fashion* of originality.

What finally freed me from these doubts and suspicions was the insistence in what was happening to me that, far from being bored and diminished and obscured to myself by my life here, I had grown more alive and more conscious than I had ever been.

I had made a significant change in my relation to the place: before, it had been mine by coincidence or accident; now it was mine by choice. My return, which at first had been hesitant and tentative, grew wholehearted and sure. I had come back to stay. I hoped to live here the rest of my life. And once that was settled I began to *see* the place with a new clarity and a new understanding and a new seriousness. Before coming back I had been willing to allow the possibility—which one of my friends insisted on—that I already knew this place as well as I ever would. But now I began to see the real abundance and richness of it. It is, I saw, inexhaustible in its history, in the details of its life, in its possibilities. I walked over it, looking, listening, smelling, touching, alive to it as never before. I listened to the talk of my kinsmen and neighbors as I never had done, alert to their knowledge of the place, and to the qualities and energies of their speech. I began more seriously than ever to learn the names of things—the wild plants and animals, the natural processes, the local places—and to articulate my observations and memories. My language increased and strengthened, and sent my mind into the place like a live root system. And so what has become the usual order of things reversed itself with me; my mind became the root of my life rather than its sublimation. I came to see myself as growing out of the earth like the other native animals and plants. I saw my body and my daily motions as brief coherences and articulations of the energy of the place, which would fall back into the earth like leaves in the autumn.

In this awakening there has been a good deal of pain. When I lived in other places I looked on their evils with the curious eye of a traveler; I was not responsible for them; it cost me nothing to be a critic, for I had not been there long, and I did not feel that I would stay. But here, now that I am both native and citizen, there is no immunity to what is wrong. It is impossible to escape the sense that I am involved in history. What I am has been to a considerable extent determined by what my forefathers were, by how they chose to treat this place while they lived in it; the lives of most of them diminished it, and limited its possibilities, and narrowed its future. And every day I am confronted by the question of what inheritance I will leave. What do I have that I am using up? For it has been our history that each generation in this place has been less welcome to it than the last. There has been less here for them. At each arrival there has been less fertility in the soil, and a larger inheritance of destructive precedent and shameful history.

I am forever being crept up on and newly startled by the realization that my people established themselves here by killing or driving out the original possessors, by the awareness that people were once bought and sold here by my people, by the sense of the violence they have done to their own kind and to each other and to the earth, by the evidence of their persistent failure to serve either the place or their own community in it. I am forced, against all my hopes and inclinations, to regard the history of my people here as the progress of the doom of what I value most in the world: the life and health of the earth, the peacefulness of human communities and households.

And so here, in the place I love more than any other and where I have chosen among all other places to live my life, I am more painfully divided within myself than I could be in any other place.

I know of no better key to what is adverse in our heritage in this place than the account of "The Battle of the Fire-Brands," quoted in Collins' *History of Kentucky* "from the autobiography of Rev. Jacob Young, a Methodist minister." The "Newcastle" referred to is the present-day New Castle, the county seat of Henry County. I give the quote in full:

> The costume of the Kentuckians was a hunting shirt, buckskin pantaloons, a leathern belt around their middle, a scabbard, and a big knife fastened to their belt; some of them wore hats and some caps. Their feet were covered with moccasins, made of dressed deer skins. They did not think themselves dressed without their powder-horn and shot-pouch, or the gun and the tomahawk. They were ready, then, for all alarms. They knew but little. They could clear ground, raise corn, and kill turkeys, deer, bears, and buffalo; and, when it became necessary, they understood the art of fighting the Indians as well as any men in the United States.
>
> Shortly after we had taken up our residence, I was called upon to assist in opening a road from the place where Newcastle now stands, to the mouth of the Kentucky

river. That country, then, was an unbroken forest; there was nothing but an Indian trail passing the wilderness. I met the company early in the morning, with my axe, three days' provisions, and my knapsack. Here I found a captain, with about 100 men, all prepared to labor; about as jovial a company as I ever saw, all good-natured and civil. This was about the last of November, 1797. The day was cold and clear. The country through which the company passed was delightful; it was not a flat country, but, what the Kentuckians called, rolling ground—was quite well stored with lofty timber, and the undergrowth was very pretty. The beautiful canebrakes gave it a peculiar charm. What rendered it most interesting was the great abundance of wild turkeys, deer, bears, and other wild animals. The company worked hard all day, in quiet, and every man obeyed the captain's orders punctually.

About sundown, the captain, after a short address, told us the night was going to be very cold, and we must make very large fires. We felled the hickory trees in great abundance; made great log-heaps, mixing the dry wood with the green hickory; and, laying down a kind of sleepers under the pile, elevated the heap and caused it to burn rapidly. Every man had a water vessel in his knapsack; we searched for and found a stream of water. By this time, the fires were showing to great advantage; so we warmed our cold victuals, ate our suppers, and spent the evening in hearing the hunter's stories relative to the bloody scenes of the Indian war. We then heard some pretty fine singing, considering the circumstances.

Thus far, well; but a change began to take place. They became very rude, and raised the war-whoop. Their shrill shrieks made me tremble. They chose two captains, divided the men into two companies, and commenced fighting with the fire-brands—the log heaps having burned down. The only law for their government was, that no man should throw a brand without fire on it—so that they might know how to dodge. They fought, for two or three hours, in perfect good nature; till brands became scarce, and they began to violate the law. Some were severely wounded, blood began to flow freely, and they were in a fair way of commencing a fight in earnest. At this moment, the loud voice of the captain rang out above the din, ordering every man to retire to rest. They dropped their weapons of warfare, rekindled the fires, and laid down to sleep. We finished our road according to directions, and returned home in health and peace.

The significance of this bit of history is in its utter violence. The work of clearing the road was itself violent. And from the orderly violence of that labor, these men turned for amusement to disorderly violence. They were men whose element was violence; the only alternatives they were aware of were those within the comprehension of main strength. And let us acknowledge that these were the truly influential men in the history of Kentucky, as well as in the history of most of the rest of America. In comparison to the fatherhood of such as these, the so-called "founding fathers" who established our political ideals are but distant cousins. It is not John Adams or Thomas Jefferson whom we see night after night in the magic mirror of the television set; we see these builders of the road from New Castle to the mouth of the Kentucky River. Their reckless violence has glamorized all our trivialities and evils. Their aggressions have simplified our complexities and problems. They have cut all our Gordian knots. They have appeared in all our disguises and costumes. They have worn all our uniforms. Their war whoop has sanctified our inhumanity and ratified our blunders of policy.

To testify to the persistence of their influence, it is only necessary for me to confess that I read the Reverend Young's account of them with delight; I yield a considerable admiration to the exuberance and extravagance of their fight with the firebrands; I take a certain pride in belonging to the same history and the same place that they belong to—though I know that they represent the worst that is in us, and in me, and that their presence in our history has been ruinous, and that their survival among us promises ruin.

"They knew but little," the observant Reverend says of them, and this is the most suggestive thing he says. It is surely understandable and pardonable, under the circumstances, that these men were ignorant by the standards of formal schooling.

But one immediately reflects that the American Indian, who was ignorant by the same standards, nevertheless knew how to live in the country without making violence the invariable mode of his relation to it; in fact, from the ecologist's or the conservationist's point of view, he did it *no* violence. This is because he had, in place of what we would call education, a fully integrated culture, the content of which was a highly complex sense of his dependence on the earth. The same, I believe was generally true of the peasants of certain old agricultural societies, particularly in the Orient. They belonged by an intricate awareness to the earth they lived on and by, which meant that they respected it, which meant that they practiced strict economies in the use of it.

The abilities of those Kentucky road builders of 1797 were far more primitive and rudimentary than those of the Stone Age people they had driven out. They could clear the ground, grow corn, kill game, and make war. In the minds and hands of men who "know but little"—or little else—all of these abilities are certain to be destructive, even of those values and benefits their use may be intended to serve.

On such a night as the Reverend Young describes, an Indian would have made do with a small shelter and a small fire. But these road builders, veterans of the Indian War, "felled the hickory trees in great abundance; made great log-heaps . . . and caused [them] to burn rapidly." Far from making a small shelter that could be adequately heated by a small fire, their way was to make no shelter at all, and heat instead a sizable area of the landscape. The idea was that when faced with abundance one should consume abundantly—an idea that has survived to become the basis of our present economy. It is neither natural nor civilized, and even from a "practical" point of view it is to the last degree brutalizing and stupid.

I think that the comparison of these road builders with the Indians, on the one hand, and with Old World peasants on the other, is a most suggestive one. The Indians and the peasants were people who belonged deeply and intricately to their places. Their ways of life had evolved slowly in accordance with their knowledge of their land, of its needs, of their own relation of dependence and responsibility to it. The road builders, on the contrary, were *placeless* people. That is why they "knew but little." Having left Europe far behind, they had not yet in any meaningful sense arrived in America, not yet having *devoted* themselves to any part of it in a way that would produce the intricate knowledge of it necessary to live in it without destroying it. Because they belonged to no place, it was almost inevitable that they should behave violently toward the places they came to. We *still* have not, in any meaningful way, arrived in America. And in spite of our great reservoir of facts and methods, in comparison to the deep earthly wisdom of established peoples we still know but little.

But my understanding of this curiously parabolic fragment of history will not be complete until I have considered more directly that the occasion of this particular violence was the building of a road. It is obvious that one who values the idea of community cannot speak against roads without risking all sorts of absurdity. It must be noticed, nevertheless, that the predecessor to this first road was "nothing but an Indian trail passing the wilderness"—a path. The Indians, then, who had the wisdom and the grace to live in this country for perhaps ten thousand years without destroying or damaging any of it, needed for their travels no more than a footpath; but their successors, who in a century and a half plundered the area of at least half its topsoil and virtually all of its forest, felt immediately that they had to have a road. My interest is not in the question of whether or not they *needed* the road, but in the fact that the road was then, and is now, the most characteristic form of their relation to the country.

The difference between a path and a road is not only the obvious one. A path is little more than a habit that comes with knowledge of a place. It is a sort of ritual of familiarity. As a form, it is a form of contact with a known landscape. It is not destructive. It is the perfect adaptation, through experience and familiarity, of movement to place; it obeys the natural contours; such obstacles as it meets it goes around. A road, on the other hand, even the most primitive road, embodies a resistance against the landscape. Its reason is not simply the necessity for movement, but haste. Its wish is to *avoid* contact with the landscape; it seeks so far as possible to go over the country, rather than through it; its aspiration, as we see clearly in the example of our modern freeways, is to be a bridge; its tendency is to translate place into space in order to traverse it with the least effort. It is destructive, seeking to remove or destroy all obstacles in its way. The primitive road advanced by the destruction of the forest; modern roads advance by the destruction of topography.

That first road from the site of New Castle to the mouth of the Kentucky River—lost now either by obsolescence or metamorphosis—is now being crossed and to some extent replaced by its modern descendant known as I-71, and I have no wish to disturb the question of whether or not *this* road was needed. I only want to observe that it bears no relation whatever to the country it passes through. It is a pure abstraction, built to serve the two abstractions that are the poles of our national life: commerce and expensive pleasure. It was built, not according to the lay of the land, but according to a blueprint. Such homes and farmlands and woodlands as happened to be in its way are now buried under it. A part of a hill near here that would have caused it to turn aside was simply cut down and disposed of as thoughtlessly as the pioneer road builders would have disposed of a tree. Its form is the form of speed, dissatisfaction, and anxiety. It represents the ultimate in engineering sophistication, but the crudest possible valuation of life in this world. It is as adequate a symbol of our relation to our country now as that first road was of our relation to it in 1797.

But the sense of the past also gives a deep richness and resonance to nearly everything I see here. It is partly the sense that what I now see other men that I have known once saw, and partly that this knowledge provides an imaginative access to what I do not know. I think of the country as a kind of palimpsest scrawled over with the comings and goings of people, the erasure of time already in process even as the marks of passage are put down. There are the ritual marks of neighborhood —roads, paths between houses. There are the domestic paths from house to barns and outbuildings and gardens, farm roads threading the pasture gates. There are the wanderings of hunters and searchers after lost stock, and the speculative or meditative or inquisitive "walking around" of farmers on wet days and Sundays. There is the spiraling geometry of the rounds of implements in fields, and the passing and returning scratches of plows across croplands. Often these have filled an interval, an opening, between the retreat of the forest from the virgin ground and the forest's return to ground that has been worn out and given up. In the woods here one often finds cairns of stones picked up out of furrows, gullies left by bad farming, forgotten roads, stone chimneys of houses long rotted away or burned.

Occasionally one stumbles into a coincidence that, like an unexpected alignment of windows, momentarily cancels out the sense of historical whereabouts, giving with an overwhelming immediacy an awareness of the reality of the past.

The possibility of this awareness is always immanent in old homesites. It may suddenly bear in upon one at the sight of old orchard trees standing in the dooryard of a house now filled with baled hay. It came to me when I looked out the attic window of a disintegrating log house and saw a far view of the cleared ridges with wooded hollows in between, and nothing in sight to reveal the date. Who was I, leaning to the window? When?

It broke upon me one afternoon when, walking in the woods on one of my family places, I came upon a gap in a fence, wired shut, but with deep-cut wagon tracks still passing through it under the weed growth and the fallen leaves. Where that thicket stands there was crop ground, maybe as late as my own time. I knew some of the men who tended it; their names and faces were instantly alive in my mind. I knew how it had been with them—how they would harness their mule teams in the early mornings in my grandfather's big barn and come to the woods-rimmed tobacco patches, the mules' feet wet with the dew. And in the solitude and silence that came upon them they would set to work, their water jugs left in the shade of bushes in the fencerows.

As a child I learned the early mornings in these places for myself, riding out in the wagons with the tobacco-cutting crews to those steep fields in the dew-wet shadow of the woods. As the day went on the shadow would draw back under the feet of the trees, and it would get hot. Little whirlwinds would cross the opening, picking up the dust and the dry "ground leaves" of the tobacco. We made a game of running with my grandfather to stand, shoulders scrunched and eyes squinched, in their middles.

Having such memories, I can acknowledge only with reluctance and sorrow that those slopes should never have been broken. Rich as they were, they were too steep. The humus stood dark and heavy over them once; the plow was its doom.

Early one February morning in thick fog and spattering rain I stood on the riverbank and listened to a towboat working its way downstream. Its engines were idling, nudging cautiously through the fog into the Cane Run bend. The end of the head barge emerged finally like a shadow, and then the second barge appeared, and then the towboat itself. They made the bend, increased power, and went thumping off out of sight into the fog again.

Because the valley was so enclosed in fog, the boat with its tow appearing and disappearing again into the muffling whiteness within two hundred yards, the moment had a curious ambiguity. It was as though I was not necessarily myself at all. I could have been my grandfather, in his time, standing there watching, as I knew he had.

II

I start down from one of the heights of the upland, the town of Port Royal at my back. It is a winter day, overcast and still, and the town is closed in itself, humming and muttering a little, like a winter beehive.

The dog runs ahead, prancing and looking back, knowing the way we are about to go. This is a walk well established with us—a route in our minds as well as on the ground. There is a sort of mystery in the establishment of these ways. Any time one crosses a given stretch of country with some frequency, no matter how wanderingly one begins, the tendency is always toward habit. By the third or fourth trip, without realizing it, one is following a fixed path, going the way one went before. After that, one may still wander, but only by deliberation, and when there is reason to hurry, or when the mind wanders rather than the feet, one returns to the old route. Familiarity has begun. One has made a relationship with the landscape, and the form and the symbol and the enactment of the relationship is the path. These paths of mine are seldom worn on the ground. They are habits of mind, directions and turns. They are as personal as old shoes. My feet are comfortable in them.

From the height I can see far out over the country, the long open ridges of the farmland, the wooded notches of the streams, the valley of the river opening beyond, and then more ridges and hollows of the same kind.

Underlying this country, nine hundred feet below the highest ridgetops, more than four hundred feet below the surface of the river, is sea level. We seldom think of it here; we are a long way from the coast, and the sea is alien to us. And yet the attraction of sea level dwells in this country as an ideal dwells in a man's mind. All our rains go in search of it and, departing, they have carved the land in a shape that is fluent and falling. The streams branch like vines, and between the branches the land rises steeply and then rounds and gentles into the long narrowing fingers of ridgeland. Near the heads of the streams even the steepest land was not too long ago farmed and kept cleared. But now it has been given up and the woods is returning. The wild is flowing back like a tide. The arable ridgetops reach out above the gathered trees like headlands into the sea, bearing their human burdens of fences and houses and barns, crops and roads.

Looking out over the country, one gets a sense of the whole of it: the ridges and hollows, the clustered buildings of the farms, the open fields, the woods, the stock ponds set like coins into the slopes. But this is a surface sense, an exterior sense, such as you get from looking down on the roof of a house. The height is a threshold from which to step down into the wooded folds of the land, the interior, under the trees and along the branching streams.

I pass through a pasture gate on a deep-worn path that grows shallow a little way beyond, and then disappears altogether into the grass. The gate has gathered thousands of passings to and fro that have divided like the slats of a fan on either side of it. It is like a fist holding together the strands of a net.

Beyond the gate the land leans always more steeply toward the branch. I follow it down, and then bear left along the crease at the bottom of the slope. I have entered the downflow of the land. The way I am going is the way the water goes. There is something comfortable and fit-feeling in this, something free in this yielding to gravity and taking the shortest way down. The mind moves through the watershed as the water moves.

As the hollow deepens into the hill, before it has yet entered the woods, the grassy crease becomes a raw gully, and along the steepening slopes on either side I can see the old scars of erosion, places where the earth is gone clear to the rock. My people's errors have become the features of my country.

It occurs to me that it is no longer possible to imagine how this country looked in the beginning, before the white people drove their plows into it. It is not possible to know what was the shape of the land here in this hollow when it was first cleared. Too much of it is gone, loosened by the plows and washed away by the rain. I am walking the route of the departure of the virgin soil of the hill. I am not looking at the same land the firstcomers saw. The original surface of the hill is as extinct as the passenger pigeon. The pristine America that the first white man saw is a lost continent, sunk like Atlantis in the sea. The thought of what was here once and is gone forever will not leave me as long as I live. It is as though I walk knee-deep in its absence.

The slopes along the hollow steepen still more, and I go in under the trees, I pass beneath the surface. I am enclosed, and my sense, my interior sense, of the country becomes intricate. There is no longer the possibility of seeing very far. The distances are closed off by the trees and the steepening walls of the hollow. One cannot grow familiar here by sitting and looking as one can up in the open on the ridge. Here the eyes become dependent on the feet. To see the woods from the inside one must look and move and look again. It is inexhaustible in its standpoints. A lifetime will not be enough to experience it all. Not far from the beginning of the woods, and set deep in the earth in the bottom of the hollow, is a rock-walled pool not a lot bigger than a bathtub. The wall is still nearly as straight and tight as when it was built. It makes a neatly turned narrow horseshoe, the open end downstream. This is a historical ruin, dug here either to catch and hold the water of the little branch, or to collect the water of a spring whose vein broke to the surface here—it is probably no longer possible to know which. The pool is filled with earth now, and grass grows in it. And the branch bends around it, cut down to the bare rock, a torrent after heavy rain, other times bone dry. All that is certain is that when the pool was dug and walled there was deep topsoil on the hill to gather and hold the water. And this high up, at least, the bottom of the hollow, instead of the present raw notch of the stream bed, wore the same mantle of soil as the slopes, and the stream was a steady seep or trickle, running most or all of the year. This tiny pool no doubt once furnished water for a considerable number of stock through the hot summers. And now it is only a lost souvenir, archaic and useless, except for the bitter intelligence there is in it. It is one of the monuments to what is lost.

Wherever one goes along the streams of this part of the country, one is apt to come upon this old stonework. There are walled springs and pools. There are the walls built in the steeper hollows where the fences cross or used to cross; the streams have drifted dirt in behind them, so that now where they are still intact they make waterfalls that have scooped out small pools at their feet. And there used to be miles of stone fences, now mostly scattered and sifted back into the ground.

Considering these, one senses a historical patience, now also extinct in the country. These walls were built by men working long days for little wages, or by slaves. It was work that could not be hurried at, a meticulous finding and fitting together, as though reconstructing a previous wall that had been broken up and scattered like puzzle pieces along the stream beds. The wall would advance only a few yards a day. The pace of it could not be borne by most modern men, even if the wages could be afforded. Those men had to move in closer accord with their own rhythms, and nature's, than we do. They had no machines. Their capacities were only those of flesh and blood. They talked as they worked. They joked and laughed. They sang. The work was exacting and heavy and hard and slow. No opportunity for pleasure was missed or slighted. The days and the years were long. The work was long. At the end of this job the next would begin. Therefore, be patient. Such pleasure as there is, is here, now. Take pleasure as it comes. Take work as it comes. The end may never come, or when it does it may be the wrong end.

Now the men who built the walls and the men who had them built have long gone underground to be, along with the buried ledges and the roots and the burrowing animals, a part of the nature of the place in the minds of the ones who come after them. I think of them lying still in their graves, as level as the sills and thresholds of their lives, as though resisting to the last the slant of the ground. And their old walls, too, re-enter nature, collecting lichens and mosses with a patience their builders never conceived of.

Like the pasture gates, the streams are great collectors of comings and goings. The streams go down, and paths always go down beside the streams. For a while I walk along an old wagon road that is buried in leaves—a fragment, beginningless and endless as the middle of a sentence on some scrap of papyrus. There is a cedar whose branches reach over this road, and under the branches I find the leavings of two kills of some bird of prey. The most recent is a pile of blue jay feathers. The other has been rained on and is not identifiable. How little we know. How little of this was intended or expected by any man. The road that has become the grave of men's passages has led to the life of the woods.

> And I say to myself: Here is your road
> without beginning or end, appearing
> out of the earth and ending in it, bearing
> no load but the hawk's kill, and the leaves
> building earth on it, something more
> to be borne. Tracks fill with earth
> and return to absence. The road was worn
> by men bearing earth along it. They have come
> to endlessness. In their passing
> they could not stay in, trees have risen
> and stand still. It is leading to the dark,
> to mornings where you are not. Here
> is your road, beginningless and endless as God.

Now I have come down within the sound of the water. The winter has been rainy, and the hill is full of dark seeps and trickles, gathering finally, along these creases, into flowing streams. The sound of them is one of the elements, and defines a zone.

When their voices return to the hill after their absence during summer and autumn, it is a better place to be. A thirst in the mind is quenched.

I have already passed the place where water began to flow in the little stream bed I am following. It broke into the light from beneath a rock ledge, a thin glittering stream. It lies beside me as I walk, overtaking me and going by, yet not moving, a thread of light and sound. And now from below comes the steady tumble and rush of the water of Camp Branch—whose nameless camp was it named for?—and gradually as I descend the sound of the smaller stream is lost in the sound of the larger.

The two hollows join, the line of the meeting of the two spaces obscured even in winter by the trees. But the two streams meet precisely as two roads. That is, the stream *beds* do; the one ends in the other. As for the meeting of the waters, there is no looking at that. The one flow does not end in the other, but continues in it, one with it, two clarities merged without a shadow.

All waters are one. This is a reach of the sea, flung like a net over the hill, and now drawn back to the sea. And as the sea is never raised in the earthly nets of fishermen, so the hill is never caught and pulled down by the watery net of the sea. But always a little of it is. Each of the gathering strands of the net carries back some of the hill melted in it. Sometimes, as now, it carries so little that the water seems to flow clear; sometimes it carries a lot and is brown and heavy with it. Whenever greedy or thoughtless men have lived on it, the hill has literally flowed out of their tracks into the bottom of the sea.

There appears to be a law that when creatures have reached the level of consciousness, as men have, they must become conscious of the creation; they must learn how they fit into it and what its needs are and what it requires of them, or else pay a terrible penalty: the spirit of the creation will go out of them, and they will become destructive; the very earth will depart from them and go where they cannot follow.

My mind is never empty or idle at the joinings of the streams. Here is the work of the world going on. The creation is felt, alive and intent on its materials, in such places. In the angle of the meeting of the two streams stands the steep wooded point of the ridge, like the prow of an upturned boat—finished, as it was a thousand years ago, as it will be in a thousand years. Its becoming is only incidental to its being. It will be because it is. It has no aim or end except to be. By being, it is growing and wearing into what it will be. The fork of the stream lies at the foot of a finished sculpture. But the stream is no dead tool; it is alive, it is still at its work. Put your hand to it to learn the health of this part of the world. It is the wrist of the hill.

Perhaps it is to prepare to hear some day the music of the spheres that I am always turning my ears to the music of streams. There is indeed a music in streams, but it is not for the hurried. It has to be loitered by and imagined. Or imagined *toward*, for it is hardly for men at all. Nature has a patient ear. To her the slowest funeral march sounds like a jig. She is satisfied to have the notes drawn out to the lengths of days or weeks or months. Small variations are acceptable to her, modulations as leisurely as the opening of a flower.

The stream is full of stops and gates. Here it has piled up rocks in its path, and pours over them into a tiny pool it has scooped at the foot of its fall. Here it has been dammed by a mat of leaves caught behind a fallen limb. Here it must force a narrow passage, here a wider one. Tomorrow the flow may increase or slacken, and the tone will shift. In an hour or a week that rock may give way, and the composition will advance by another note. Some idea of it may be got by walking slowly along and noting the changes as one passes from one little fall or rapid to another. But this is a highly simplified and diluted version of the real thing, which is too complex and widespread ever to be actually heard by us. The ear must imagine an impossible patience in order to grasp even the unimaginableness of such music.

But the creation is musical, and this is a part of its music, as bird song is, or the words of poets. The music of the streams is the music of the shaping of the earth, by which the rocks are pushed and shifted downward toward the level of the sea.

And now I find lying in the path an empty beer can. This is the track of the ubiquitous man Friday of all our woods. In my walks I never fail to discover some sign that he has preceded me. I find his empty shotgun shells, his empty cans and bottles, his sandwich wrappings. In wooded places along roadsides one is apt to find, as well, his overtraveled bedsprings, his outcast refrigerator, and heaps of the imperishable refuse of his modern kitchen. A year ago, almost in this same place where I have found this beer can, I found a possum that he had shot dead and left lying, in celebration of his manhood. He is the true American pioneer, perfectly at rest in his assumption that he is the first and the last whose inheritance and fate this place will ever be. Going forth, as he may think, to sow, he only broadcasts his effects.

As I go on down the path alongside Camp Branch, I walk by the edge of croplands abandoned only within my own lifetime. On my left are the south slopes where the woods is old, long undisturbed. On my right, the more fertile north slopes are covered with patches of briars and sumacs and a lot of young walnut trees. Tobacco of an extraordinary quality was once grown here, and then the soil wore thin, and these places were given up for the more accessible ridges that were not steep, where row cropping made better sense anyway. But now, under the thicket growth, a mat of bluegrass has grown to testify to the good nature of this ground. It was fine dirt that lay here once, and I am far from being able to say that I could have resisted the temptation to plow it. My understanding of what is best for it is the tragic understanding of hindsight, the awareness that I have been taught what was here to be lost by the loss of it.

We have lived by the assumption that what was good for us would be good for the world. And this has been based on the even flimsier assumption that we could know with any certainty what was good even for us. We have fulfilled the danger of this by making our personal pride and greed the standard of our behavior toward the world—to the incalculable disadvantage of the world and every living thing in it. And now, perhaps very close to too late, our great error has become clear. It is not only our own creativity—our own capacity for life—that is stifled by our arrogant assumption; the creation itself is stifled.

We have been wrong. We must change our lives, so that it will be possible to live by the contrary assumption that what is good for the world will be good for us. And that requires that we make the effort to *know* the world and to learn what is good for it. We must learn to co-operate in its processes, and to yield to its limits. But even more important, we must learn to acknowledge that the creation is full of mystery; we will never entirely understand it. We must abandon arrogance and stand in awe. We must recover the sense of the majesty of creation, and the ability to be worshipful in its presence. For I do not doubt that it is only on the condition of humility and reverence before the world that our species will be able to remain in it.

Standing in the presence of these worn and abandoned fields, where the creation has begun its healing without the hindrance or the help of man, with the voice of the stream in the air and the woods standing in silence on all the slopes around me, I am deep in the interior not only of my place in the world, but of my own life, its sources and searches and concerns. I first came into these places following the men to work when I was a child. I knew the men who took their lives from such fields as these, and their lives to a considerable extent made my life what it is. In what came to me from them there was both wealth and poverty, and I have been a long time discovering which was which.

It was in the woods here along Camp Branch that Bill White, my grandfather's Negro hired hand, taught me to hunt squirrels. Bill lived in a little tin-roofed house on up nearer the head of the hollow. And this was, I suppose more than any other place, his hunting ground. It was the place of his freedom, where he could move without subservience, without considering who he was or who anybody else was. On late summer mornings, when it was too wet to work, I would follow him into the woods. As soon as we stepped in under the trees he would become silent and absolutely attentive to the life of the place. He was a good teacher and an exacting one. The rule seemed to be that if I wanted to stay with him, I had to make it possible for him to forget I was there. I was to make no noise. If I did he would look back and make a downward emphatic gesture with his hand, as explicit as writing: Be quiet, or go home. He would see a squirrel crouched in a fork or lying along the top of a branch, and indicate with a grin and a small jerk of his head where I should look; and then wait, while I, conscious of being watched and demanded upon, searched it out for myself. He taught me to look and to listen and to be quiet. I wonder if he knew the value of such teaching or the rarity of such a teacher.

In the years that followed I hunted often here alone. And later in these same woods I experienced my first obscure dissatisfactions with hunting. Though I could not have put it into words then, the sense had come to me that hunting as I knew it—the eagerness to kill something I did not need to eat—was an artificial relation to the place, when what I was beginning to need, just as inarticulately then, was a relation that would be deeply natural and meaningful. That was a time of great uneasiness and restlessness for me. It would be the fall of the year, the leaves would be turning, and ahead of me would be another year of school. There would be confusions about girls and ambitions, the wordless hurried feeling that time and events and my own nature were pushing me toward what I was going to be—and I had no notion what it was, or how to prepare.

And then there were years when I did not come here at all—when these places and their history were in my mind, and part of me, in places thousands of miles away. And now I am here again, changed from what I was, and still changing. The future is no more certain to me now than it ever was, though its risks are clearer, and so are my own desires: I am the father of two young children whose lives are hostages given to the future. Because of them and because of events in the world, life seems more fearful and difficult to me now than ever before. But it is also more inviting, and I am constantly aware of its nearness to joy. Much of the interest and excitement that I have in my life now has come from the deepening, in the years since my return here, of my relation to this countryside that is my native place. For in spite of all that has happened to me in other places, the great change and the great possibility of change in my life has been in my sense of this place. The major difference is perhaps only that I have grown able to be wholeheartedly present here. I am able to sit and be quiet at the foot of some tree here in this woods along Camp Branch, and feel a deep peace, both in the place and in my awareness of it, that not too long ago I was not conscious of the possibility of. This peace is partly in being free of the suspicion that pursued me for most of my life, no matter where I was, that there was perhaps another place I *should* be, or would be happier or better in; it is partly in the increasingly articulate consciousness of being here, and of the significance and importance of being here.

After more than thirty years I have at last arrived at the candor necessary to stand on this part of the earth that is so full of my own history and so much damaged by it, and ask: What *is* this place? What is in it? What is its nature? How should men live in it? What must I do?

I have not found the answers, though I believe that in partial and fragmentary ways they have begun to come to me. But the questions are more important than their answers. In the final sense they *have* no answers. They are like the questions—they are perhaps the same questions—that were the discipline of Job. They are a part of the necessary enactment of humility, teaching a man what his importance is, what his responsibility is, and what his place is, both on the earth and in the order of things. And though the answers must always come obscurely and in fragments, the questions must be persistently asked. They are fertile questions. In their implications and effects, they are moral and aesthetic and, in the best and fullest sense, practical. They promise a relationship to the world that is decent and preserving.

They are also, both in origin and effect, religious. I am uneasy with the term, for such religion as has been openly practiced in this part of the world has promoted and fed upon a destructive schism between body and soul, heaven and earth. It has encouraged people to believe that the world is of no importance, and that their only obligation in it is to submit to certain churchly formulas in order to get to heaven. And so the people who might have been expected to care most selflessly for the world have had their minds turned elsewhere—to a pursuit of "salvation" that was really only another form of gluttony and self-love, the desire to perpetuate their own small lives beyond the life of the world. The heaven-bent have abused the earth thoughtlessly, by inattention, and their negligence has permitted and encouraged others to abuse it deliberately. Once the creator was removed from the creation, divinity became only a remote abstraction, a social weapon in the hands of the religious institutions. This split in public values produced or was accompanied by, as it was bound to be, an equally artificial and ugly division in people's lives, so that a man, while pursuing heaven with the sublime appetite he thought of as his soul, could turn his heart against his neighbors and his hands against the world. For these reasons, though I know that my questions *are* religious, I dislike having to *say* that they are.

But when I ask them my aim is not primarily to get to heaven. Though heaven is certainly more important than the earth if all they say about it is true, it is still morally incidental to it and dependent on it, and I can only imagine it and desire it in terms of what I know of the earth. And so my questions do not aspire beyond the earth. They aspire *toward* it and *into* it. Perhaps they aspire *through* it. They are religious because they are asked at the limit of what I know; they acknowledge mystery and honor its presence in the creation; they are spoken in reverence for the order and grace that I see, and that I trust beyond my power to see.

The stream has led me down to an old barn built deep in the hollow to house the tobacco once grown on those abandoned fields. Now it is surrounded by the trees that have come back on every side—a relic, a fragment of another time, strayed out of its meaning. This is the last of my historical landmarks. To here, my walk has had insistent overtones of memory and history. It has been a movement of consciousness through knowledge, eroding and shaping, adding and wearing away. I have descended like the water of the stream through what I know of myself, and now that I have there is a little more to know. But here at the barn, the old roads and the cow paths—the formal connections with civilization—come to an end.

I stoop between the strands of a barbed-wire fence, and in that movement I go out of time into timelessness. I come into a wild place. I walk along the foot of a slope that was once cut bare of trees, like all the slopes of this part of the country—but long ago; and now the woods is established again, the ground healed, the trees grown big, their trunks rising clean, free of undergrowth. The place has a serenity and dignity that one feels immediately; the creation is whole in it and unobstructed. It is free of the strivings and dissatisfactions, and partialities and imperfections of places under the mechanical dominance of men. Here, what to a housekeeper's eye might seem disorderly is nonetheless orderly and within order; what might seem arbitrary or accidental is included in the design of the whole as if by intention; what might seem evil or violent is a comfortable member of the household. Where the creation is whole nothing is extraneous. The presence of the creation here makes this a holy place, and it is as a pilgrim that I have come—to give the homage of awe and love, to submit to mystification. It is the creation that has attracted me, its perfect interfusion of life and design. I have made myself its follower and its apprentice.

One early morning last spring, I came and found the woods floor strewn with bluebells. In the cool sunlight and the lacy shadows of the spring woods the blueness of those flowers, their elegant shape, their delicate fresh scent kept me standing and looking. I found a rich delight in them that I cannot describe and that I will never forget. Though I had been familiar for years with most of the spring woods flowers, I had never seen these and had not known they grew here. Looking at them, I felt a strange feeling of loss and sorrow that I had never seen them before. But I was also exultant that I saw them now—that they were here.

For me, in the thought of them will always be the sense of the joyful surprise with which I found them—the sense that came suddenly to me then that the world is blessed beyond my understanding, more abundantly than I will ever know. What lives are still ahead of me here to be discovered and exulted in, tomorrow, or in twenty years? What wonder will be found here on the morning after my death? Though as a man I inherit great evils and the possibility of great loss and suffering, I know that my life is blessed and graced by the yearly flowering of the bluebells. How perfect they are! In their presence I am humble and joyful. If I were given all the learning and all the methods of my race I could not make one of them, or even imagine one. Solomon in all his glory was not arrayed like one of these. It is the privilege and the labor of the apprentice of creation to come with his imagination into the unimaginable, and with his speech into the unspeakable.

III

On weekends the Air Force reservists practice war. Their jet fighters come suddenly over the crest of the hill into the opening of the valley, very low, screeching insistently into our minds —perfecting deadliness. We can see the rockets nestling under their wings. They do not represent anything I understand as my own or that I identify with. Unable to imagine myself flying one or employing one for any of my purposes, I am left with the alternative of imagining myself their target. They strike fear into me. I fear them, I think, with the same fear—though not, certainly, with the same intensity—with which they are feared in the villages and the rice paddies of Vietnam. I am afraid that nothing I value can withstand them. I am unable to believe that what I most hope for can be served by them.

On weekends, too, the hunters come out from the cities to kill for pleasure such game as has been able to survive the constraints and destructions of the human economy, the highway traffic, the poison sprays. They come weighted down with expensive clothes and equipment, all purchased for the sake of a few small deaths in a country they neither know nor care about. We hear their guns, and see their cars parked at lane ends and on the roadsides.

Sometimes I can no longer think in the house or in the garden or in the cleared fields. They bear too much resemblance to our failed human history—failed, because it has led to this human present that is such a bitterness and a trial. And so I go to the woods. As I enter in under the trees, dependably, almost at once, and by nothing I do, things fall into place. I enter an order that does not exist outside, in the human spaces. I feel my life take its place among the lives—the trees, the annual plants, the animals and birds, the living of all these and the dead—that go and have gone to make the life of the earth. I am less important than I thought, the human race is less important than I thought. I rejoice in that. My mind loses its urgings, senses its nature, and is free. The forest grew here in its own time, and so I will live, suffer and rejoice, and die in my own time. There is nothing that I may decently hope for that I cannot reach by patience as well as by anxiety. The hill, which is a part of America, has killed no one in the service of the American government. Then why should I, who am a fragment of the hill? I wish to be as peaceable as my land, which does no violence, though it has been the scene of violence and has had violence done to it.

How, having a consciousness, an intelligence, a human spirit —all the vaunted equipment of my race—can I humble myself before a mere piece of the earth and speak of myself as its fragment? Because my mind transcends the hill only to be filled with it, to comprehend it a little, to know that it lives on the hill in time as well as in place, to recognize itself as the hill's fragment.

The false and truly belittling transcendence is ownership. The hill has had more owners than its owners have had years —they are grist for its mill. It has had few friends. But I wish to be its friend, for I think it serves its friends well. It tells them they are fragments of its life. In its life they transcend their years.

The most exemplary nature is that of the topsoil. It is very Christ-like in its passivity and beneficence, and in the penetrating energy that issues out of its peaceableness. It increases by experience, by the passage of seasons over it, growth rising out of it and returning to it, not by ambition or aggressiveness. It is enriched by all things that die and enter into it. It keeps the past, not as history or as memory, but as richness, new possibility. Its fertility is always building up out of death into promise. Death is the bridge or the tunnel by which its past enters its future.

Ownership presumes that only man lives in time, and that the hill is merely present, merely a place. But the poet who lives, heeded or not, within the owner knows that the hill is also taking place. It is as alive as its owner. The two lives go side by side in time together, and ultimately they are the same.

To walk in the woods, mindful only of the *physical* extent of it, is to go perhaps as owner, or as knower, confident of one's own history and of one's own importance. But to go there, mindful as well of its temporal extent, of the age of it, and of all that led up to the present life of it, and of all that will probably follow it, is to feel oneself a flea in the pelt of a great living thing, the discrepancy between its life and one's own so great that it cannot be imagined. One has come into the presence of mystery.

After all the trouble one has taken to be a modern man, one has come back under the spell of a primitive awe, wordless and humble.

In the centuries before its settlement by white men, among the most characteristic and pleasing features of the floor of this valley, and of the stream banks on its slopes, were the forests and the groves of great beech trees. With their silver bark and their light graceful foliage, turning gold in the fall, they were surely as lovely as any forests that ever grew on earth. I think so because I have seen their diminished descendants, which have returned to stand in the wasted places that we have so quickly misused and given up. But those old forests are all gone. We will never know them as they were. We have driven them beyond the reach of our minds, only a vague hint of their presence returning to haunt us, as though in dreams—a fugitive rumor of the nobility and beauty and abundance of the squandered maidenhood of our world—so that, do what we will, we will never quite be satisfied ever again to be here.

The country, as we have made it by the pretense that we can do without it as soon as we have completed its metamorphosis into cash, no longer holds even the possibility of such forests, for the topsoil that they made and stood upon, like children piling up and trampling underfoot the fallen leaves, is no longer here.

It is thought that the beech forests of the Midwest, in the time before the white invasion, produced an annual nut crop of a billion bushels. Suppose that early in our history we had learned, like the Indians, to make use of this bounty, and of the rest of the natural produce of the country. We would have been unimaginably different, and would have been less in need of forgiveness. Instead of asking what was already here that might be of use to us, we hastened to impose on the face of the new country, like the scraps and patches of a collage, the fields and the crop rows and the fences of Europe. We destroyed the abundance that lay before us simply by being unwilling or unable to acknowledge that it was there. We did not know where we were, and to avoid the humility and the labor of our ignorance, we pretended to be where we had come from. And so there is a sense in which we are still not here. Because we have ignored the place we have come to, our presence here remains curiously accidental, as though we came by misapprehension or mistake, like birds driven out to sea by a storm.

We still want to be Europeans. We have abandoned Europe only reluctantly and only by necessity. We tried for two hundred years to grow the European grape here, without success, before turning to the development of the native vines that abounded in variety and in health in all parts of the country.

There is an ominous—perhaps a fatal—presumptuousness, like the *hybris* of the ancient kings, in living in a place by the *imposition* on it of one's preconceived or inherited ideas and wishes. And that is the way we white people have lived in America throughout our history, and it is the way our history now teaches us to live here.

Surely there could be a more indigenous life than we have. There could be a consciousness that would establish itself on a place by understanding its nature and learning what is potential in it. A man ought to study the wilderness of a place before applying to it the ways he learned in another place. Thousands of acres of hill land, here and in the rest of the country, were wasted by a system of agriculture that was fundamentally alien to it. For more than a century, here, the steepest hillsides were farmed, by my forefathers and their neighbors, as if they were flat, and as if this was not a country of heavy rains. And that symbolizes well enough how alien we remain, in our behavior and in our thoughts, to our country. We haven't yet, in any meaningful sense, arrived in these places that we declare we own. We undertook the privilege of the virgin abundance of this land without any awareness at all that we undertook at the same time a responsibility toward it. That responsibility has never yet impressed itself upon our character; its absence in us is signified on the land by scars.

Until we understand what the land is, we are at odds with everything we touch. And to come to that understanding it is necessary, even now, to leave the regions of our conquest—the cleared fields, the towns and cities, the highways—and re-enter the woods. For only there can a man encounter the silence and the darkness of his own absence. Only in this silence and darkness can he recover the sense of the world's longevity, of its ability to thrive without him, of his inferiority to it and his dependence on it. Perhaps then, having heard that silence and seen that darkness, he will grow humble before the place and begin to take it in—to learn *from it* what it is. As its sounds come into his hearing, and its lights and colors come into his vision, and its odors come into his nostrils, then he may come into *its* presence as he never has before, and he will arrive in his place and will want to remain. His life will grow out of the ground like the other lives of the place, and take its place among them. He will be *with* them—neither ignorant of them, nor indifferent to them, nor against them—and so at last he will grow to be native-born. That is, he must re-enter the silence and the darkness, and be born again.

One winter night nearly twenty years ago I was in the woods with the coon hunters, and we were walking toward the dogs, who had moved out to the point of the bluff where the valley of Cane Run enters the valley of the river. The footing was difficult, and one of the hunters was having trouble with his lantern. The flame would "run up" and smoke the globe, so that the light it gave obscured more than it illuminated, an obstacle between his eyes and the path. At last he cursed it and flung it down into a hollow. Its little light went looping down through the trees and disappeared, and there was a distant tinkle of glass as the globe shattered. After that he saw better and went along the bluff easier than before, and lighter, too.

Not long ago, walking up there, I came across his old lantern lying rusted in the crease of the hill, half buried already in the siftings of the slope, and I let it lie. But I've kept the memory that it renewed. I have made it one of my myths of the hill. It has come to be truer to me now than it was then.

For I have turned aside from much that I knew, and have given up much that went before. What will not bring me, more certainly than before, to where I am is of no use to me. I have stepped out of the clearing into the woods. I have thrown away my lantern, and I can see the dark.

In order to know the hill it is necessary to slow the mind down, imaginatively at least, to the hill's pace. For the hill, like Valéry's sycamore, is a voyager standing still. Never moving a step, it travels through years, seasons, weathers, days and nights. These are the measures of its time, and they alter it, marking their passage on it as on a man's face. The hill has never observed a Christmas or an Easter or a Fourth of July. It has nothing to do with a dial or a calendar. Time is told in it mutely and immediately, with perfect accuracy, as it is told by the heart in the body. Its time is the birth and the flourishing and the death of the many lives that are its life.

The hill is like an old woman, all her human obligations met, who sits at work day after day, in a kind of rapt leisure, at an intricate embroidery. She has time for all things. Because she does not expect ever to be finished, she is endlessly patient with details. She perfects flower and leaf, feather and song, adorning the briefest life in great beauty as though it were meant to last forever.

In the early spring I climb up through the woods to an east-facing bluff where the bloodroot bloom in scattered colonies around the foot of the rotting monument of a tree trunk. The sunlight is slanting, clear, through the leafless branches. The flowers are white and perfect, delicate as though shaped in air and water. There is a fragility about them that communicates how short a time they will last. There is some subtle bond between them and the dwindling great trunk of the dead tree. There comes on me a pressing wish to imitate them and so preserve them. But I know that what draws me to them would not pass over into anything I can *do*. They will be lost. In a few days none will be here. Suddenly I want to be, I don't know which, a worshiper or a god.

Coming upon a mushroom growing out of a pad of green moss between the thick roots of an oak, the sun and the dew still there together, I have felt my mind irresistibly become small, to inhabit that place, leaving me standing vacant and be-wildered, like a boy whose captured field mouse has just leaped out of his hand.

As I slowly fill with the knowledge of this place, and sink into it, I come to the sense that my life here is inexhaustible, that its possibilities lie rich behind and ahead of me, that when I am dead it will not be used up.

Too much that we do is done at the expense of something else, or somebody else. There is some intransigent destructiveness in us. My days, though I think I know better, are filled with a thousand irritations, worries, regrets for what has happened and fears for what may, trivial duties, meaningless torments—as destructive of my life as if I wanted to be dead. Take today for what it is, I counsel myself. Let it be enough.

And I dare not, for fear that if I do, yesterday will infect tomorrow. We are in the habit of contention—against the world, against each other, against ourselves.

It is not from ourselves that we will learn to be better than we are.

In spite of all the talk about the law of tooth and fang and the struggle for survival, there is in the lives of the animals and birds a great peacefulness. It is not all fear and flight, pursuit and killing. That is part of it, certainly; and there is cold and hunger; there is the likelihood that death, when it comes, will be violent. But there is peace, too, and I think that the intervals of peace are frequent and prolonged. These are the times when the creature rests, communes with himself or with his kind, takes pleasure in being alive.

This morning while I wrote I was aware of a fox squirrel hunched in the sunlight on a high elm branch beyond my window. The night had been frosty, and now the warmth returned. He stayed there a long time, warming and grooming himself. Was he not at peace? Was his life not pleasant to him then?

I have seen the same peacefulness in a flock of wood ducks perched above the water in the branches of a fallen beech, preening and dozing in the sunlight of an autumn afternoon. Even while they dozed they had about them the exquisite alertness of wild things. If I had shown myself they would have been instantly in the air. But for the time there was no alarm among them, and no fear. The moment was whole in itself, deeply satisfying both to them and to me.

Or the sense of it may come with watching a flock of cedar waxwings eating wild grapes in the top of the woods on a

November afternoon. Everything they do is leisurely. They pick the grapes with a curious deliberation, comb their feathers, converse in high windy whistles. Now and then one will fly out and back in a sort of dancing flight full of whimsical flutters and turns. They are like farmers loafing in their own fields on Sunday. Though they have no Sundays, their days are full of sabbaths.

One clear fine morning in early May, when the river was flooded, my friend and I came upon four rough-winged swallows circling over the water, which was still covered with frail wisps and threads of mist from the cool night. They were bathing, dipping down to the water until they touched the still surface with a little splash. They wound their flight over the water like the graceful falling loops of a fine cord. Later they perched on a dead willow, low to the water, to dry and groom themselves, the four together. We paddled the canoe almost within reach of them before they flew. They were neat, beautiful, gentle birds. Sitting there preening in the sun after their cold bath, they communicated a sense of domestic integrity, the serenity of living within order. We didn't belong within the order of the events and needs of their day, and so they didn't notice us until they had to.

But there is not only peacefulness, there is joy. And the joy, less deniable in its evidence than the peacefulness, is the confirmation of it. I sat one summer evening and watched a great blue heron make his descent from the top of the hill in to the valley. He came down at a measured deliberate pace, stately as always, like a dignitary going down a stair. And then, at a point I judged to be midway over the river, without at all varying his wingbeat he did a backward turn in the air, a loop-the-loop. It could only have been a gesture of pure exuberance, of joy—a speaking of his sense of the evening, the day's fulfillment, his descent homeward. He made just the one slow turn, and then flew on out of sight in the direction of a slew farther down in the bottom. The movement was incredibly beautiful, at once exultant and stately, a benediction on the evening and on the river and on me. It seemed so perfectly to confirm the presence of a free nonhuman joy in the world—a joy I feel a great need to believe in—that I had the skeptic's impulse to doubt that I had seen it. If I had, I thought, it would be a sign of the presence of something heavenly in the earth. And then, one evening a year later, I saw it again.

Every man is followed by a shadow which is his death—dark, featureless, and mute. And for every man there is a place where his shadow is clarified and is made his reflection, where his face is mirrored in the ground. He sees his source and his destiny, and they are acceptable to him. He becomes the follower of what pursued him. What hounded his track becomes his companion.

That is the myth of my search and my return.

I have been walking in the woods, and have lain down on the ground to rest. It is the middle of October, and around me, all through the woods, the leaves are quietly sifting down. The newly fallen leaves make a dry, comfortable bed, and I lie easy, coming to rest within myself as I seem to do nowadays only when I am in the woods.

And now a leaf, spiraling down in wild flight, lands on my shirt front at about the third button below the collar. At first I am bemused and mystified by the coincidence—that the leaf should have been so hung, weighted and shaped, so ready to fall, so nudged loose and slanted by the breeze, as to fall where I, by the same delicacy of circumstance, happened to be lying. The event, among all its ramifying causes and considerations, and finally its mysteries, begins to take on the magnitude of history. Portent begins to dwell on it.

And suddenly I apprehend in it the dark proposal of the ground. Under the fallen leaf my breastbone burns with imminent decay. Other leaves fall. My body begins its long shudder into humus. I feel my substance escape me, carried into the mold by beetles and worms. Days, winds, seasons pass over me as I sink under the leaves. For a time only sight is left to me, a passive awareness of the sky overhead, birds crossing, the mazed interreaching of the treetops, the leaves falling— and then that, too, sinks away. It is acceptable to me, and I am at peace.

When I move to go, it is as though I rise up out of the world.

THE HIDDEN WOUND
(1970)

FROM

THE HIDDEN WOUND
(1970)

Chapter 4

WHEN I WAS three years old Nick Watkins, a black man, came to work for my Grandfather Berry. I don't remember when he came, which is to say that I don't remember not knowing him. When I was older and Nick and I would reminisce about the beginnings of our friendship, he used to laugh and tell me that when he first came I would follow him around calling him Tommy. Tommy was the hand who had lived there just before Nick. It was one of those conversations that are repeated ritually between friends. I would ask Nick to tell how it had been when he first came, and he would always tell about me calling him Tommy, and he would laugh. When I was eight or nine the story was important to me because it meant that Nick and I had known each other since way back, and were old buddies.

I have no idea of Nick's age when I first knew him. He must have been in his late fifties, and he worked for us until his death in, I believe, 1945—a period of about eight years. During that time one of my two or three chief ambitions was to be with him. With my brother or by myself, I dogged his steps. So faithful a follower, and so young and self-important and venturesome as I was, I must have been a trial to him. But he never ran out of patience.

From something philosophical and serene in that patience, and from a few things he said to me, I know that Nick had worked hard ever since his childhood. He told me that when he was a small boy he had worked for a harsh white woman, a widow or a spinster. When he milked, the cow would often kick the bucket over, and he would have to carry it back to the house empty, and the white woman would whip him. He had worked for hard bosses. Like thousands of others of his race he had lived from childhood with the knowledge that his fate was to do the hardest of work for the smallest of wages, and that there was no hope of living any other way.

White people thought of Nick as "a good nigger," and within the terms of that designation he had lived his life. But in my memory of him, and I think in fact, he was possessed of a considerable dignity. I think this was because there was a very conscious peace and faithfulness that he had made between himself and his lot. When there was work to be done, he did it dependably and steadily and well, and thus escaped the indignity of being bossed. I do not remember seeing him servile or obsequious. My grandfather, within the bounds of the racial bias, thought highly of him. He admired him particularly, I remember, as a teamster, and was always pointing him out to me as an example: "Look a yonder how old Nick sets up to drive his mules. Look how he takes hold of the lines. Remember that, and you'll know something."

In the eight or so years that Nick lived on the place, he and my grandfather spent hundreds and hundreds of work days together. When Nick first came there my grandfather was already in his seventies. Beyond puttering around and "seeing to things," which he did compulsively as long as he could stand up, he had come to the end of his working time. But despite the fact that my father had quietly begun to make many of the decisions in the running of the farm, and had assumed perhaps most of the real worries of running it, the old man still thought of himself as the sovereign ruler there, and it could be a costly mistake to attempt to deal with him on the assumption that he was not. He still got up at four o'clock as he always had, and when Nick and the other men on the place went to work he would be with them, on horseback, following the mule teams to the field. He rode a big bay mare named Rose; he would continue to ride her past the time when he could get into the saddle by himself. Through the long summer days he would stay with Nick, sitting and watching and talking, reminiscing, or riding behind him as he drove the rounds of a pasture on a mowing machine. When there was work that he could do, he would be into it until he tired out, and then he would invent an errand so he could get away with dignity.

Given Nick's steadiness at work, I don't think my grandfather stayed with him to boss him. I think he stayed so close because he couldn't stand not to be near what was going on, and because he needed the company of men of his own kind, working men. I have the clearest memory of the two of them passing again and again in the slowly shortening rounds of a big pasture, Nick driving a team of good black mules hitched to a mowing machine, my grandfather on the mare always only two or three steps behind the cutter bar. I don't know where I am in the memory, perhaps watching from the shade of some bush in a fencerow. In the bright hot sun of the summer day they pass out of sight and the whole landscape falls quiet. And then I hear the chuckling of the machine again, and then I see the mules' ears and my grandfather's hat appear over the top of the ridge, and they all come back into sight and pass around again. Within the steady monotonous racket of the machine, they keep a long silence, rich, it seems to me, with the deep camaraderie of men who have known hard work all their lives. Though their long labor in barns and fields had been spent in radically different states of mind, with radically different expectations, it was a common ground and a bond between them— never by men of their different colors, in that time and place, to be openly acknowledged or spoken of. Nick drives on and on into the day, deep in his silence, erect, alert and solemn faced with the patience that has kept with him through thousands of such days before, the elemental reassurance that dinnertime will come, and then quitting time, supper and rest. Behind him as the day lengthens, my grandfather dozes on the mare; when he sways in the saddle the mare steps under him, keeping him upright. Nick would claim that the mare did this out of a conscious sense of responsibility, and maybe she did.

On those days I know that Nick lived in constant fear that the mare too would doze and step over the cutter bar, and would be cut and would throw her rider before the mules could be stopped. Despite my grandfather's unshakable devotion to the idea that he was still in charge of things, it was clearly Nick who bore the great responsibility of those days. Because of childishness or whatever, the old man absolutely refused to accept the limits of age. He was fiercely headstrong in everything, and so was constantly on the verge of doing some damage to himself. I can see Nick working along, pretending not to watch him, but watching him all the same out of the corner of his eye, and then hustling anxiously to the rescue: "Whoa, boss. Whoa. Wait, boss." When he had my brother and me, and maybe another boy or two, to look after as well, Nick must have been driven well nigh out of his mind, but he never showed it.

When they were in the mowing or other such work, Nick

and my grandfather were hard to associate with. Of course we could get on horseback ourselves and ride along behind the old man's mare, but it was impossible to talk and was consequently boring. But there was other work, such as fencing or the handwork in the crops, that allowed the possibility of conversation, and whenever we could we got into that—in everybody's way, whether we played or tried to help, often getting scolded, often aware when we were not being scolded that we were being stoically put up with, but occasionally getting the delicious sense that we were being kindly indulged and catered to for all our sakes, or more rarely that we were being of use.

I remember one fine day we spent with Nick and our grandfather, cutting a young sassafras thicket that had grown up on the back of the place. Nick would fell the little trees with his ax, cutting them off about waist high, so that when they sprouted the cattle would browse off the foliage and so finally kill them. We would pile the trees high on the sled, my brother and I would lie on the mass of springy branches, in the spiced sweetness of that foliage, among the pretty leaves and berries, and Nick would drive us down the hill to unload the sled in a wash that our grandfather was trying to heal.

That was quiet slow work, good for talk. At such times the four of us would often go through a conversation about taking care of Nick when he got old. I don't remember how this conversation would start. Perhaps Nick would bring the subject up out of some anxiety he had about it. But our grandfather would say, "Don't you worry, Nick. These boys'll take care of you."

And one of us would say, "Yessir, Nick, we sure will."

And our grandfather would shake his head in sober emphasis and say, "By God, they'll do it!"

Usually, then, there would follow an elaborately detailed fantasy in which Nick would live through a long carefree old age, with good foxhounds and time to hunt, looked after by my brother and me who by then would have grown up to be lawyers or farmers.

Another place we used to talk was in the barn. Usually this would be on a rainy day, or in the late evening after work. Nick and the old man would sit in the big doorway on upturned buckets, gazing out into the lot. They would talk about old times. Or we would all talk about horses, and our grandfather would go through his plan to buy six good colts for my brother and me to break and train. Or we would go through the plan for Nick's old age.

Or our grandfather would get into a recurrent plan all his own about buying a machine, which was his word for automobile. According to the plan, he would buy a good new machine, and Nick would drive it, and they would go to town and to "Louis-ville" and maybe other places. The intriguing thing about this plan was that it was based on the old man's reasoning that since Nick was a fine teamster he was therefore a fine automobile driver. Which Nick wasn't; he couldn't drive an automobile at all. But as long as Nick lived, our grandfather clung to that dream of buying a good machine. Under the spell of his own talk about it, he always believed that he was right on the verge of doing it.

We also talked about the war that was being fought "across the waters." The two men were deeply impressed with the magnitude of the war and with the ominous new weapons that were being used in it. I remember sitting there in the barn door one day and hearing our grandfather say to Nick: "They got cannons now that'll shoot clean across the water. Good God Amighty!" I suppose he meant the English Channel, but I thought then that he meant the ocean. It was one of the ways the war and modern times became immediate to my imagination.

A place I especially liked to be with Nick was at the woodpile. At his house and at my grandparents' the cooking was still done on wood ranges, and Nick had to keep both kitchens supplied with stove wood. The poles would be laid up in a sawbuck and sawed to the proper lengths with a crosscut saw, and then the sawed lengths would be split on the chopping block with an ax. It was a daily thing throughout the year, but more wood was needed of course in the winter than in the summer. When I was around I would often help Nick with the sawing, and then sit up on the sawbuck to watch while he did the splitting, and then I would help him carry the wood in to the woodbox in the kitchen. Those times I would carry on long conversations, mostly by myself; Nick, who needed his breath for the work, would reply in grunts and monosyllables.

Summer and winter he wore two pairs of pants, usually an old pair of dress pants with a belt under a pair of bib overalls, swearing they kept him cool in hot weather and warm in cold. Like my grandfather, he often wore an old pair of leather puttees, or he would have his pants legs tied snugly above his shoe tops with a piece of twine. This, he liked to tell us, was to keep the snakes and mice from running up his britches legs. He wore old felt hats that were stained and sweaty and shaped to his character. He had a sober open dignified face and gentle manners, was quick to smile and to laugh. His teeth were ambeer-stained from chewing tobacco. His hands were as hard as leather; one of my hopes was someday to have hands as hard as his. He seemed instinctively to be a capable handler of stock. He could talk untiringly of good saddle horses and good work teams that he had known. He was an incurable fox hunter and was never without a hound or two. I think he found it easy to be solitary and quiet.

I heard my grandfather say to him one day: "Nick, you're the first darkie I ever saw who didn't sing while he worked."

But there were times, I knew, when Nick did sing. It was only one little snatch of a song that he sang. When the two of us would go on horseback to the store or to see about some stock—Nick on my grandfather's mare, I on a pony—and we had finished our errand and started home, Nick would often sing: "Get along home, home, Cindy, get along home!" And he would laugh.

"Sing it again," I would say.

And he would sing it again.

Chapter 5

BUT before I can tell any more about Nick, I will have to tell about Aunt Georgie. She came to live in the little two-room house with Nick in perhaps the third year he worked for us. Why did we call her Aunt when we never called Nick Uncle? I suppose it was because Nick was informal and she was formal. I remember my grandparents insisting that my brother and I should say Uncle Nick, but we would never do it. Our friendship was somehow too democratic for that. They might as well have insisted that we boys call each other mister. But we used the title Aunt Georgie from the first. She was a woman of a rather stiff dignity and a certain aloofness, and the term of respect was clearly in order.

She was short and squat, bowlegged, bent, her hands crooked with arthritis, her two or three snaggling front teeth stained with snuff. I suppose she was ugly—though I don't believe I ever made that judgment in those days. She looked like Aunt Georgie, who looked like nobody else.

Her arrival at Nick's house suddenly made that one of the most intriguing places that my brother and I had ever known. We began to spend a lot of our time there just sitting and talking. She would always greet us and make us at home with a most gracious display of pleasure in seeing us. Though she could cackle with delight like an old child, though she would periodically interrupt her conversation to spit ambeer into a coffee can she kept beside her chair, there was always a reserve about her, an almost haughty mannerliness that gave a peculiar sense of *occasion* to these visits. When she would invite us to eat supper, as she sometimes did, her manner would impose a curious self-consciousness on us—not a *racial* self-consciousness, but the demanding self-consciousness of a child who has been made, in the fullest sense, the guest of an adult, and of whom therefore a certain dignity is expected. Her house was one of the few places I visited as a child where I am certain I always behaved well. There was something in her presence that kept you always conscious of how you were acting; in response to her you became capable of social delicacy.

Thinking of her now, in spite of my rich experience of her, I realize how little I really *know* about her. The same is true of Nick. I knew them as a child knows people, as they revealed themselves to a child. What they were in themselves, as they spoke to each other, or thought in solitude, I can only surmise from that child's knowledge. I will never know.

Aunt Georgie had lived in a small rural community thirty or forty miles away. She had also lived in Louisville, where I believe she had relatives. She was a great reader of the Bible, and I assume of tracts of various sorts, and I don't know what else.

The knowledge that came out in her talk—fantastical, superstitious, occult, theological, Biblical, autobiographical, medical,
historical—was amazing in variety and extent. She was one of
the most intricate and powerful characters I have ever known.
Perhaps not twenty-five percent of her knowledge was subject
to any kind of proof—a lot of it was the stuff of unwritten fairy
tales and holy books, a lot I think she had made up herself—but
all of it, every last occult or imaginary scrap of it, was caught up
and held together in the force of her personality. Nothing was
odd to her, nothing she knew stood aside from her unassimi-
lated into the restless wayward omnivorous force of her mind. I
write of her with fear that I will misrepresent or underestimate
her. I believe she had great intelligence, which had been forced
to grow and form itself on the strange straggling wildly hetero-
geneous bits of information that sifted down to her through
various leaks in the stratification of white society.

The character of Aunt Fanny in my book *A Place on Earth*
is to some extent modeled on Aunt Georgie: "She's an accomplished seamstress, and the room is filled with her work: quilts,
crocheted doilies, a linen wall-hanging with the Lord's Prayer
embroidered on it in threads of many colors. In the house she's
nearly always occupied with her needle, always complaining of
her dim eyesight and arthritic hands. Her dark hands, though
painfully crooked and drawn by the disease, are still . . . dex-
terous and capable. She's extremely attentive to them, always
anointing them with salves and ointments of her own making.
The fingers wear rings made of copper wire, which she believes
to have the power of prevention and healing. She's an excellent
persistent canny gardener. The garden beside the house is her
work. She makes of its small space an amplitude . . . rows of
vegetables and flowers—and herbs, for which she knows the
recipes and the uses." If that is her daylight aspect, there is also
an aspect of darkness: ". . . the intricate quilts which are always prominent in the room, on the bed or the quilting frame
. . . always seem threatened, like earthquake country, by an
ominous nearness of darkness in the character of their creator.
Aunt Fanny has seen the Devil, not once but often, especially
in her youth, and she calls him familiarly by his name: Red Sam.
Her obsessions are Hell and Africa, and she has the darkest,
most fire-lit notions of both. Her idea of Africa is a hair-raising

blend of lore and hearsay and imagination. She thinks of it with nostalgia and longing—a kind of earthly Other Shore, Eden and Heaven—and yet she fears it because of its presumed darkness, its endless jungles, its stock of deliberately malevolent serpents and man-eating beasts. And by the thought of Hell she's held as endlessly fascinated as if her dearest ambition is to go there. She can talk at any length about it, cataloguing its tortures and labyrinths in almost loving detail. . . . Stooped in the light of a coal-oil lamp at night, following her finger down some threatening page of the Bible, her glasses opaquely reflecting the yellow of the lamp, her pigtails sticking out like compass points around her head, she looks like a black Witch of Endor.

"She possesses a nearly inexhaustible lore of snakes and deaths, and bottomless caves and pools, and mysteries and ghosts and wonders."

But the purpose of Aunt Fanny in that book is not to represent Aunt Georgie, and she comes off as a much simpler character. Aunt Fanny had spent all her life in the country, but Aunt Georgie had lived for some time in the city, and her mind, in its way of rambling and sampling, had become curiously cosmopolitan. Like Aunt Fanny, she had an obsession with Africa, but I think it must have started with her under the influence of the Back to Africa movement. She must at some time or other have heard speakers involved in this movement, for I remember her quoting someone who said, "Don't let them tell you they won't know you when you go back. They know their own people, and they'll *welcome* you." There was much more of this talk, but I had no context in which to place it and understand it, and so I have lost the memory. She used to tell a lot of stories about Africa, and I remember only one of them: a story of a woman and her small child who somehow happened to be passing through the jungle alone. A lion was following them, and the woman was terrified. They ran on until they were exhausted and could go no farther. Not knowing what else to do, the woman resigned herself and sat down to wait for death. And the lion came up to them. But instead of attacking, he walked calmly up and laid his paw in the woman's lap. She saw that the paw had a thorn in it that had made a very sore wound. She removed the thorn and treated the wound, and the lion became her protector, driving away the other large animals that threatened them in the jungle, hunting and providing for them, and in all ways taking care of them. I suppose I remember that story among all the others she told because I visualized it very clearly as she told it; it has stayed in my mind all these years with the straightforwardness and innocence of a Rousseau painting, but darker, the foliage ranker and more blurred in detail. But also I have remembered it, I think, because of the sense that began then, and that remains poignant in the memory, that the lion had become the woman's husband.

It was from Aunt Georgie, sitting and listening to her by the hour when I was seven and eight and nine and ten years old, that I first heard talk of the question of civil rights for Negroes. Again, I could supply no context for what I heard, and so have forgotten most of it. But one phrase has stuck in my mind along with her manner of saying it. She said that many times the white people had promised the Negro people "a right to the flag," and they never had given it to them. The old woman was capable of a moving eloquence, and I was deeply disturbed by what she said. I remember that, and I remember the indignation of my white elders when I would try to check the point with them. After Aunt Georgie moved away it was probably ten years before I paid attention to any more talk about civil rights, and it was longer than that before I felt again anything like the same disturbed sense of personal responsibility that she made me feel.

She used to tell a story about the end of the First World War, of people dancing wildly through the streets, carrying the Kaiser's head impaled on the end of a fence rail. I keep a clear image of that scene, too. Was it some celebration she had seen in Louisville, at which the Kaiser had been mutilated in effigy? Was it something she had read or imagined? I don't know. I assumed then that it was really the Kaiser's head, that she had seen that barbaric celebration herself, and that it was one of the central events of the history of the world.

Germany was also a sort of obsession with her, I suppose because of the two world wars, and she had nothing good to say for it. If to her Africa was a darkish jungly place of marvels that enticed the deepest roots of her imagination, Germany was a medieval torture chamber, a place of dire purposes and devices, pieces of diabolical machinery. In her sense of it the guillotine had come to be its emblem. It was a sort of earthly hell where people were sent to be punished for such crimes as public drunkenness. According to her, if you were caught drunk once you were *warned*, if you were caught twice you were put in jail, but the *third* time you were put on the train and sent to Germany where the Kaiser would cut your head off. She told of a man and his son in the town of Finchville, Kentucky, who had been caught drunk for the third time. The people of the town went down to the railroad station to witness their condemned neighbors' departure for Germany. Through the coach windows the doomed father and son could be seen finding an empty seat and sitting down together to begin their awesome journey. Over the condemned men and over the watching crowd there hung a great heaviness of finality and fate and horror. She told it as one who had been there and had seen it. And I remember it as if *I* had seen it; it is as vivid in my mind as anything that ever happened to *me*. The first time I heard her tell that story I believe I must have spent an hour or two cross-examining her, trying to find some small mitigation of the implacable finality of it, and she would not yield me the tiniest possibility of hope. Thinking about it now I feel the same contraction of despair that I felt then.

She told about awful sicknesses, and acts of violence, and terrible deaths. There was once a lion tamer who used to put his head in a lion's mouth, and one morning he nicked himself shaving and when he put his head into the lion's mouth that day, the lion tasted his blood and bit his head off. She talked about burials, bodies lying in the ground, dug-up bodies, ghosts.

And snakes. She believed that woman should bruise the head of the serpent with her heel. She would snick them up into little pieces with her sharp hoe. But they fascinated her, too, and they lived in her mind with the incandescence that her imagination gave to everything it touched. She was full of the lore of snakes—little and big snakes, deadly poisonous snakes, infernal snakes, *blue* snakes, snakes with yellow stripes and bright red heads, snakes that sucked cows, snakes that swallowed large animals such as people, hoop snakes. Once there was a woman who was walking in the woods, and her feet got tired and she sat down on a log to rest, and she took off her shoe and beat it on the log to dislodge a rock that was in it, and slowly that log began to *move*. My goodness gracious sakes alive!

The most formidable snake of all was the hoop snake. The hoop snake traveled by the law of gravity—a concept as staggering to the mind as atomic energy or perpetual motion. The hoop snake could travel on his belly like any other snake, but when he found himself on the top of a hill he was apt to whip himself up into the shape of a wheel, and go flying off down the slope at a dazzling speed. Once he had started his descent, of course, he had no sense of direction, and no way to stop or to turn aside and he had a deadly poisonous sharp point on the end of his tail that killed whatever it touched. Trees, horses and cattle, cats, dogs, men, women, and even *boys* were all brought down, like grass before the breath of the Lord, in the furious free-wheeling descent of the hoop snake, quicker than the eye. How could you know when he was coming? You couldn't. How could you get out of his way? You couldn't.

It has occurred to me to wonder if there wasn't a degree of conscious delight that Aunt Georgie took in scaring the wits out of two gullible little white boys. I expect there was. There was undoubtedly some impulse of racial vengeance in it; and it was bound to have given her a sense of power. But there was no cynicism in it; she believed what she said. I think, in fact, that she was motivated somewhat by a spirit of evangelism; she was instructing us, warning us against human evils, alerting us to the presence of ominous powers. Also she must have been lonely; we were company for her, and these things that she had on her mind were what she naturally talked to us about. And probably beyond any other reasons was that, like all naturally talkative people, she loved to listen to herself and to provide herself with occasions for eloquence.

But she was also a tireless loving gardener, a rambler in the fields, a gatherer of herbs and mushrooms, a raiser of chickens. And she talked well about all these things. She was a marvelous teller of what went on around the farm: How the mules ran the goat and the goat jumped up in the loading chute, safe, and looked back at the mules and wagged his head as if to say, "Uh *huh*!" How the hawk got after her chickens, and how she got her old self out of the house just in time and said, "Whoooeee! Hi! Get *out* of here, you *devil* you, and don't you come *back*!"

She knew about healing herbs and tonics and poultices and ointments. In those years I was always worrying about being skinny. I knew from the comic books that a ninety-eight-pound weakling was one of the worst things you could grow up to be; I foresaw my fate, and dreaded it. I confided my worry to Aunt Georgie, as I did to nearly everybody who seemed willing to listen, and she instantly prescribed a concoction known as Dr. Bell's Pine Tar Honey. By propounding the cure she ratified the ill; no one would ever persuade me again that I was not disastrously skinny. But Dr. Bell's Pine Tar Honey was nowhere to be found. No drugstore had it. The more I failed to find it, the more I believed in its power. It could turn a skinny boy into a *normal* boy. It was the elusive philosopher's stone of my childhood. And Aunt Georgie stood by her truth. She never quit recommending it. She would recommend no substitute.

She and Nick never married, though I believe she never abandoned the hope that they would. She always referred to Nick, flirtatiously and a little wistfully, as Nickum-Nackum. But Nick, with what I believe to have been a very pointed discretion, always called her *Miss* Georgie.

Chapter 6

IF Aunt Georgie was formal and austere, Nick was casual and familiar. If the bent of her mind was often otherworldly, Nick's belonged incorrigibly to this world; he was a man of the fields and the barns and the long nighttime courses of fox hunts. When he spoke of the Lord he called him, as my grandfather did, the Old Marster, by which they meant a god of mystery, the maker of weather and seasons, of abundance and dearth, of growth and death—a god far more remote, and far less talkative, than the god of the churches.

At night after I had finished supper in my grandmother's kitchen I would often walk down across the field to where Nick's house sat at the corner of the woods. He would come out, lighting his pipe, and we would sit down on the stones of the doorstep, and Nick would smoke and we would talk while it got dark and the stars came out. Up the hill we could see the lighted windows of my grandparents' house. In front of us the sloping pasture joined the woods. The woods would become deep and softly massive in its darkness. At dusk a toad who lived under the doorstep would come out, and always we would notice him and Nick would comment that he lived there and that he came out every night. Often as we sat there, comfortable with all the outdoors and the night before us, Aunt Georgie would be sitting by the lamp in the house behind us, reading aloud from the Bible, or trying to lecture to Nick on the imminence of eternity, urging him to think of the salvation of his soul. Nick's reluctance to get disturbed over such matters was always a worry to her. Occasionally she would call out the door: "Nickum-Nackum? Are you listening to me?" and Nick would serenely interrupt whatever he was telling me—"Yessum, Miss Georgie"—and go on as before.

What we would often be talking about was a fine foxhound named Waxy that Nick had owned a long time ago. He would tell of the old fox hunts, saying who the hunters had been and what kind of dogs they'd had. He would tell over the whole course of some hunt. By the time the race was over his Waxy would always have far outdistanced all the other dogs, and everybody would have exclaimed over what a great, fleet, wise hound she was, and perhaps somebody would have tried to buy her from Nick at a high price, which Nick would never take. Thinking about it since, I have had the feeling that those dogs so far outhunted and outrun by Nick's Waxy were white men's dogs. But I don't know for sure. In addition to telling how well Waxy had performed in some fox race or other, Nick would always tell how she looked, how she was marked and made. And he would frequently comment that Waxy was a fine name for a foxhound. He had thought a lot about how things ought to be named. In his mind he had lists of the best names for milk cows and horses and dogs. Blanche, I remember, he thought to be the prettiest name for a certain kind of light-coated Jersey cow. As long as he was with us I don't think he ever had the luck to give that name to a cow. Among others whose names I have forgotten, I remember that he milked one he called Mrs. Williams.

There were certain worldly ideals that always accompanied him, as hauntingly, I think, as Aunt Georgie was accompanied by devils and angels and snakes and ghosts. There was the ideal foxhound and the ideal team of mules and the ideal saddle horse, and he could always name some animal he had known that had not been quite perfect but had come close. Someday he would like to own a foxhound like Waxy but just a little better. Someday he would like to work a team of gray mare mules each just a little better than the one named Fanny that my grandfather owned.

We talked a great deal about the ideal saddle horse, because I persistently believed that I was going to pick out and buy one of my own. In the absence of any particular horse that I intended to buy, the conversational possibilities of this subject were nearly without limit. Listing points of color, conformation, breeding, disposition, size, and gait, we would arrive within a glimpse of the elusive outlines of the ideal. And then, supposing some divergence from one of the characteristics we had named, we might prefigure a horse of another kind. We were like those experts who from a track or a single bone can reconstruct an extinct animal: give us a color or a trait of conformation or character, and we could produce a horse to go with it. And once we had the horse, we had to settle on a name and on the way it should be broke and fed and kept.

But our *great* plan, the epic of our conversation, was to go camping and hunting in the mountains. I don't think that either Nick or my brother and I knew very much about the mountains. None of us had been there. My own idea seems to have blended what I had read of the virgin Kentucky woods of Boone's time with Aunt Georgie's version of Africa. My brother and I believed there was a great tract of wilderness in the mountains, thickly populated with deer and bears and black panthers and mountain lions. We thought we would go up there and kill some of those beasts and eat their flesh and dress in their skins and make ornaments and weapons out of their long teeth, and live in the untouched maidenhood of history.

I don't know what idea Nick may have had of the mountains; he always seemed more or less to go along with our idea of them; whether or not he contributed to the notions we had, I don't remember. Though he was an ignorant man, he was knowing and skilled in the realities that had been available to him, and so he was bound to have known that we would never make any such trip. But he elaborated this plan with us year after year; it was, in fact, in many ways more his creature than ours. It was generous—one of the most generous things anybody ever did for us. And yet it was more than generous, for I think Nick believed in that trip as a novelist believes in his novel: his imagination was touched by it; he couldn't resist it.

We were going to get new red tassels to hang on the mules' harness, and polish the brass knobs on their hames. And then one night we were going to load the wagon—we had told over many times what all we were going to put in it. And the next morning early, way before daylight so my grandfather wouldn't catch us, we would slip off and go to the mountains. We had a highly evolved sense of the grandeur of the spectacle we would make as we went through New Castle, where everybody who saw us would know where we were going and what a fine adventure it was. Here we would come through the town just as everybody was getting up and out into the street. Nick would be driving, my brother and I sitting on either side of him, his ready henchmen. We would have the mules in a spanking trot, the harness jangling, the red tassels swinging, and everything fine—right on through town past the last house. From there to the tall wilderness where we would make our camp we had no plans. The trip didn't interest us after people we knew quit watching.

When we got to the mountains we were going to build a log cabin, and we talked a lot about the proper design and construction of that. And there were any number of other matters, such as cooking, and sewing the clothes we would make out of the skins of animals. Sometimes, talking of the more domestic aspects of our trip, we would get into a bind and have to plan to take Aunt Georgie. She never could work up much enthusiasm for the project; it was too mundane for her.

There were several times when we actually named the day of our departure, and my brother and I went to bed expecting to leave in the morning. We had agreed that to wake us up Nick would rattle a few tin cans in the milk bucket. But we never woke up until too late. I remember going out one morning and finding the very tin cans that Nick had rattled floating in the rain barrel. Later I realized that of course Nick had never rattled those cans or any others. But in the years since I have often imagined him taking the time to put those cans there for me to see, and then slipping off to his day's work—conspiring with himself like Santa Claus to deceive and please me.

In the present time, when the working man in the old sense of the phrase is so rapidly disappearing from the scene, his conversation smothered in the noise of machinery, it may be necessary to point out that such talk as we carried on with Nick was by no means uncommon among the farming people of that place, and I assume of other places as well. It was one of the natural products and pleasures of the life there. With the hands occupied in the work, the mind was set free; and since there was often considerable misery in the work, the mind turned naturally to what would be desirable or pleasant. When one was at work with people with whom one had worked many days or years before, one naturally spoke of the good things that were on one's mind. And once the conversation was started, it began to be a source of pleasure in itself, producing and requiring an energy that propelled it beyond the bounds of reality or practicality. As long as one was talking about what would be good, why not go ahead and talk about what would be *really* good? My childhood was surrounded by a communal daydreaming, the richest sort of imaginative talk, that began in this way—in work, in the misery of work, to make the work bearable and even pleasant. Such talk ranged all the way from a kind of sensuous realism to utter fantasy, but because the bounds of possibility were almost always ignored I would say that the impetus was always that of fantasy. I have heard crews of men, weary and hungry and hot near the end of a day's work, construct long elaborate conversations on the subject of what would be good to eat and drink, dwelling at length and with subtlety on the taste and the hotness or the coldness of various dishes and beverages, and on combinations of dishes and beverages, the menu lengthening far beyond the capacity of any living stomach. I knew one man who every year got himself through the ordeal of the tobacco harvest by elaborating from one day to the next the fantasy of an epical picnic and celebration which was always to occur as soon as the work was done—and which never did occur except, richly, marvelously, in the minds of his listeners. Such talk could be about anything: where you would like to go, what you would like to do when you got there, what you would like to be able to buy, what would be a good kind of a farm, what science was likely to do, and so on.

Nick often had fox hunting on his mind, and he always knew where there was a fox, and he was always going to hunt as soon as he had time. Occasionally he *would* have time—a wet afternoon or a night when he didn't feel too tired—and he would go, and would sometimes take one of us. I like the thought of Nick, alone as he often was, out in the woods and the fields at night with his dogs. I think he transcended his lot then, and was free in the countryside and in himself—beyond anybody's knowledge of where he was and anybody's notion of what he ought to do and anybody's claim on his time, his dogs mouthing out there in the dark, and all his senses with them, discovering the way the fox had gone. I don't remember much about any hunt I ever made with him, but I do remember the sense of accompanying him outside the boundaries of his life as a servant, into that part of his life that clearly belonged to him, in which he was competent and knowing. And I remember how it felt to be riding sleepily along on the pony late at night, the country all dark around us, not talking, not sure where we were going. Nick would be ahead of me on old Rose, and I could hear her footsteps soft on the grass, or hard on the rocks when we crossed a creek. I knew that Nick knew where we were, and I felt comfortable and familiar with the dark because of that.

When Joe Louis fought we would eat supper and then walk across the fields to the house of the man who was raising my grandfather's tobacco; we would have been planning for days to go there to hear the fight on the radio. My grandparents had a radio, too, but Nick would never have thought of going there. That he felt comfortable in going to another household of white people and sitting down by the radio with them was always a little embarrassing to me, and it made me a little jealous. And yet I understood it, and never questioned it. It was one of those complex operations of race consciousness that a small child will comprehend with his feelings, and yet perhaps never live long enough to unravel with his mind. The prowess of Joe Louis, anyhow, was something Nick liked to talk about; it obviously meant a great deal to him. And so the championship matches of those days were occasions when I felt called on to manifest my allegiance to Nick. Though I could have listened to the fight in my grandparents' living room, I would go with Nick, and the more my grandmother declared she didn't know why I had to do that the more fierce and clear my loyalty to Nick became.

The people of that house had a boy a little older than I was, and so I also had my own reasons for going there. We boys would listen to the fight and then, while the grownups talked, go out to play, or climb the sugar pear trees and sit up in the branches, eating. And then holding Nick's hand, I would walk with him back home across the dark fields.

One winter there suddenly appeared in the neighborhood a gaunt possessed-looking white man who was, as people said, "crazy on religion" or "religious crazy." He was, I believe, working on a nearby farm. Somehow he discovered Aunt Georgie, and found that she could quote the Bible and talk religion as long as he could. He called Nick Mix, and so in derision and reprisal Nick and my brother and I called *him* Mix, and that was the only name we ever knew him by. We would see him coming over the hill, his ragged clothes flapping in the cold wind, and see him go into the little house. And then after a while we would go down, too, and listen. The tone of their conversations was very pious and oratorical; they more or less preached to each other, and the sessions would go on for hours.

Of it all, I remember only a little fragment: Mix, sitting in a chair facing Aunt Georgie, sticks his right hand out into the air between them, dramatically, and asks her to put hers out there beside it. She does, and then he raises his left hand like a Roman orator and, glaring into her eyes, says: "There they are, Auntie! *There* they are! Look at 'em! One of 'em's white, and one of 'em's black! But *inside*, they both white!"

What impresses me now, as I remember those times, is that

while Mix was never anything but crazy and absurd, ranting mad, Aunt Georgie always seemed perfectly dignified and sane. It seems to me that this testifies very convincingly to the strength of her intelligence: she had *comprehended* the outlandishness of her own mind, and so she could keep a sort of grace before the unbridled extravagance of his. In her way she had faced the worst. The raging of Lear himself would not have perturbed her; she would have told him stories worse than his, and referred him to the appropriate Scripture. To Nick, who took a very detached and secular view of most things, these sessions were a source of ironic amusement. He liked to look over at us through the thick of the uproar and wink.

Chapter 7

I suppose that it is usually the aim of a writer to produce a definitive statement, one that will prove him to be the final authority on what he has said. But though my aim here is to tell the truth as nearly as I am able, I am aware that the truth I am telling may be a very personal one, the truth, that is, as distorted and qualified by my own heritage and personality. I am, after all, writing about people of another race and a radically different heritage, whom I knew only as a child, and whose lives parted from mine nearly a quarter of a century ago. As I write I can hardly help but think of the possibility that if Nick and Aunt Georgie were alive to read this, they might not recognize themselves. And in the face of the extreme racial sensitivity of the present time, I can hardly ignore the possibility that my black contemporaries may find some of my assumptions highly objectionable. And so I write with the feeling that the truth I may tell will not be definitive or objective or even demonstrable, but in the strictest sense subjective, relative to the peculiar self-consciousness of a diseased man struggling toward a cure. I am trying to establish the outlines of an understanding of myself in regard to what was fated to be the continuing crisis of my life, the crisis of racial awareness—the sense of being doomed by my history to be, if not always a racist, then a man always limited by the inheritance of racism, condemned to be always conscious of the necessity *not* to be a racist, to be always dealing deliberately with the reflexes of racism that are embedded in my mind as deeply at least as the language I speak.

(In the Western tradition of individualism there is the assumption that art can grow out of a personal or a cultural disease, and triumph over it. I no longer believe that. It is related to the idea that a man can achieve personal immortality in a work of art, which I also no longer believe. Though I believe that the liveliest art is suffused with the energy of the creation, and in that sense participates in immortality, I do not believe that any one work of art is immortal any more than I believe that a grove of trees or a nation is immortal. A man cannot be immortal except by saving his soul, and he cannot save his soul except by freeing his body and mind from the destructive forces in his history. A work of art that grows out of a diseased culture has not only the limits of art but the limits of the disease—if it is not an affirmation of the disease, it is a reaction against it. The art of a man divided within himself and against his neighbors, no matter how sophisticated its techniques or how beautiful its forms and textures, will never have the communal *power* of the simplest tribal song.)

There is an inescapable tentativeness in writing of one's own formative experiences: as long as the memory of them stays alive they *remain* formative; the power of growth and change remains in them, and they never become quite predictable in their influence. If memory is in a way the ancestor of consciousness, it yet remains dynamic within consciousness. And so, though I can write about Nick and Aunt Georgie as two of the significant ancestors of my mind, I must also deal with their memory as a live resource, a *power* that will live and change in me as long as I live. To fictionalize them, as I did Aunt Georgie to some extent in *A Place on Earth*, would be to give them an imaginative stability at the cost of oversimplifying them. To attempt to tell the "truth" about them as they really were is to resign oneself to enacting a small fragment of an endless process. Their truth is inexhaustible both in their lives as they were, and in my life as I think they were.

The peculiar power of my memory of them comes, I think, from the fact that all my association with them occurred within a tension between the candor and openness of a child's view of things, and the racial contrivances of the society we lived in. As I have already suggested, there were times when I was inescapably aware of the conflict between what I felt about them, in response to what I knew of them, and the feelings that were prescribed to me by the society's general prejudice against their race. Being with them, it was hard to escape for very long from the sense of racial difference prepared both in their minds and in my own.

The word *nigger* might be thought of as rattling around, with devastating noise and impact, within the silence, that black-manshaped hollow, inside our language. When the word was spoken abstractly, I believe that it seemed as innocuous and casual to me as any other word. I used it that way myself, in the *absence* of black people, without any consciousness that I was participating in a judgment and a condemnation; certainly I used it without any feeling that my use of it manifested anything that was wrong with *me*. But when it was used with particular reference to a person one cared for, as a child cares, it took on a tremendous force; its power reached ominously over one's sense of things. I remember the shock and confusion I felt one night when, saying good-bye to Nick, I impulsively kissed his hand, which I had held as we brought the milk to the house, and then came into the kitchen to hear: "Lord, child, how can you kiss that old nigger's hand?"

And I remember sitting down at Nick's house one day when he and Aunt Georgie had company. There was a little boy, four or five years old, who was rambunctious and continually on the verge of mischief. To keep him in hand Nick would lean over and rattle his old leather leggings that he had taken off and put behind the drum stove, and he would say to the boy: "John's going to get you. If you don't quit that, John's going to get you." All this—the rattling of the leggings, the threat, the emphatic result—impressed me a good deal, and I wondered about it. Was this John-the-Leggings somebody Nick had thought up on the spur of the moment? Was he one of the many associates of Red Sam? And then it hit me. That John that my friend Nick was speaking of in so formidable a tone was my father! The force of a realization like that is hardly to be measured; it's the sort of thing that can initiate a whole epoch of the development of a mind, and yet remain on as a force. It gave me the strongest sort of a hint of the existence of something large and implacable and rigid that I had been born into, and lived in—something I have been trying to get out of ever since. In justice to my father, I must say that I don't believe his name was used in this way because of anything he had *done*, but because of his place in the system. In spite of my grandfather's insistences to the contrary, Nick knew as I did that my father had become the man really in charge of things. Thus he had entered the formidable role of "boss man": whoever he was, whatever he did, he had the power and the austerity of that role; the society assigned it to him, as it assigned to Nick the role of "nigger."

At times in my childhood I was made to feel the estranging power of this role myself—though I believe always by white men. When I would be playing where the men were working in the fields—I suppose in response to some lapse of tact on my part, some unconscious display of self-importance or arrogance—one of them would sing or quote:

> He ain't the boss, he's the boss's son,
> But he'll be the boss when the boss is done.

A complementary rhyme, which might have been made up as a companion to that, goes this way:

> A naught's a naught and a figger's a figger.
> All for the white man, none for the nigger.

Between them these two little rhymes make a surprisingly complete definition of the values and the class structure of the place and the time of my childhood, and they define its most erosive psychic tensions.

That sense of difference, given the candor and the affection and the high spirits of children, could only beget in us its opposite: a very strong and fierce sense of allegiance. Whenever there was a conflict of interest between Nick and Aunt Georgie and our family, neither my brother nor I ever hesitated to take the side of Nick and Aunt Georgie. We had an uncle who would make a game of insulting Nick in our presence because it tickled him the way we would come to the rescue; we would attack him and fight him as readily as if he had been another child. And there were winter Saturdays that we spent with our goat hitched to a little sled we had made, hauling coal from our grandfather's pile down to Nick's. That was stealing, and we knew it; it had the moral flair and the dangerous loyalty of the adventures of Robin Hood.

But the clearest of all my own acts of taking sides happened at a birthday party my grandmother gave for me when I must have been nine or ten years old. As I remember, she invited all the family and perhaps some of the neighbors. I issued one invitation of my own, to Nick. I believe that in my eagerness to have him come, and assuming that as my friend he ought to be there, I foresaw none of the social awkwardness that I created. But I had, in fact, surrounded us all with the worst sort of discomfort. Nick, trying to compromise between his wish to be kind to me and his embarrassment at my social mis-conception, quit work at the time of the party and came and sat on the cellar wall behind the house. By that time even I had begun to sense the uneasiness I had created: I had done a thing more powerful than I could have imagined at the time; I had scratched the wound of racism, and all of us, our heads beclouded in the social dream that all was well, were feeling the pain. It was suddenly evident to me that Nick neither would nor could come into the house and be a member of the party. My grandmother, to her credit, allowed me to follow my in-stincts in dealing with the situation, and I did. I went out and spent the time of the party sitting on the cellar wall with Nick.

It was obviously the only decent thing I could have done; if I had thought of it in moral terms I would have had to see it as my duty. But I didn't. I didn't think of it in moral terms at all. I did simply what I *preferred* to do. If Nick had no place at my party, then I would have no place there either; my place would be where he was. The cellar wall became the place of a definitive enactment of our friendship, in which by the grace of a child's honesty and a man's simple-hearted generosity, we transcended our appointed roles. I like the thought of the two of us sitting out there in the sunny afternoon, eating ice cream and cake, with all my family and my presents in there in the house without me. I was full of a sense of loyalty and love that clarified me to myself as nothing ever had before. It was a time I would like to live again.

Chapter 8

ONE day when we were all sitting in the barn door, resting and talking, my grandfather sat gazing out a while in silence, and then he pointed to a far corner of the lot and said: "When I'm dead, Nick, I want you to bury me there."

And Nick laughed and said, "Boss, you got to go farther away from here than that."

My grandfather's words bore the forlorn longing to remain in the place, in the midst of the work and the stirring, that had interested him all his life. It was also a characteristic piece of contrariness: if he didn't want to go, then by God he *wouldn't*. Nick's rejoinder admitted how tough it would be on everybody to have such a stubborn old boss staying on there forever, and it told a plain truth: there wasn't to be any staying; good as it might be to be there, we were all going to have to leave.

Those words stand in my memory in high relief; strangely powerful then (it was an exchange that we younger ones quoted over and over in other conversations) they have assumed a power in retrospect that is even greater. That whole scene—the two old men, two or three younger ones, two boys, all looking out that big doorway at the world, and beyond the world—seems to me now not to belong to my childhood at all, but to stand like an illuminated capital at the head of the next chapter. It presages a series of deaths and departures and historical changes that would put an end both to my childhood and to the time and the way of life I knew as a child. Within two or three years of my tenth summer both Nick and my grandfather died, and after Nick's death Aunt Georgie moved away. Also the war ended, and our part of the country moved rapidly into the era of mechanized farming. People, especially those who had worked as hired hands in the fields, began to move to the cities, and the machines moved from the cities out into the fields. Soon nearly everybody had a tractor; the few horses and mules still left in the country were being kept only for old time's sake.

It was a school day in the winter and I was at home in New Castle eating breakfast, when my grandmother telephoned from the farm to tell us that Nick had had a stroke. He had eaten breakfast and was starting to go up to the barn when he fell down in the doorway. He had opened the door before he fell and Aunt Georgie was unable to move him, and so he had lain in the cold for some time before help came. Until he died a few days later he never moved or spoke. I wasn't allowed to go to see him, but afterward my grandmother told me how he had been. He lay there on his back, she said, unable to move, looking up at the ceiling. When she would come in he would recognize her; she could tell that by his eyes. And I am left with that image of him: lying there still as death, his life showing only in his eyes. As if I had been there, I am aware of the intelligence and the gentleness and the sorrow of his eyes. He had fallen completely into the silence that had so nearly surrounded him all his life. It was not the silence of death, which men may speak of with words, and so know each other. It was the racial silence in the speech of white men, the wound of their history, formed three hundred years before my birth to stand between him and me, so that when I think of him now, as important as his memory is to me, it must be partly to wonder if I knew him.

One night not long after Nick's death I went with my father down to the little house to see Aunt Georgie. She had a single lamp lighted in the bedroom, and we went in and sat down. For a time, while they talked, I sat there without hearing, saddened and bewildered by the heavy immanence of change I felt in the room and by Nick's absence from it. And then I began to listen to what was being said. Aunt Georgie's voice was without the readiness of laughter I had always known in her. She spoke quietly and deliberately, avoiding betrayal of her grief. My father was talking, I remember, very gently and generously to her, helping her to get her affairs in order, discussing an insurance policy that Nick had, determining what her needs were. And all at once I realized that Aunt Georgie was going to leave. I hadn't expected it; she was deeply involved in my history and my affections; she had been there what seemed to me a long long time. Perhaps because of those conversations we used to have about taking care of Nick when he got old, I had supposed that such permanent arrangements were made as a matter of course. I interrupted and said something to the effect that I thought Aunt Georgie ought to stay, I didn't want her to go. And they turned to me and explained that it was not possible. She had relatives in Louisville; she would go to live with them. Their few words made it clear to me that there were great and demanding realities that I had never considered: realities of allegiance, realities of economics. I waited in silence then until their talk was finished, and we went out into the winter night and walked back up the hill.

I don't know if I saw Aunt Georgie again before she left. That is the last memory I have of her: strangely detached and remote from the house as though she already no longer lived there, an old woman standing in the yellow lamplight among the shadowy furnishings of the room, stooped and slow as if newly aware of the heaviness of her body.

And so it ended.

A CONTINUOUS HARMONY
(1972)

Think Little

FIRST THERE WAS Civil Rights, and then there was the War, and now it is the Environment. The first two of this sequence of causes have already risen to the top of the nation's consciousness and declined somewhat in a remarkably short time. I mention this in order to begin with what I believe to be a justifiable skepticism. For it seems to me that the Civil Rights Movement and the Peace Movement, as popular causes in the electronic age, have partaken far too much of the nature of fads. Not for all, certainly, but for too many they have been the fashionable politics of the moment. As causes they have been undertaken too much in ignorance; they have been too much simplified; they have been powered too much by impatience and guilt of conscience and short-term enthusiasm, and too little by an authentic social vision and long-term conviction and deliberation. For most people those causes have remained almost entirely abstract; there has been too little personal involvement, and too much involvement in organizations that were insisting that *other* organizations should do what was right.

There is considerable danger that the Environmental Movement will have the same nature: that it will be a public cause, served by organizations that will self-righteously criticize and condemn other organizations, inflated for a while by a lot of public talk in the media, only to be replaced in its turn by another fashionable crisis. I hope that will not happen, and I believe that there are ways to keep it from happening, but I know that if this effort is carried on solely as a public cause, if millions of people cannot or will not undertake it as a *private* cause as well, then it is *sure* to happen. In five years the energy of our present concern will have petered out in a series of public gestures—and no doubt in a series of empty laws—and a great, and perhaps the last, human opportunity will have been lost.

It need not be that way. A better possibility is that the movement to preserve the environment will be seen to be, as I think it has to be, not a digression from the civil rights and peace movements, but the logical culmination of those movements.

For I believe that the separation of these three problems is artificial. They have the same cause, and that is the mentality of greed and exploitation. The mentality that exploits and destroys the natural environment is the same that abuses racial and economic minorities, that imposes on young men the tyranny of the military draft, that makes war against peasants and women and children with the indifference of technology. The mentality that destroys a watershed and then panics at the threat of flood is the same mentality that gives institutionalized insult to black people and then panics at the prospect of race riots. It is the same mentality that can mount deliberate warfare against a civilian population and then express moral shock at the logical consequence of such warfare at My Lai. We would be fools to believe that we could solve any one of these problems without solving the others.

To me, one of the most important aspects of the Environmental Movement is that it brings us not just to another public crisis, but to a crisis of the protest movement itself. For the environmental crisis should make it dramatically clear, as perhaps it has not always been before, that there is no public crisis that is not also private. To most advocates of civil rights, racism has seemed mostly the fault of someone else. For most advocates of peace, the war has been a remote reality, and the burden of the blame has seemed to rest mostly on the government. I am certain that these crises have been more private, and that we have each suffered more from them and been more responsible for them, than has been readily apparent, but the connections have been difficult to see. Racism and militarism have been institutionalized among us for too long for our personal involvement in those evils to be easily apparent to us. Think, for example, of all the Northerners who assumed—until black people attempted to move into *their* neighborhoods—that racism was a Southern phenomenon. And think how quickly—one might almost say how naturally—among some of its members the Peace Movement has spawned policies of deliberate provocation and violence.

But the environmental crisis rises closer to home. Every time we draw a breath, every time we drink a glass of water, we are suffering from it. And more important, every time we indulge in, or depend on, the wastefulness of our economy—and our economy's first principle is waste—we are *causing* the crisis. Nearly every one of us, nearly every day of his life, is contributing *directly* to the ruin of this planet. A protest meeting on the issue of environmental abuse is not a convocation of accusers, it is a convocation of the guilty. That realization ought to clear the smog of self-righteousness that has almost conventionally hovered over these occasions and let us see the work that is to be done.

In this crisis it is certain that every one of us has a public responsibility. We must not cease to bother the government and the other institutions to see that they never become comfortable with easy promises. For myself, I want to say that I hope never again to go to Frankfort to present a petition to the governor on an issue so vital as that of strip mining, only to be dealt with by some ignorant functionary—as several of us were not so long ago, the governor himself being "too busy" to receive us. Next time I will go prepared to wait as long as necessary to see that the petitioners' complaints and their arguments are heard *fully*—and by the governor. And then I will hope to find ways to keep those complaints and arguments from being forgotten until something is done to relieve them. The time is past when it was enough merely to elect our officials. We will have to elect them and then go and *watch* them and keep our hands on them, the way the coal companies do. We have made a tradition in Kentucky of putting self-servers, and worse, in charge of our vital interests. I am sick of it. And I think that one way to change it is to make Frankfort a less comfortable place. I believe in American political principles, and I will not sit idly by and see those principles destroyed by sorry practice. I am ashamed that American government should have become the chief cause of disillusionment with American principles.

And so when the government in Frankfort again proves too stupid or too blind or too corrupt to see the plain truth and to act with simple decency, I intend to be there, and I trust that I won't be alone. I hope, moreover, to be there, not with a sign or a slogan or a button, but with the facts and the arguments. A crowd whose discontent has risen no higher than the level of slogans is *only* a crowd. But a crowd that understands the reasons for its discontent and knows the remedies is a vital community, and it will have to be reckoned with. I would rather go before the government with two men who have a competent understanding of an issue, and who therefore deserve a hearing, than with two thousand who are vaguely dissatisfied.

But even the most articulate public protest is not enough. We don't live in the government or in institutions or in our public utterances and acts, and the environmental crisis has its roots in our *lives*. By the same token, environmental health will also be rooted in our lives. That is, I take it, simply a fact, and in the light of it we can see how superficial and foolish we would be to think that we could correct what is wrong merely by tinkering with the institutional machinery. The changes that are required are fundamental changes in the way we are living.

What we are up against in this country, in any attempt to invoke private responsibility, is that we have nearly destroyed private life. Our people have given up their independence in return for the cheap seductions and the shoddy merchandise of so-called "affluence." We have delegated all our vital functions and responsibilities to salesmen and agents and bureaus and experts of all sorts. We cannot feed or clothe ourselves, or entertain ourselves, or communicate with each other, or be charitable or neighborly or loving, or even respect ourselves, without recourse to a merchant or a corporation or a public service organization or an agency of the government or a stylesetter or an expert. Most of us cannot think of dissenting from the opinions or the actions of one organization without first forming a new organization. Individualism is going around these days in uniform, handing out the party line on individualism. Dissenters want to publish their personal opinions over a thousand signatures.

The Confucian *Great Digest* says that the "chief way for the production of wealth" (and it is talking about real goods, not money) is "that the producers be many and that the mere consumers be few. . . ." But even in the much-publicized rebellion of the young against the materialism of the affluent society, the consumer mentality is too often still intact: The standards of behavior are still those of kind and quantity, the security sought is still the security of numbers, and the chief motive is still the consumer's anxiety that he is missing out on what is "in." In this state of total consumerism—which is to say a state of helpless dependence on things and services and ideas and motives that we have forgotten how to provide ourselves—all meaningful contact between ourselves and the earth is broken. We do not understand the earth in terms either of what it offers us or of what it requires of us, and I think it is the rule that people inevitably destroy what they do not understand. Most of us are not directly responsible for strip mining and extractive agriculture and other forms of environmental abuse. But we are guilty nevertheless, for we connive in them by our ignorance. We are ignorantly dependent on them. We do not know enough about them; we do not have a particular enough sense of their danger. Most of us, for example, not only do not know how to produce the best food in the best way—we don't know how to produce any kind in any way. Our model citizen is a sophisticate who before puberty understands how to produce a baby, but who at the age of thirty will not know how to produce a potato. And for this condition we have elaborate rationalizations, instructing us that dependence for everything on somebody else is efficient and economical and a scientific miracle. I say, instead, that it is madness, mass produced. A man who understands the weather only in terms of golf is participating in a public insanity that either he or his descendants will be bound to realize as suffering. I believe that the death of the world is breeding in such minds much more certainly and much faster than in any political capital or atomic arsenal.

For an index of our loss of contact with the earth we need only look at the condition of the American farmer—who must enact our society's dependence on the land. In an age of unparalleled affluence and leisure, the American farmer is harder pressed and harder worked than ever before; his margin of profit is small, his hours are long; his outlays for land and equipment and the expenses of maintenance and operation are growing rapidly greater; he cannot compete with industry for labor; he is being forced more and more to depend on the use of destructive chemicals and on the wasteful methods of haste. As a class, farmers are one of the despised minorities. So far as I can see, farming is considered marginal or incidental to the economy of the country, and farmers, when they are thought of at all, are thought of as hicks and yokels, whose lives do not fit into the modern scene. The average American farmer is now an old man whose children have moved away to the cities. His knowledge, and his intimate connection with the land, are about to be lost. The small independent farmer is going the way of the small independent craftsmen and storekeepers. He is being forced off the land into the cities, his place taken by absentee owners, corporations, and machines. Some would justify all this in the name of efficiency. As I see it, it is an enormous social and economic and cultural blunder. For the small farmers who lived on their farms *cared* about their land. And given their established connection to their land—which was often hereditary and traditional as well as economic—they could have been encouraged to care for it more competently than they have so far. The corporations and machines that replace them will never be bound to the land by the sense of birthright and continuity, or by the love that enforces care. They will be bound by the rule of efficiency, which takes thought only of the volume of the year's produce, and takes no thought of the life of the land, not measurable in pounds or dollars, which will assure the livelihood and the health of the coming generations.

If we are to hope to correct our abuses of each other and of other races and of our land, and if our effort to correct these abuses is to be more than a political fad that will in the long run be only another form of abuse, then we are going to have to go far beyond public protest and political action. We are going to have to rebuild the substance and the integrity of private life in this country. We are going to have to gather up the fragments of knowledge and responsibility that we have parceled out to the bureaus and the corporations and the specialists, and put those fragments back together in our own minds and in our families and households and neighborhoods. We need better government, no doubt about it. But we also need better minds, better friendships, better marriages, better communities. We need persons and households that do not have to wait upon organizations, but can make necessary changes in themselves, on their own.

For most of the history of this country our motto, implied or spoken, has been Think Big. A better motto, and an essential one now, is Think Little. That implies the necessary change of thinking and feeling, and suggests the necessary work. Thinking Big has led us to the two biggest and cheapest political dodges of our time: plan-making and law-making. The lotus-eaters of this era are in Washington, D.C., Thinking Big. Somebody perceives a problem, and somebody in the government comes up with a plan or a law. The result, mostly, has been the persistence of the problem, and the enlargement and enrichment of the government.

But the discipline of thought is not generalization; it is detail, and it is personal behavior. While the government is "studying" and funding and organizing its Big Thought, nothing is being done. But the citizen who is willing to Think Little, and, accepting the discipline of that, to go ahead on his own, is already solving the problem. A man who is trying to live as a neighbor to his neighbors will have a lively and practical understanding of the work of peace and brotherhood, and let there be no mistake about it—he is *doing* that work. A couple who make a good marriage, and raise healthy, morally competent children, are serving the world's future more directly and surely than any political leader, though they never utter a public word. A good farmer who is dealing with the problem of soil erosion on an acre of ground has a sounder grasp of that problem and *cares* more about it and is probably doing more to solve it than any bureaucrat who is talking about it in general. A man who is willing to undertake the discipline and the difficulty of mending his own ways is worth more to the conservation movement than a hundred who are insisting merely that the government and the industries mend *their* ways.

If you are concerned about the proliferation of trash, then by all means start an organization in your community to do something about it. But before—*and while*—you organize, pick up some cans and bottles yourself. That way, at least, you will assure yourself and others that you mean what you say. If you are concerned about air pollution, help push for government controls, but drive your car less, use less fuel in your home. If you are worried about the damming of wilderness rivers, join the Sierra Club, write to the government, but turn off the lights you're not using, don't install an air conditioner, don't be a sucker for electrical gadgets, don't waste water. In other words, if you are fearful of the destruction of the environment, then learn to quit being an environmental parasite. We all are, in one way or another, and the remedies are not always obvious, though they certainly will always be difficult. They require a new kind of life—harder, more laborious, poorer in luxuries and gadgets, but also, I am certain, richer in meaning and more abundant in real pleasure. To have a healthy environment we will all have to give up things we like; we may even have to give up things we have come to think of as necessities. But to be fearful of the disease and yet unwilling to pay for the cure is not just to be hypocritical; it is to be doomed. If you talk a good line without being changed by what you say, then you are not just hypocritical and doomed; you have become an agent of the disease. Consider, for an example, President Nixon, who advertises his grave concern about the destruction of the environment, and who turns up the air conditioner to make it cool enough to build a fire.

Odd as I am sure it will appear to some, I can think of no better form of personal involvement in the cure of the environment than that of gardening. A person who is growing a garden, if he is growing it organically, is improving a piece of the world. He is producing something to eat, which makes him somewhat independent of the grocery business, but he is also enlarging, for himself, the meaning of food and the pleasure of eating. The food he grows will be fresher, more nutritious, less contaminated by poisons and preservatives and dyes than what he can buy at a store. He is reducing the trash problem; a garden is not a disposable container, and it will digest and reuse its own wastes. If he enjoys working in his garden, then he is less dependent on an automobile or a merchant for his pleasure. He is involving himself directly in the work of feeding people.

If you think I'm wandering off the subject, let me remind you that most of the vegetables necessary for a family of four can be grown on a plot of forty by sixty feet. I think we might see in this an economic potential of considerable importance, since we now appear to be facing the possibility of widespread famine. How much food could be grown in the dooryards of cities and suburbs? How much could be grown along the extravagant right-of-ways of the interstate system? Or how much could be grown, by the intensive practices and economics of the garden or small farm, on so-called marginal lands? Louis Bromfield liked to point out that the people of France survived crisis after crisis because they were a nation of gardeners, who in times of want turned with great skill to their own small plots of ground. And F. H. King, an agriculture professor who traveled extensively in the Orient in 1907, talked to a Chinese farmer who supported a family of twelve, "one donkey, one cow . . . and two pigs on 2.5 acres of cultivated land"—and who did this, moreover, by agricultural methods that were sound enough to have maintained his land in prime fertility through several thousand years of such use. These are possibilities that are readily apparent and attractive to minds that are prepared to Think Little. To Big Thinkers—the bureaucrats and businessmen of agriculture—they are invisible. But intensive, organic agriculture kept the farms of the Orient thriving for thousands of years, whereas extensive—which is to say, exploitive or extractive—agriculture has critically reduced the fertility of American farmlands in a few centuries or even a few decades.

A person who undertakes to grow a garden at home, by practices that will preserve rather than exploit the economy of the soil, has set his mind decisively against what is wrong with us. He is helping himself in a way that dignifies him and that is rich in meaning and pleasure. But he is doing something else that is more important: He is making vital contact with the soil and the weather on which his life depends. He will no longer look upon rain as a traffic impediment, or upon the sun as a holiday decoration. And his sense of humanity's dependence on the world will have grown precise enough, one would hope, to be politically clarifying and useful.

What I am saying is that if we apply our minds directly and competently to the needs of the earth, then we will have begun to make fundamental and necessary changes in our minds. We will begin to understand and to mistrust *and to change* our wasteful economy, which markets not just the produce of the earth, but also the earth's ability to produce. We will see that beauty and utility are alike dependent upon the health of the world. But we will also see through the fads and the fashions of protest. We will see that war and oppression and pollution are not separate issues, but are aspects of the same issue. Amid the outcries for the liberation of this group or that, we will know that no person is free except in the freedom of other persons, and that our only real freedom is to know and faithfully occupy our place—a much humbler place than we have been taught to think—in the order of creation.

But the change of mind I am talking about involves not just a change of knowledge, but also a change of attitude toward our essential ignorance, a change in our bearing in the face of mystery. The principles of ecology, if we will take them to heart, should keep us aware that our lives depend upon other lives and upon processes and energies in an interlocking system that, though we can destroy it, we can neither fully understand nor fully control. And our great dangerousness is that, locked in our selfish and myopic economy, we have been willing to change or destroy far beyond our power to understand. We are not humble enough or reverent enough.

Some time ago, I heard a representative of a paper company refer to conservation as a "no-return investment." This man's thinking was exclusively oriented to the annual profit of his industry. Circumscribed by the demand that the profit be great, he simply could not be answerable to any other demand—not even to the obvious needs of his own children.

Consider, in contrast, the profound ecological intelligence of Black Elk, "a holy man of the Oglala Sioux," who in telling his story said that it was not his own life that was important to him, but what he had shared with all life: "It is the story of all life that is holy and it is good to tell, and of us two-leggeds sharing in it with the four-leggeds and the wings of the air and all green things. . . ." And of the great vision that came to him when he was a child he said: "I saw that the sacred hoop of my people was one of many hoops that made one circle, wide as daylight and as starlight, and in the center grew one mighty flowering tree to shelter all the children of one mother and father. And I saw that it was holy."

Discipline and Hope

I: INTRODUCTORY NOTE

I BEGIN WITH what I believe is a safe premise, at least in the sense that most of the various sides of the current public argument would agree, though for different reasons: that we have been for some time in a state of general cultural disorder, and that this disorder has now become critical. My interest here has been to examine to what extent this disorder is a failure of discipline—specifically, a failure of those disciplines, both private and public, by which desired ends might be reached, or by which the proper means to a desired end might be determined, or by which it might be perceived that one apparently desirable end may contradict or forestall another more desirable. Thus we have not asked how the "quality of life," as we phrase it, may be fostered by social and technological means that are sensitive only to quantitative measures; we have not really questioned the universal premise of power politics that peace (among the living) is the natural result of war; we have hardly begun to deal with the fact that an economy of waste is not compatible with a healthy environment.

I realize that I have been rather severely critical of several, perhaps all, sides of the present disagreement, having in effect made this a refusal to be a partisan of *any* side. If critical intelligence has a use it is to prevent the coagulation of opinion in social or political cliques. My purpose has been to invoke the use of principle, rather than partisanship, as a standard of behavior, and to clear the rhetoric of the various sides from what ought to be the ground of personal experience and common sense.

It appears to me that the governing middle, or the government, which supposedly represents the middle, has allowed the extremes of left and right to force it into an extremism of its own. These three extremes of left, right, and middle, egged on by and helplessly subservient to each other's rhetoric, have now become so self-righteous and self-defensive as to have no social use. So large a ground of sanity and good sense and decency has been abandoned by these extremes that it becomes possible now to think of a New Middle made up of people conscious and knowledgeable enough to despise the blandishments and oversimplifications of the extremes—and roomy and diverse enough to permit a renewal of intelligent cultural dialogue. That is what I hope for: a chance to live and speak as a person, not as a function of some political bunch.

2: THE POLITICS OF KINGDOM COME

Times of great social stress and change, when realities become difficult to face and to cope with, give occasion to forms of absolutism, demanding perfection. We are in such a time now, and it is producing the characteristic symptoms. It has suddenly become clear to us that practices and ambitions that we have been taught from the cradle to respect have made us the heirs apparent of a variety of dooms; some of the promised solutions, on which we have been taught to depend, are not working, are probably not going to work. As a result the country is burdened with political or cultural perfectionists of several sorts, demanding that the government or the people create *right now* one or another version of the ideal state. The air is full of dire prophecies, warnings, and threats of what will happen if the Kingdom of Heaven is not precipitately landed at the nearest airport.

It is important that we recognize the childishness of this. Its ancestor is the kicking fit of childhood, a sort of behavioral false rhetoric that offers the world two absolute alternatives: "If I can't have it, I'll tear it up." Its cultural model is the fundamentalist preacher, for whom there are no degrees of behavior, who cannot tell the difference between a shot glass and a barrel. The public *demand* for perfection, as opposed to private striving for it, is almost always productive of violence, and is itself a form of violence. It is totalitarian in impulse, and often in result.

The extremes of public conviction are always based upon rhetorical extremes, which is to say that their words—and their actions—have departed from facts, causes, and arguments, and have begun to follow the false logic of a feud in which nobody remembers the cause but only what was last said or done by the other side. Language and behavior become purely negative in function. The opponents no longer speak in support of their vision or their arguments or their purposes, but only in opposition to each other. Language ceases to bind head to heart, action to principle, and becomes a weapon in a contention deadly as war, shallow as a game.

But it would be an oversimplification to suggest that the present contention of political extremes involves only the left and the right. These contend not so much against each other as against the middle—the administration in power—which each accuses of *being* the other. In defending itself, the middle characteristically adopts the tactics of the extremes, corrupting its language by a self-congratulatory rhetoric bearing no more kinship to the truth and to honest argument than expediency demands, and thus it becomes an extreme middle. Whereas the extreme left and right see in each other the imminence of Universal Wrong, the extreme middle appears to sense in itself the imminence of the Best of All Possible Worlds, and therefore looks upon all critics as traitors. The rhetoric of the extreme middle equates the government with the country, loyalty to the government with patriotism, the will of the chief executive with the will of the people. It props itself with the tone of divine good will and infallibility, demanding an automatic unquestioning faith in its actions, upholding its falsehoods and errors with the same unblinking piety with which it obscures its truths and its accomplishments.

Because the extreme middle is characteristically in power, its characteristic medium is the one that is most popular—television. How earnestly and how well this middle has molded itself to the demands of television is apparent when one considers how much of its attention is given to image making, or remaking, and to public relations. It has given up almost altogether the disciplines of political discourse (considerations of fact and of principle and of human and historical limits and possibilities), and has taken up the cynical showmanship of those who have cheap goods for sale. Its catch phrases do not rise from any viable political tradition; their next of kin are the TV jingles of soup and soap. It is a politics of illusion, and its characteristic medium is pre-eminently suited—as it is almost exclusively limited—to the propagation of illusion.

Of all the illusions of television, that of its much-touted "educational value" is probably the first. Because of its utter transience as a medium and the complete passivity of its audience, television is doomed to have its effect within the limits of the most narrow and shallow definition of entertainment—that is, entertainment as diversion. The watcher sees the program at the expense of no effort at all; he is inert. All the live connections are broken. More important, a TV program can be seen only once; it cannot be re-examined or judged upon the basis of study, as even a movie can be; a momentous event or a serious drama slips away from us among the ordinary furniture of our lives, as transient and fading as the most commonplace happenings of every day. For these reasons a political speech on television has to be first and last a show, simply because it has no chance to become anything else. The great sin of the medium is not that it presents fiction as truth, as undoubtedly it sometimes does, but that it cannot help presenting the truth as fiction, and that of the most negligible sort—a way to keep awake until bedtime.

In depending so much upon a medium that will not permit scrutiny, the extreme middle has perhaps naturally come to speak a language that will not *bear* scrutiny. It has thus abased its own part in the so-called political dialogue to about the level of the slogans and chants and the oversimplified invective of the extreme left and right—a fact that in itself might sufficiently explain the obsession with image-making. Hearing the televised pronouncements of the political leaders of our time upon the great questions of human liberty, community obligation, war and peace, poverty and wealth, one might easily forget that such as John Adams and Thomas Jefferson once spoke here upon those questions. Indeed, our contemporary men of power have produced in their wake an industry of journalistic commentary and interpretation, because it is so difficult to determine *what* they have said and whether or not they meant it. Thus one sees the essential contradiction in the expedient doctrine that the end may justify the means. Corrupt or false means must inevitably corrupt or falsify the end. There is an important sense in which the end *is* the means.

What is disturbing, then, about these three "sides" of our present political life is not their differences but their similarities.

They have all abandoned discourse as a means of clarifying and explaining and defending and implementing their ideas. They have taken almost exclusively to the use of the rhetoric of ad-writers: catch phrases, slogans, clichés, euphemisms, flatteries, falsehoods, and various forms of cheap wit. This has led them —as such rhetoric must—to the use of power and the use of violence against each other. But however their ideological differences might be graphed, they are, in effect, all on the same side. They are on the side of their quarrel, and are against all other, and all better, possibilities. There is a political and social despair in this that is the greatest peril a country can come to, short of the inevitable results of such despair should it continue very long. "We are fatalists," Edward Dahlberg wrote, "only when we cease telling the truth, but, so long as we communicate the truth, we move ourselves, life, history, men. There is no other way."

Our present political rhetoric is the desperation of argument. It is like a weapon in its inflexibility, in its insensitivity to circumstance, and in its natural inclination toward violence. It is by this recourse to loose talk—this willingness to say whatever will be easiest to say and most willingly heard—that the left permits its methods to contradict its avowed aims, as when it contemplates violence as a means of peace or permits arguments to shrink into slogans. By this process the right turns from its supposed aim of conserving the best of the past and undertakes the defense of economic privilege and the deification of symbols. And by this process the middle abandons its obligation to lead and enlighten the majority it claims as its constituency, and takes to the devices of a sterile showmanship, by which it hopes to elude criticism and obscure its failures.

The political condition in this country now is one in which the means or the disciplines necessary to the achievement of professed ends have been devalued or corrupted or abandoned altogether. We are offered peace without forbearance or tolerance or love, security without effort and without standards, freedom without risk or responsibility, abundance without thrift. We are asked repeatedly by our elected officials to console ourselves with that most degenerate of political arguments: Though we are not doing as well as we might, we could do worse, and we are doing better than some.

3: THE KINGDOM OF EFFICIENCY AND SPECIALIZATION

But this political indiscipline is exemplary of a condition that is widespread and deeply rooted in almost all aspects of our life. Nearly all the old standards, which implied and required rigorous disciplines, have now been replaced by a new standard of efficiency, which requires not discipline, not a mastery of means, but rather a carelessness of means, a relentless subjection of means to immediate ends. The standard of efficiency displaces and destroys the standards of quality because, by definition, it cannot even consider them. Instead of asking a man what he can do well, it asks him what he can do fast and cheap. Instead of asking the farmer to practice the best husbandry, to be a good steward and trustee of his land and his art, it puts irresistible pressures on him to produce more and more food and fiber more and more cheaply, thereby destroying the health of the land, the best traditions of husbandry, and the farm population itself. And so when we examine the principle of efficiency as we now practice it, we see that it is not really efficient at all. As we use the word, efficiency means no such thing, or it means short-term or temporary efficiency, which is a contradiction in terms. It means cheapness at any price. It means hurrying to nowhere. It means the profligate waste of humanity and of nature. It means the greatest profit to the greatest liar. What we have called efficiency has produced among us, and to our incalculable cost, such unprecedented monuments of destructiveness and waste as the strip-mining industry, the Pentagon, the federal bureaucracy, and the family car.

Real efficiency is something entirely different. It is neither cheap (in terms of skill and labor) nor fast. Real efficiency is long-term efficiency. It is to be found in means that are in keeping with and preserving of their ends, in methods of production that preserve the sources of production, in workmanship that is durable and of high quality. In this age of planned obsolescence, frivolous horsepower and surplus manpower, those salesmen and politicians who talk about efficiency are talking, in reality, about spiritual and biological death.

Specialization, a result of our nearly exclusive concern with the form of exploitation that we call efficiency, has in its turn become a destructive force. Carried to the extent to which we have carried it, it is both socially and ecologically destructive. That specialization has vastly increased our knowledge, as its defenders claim, cannot be disputed. But I think that one might reasonably dispute the underlying assumption that knowledge per se, undisciplined knowledge, is good. For while specialization has increased knowledge, it has fragmented it. And this fragmentation of knowledge has been accompanied by a fragmentation of discipline. That is, specialization has tended to draw the specialist toward the discipline that will lead to the discovery of new facts or processes within a narrowly defined area, and it has tended to lead him away from or distract him from those disciplines by which he might consider the *effects* of his discovery upon human society or upon the world.

Nowhere are these tendencies more apparent than in agriculture. For years now the agricultural specialists have tended to think and work in terms of piecemeal solutions and in terms of annual production, rather than in terms of a whole and coherent system that would maintain the fertility and the ecological health of the land over a period of centuries. Focused nearly exclusively upon so-called efficiency with respect to production, as if the only discipline pertinent to agriculture were that of economics, they have eagerly abetted a rapid industrialization of agriculture that is potentially catastrophic, both in the ecological deterioration of farm areas and in the dispossession and displacement of the rural population.

Ignoring the ample evidence that a healthy, ecologically sound agriculture is highly diversified, using the greatest possible variety of animals and plants, and that it returns all organic wastes to the soil, the specialists of the laboratories have promoted the specialization of the farms, encouraging one-crop agriculture and the replacement of humus by chemicals. And as the pressures of urban populations upon the land have grown, the specialists have turned more and more not to the land, but to the laboratory.

Ignoring the considerable historical evidence that to have a productive agriculture over a long time it is necessary to have a stable, prosperous rural population closely bound in sympathy and association to the land, the specialists have either connived in the dispossession of small farmers by machinery and technology, or have actively encouraged their migration into the cities.

The result of the short-term vision of these experts is a whole series of difficulties that together amount to a rapidly building ecological and social disaster, which there is little disposition at present to regret, much less to correct. The organic wastes of our society, for which our land is starved and which in a sound agricultural economy would be returned to the land, are flushed out through the sewers to pollute the streams and rivers and, finally, the oceans; or they are burned and the smoke pollutes the air; or they are wasted in other ways. Similarly, the small farmers who in a healthy society ought to be the mainstay of the country—whose allegiance to their land, continuing and deepening in association from one generation to another, would be the motive and guarantee of good care—are forced out by the homicidal economics of efficiency, to become emigrants and dependents in the already overcrowded cities. In both instances, by the abuse of knowledge in the name of efficiency, assets have been converted into staggering problems.

The metaphor governing these horrendous distortions has been that of the laboratory. The working assumption has been that nature and society, like laboratory experiments, can be manipulated by processes that are for the most part comprehensible toward ends that are for the most part foreseeable. But the analogy, as any farmer knows, is too simple, for both nature and humanity are unpredictable and ultimately mysterious. Sir Albert Howard was speaking to this problem in *An Agricultural Testament*: "Instead of breaking up the subject into fragments and studying agriculture in piecemeal fashion by the analytical methods of science, appropriate only to the discovery of new facts, we must adopt a synthetic approach and look at the wheel of life as one great subject and not as if it were a patchwork of unrelated things." A much more appropriate model for the agriculturist, scientist, or farmer is the forest, for the forest, as Howard pointed out, "manures itself" and is therefore self-renewing; it has achieved that "correct relation between the processes of growth and the processes of decay that is the first principle of successful agriculture." A healthy agriculture can take place only within nature, and in cooperation with its processes, not in spite of it and not by "conquering" it. Nature, Howard points out, in elaboration of his metaphor, "never attempts to farm without livestock; she always raises mixed crops; great pains are taken to preserve the soil and to prevent erosion; the mixed vegetable and animal wastes are converted into humus; *there is no waste* [my emphasis]; the processes of growth and the processes of decay balance one another; ample provision is made to maintain large reserves of fertility; the greatest care is taken to store the rainfall; both plants and animals are left to protect themselves against disease."

The fact is that farming is not a laboratory science, but a science of practice. It would be, I think, a good deal more accurate to call it an art, for it grows not only out of factual knowledge but out of cultural tradition; it is learned not only by precept but by example, by apprenticeship; and it requires not merely a competent knowledge of its facts and processes, but also a complex set of attitudes, a certain culturally evolved stance, in the face of the unexpected and the unknown. That is to say, it requires *style* in the highest and richest sense of that term.

One of the most often repeated tenets of contemporary optimism asserts that "a nation that can put men on the moon certainly should be able to solve the problem of hunger." This proposition seems to me to have three important flaws, which I think may be taken as typical of our official view of ourselves:

1—It construes the flight to the moon as an historical event of a complete and coherent significance, when in fact it is a fragmentary event of very uncertain significance. Americans have gone to the moon as they came to the frontiers of the New World: with their minds very much upon getting there, very little upon what might be involved in *staying* there. I mean that, because of our history of waste and destruction here, we have no assurance that we can survive in America, much less on the moon. And until we can bring into balance the processes of growth and decay, the white man's settlement of this continent will remain an incomplete event. When a Japanese peasant went to the fields of his tiny farm in the preindustrial age, he worked in the governance of an agricultural tradition that had sustained the land in prime fertility for thousands of years, in spite of the pressures of a population that in 1907 had reached a density, according to F. H. King's *Farmers of Forty Centuries*, of "more than three people to each acre." Such a farmer might look upon his crop year as a complete and coherent historical event, suffused and illuminated with a meaning and mystery that were both its own and the world's, because in his mind and work agricultural process had come into an enduring and preserving harmony with natural process. To him the past confidently promised a future. What are we to say, by contrast, of a society that places no value at all upon such a tradition or such a man, that instead works the destruction of such imperfect agricultural tradition as it has, that replaces the farm people with machines, that values the techniques of production far above the techniques of land maintenance, and that has espoused as an ideal a depopulated countryside farmed by a few technicians for the supposedly greater benefit of hundreds of millions crowded into cities and helpless to produce food or any other essential for themselves?

2—The agricultural optimism that bases itself upon the moon landings assumes that there is an equation between agriculture and technology, or that agriculture is a kind of technology. This grows out of the much-popularized false assumptions of the agricultural specialists, who have gone about their work as if agriculture was answerable only to the demands of economics, not to those of ecology or human culture, just as most urban consumers conceive eating to be an activity associated with economics but not with agriculture. The ground of agricultural thinking is so narrowly circumscribed, one imagines, to fit the demands of laboratory science, as well as the popular prejudice that prefers false certainties to honest doubts. The discipline proper to eating, of course, is not economics but agriculture. The discipline proper to agriculture, which survives not just by production but also by the return of wastes to the ground, is not economics but ecology. And ecology may well find its proper disciplines in the arts, whose function is to refine and enliven *perception*, for ecological principle, however publicly approved, can be enacted only upon the basis of each man's perception of his relation to the world.

Under the governance of the laboratory analogy, the *device*, which is simple and apparently simplifying, becomes the focal point and the standard rather than the human need, which is complex. Thus an agricultural specialist, prescribing the best conditions for the use of a harvesting machine, thinks only of the machine, not its cultural or ecological effects. And because of the fixation on optimum conditions, big-farm technology has come to be highly developed, whereas the technology of the family farm, which must still involve methods and economies that are "old-fashioned," has been neglected. For this reason, and others perhaps more pressing, small-farm technology is rapidly passing from sight, along with the small farmers. As a result we have an increasing acreage of supposedly "marginal" but potentially productive land for the use of which we have neither methods nor people—an alarming condition in view of the likelihood that someday we will desperately need to farm these lands again.

The drastic and incalculably dangerous assumption is that farming can be considered apart from farmers, that the land may be conceptually divided in its use from human need and human care. The assumption is that moving a farmer into a factory is as simple a cultural act as moving a worker from one factory to another. It is inconceivably more complicated, and more final. American agricultural tradition has been for the most part inadequate from the beginning, and we have an abundance of diminished land to show for it. But American farmers are nevertheless an agricultural population of long standing. Most settlers who farmed in America farmed in Europe. The farm population in this country therefore embodies a knowledge and a set of attitudes and interests that have been thousands of years in the making. This mentality is, or was, a great resource upon which we might have built a truly indigenous agriculture, fully adequate to the needs and demands of American regions. Ancient as it is, it is destroyed in a generation in every family that is forced off the farm into the city—or in less than a generation, for the farm mentality can survive only in sustained vital contact with the land.

A truer agricultural vision would look upon farming not as a function of the economy or even of the society, but as a function of the land; and it would look upon the farm population as an indispensable and inalienable part of the ecological system. Among the Incas, according to John Collier (*Indians of the Americas*), the basic social and economic unit was the tribe, or *ayllu*, but "the *ayllu* was not merely its people, and not merely the land, but people and land wedded through a mystical bond." The union of the land and the people was indissoluble, like marriage or grace. Chief Rekayi of the Tangwena tribe of Rhodesia, in refusing to leave his ancestral home, which had been claimed by the whites, is reported in recent newspaper accounts to have said: "I am married to this land. I was put here by God . . . and if I am to leave, I must be removed by God who put me here." This altogether natural and noble sentiment was said by the Internal Affairs Minister to have been "Communist inspired."

3—The notion that the moon voyages provide us assurance of enough to eat exposes the shallowness of our intellectual confidence, for it is based upon our growing inability to distinguish between training and education. The fact is that a man can be made an astronaut much more quickly than he can be made a good farmer, for the astronaut is produced by training and the farmer by education. Training is a process of conditioning, an orderly and highly efficient procedure by which a man learns a prescribed pattern of facts and functions. Education, on the other hand, is an obscure process by which a person's experience is brought into contact with his place and his history. A college can train a person in four years; it can barely begin his education in that time. A person's education begins before his birth in the making of the disciplines, traditions, and attitudes of mind that he will inherit, and it continues until his death under the slow, expensive, uneasy tutelage of his experience. The process that produces astronauts may produce soldiers and factory workers and clerks; it will never produce good farmers or good artists or good citizens or good parents.

White American tradition, so far as I know, contains only one coherent social vision that takes such matters into consideration, and that is Thomas Jefferson's. Jefferson's public reputation seems to have dwindled to that of Founding Father and advocate of liberty, author of several documents and actions that have been enshrined and forgotten. But in his thinking, democracy was not an ideal that stood alone. He saw that it would have to be secured by vigorous disciplines or its public offices would become merely the hunting grounds of mediocrity and venality. And so those who associate his name only with his political utterances miss both the breadth and depth of his wisdom. As Jefferson saw it, two disciplines were indispensable to democracy: on the one hand, education, which was to produce a class of qualified leaders, an aristocracy of "virtue and talents" drawn from all economic classes; and on the other hand, widespread land ownership, which would assure stable communities, a tangible connection to the country, and a permanent interest in its welfare. In language that recalls Collier's description of the *ayllu* of the Incas, and the language of Chief Rekayi of the Tangwenans, Jefferson wrote that farmers "are tied to their country, and wedded to its liberty and interests, by the most lasting bonds." And: ". . . legislators cannot invent too many devices for subdividing property. . . ." And: ". . . it is not too soon to provide by every possible means that as few as possible shall be without a little portion of land. The small landholders are the most precious part of a state." For the discipline of education of the broad and humane sort that Jefferson had in mind, to produce a "natural aristocracy . . . for the instruction, the trusts, and government of society," we have tended more and more to substitute the specialized training that will most readily secure the careerist in his career. For the ownership of "a little portion of land" we have, and we apparently wish, to substitute the barbarous abstraction of nationalism, which puts our minds within the control of whatever demagogue can soonest rouse us to self-righteousness.

On September 10, 1814, Jefferson wrote to Dr. Thomas Cooper of the "condition of society" as he saw it at that time: ". . . we have no paupers, the old and crippled among us, who possess nothing and have no families to take care of them, being too few to merit notice as a separate section of society. . . . The great mass of our population is of laborers; our rich . . . being few, and of moderate wealth. Most of the laboring class possess property, cultivate their own lands . . . and from the demand for their labor are enabled . . . to be fed abundantly, clothed above mere decency, to labor moderately. . . . The wealthy . . . know nothing of what the Europeans call luxury." This has an obvious kinship with the Confucian formula: ". . . that the producers be many and that the mere consumers be few; that the artisan mass be energetic and the consumers temperate. . . ."

In the loss of that vision, or of such a vision, and in the abandonment of that possibility, we have created a society characterized by degrading urban poverty and an equally degrading affluence—a society of undisciplined abundance, which is to say a society of waste.

4: THE KINGDOM OF CONSUMPTION

The results have become too drastic to be concealed by our politicians' assurances that we have built a "great society" or that we are doing better than India. Official pretense has begun to break down under the weight of the obvious. In the last decade we have become unable to condition our children's minds to approve or accept our errors. Our history has created in the minds of our young people a bitter division between official pretense and social fact, and we have aggravated this division by asking many of them to fight and die in support of official pretense. In this way we have produced a generation whose dissidence and alienation are without precedent in our national experience.

The first thing to be said about this rebelliousness is that it is understandable, and that it deserves considerate attention. Many of this generation have rejected values and practices that they believe to be destructive, and they should do so. Many of them have begun to search for better values and forms of life, and they should do so. But the second thing to be said is that this generation is as subject as any other to intelligent scrutiny and judgment, and as deserving of honest criticism. It has received much approbation and condemnation, very little criticism.

One of its problems is that it has been isolated in its youthfulness, cut off from the experience and the counsel of older people, as probably no other generation has ever been. It is true that the dissident young have had their champions among the older people, but it is also true, I think, that these older people have been remarkably uncritical of the young, and so have abdicated their major responsibility to them. Some appear to have *joined* the younger generation, buying their way in by conniving in the notion that idealistic youth can do no wrong —or that one may reasonably hope to live without difficulty or effort or tragedy, or that surfing is "a life." The uncritical approval of a band of senior youth freaks is every bit as isolating and every bit as destructive as the uncritical condemnation of those who have made hair length the foremost social issue of the time.

And so a number of the problems of the young people now are problems that have always attended youthfulness, but which isolation has tended to aggravate in the present generation: impetuousness, a haste to undertake work that one is not yet prepared for; a tendency to underestimate difficulty and overestimate possibility, which is apt, through disillusionment, to lead to the overestimation of difficulty and the underestimation of possibility; oversimplification, as when rejection takes the place of evaluation; and, finally, naïve prejudice, as when people who rightly condemn the use of such terms as "nigger" or "greaser" readily use such terms as "pig" and "redneck."

Another of its problems, and a much larger one, is that the propaganda both of the "youth culture" and of those opposed to it has inculcated in many minds, both young and old, the illusion that this is a wholly new generation, a generation free of history. The proposition is dangerously silly. The present younger generation is, as much as any other, a product of the past; it would not be as it is if earlier generations had not been as they were. Like every other young generation, this one bears the precious human burden of new possibility and new hope, the opportunity to put its inheritance to better use. And like every other, it also bears the germ of historical error and failure and weakness—which it rarely forgives in its predecessors, and seldom recognizes in itself. In the minds of those who do not know it well, and who have not mastered the disciplines of self-criticism, historical error is a subtle virus indeed. It is of the greatest importance that we recognize in the youth culture the persistence in new forms of the mentality of waste, certain old forms of which many of the young have rightly repudiated.

Though it has forsworn many of the fashions and ostentations of the "affluent society," the youth culture still supports its own forms of consumerism, the venerable American doctrine which holds that if enough is good, too much is better. As an example, consider the present role of such drugs as marijuana and the various hallucinogens. To deal sensibly with this subject, it is necessary to say at the outset that the very concept of drug abuse implies the possibility of drug use that is *not* abusive. And it is, in fact, possible to produce examples of civilizations that have employed drugs in disciplines and ceremonies that have made them culturally useful and prevented their abuse.

Tobacco, for instance, is a drug that we have used so massively and thoughtlessly that we have, in typical fashion, come to be endangered by it. This is the pattern of the consumer economy and it applies not just to drugs, but to such commodities as the automobile and electrical power. But American Indians attached to tobacco a significance that made it more valuable to them than it has ever been to us, and at the same time kept them from misusing it as we have. In the Winnebago Origin Myth tobacco had a ceremonial and theological role. According to this myth, Paul Radin wrote in his introduction to *The Road of Life and Death*, man "is not to save himself or receive the wherewithal of life through the accidental benefactions of culture-heroes. On the contrary, he is to be in dire straits and saved. Earthmaker is represented as withholding tobacco from the spirits in order to present it to man and as endowing these same spirits with an overpowering craving for it. In short, it is to be the mechanism for an exchange between man and the deities. He will give them tobacco; they will give him powers to meet life and overcome obstacles."

The use of alcohol has had, I believe, a similar history: a decline and expansion from ceremonial use to use as a commodity and extravagance, from cultural usefulness to cultural liability.

The hallucinogenic drugs have now also run this course of cultural diminishment, and at the hands not of the salesmen of the corporations and the advertising agencies, but of the self-proclaimed enemies of those salesmen. Most of these drugs have been used by various cultures in association with appropriate disciplines and ceremonies. Anyone who reads an account of the Peyote Meeting of the Native American Church will see that it resembles very much the high ritual and art of other cultures but very little indeed the usual account of the contemporary "dope scene."

A very detailed and well-understood account of the disciplined use of such drugs is in Carlos Castaneda's *The Teachings of Don Juan: A Yaqui Way of Knowledge*. Don Juan, a medicine man and sorcerer, a Yaqui Indian from Sonora, Mexico, undertakes to teach Castaneda the uses of jimson weed, peyote, and the psilocybe mushrooms. The book contains some remarkable accounts of the author's visions under the influence of these drugs, but equally remarkable is the rigor of the disciplines and rituals by which his mentor prepared him for their use. At one point early in their association the old Indian said to him: "A man goes to knowledge as he goes to war, wide awake, with fear, with respect, and with absolute assurance. Going to knowledge or going to war in any other manner is a mistake, and whoever makes it will live to regret his steps."

The cultural role of both hallucinogens and intoxicants, in societies that have effectively disciplined their use, has been strictly limited. They have been used either for the apprehension of religious or visionary truth or, a related function, to induce in conditions prescribed by ceremony and festivity a state of self-abandon in which one may go free for a limited time of the obscuring and distorting preoccupation with one's own being. Other cultures have used other means—music or dance or poetry, for examples—to produce these same ends, and although the substance of Don Juan's teaching may be somewhat alien to the mainstream of our tradition, the terms of its discipline are not.

By contrast, the youth culture tends to use drugs in a way very similar to the way its parent culture uses alcohol: at random, as a social symbol and crutch, and with the emphasis upon the fact and quantity of use rather than the quality and the content of the experience. It would be false to say that these drugs have come into contemporary use without any of their earlier cultural associations. Indeed, a good deal of importance has been assigned to the "religious" aspect of the drug experience. But too often, it seems to me, the tendency has been to make a commodity of religion, as if in emulation of some churches. Don Juan looked upon drugs as a way to knowledge, difficult and fearful as other wise men have conceived other ways to knowledge, and therefore to be rigorously prepared for and faithfully followed; the youth culture, on the other hand, has tended to look upon drugs as a sort of instant Holy Truth, of which one need not become worthy. When they are inadequately prepared for the use of drugs, that is to say, people "consume" and waste them.

The way out of this wastefulness obviously cannot lie simply in a shift from one fashionable commodity to another commodity equally fashionable. The way out lies only in a change of mind by which we will learn not to think of ourselves as "consumers" in any sense. A consumer is one who uses things up, a concept that is alien to the creation, as are the concepts of waste and disposability. A more realistic and accurate vision of ourselves would teach us that our ecological obligations are to use, not use up; to use by the standard of real need, not of fashion or whim; and then to relinquish what we have used in a way that returns it to the common ecological fund from which it came.

The key to such a change of mind is the realization that the first and final order of creation is not such an order as men can impose on it, but an order in the creation itself by which its various parts and processes sustain each other, and which is only to some extent understandable. It is, moreover, an order in which things find their places and their values not according to their inert quantities or substances but according to their energies, their powers, by which they cooperate or affect and influence each other. The order of the creation, that is to say, is closer to that of drama than to that of a market.

This relation of power and order is another of the major concerns of the Winnebago Creation Myth. "Having created order within himself and established it for the stage on which man is to play," Paul Radin says, "Earthmaker proceeds to create the first beings who are to people the Universe, the spirits and deities. To each one he assigns a fixed and specific amount of power, to some more, to some less. . . . This principle of gradation and subordination is part of the order that Earthmaker is represented as introducing into the Universe. . . . The instant it is changed there is danger and the threat of disruption." The principle is dramatized, according to Radin, in the legend of Morning Star: ". . . one of the eight great Winnebago deities, Morning Star, has been decapitated by his enemy, a water-spirit. The body of the hero still remains alive and is being taken care of by his sister. The water-spirit, by keeping the head of Morning Star, has added the latter's power to his own. So formidable is this combination of powers that none of the deities [is] a match for him now. In fact only Earthmaker is his equal. Here is a threat of the first magnitude to the order ordained by Earthmaker and it must be met lest destruction overtake the world."

The point is obvious: To take and keep, to consume the power of another creature is an act profoundly disordering, contrary to the nature of the creation. And equally obvious is its applicability to our own society, which sees its chief function in such accumulations of power. A man grows rich by strip mining, adding the power of a mountain to himself in such a way that he cannot give it back. As a nation, we have so far grown rich by adding the power of the continent to ourselves in such a way that we cannot give it back. "Here is a threat of the first magnitude to the order ordained by Earthmaker and it must be met lest destruction overtake the world."

Though we generally concede that a man may have more of the world's goods than he deserves, I think that we have never felt that a man may have more light than he deserves. But an interesting implication of the Winnebago doctrine of power and order is that a man must not only become worthy of enlightenment, but has also an ethical obligation to make himself worthy of the world's goods. He can make himself worthy of them only by using them carefully, preserving them, relinquishing them in good order when he has had their use. That a man shall find his life by losing it is an ethical concept that applies to the body as well as the spirit.

An aspect of the consumer mentality that has cropped up with particular virulence in the youth culture is an obsessive fashionableness. The uniformity of dress, hair style, mannerism, and speech is plain enough. But more serious, because less conscious and more pretentious, is an intellectual fashionableness pinned up on such shibboleths as "the people" (the most procrustean of categories), "relevance" (the most reactionary and totalitarian of educational doctrines), and "life style."

This last phrase furnishes a particularly clear example of the way poor language can obscure both a problem and the possibility of a solution. Compounded as "alternate life style," the phrase becomes a part of the very problem it aspires to solve. There are, to begin with, two radically different, even opposed meanings of style: style as fashion, an imposed appearance, a gloss upon superficiality; and style as the signature of mastery, the efflorescence of long discipline. It is obvious that the style of mastery can never become the style of fashion, simply because every master of a discipline is different from every other; his mastery is suffused with his own character and his own materials. Cézanne's paintings could not have been produced by a fad, for the simple reason that they could not have been produced by any other person. As a popular phrase, "life style" necessarily has to do only with what is imitable in another person's life, its superficial appearances and trappings; it cannot touch its substances, disciplines, or devotions. More important is the likelihood that a person who has identified his interest in another person as an interest in his "life style" will be *aware* of nothing but appearances. The phrase "alternate life style" attempts to recognize our great need to change to a kind of life that is not wasteful and destructive, but stifles the attempt, in the same breath, by infecting it with that superficial concept of style. An essential recognition is thus obscured at birth by the old lie of advertising and public relations: that you can alter substance by altering appearance. "Alternate life style" suggests, much in the manner of the fashion magazines, that one can change one's life by changing one's clothes.

Another trait of consumption that thrives in the youth culture is that antipathy to so-called "drudgery" that has made us, with the help of salesmen and advertisers, a nation of suckers. This is the pseudoaristocratic notion, early popularized in America, that one is too good for the fundamental and recurring tasks of domestic order and biological necessity; to dirty one's hands in the soil or to submerge them for very long in soapy water is degrading and brutalizing. With one's hands thus occupied, the theory goes, one is unlikely to reach those elusive havens of "self-discovery" and "self-fulfillment"; but if one can escape such drudgery, one then has a fair chance of showing the world that one is *really* better than all previous evidence would have indicated. In every drudge there is an artist or a tycoon yearning to breathe free.

The entire social vision, as I understand it, goes something like this: Man is born into a fallen world, doomed to eat bread in the sweat of his face. But there is an economic redemption. He should go to college and get an education—that is, he should acquire the "right" certificates and meet the "right" people. An education of this sort should enable him to get a "good" job—that is, short hours of work that is either easy or prestigious for a lot of money. Thus he is saved from the damnation of drudgery, and is presumably well on the way to proving the accuracy of his early suspicion that he is *really* a superior person.

Or, in a different version of the same story, the farmer at his plow or the housewife at her stove dreams of the neat outlines and the carefree boundaries of a factory worker's eight-hour day and forty-hour week, and his fat, unworried paycheck. They will leave their present drudgery to take the bait, in this case, of leisure, time, and money to enjoy the "good things of life."

In reality, this despised drudgery is one of the constants of life, like water only changing its form in response to changes of atmosphere. Our aversion to the necessary work that we call drudgery and our strenuous efforts to avoid it have not diminished it at all, but only degraded its forms. The so-called drudgery has to be done. If one is "too good" to do it for oneself, then it must be done by a servant, or by a machine manufactured by servants. If it is not done at home, then it must be done in a factory, which degrades both the conditions of work and the quality of the product. If it is not done well by the hands of one person, then it must be done poorly by the hands of many. But somewhere the hands of someone must be soiled with the work. Our aversion to this was once satisfied by slavery, or by the abuse of a laboring class; now it is satisfied by the assembly line, or by similar redundancy in bureaus and offices. For decades now our people have streamed into the cities to escape the drudgery of farm and household. Where do they go to escape the drudgery of the city? Only home at night, I am afraid, to the spiritual drudgery of factory-made suppers and TV. On weekends many of them continue these forms of urban drudgery in the country.

The youth culture has accepted, for the most part uncritically, the conviction that all recurring and necessary work is drudgery, even adding to it a uniquely gullible acquiescence in the promoters' myth that the purpose of technology is to free mankind for spiritual and cultural pursuits. But to the older idea of economic redemption from drudgery, the affluent young have added the even more simple-minded idea of redemption by spontaneity. Do what you feel like, they say—as if every day one could "feel like" doing what is necessary. Any farmer or mother knows the absurdity of this. Human nature is such that if we waited to do anything until we felt like it, we would do very little at the start, even of those things that give us pleasure, and would do less and less as time went on. One of the common experiences of people who regularly do hard work that they enjoy is to find that they begin to "feel like it" only after the task is begun. And one of the chief uses of discipline is to assure that the necessary work gets done even when the worker *doesn't* feel like it.

Because of the prevalence of the economics and the philosophy of laborsaving, it has become almost a heresy to speak of hard work, especially manual work, as an inescapable human necessity. To speak of such work as good and ennobling, a source of pleasure and joy, is almost to declare oneself a pervert. Such work, and any aptitude or taste for it, are supposedly mere relics of our rural and primitive past—a past from which it is the business of modern science and technology to save us.

Before one can hope to use any intelligence in this matter, it is necessary to resurrect a distinction that was probably not necessary before the modern era, and that has so far been made only by a few eccentrics and renegades. It is a distinction not made in business and government, and very little made in the universities. I am talking about the distinction between work that is necessary and therefore meaningful, and work that is unnecessary or devoid of meaning. There is no intelligent defense of what Thoreau called "the police of meaningless labor." The unnecessary work of producing notions or trinkets or machines intended to be soon worn out, or necessary work the meaning of which has been destroyed by mechanical process, is as degrading as slavery. And the purpose of such slavery, according to the laborsaving philosophy, is to set men free from work. Freed from work, men will presumably take to more "worthy" pursuits such as "culture." Noting that there have always been some people who, when they had leisure, studied literature and painting and music, the prophets of the technological paradise have always assured us that once we have turned all our work over to machines we will become a nation of artists or, at worst, a nation of art critics. This notion seems to me highly questionable on grounds both of fact and of principle.

In fact, we already know by experience what the "leisure" of most factory and office workers usually is, and we may reasonably predict that what it is it is likely to continue to be. Their leisure is a frantic involvement with salesmen, illusions, and machines. It is an expensive imitation of their work—anxious, hurried, unsatisfying. As their work offers no satisfactions in terms of work but must always be holding before itself the will-o'-the-wisp of freedom from work, so their leisure has no leisurely goals but must always be seeking its satisfaction outside itself, in some activity or some thing typically to be provided by a salesman. A man doing wholesome and meaningful work that he is pleased to do well is three times more at rest than the average factory or office worker on vacation. A man who does meaningless work does not have his meaning at hand. He must go anxiously in search of it—and thus fail to find it. The farmer's Sunday afternoon of sitting at home in the shade of a tree has been replaced by the "long weekend" of a thousand miles. The difference is that the farmer was where he wanted to be, understood the value of being there, and therefore when he had no work to do could sit still. How much have we spent to obscure so simple and obvious a possibility? The point is that there is an indissoluble connection and dependence between work and leisure. The freedom from work must produce not leisure, but an ever more frantic search for something to do.

The principle was stated by Thoreau in his *Journal*: "Hard and steady and engrossing labor with the hands, especially out of doors, is invaluable to the literary man and serves him directly. Here I have been for six days surveying in the woods, and yet when I get home at evening, somewhat weary at last . . . I find myself more susceptible than usual to the finest influences, as music and poetry." That is, certainly, the testimony of an exceptional man, a man of the rarest genius, and it will be asked if such work could produce such satisfaction in an ordinary man. My answer is that we do not have to look far or long for evidence that all the fundamental tasks of feeding and clothing and housing—farming, gardening, cooking, spinning, weaving, sewing, shoemaking, carpentry, cabinetwork, stonemasonry—were once done with consummate skill by ordinary people, and as that skill indisputably involved a high measure of pride, it can confidently be said to have produced a high measure of satisfaction.

We are being saved from work, then, for what? The answer can only be that we are being saved from work that is meaningful and ennobling and comely in order to be put to work that is unmeaning and degrading and ugly. In 1930, the Twelve Southerners of *I'll Take My Stand* issued as an introduction to their book "A Statement of Principles," in which they declared for the agrarian way of life as opposed to the industrial. The book, I believe, was never popular. At the time, and during the three decades that followed, it might have been almost routinely dismissed by the dominant cultural factions as an act of sentimental allegiance to a lost cause. But now it has begun to be possible to say that the cause for which the Twelve Southerners spoke in their introduction was not a lost but a threatened cause: the cause of human culture. "The regular act of applied science," they said, "is to introduce into labor a labor-saving device or a machine. Whether this is a benefit depends on how far it is advisable to save the labor. The philosophy of applied science is generally quite sure that the saving of labor is a pure gain, and that the more of it the better. This is to assume that labor is an evil, that only the end of labor or the material product is good. On this assumption labor becomes mercenary and servile. . . . The act of labor as one of the happy functions of human life has been in effect abandoned. . . .

"Turning to consumption, as the grand end which justifies the evil of modern labor, we find that we have been deceived. We have more time in which to consume, and many more products to be consumed. But the tempo of our labors communicates itself to our satisfactions, and these also become brutal and hurried. The constitution of the natural man probably does not permit him to shorten his labor-time and enlarge his consuming-time indefinitely. He has to pay the penalty in satiety and aimlessness."

The outcry in the face of such obvious truths is always that if they were implemented they would ruin the economy. The peculiarity of our condition would appear to be that the implementation of *any* truth would ruin the economy. If the Golden Rule were generally observed among us, the economy would not last a week. We have made our false economy a false god, and it has made blasphemy of the truth. So I have met the economy in the road, and am expected to yield it right of way. But I will not get over. My reason is that I am a man, and have a better right to the ground than the economy. The economy is no god for me, for I have had too close a look at its wheels. I have seen it at work in the strip mines and coal camps of Kentucky, and I know that it has no moral limits. It has emptied the country of the independent and the proud, and has crowded the cities with the dependent and the abject. It has always sacrificed the small to the large, the personal to the impersonal, the good to the cheap. It has ridden to its questionable triumphs over the bodies of small farmers and tradesmen and craftsmen. I see it, still, driving my neighbors off their farms into the factories. I see it teaching my students to give themselves a price before they can give themselves a value. Its principle is to waste and destroy the living substance of the world and the birthright of posterity for a monetary profit that is the most flimsy and useless of human artifacts.

Though I can see no way to defend the economy, I recognize the need to be concerned for the suffering that would be produced by its failure. But I ask if it is necessary for it to fail in order to change; I am assuming that if it does not change it must sooner or later fail, and that a great deal that is more valuable will fail with it. As a deity the economy is a sort of egotistical French monarch, for it apparently can see no alternative to itself except chaos, and perhaps that is its chief weakness. For, of course, chaos is not the only alternative to it. A better alternative is a better economy. But we will not conceive the possibility of a better economy, and therefore will not begin to change, until we quit deifying the present one.

A better economy, to my way of thinking, would be one that would place its emphasis not upon the *quantity* of notions and luxuries but upon the *quality* of necessities. Such an economy would, for example, produce an automobile that would last at least as long, and be at least as easy to maintain, as a horse.* It would encourage workmanship to be as durable as its materials; thus a piece of furniture would have the durability not of glue but of wood. It would substitute for the pleasure of frivolity a pleasure in the high quality of essential work, in the use of good tools, in a healthful and productive countryside. It would encourage a migration from the cities back to the farms, to assure a work force that would be sufficient not only to the production of the necessary quantities of food, but to the production of food of the best *quality* and to the maintenance of the land at the highest fertility—work that would require a great deal more personal attention and care and hand labor than the present technological agriculture that is focused so exclusively upon production. Such a change in the economy would not involve large-scale unemployment, but rather large-scale changes and shifts of employment.

"You are tilting at windmills," I will be told. "It is a hard world, hostile to the values that you stand for. You will never enlist enough people to bring about such a change." People who talk that way are eager to despair, knowing how easy despair is. The change I am talking about appeals to me precisely because it need not wait upon "other people." Anybody who wants to do so can begin it in himself and in his household as soon as he is ready—by becoming answerable to at least some of his own needs, by acquiring skills and tools, by learning what his real needs are, by refusing the glamorous and the frivolous. When a person learns to *act* on his best hopes he enfranchises and validates them as no government or public policy ever will. And by his action the possibility that other people will do the same is made a likelihood.

But I must concede that there is also a sense in which I *am* tilting at windmills. While we have been preoccupied by various ideological menaces, we have been invaded and nearly overrun by windmills. They are drawing the nourishment from our soil and the lifeblood out of our veins. Let us tilt against the windmills. Though we have not conquered them, if we do not keep going at them they will surely conquer us.

---

*If automobiles are not more durable and economic than horses, then obviously a better economy would replace them with horses. It would be progressive to do so.

5: THE KINGDOM OF ABSTRACTION AND ORGANIZATION

I do not wish to discount the usefulness of either abstraction or organization, but rather to point out that we have given them such an extravagant emphasis and such prodigal subsidies that their *functioning* has come to overbear and obscure and even nullify their usefulness. Their ascendancy no doubt comes naturally enough out of the need to deal with the massive populations of an urban society. But their disproportionate, their almost exclusive, importance among us can only be explained as a disease of the specialist mentality that has found a haven in the government bureaus and the universities.

The bureaucrat who has formulated a plan, the specialist who has discovered a new fact or process, and the student who has espoused a social vision or ideal, all are of a kind in the sense that they all tend to think that they are at the end of a complete disciplinary process when in fact they have little more than reached the beginning of one. And this is their weakness: that they conceive abstractions to be complete in themselves, and therefore have only the simplest and most mechanical notions of the larger processes within which the abstractions will have their effect—processes that are apt, ultimately, to be obscure or mysterious in their workings and are therefore alien to the specialist mentality in the first place.

Having produced or espoused an abstraction, they next seek to put it to use by means of another abstraction—that is, an organization. But there is a sense in which organization is not a means of implementation, but rather a way of clinging to the clear premises and the neat logic of abstraction. The specialist mentality, unable by the terms of its narrow discipline to relinquish the secure order of abstraction, is prevented by a sort of Zeno's law from ever reaching the real ground of proof in the human community or in the world; it never *meets* the need it purports to answer. Demanding that each step toward the world be a predictable one, the specialist is by that very token not moving in the direction of the world at all, but on a course parallel to it. He can reach the world, not by any organizational process, but only by a reverse leap of faith from the ideal realm of the laboratory or theoretical argument onto the obscure and clumsifying ground of experience, where other and larger disciplines are required.

The man who must actually put the specialist's abstraction to use and live with its effects is never a part of the specialist organization. The organization can only deliver the abstraction to him and, of necessity, largely turn him loose with it. The farmer is not a part of the college of agriculture and the extension service; he is, rather, their object. The impoverished family is not a part of the welfare structure, but its object. The abstraction handed to these object-people is either true only in theory or it has been tried only under ideal (laboratory) conditions. For the bureaucrat, social planning replaces social behavior; for the agricultural scientist, chemistry and economics replace culture and ecology; for the political specialist (student or politician), theory replaces life, or tries to. Thus we institutionalize an impasse between the theoretical or ideal and the real, between the abstract and the particular; the specialist maintains a sort of esthetic distance between himself and the ground of proof and responsibility; and we delude ourselves that precept can have life and useful force without example.

Abstractions move toward completion only in the particularity of enactment or of use. Their completion is only in that mysterious whole that Sir Albert Howard and others have called the wheel of life. A vision or a principle or a discovery or a plan is therefore only *half* a discipline, and, practically speaking, it is the least important half. Black Elk, the holy man of the Sioux, said in his autobiography, *Black Elk Speaks*: "I think I have told you, but if I have not, you must have understood, that a man who has a vision is not able to use the power of it until after he has performed the vision on earth for the people to see." And only a few years later another American, William Carlos Williams, said much the same: "No ideas but in things." The difference of which both men spoke is that between knowledge and the *use* of knowledge. Similarly, one may speak of the difference between the production of an idea or a thing and its use. The disciplines of production are always small and specialized. The disciplines of use and continuity are both different and large. A man who produces a fact or an idea has not completed his responsibility to it until he sees that it is well used in the world. A man may grow potatoes as a specialist of sorts, but he falsifies himself and his potatoes too if he eats them and fails to live as a man.

If the culture fails to provide highly articulate connections between the abstract and the particular, the organizational and the personal, knowledge and behavior, production and use, the ideal and the world—that is, if it fails to bring the small disciplines of each man's work within the purview of those larger disciplines implied by the conditions of our life in the world —then the result is a profound disorder in which men release into their community and dwelling place powerful forces the consequences of which are unconsidered or unknown. New knowledge, political ideas, technological innovations, all are injected into society merely on the ground that to the specialists who produce them they appear to be good in themselves. A "labor-saving" device that does the work it was intended to do is thought by its developers to be a success: In terms of their discipline and point of view it *works*. That, in working, it considerably lowers the quality of a product and makes obsolete a considerable number of human beings is, to the specialists, merely an opportunity for other specialists.

If this attitude were restricted to the elite of government and university it would be bad enough; but it has been so popularized by their propaganda and example that the general public is willing to attribute to declarations, promises, mere words, the force of behavior. We have allowed and even encouraged a radical disconnection between our words and our deeds. Our speech has drifted out of the world into a realm of fantasy in which whatever we say is true. The President of the republic* openly admits that there is no connection between what he says and what he does—this in spite of his evident wish to be re-elected on the strength of what he says. We find it not extraordinary that lovers of America are strip mining in Appalachia, that lovers of peace are bombing villages in Southeast Asia, that lovers of freedom are underwriting dictatorships. If we *say* we are lovers of America and peace and freedom, then this must be what lovers of America and peace and freedom *do*. Having no need to account for anything they have done, our politicians do not find it necessary to trouble

*Nixon.

us with either evidence or argument, or to confess their errors, or to subtract their losses from their gains; they speak like the gods of Olympus, assured that if they *say* they are our servants anything they do in their own interest is right. Our public discourse has been reduced to the manipulation of uprooted symbols: good words, bad words, the names of gods and devils, emblems, slogans, flags. For some the flag no longer stands for the country, it *is* the country; they plant their crops and bury their dead in it.

There is no better example of this deterioration of language than in the current use of the word "freedom." Across the whole range of current politics this word is now being mouthed as if its devotees cannot decide whether it should be kissed or eaten, and this adoration has nothing to do with its meaning. The government is protecting the freedom of people by killing them or hiding microphones in their houses. The government's opponents, left and right, wish to set people free by telling them exactly what to do. All this is for the sake of the political power the word has come to have. The up-to-date politician no longer pumps the hand of a prospective constituent; he offers to set him—or her—free. And yet it seems to me that the word has no political meaning at all; the government cannot serve freedom except negatively—"by the alacrity," in Thoreau's phrase, "with which it [gets] out of its way."*

The going assumption seems to be that freedom can be granted only by an institution, that it is the gift of the government to its people. I think it is the other way around. Free men are not set free by their government; they have set their government free of themselves; they have made it unnecessary. Freedom is not accomplished by a declaration. A declaration of freedom is either a futile and empty gesture, or it is the statement of a finished fact. Freedom is a personal matter; though we may be enslaved as a group, we can be free only as persons. We can set each other free only as persons. It is a matter of discipline. A person can free himself of a bondage that has been imposed on him only by accepting another bondage that he

---

*And—still negatively—by keeping the selfish or vicious intentions of people out of its way.

has chosen. A man who would not be the slave of other men must be the master of himself—that is the real meaning of self-government. If we all behaved as honorably and honestly and as industriously as we expect our representatives to behave, we would soon put the government out of work.

A person dependent on somebody else for everything from potatoes to opinions may declare that he is a free man, and his government may issue a certificate granting him his freedom, but he will not be free. He is that variety of specialist known as a consumer, which means that he is the abject dependent of producers. How can he be free if he can do nothing for himself? What is the First Amendment to him whose mouth is stuck to the tit of the "affluent society"? Men are free precisely to the extent that they are equal to their own needs. The most able are the most free.

6: DISCIPLINE AND HOPE, MEANS AS ENDS

The various problems that I have so far discussed can best be understood, I think, as failures of discipline caused by a profound confusion as to the functions and the relative values of means and ends. I do not suggest simply that we fall with the ease of familiarity into the moral expedient of justifying means by ends, but that we have also come to attribute to ends a moral importance that far outweighs that which we attribute to means. We expect ends not only to justify means, but to rectify them as well. Once we have reached the desired end, we think, we will turn back to purify and consecrate the means. Once the war that we are fighting for the sake of peace is won, then the generals will become saints, the burned children will proclaim in heaven that their suffering is well repaid, the poisoned forests and fields will turn green again. Once we have peace, we say, or abundance or justice or truth or comfort, everything will be all right. It is an old dream.

It is a vicious illusion. For the discipline of ends is no discipline at all. The end is preserved in the means; a desirable end may perish forever in the wrong means. Hope lives in the means, not the end. Art does not survive in its revelations, or agriculture in its products, or craftsmanship in its artifacts, or civilization in its monuments, or faith in its relics.

That good ends are destroyed by bad means is one of the dominant themes of human wisdom. The *I Ching* says: "If evil is branded it thinks of weapons, and if we do it the favor of fighting against it blow for blow, we lose in the end because thus we ourselves get entangled in hatred and passion. Therefore it is important to begin at home, to be on guard in our own persons against the faults we have branded. . . . For the same reasons we should not combat our own faults directly. . . . As long as we wrestle with them, they continue victorious. Finally, the best way to fight evil is to make energetic progress in the good." Confucius said of riches that "if not obtained in the right way, they do not last." In the Sermon on the Mount, Jesus said: "Ye have heard that it hath been said, An eye for an eye, and a tooth for a tooth: But I say unto you, That ye resist not evil. . . ." And for that text Ken Kesey supplies the modern exegesis: "As soon as you resist evil, as soon as it's gone, you fold, because it's what you're based on." In 1931, Judge Lusk of the Chattanooga criminal court handed down a decision in which he wrote: "The best way, in my judgment, to combat Communism, or any other movement inimical to our institutions, is to show, if we can, that the injustices which they charge against us are, in fact, nonexistent." And a friend of mine, a graduate of the University of Emily's Run, was once faced with the argument that he could "make money" by marketing some inferior lambs; he refused, saying that his purpose was the production of *good* lambs, and he would sell no other kind. He meant that his disciplines had to be those of a farmer, and that he would be diminished as a farmer by adopting the disciplines of a money-changer. It is a tragedy of our society that it neither pays nor honors a man for this sort of integrity —though it depends on him for it.

It is by now a truism that the great emphasis of our present culture is upon things, things as things, things in quantity without respect to quality; and that our predominant techniques and attitudes have to do with production and acquisition. We persist in the belief—against our religious tradition, and in the face of much evidence to the contrary—that if we leave our children wealthy we will assure their happiness. A corollary of this is the notion, rising out of the work of the geneticists, that we can assure a brighter future for the world by *breeding* a more intelligent race of humans—even though the present problems of the world are the result, not of human stupidity, but of human intelligence without adequate cultural controls. Both ideas are typical of the materialist assumption that human destiny can be improved by being constantly tinkered at, as if it were a sort of balky engine. But we can do nothing for the human future that we will not do for the human present. For the amelioration of the future condition of our kind we must look, not to the wealth or the genius of the coming generations, but to the quality of the disciplines and attitudes that we are preparing now for their use.

We are being virtually buried by the evidence that those disciplines by which we manipulate *things* are inadequate disciplines. Our cities have become almost unlivable because they have been built to be factories and vending machines rather than communities. They are conceptions of the desires for wealth, excitement, and ease—all illegitimate motives from the standpoint of community, as is proved by the fact that without the community disciplines that make for a stable, neighborly population, the cities have become scenes of poverty, boredom, and dis-ease.

The rural community—that is, the land and the people—is being degraded in complementary fashion by the specialists' tendency to regard the land as a factory and the people as spare parts. Or, to put it another way, the rural community is being degraded by the fashionable premise that the exclusive function of the farmer is production and that his major discipline is economics. On the contrary, both the function and the discipline of the farmer have to do with provision: He must provide, he must look ahead. He must look ahead, however, not in the economic-mechanistic sense of anticipating a need and fulfilling it, but the sense of using methods that preserve the source. In his work sound economics becomes identical with sound ecology. The farmer is not a factory worker, he is the trustee of the life of the topsoil, the keeper of the rural community. In precisely the same way, the dweller in a healthy city is not an office or a factory worker, but part and preserver of the urban community. It is in thinking of the whole citizenry as factory workers—as readily interchangeable parts of an entirely mechanistic and economic order—that we have reduced our people to the most abject and aimless of nomads, and displaced and fragmented our communities.

An index of the health of a rural community—and, of course, of the urban community, its blood kin—might be found in the relative acreages of field crops and tree crops. By tree crops I mean not just those orchard trees of comparatively early bearing and short life, but also the fruit and nut and timber trees that bear late and live long. It is characteristic of an unsettled and anxious farm population—a population that feels itself, because of economic threat or the degradation of cultural value, to be ephemeral—that it farms almost exclusively with field crops, within economic and biological cycles that are complete in one year. This has been the dominant pattern of American agriculture. Stable, settled populations, assured both of an economic sufficiency in return for their work and of the cultural value of their work, tend to have methods and attitudes of a much longer range. Though they have generally also farmed with field crops, established farm populations have always been planters of trees. In parts of Europe, according to J. Russell Smith's important book, *Tree Crops*, steep hillsides were covered with orchards of chestnut trees, which were kept and maintained with great care by the farmers. Many of the trees were ancient, and when one began to show signs of dying, a seedling would be planted beside it to replace it. Here is an agricultural discipline that could develop only among farmers who felt secure —as individuals and also as families and communities—in their connection to their land. Such a discipline depends not just on the younger men in the prime of their workdays but also on the old men, the keepers of tradition. The model figure of this agriculture is an old man planting a young tree that will live longer than a man, that he himself may not live to see in its first bearing. And he is planting, moreover, a tree whose worth lies beyond any conceivable market prediction. He is planting it because the good sense of doing so has been clear to men of his place and kind for generations. The practice has been continued because it is ecologically and agriculturally sound; the economic soundness of it must be assumed. While the planting of a field crop, then, may be looked upon as a "short-term investment," the planting of a chestnut tree is a covenant of faith.

An urban discipline that in good health is closely analogous to healthy agriculture is teaching. Like a good farmer, a good teacher is the trustee of a vital and delicate organism: the life of the mind in his community. The standard of his discipline is his community's health and intelligence and coherence and endurance. This is a high calling, deserving of a life's work. We have allowed it to degenerate into careerism and specialization. In education as in agriculture we have discarded the large and enlarging disciplines of community and place, and taken up in their stead the narrow and shallow discipline of economics. Good teaching is an investment in the minds of the young, as obscure in result, as remote from immediate proof as planting a chestnut seedling. But we have come to prefer ends that are entirely foreseeable, even though that requires us to shorten our vision. Education is coming to be not a long-term investment in young minds and in the life of the community, but a short-term investment in the economy. We want to be able to tell how many dollars an education is worth and how soon it will begin to pay.

To accommodate these frivolous desires, education becomes training and specialization, which is to say, it institutionalizes and justifies ignorance and intellectual irresponsibility. It produces a race of learned mincers, whose propriety and pride it is to keep their minds inside their "fields," as if human thoughts were a kind of livestock to be kept out of the woods and off the roads. Because of the obsession with short-term results that may be contained within the terms and demands of a single life, the interest of community is displaced by the interest of career. The careerist teacher judges himself, and is judged by his colleagues, not by the influence he is having upon his students or the community, but by the number of his publications, the size of his salary and the status of the place to which his career has taken him thus far. And in ambition he is where he is only temporarily; he is on his way to a more lucrative and prestigious place. Because so few stay to be aware of the *effects* of their work, teachers are not judged by their teaching, but by the short-term incidentals of publication and "service." That teaching is a long-term service, that a teacher's best work may be published in the children or grandchildren of his students, cannot be considered, for the modern educator, like his "practical" brethren in business and industry, will honor nothing that he cannot see. That is not to say that books do not have their progeny in the community, or that a legitimate product of a teacher's life may not be a book. It *is* to say that if *good* books are to be written, they will be written out of the same resources of talent and discipline and character and delight as always, and not by institutional coercion.

It is not from the standpoint of the university itself that we will see its faults, but from the standpoint of the whole community. Looking only at the university, one might perhaps believe that its first obligation is to become a better exemplar of its species: a *bigger* university, with more prestigious professors publishing more books and articles. But look at the state of Kentucky—whose land is being vandalized and whose people are being impoverished by the absentee owners of coal; whose dispossessed are refugees in the industrial cities to the north; whose farm population and economy are under the heaviest threat of their history; whose environment is generally deteriorating; whose public schools have become legendary for their poor quality; whose public offices are routinely filled by the morally incompetent. Look at the *state* of Kentucky, and it is clear that, more than any publication of books and articles, or any research, we need an annual increment of several hundred competently literate *graduates* who have some critical awareness of their inheritance and a sense of their obligation to it, and who know the use of books.

That, and that only, is the disciplining idea of education, and the methods must be derived accordingly. It has nothing to do with number or size. It would be impossible to value economically; it is the antithesis of that false economy which thrives upon the exploitation of stupidity. It stands forever opposed to the assumption that you can produce a good citizen by subjecting a moral simpleton to specialized training or expert advice.

It is the obsession with immediate ends that is degrading, that destroys our disciplines, and that drives us to our inflexible concentration upon number and price and size. I believe that the closer we come to correct discipline, the less concerned we are with ends, and with questions of futurity in general.

Correct discipline brings us into alignment with natural process, which has no explicit or deliberate concern for the future. We do not eat, for instance, because we want to live until tomorrow, but because we are hungry today and it *satisfies* us to eat. Similarly, a good farmer plants, not because of the abstractions of demand or market or his financial condition, but because it is planting time and the ground is ready—that is, he plants in response to his discipline and to his place. And the real teacher does not teach with reference to the prospective job market or some program or plan for the society's future; he teaches because he has something to teach and because he has students. A poet could not write a poem in order to earn a place in literary history. His place in literary history is another subject, and as such a distraction. He writes because he has a poem to write, he knows how, the work pleases him, and he has forgotten all else. "Take therefore no thought for the morrow: for the morrow shall take thought for the things of itself." This passage rests, of course, on the fact that we do not know what tomorrow will be, and are therefore strictly limited in our ability to take thought for it. But it also rests upon the assumption of correct disciplines. The man who works and behaves well today *need* take no thought for the morrow; he has discharged today's only obligation to the morrow.

7: THE ROAD AND THE WHEEL

There are, I believe, two fundamentally opposed views of the nature of human life and experience in the world: one holds that though natural processes may be cyclic, there is within nature a human domain the processes of which are linear; the other, much older, holds that human life is subject to the same cyclic patterns as all other life. If the two are contradictory that is not so much because one is wrong and the other right as because one is partial and the other complete. The linear idea, of course, is the doctrine of progress, which represents man as having moved across the oceans and the continents and into space on a course that is ultimately logical and that will finally bring him to a man-made paradise. It also sees him as moving through time in this way, discarding old experience as he encounters new. The cyclic vision, on the other hand, sees our life ultimately not as a cross-country journey or a voyage of discovery, but as a circular dance in which certain basic *and necessary* patterns are repeated endlessly. This is the religious and ethical basis of the narrative of Black Elk: "Everything the Power of the World does is done in a circle. The sky is round, and I have heard that the earth is round like a ball, and so are all the stars. The wind, in its greatest power, whirls. Birds make their nests in circles, for theirs is the same religion as ours. The sun comes forth and goes down again in a circle. The moon does the same, and both are round. Even the seasons form a great circle in their changing, and always come back again to where they were. The life of a man is a circle from childhood to childhood, and so it is in everything where power moves. Our tepees were round like the nests of birds, and these were always set in a circle, the nation's hoop, a nest of many nests, where the Great Spirit meant for us to hatch our children." The doctrine of progress suggests that the fluctuations of human fortune are a series of ups and downs in a road tending generally upward toward the earthly paradise. To Black Elk earthly blessedness did not lie ahead or behind; it was the result of harmony within the circle of the people and between the people and the world. A man was happy or sad, he thought, in proportion as he moved toward or away from "the sacred hoop of [his] people [which] was one of many hoops that made one circle, wide as daylight and as starlight. . . ."

Characteristic of the linear vision is the idea that anything is justifiable only insofar as it is immediately and obviously good for something else. The linear vision tends to look upon everything as a cause, and to require that it proceed directly and immediately and obviously to its effect. What is it good for? we ask. And only if it proves immediately to be good *for* something are we ready to raise the question of value: How much is it worth? But we mean how much money, for if it can only be good for something else then obviously it can only be *worth* something else. Education becomes training as soon as we demand, in this spirit, that it serve some immediate purpose and that it be worth a predetermined amount. Once we accept so specific a notion of utility, all life becomes subservient to its use; its value is drained into its use. That is one reason why these are such hard times for students and old people: They are living either before or after the time of their social utility. It is also the reason why so many non-human species are threatened with extinction. Any organism that is not contributing obviously and directly to the workings of the economy is now endangered—which means, as the ecologists are showing, that human society is to the same extent endangered. The cyclic vision is more accepting of mystery and more humble. Black Elk *assumes* that all things have a use—that is the condition of his respect for all things—but he does not know what all their uses are. Because he does not value them for their uses, he is free to value them for their own sake: "The Six Grandfathers have placed in this world many things, all of which should be happy. Every little thing is sent for something, and in that thing there should be happiness and the power to make happy." It should be emphasized that this is ecologically sound. The ecologists recognize that the creation is a great union of interlocking lives and processes and substances, all dependent on each other; and because they cannot discover the whole pattern of interdependence they recognize the need for the greatest possible care in the use of the world. Black Elk and his people, however, were further advanced, for they possessed the cultural means for the enactment of a ceremonious respect for and delight in the lives with which they shared the world, and that respect and delight afforded those other lives an effective protection.

The linear vision looks fixedly straight ahead. It never looks back, for its premise is that there is no return. The doctrine of possession is complemented by no doctrine of relinquishment. Our shallow concept of use does not imply good use or preservation; thus quantity depresses quality, and we arrive at the concepts of waste and disposability. Similarly, life is lived without regard or respect for death. Death thus becomes accidental, the chance interruption of a process that might otherwise go on forever—therefore, always a surprise and always feared. Dr. Leon R. Kass, of the National Academy of Sciences, recently said that "medicine seems to be sharpening its tools to do battle with death as though death were just one more disease." The cyclic vision, at once more realistic and more generous, recognizes in the creation the essential principle of return: What is here will leave to come again; if there is to be having there must also be giving up. And it sees death as an integral and indispensable part of life. In one of the medicine rites of the Winnebago, according to Paul Radin, an old woman says: "Our father has ordained that my body shall fall to pieces. I am the earth. Our father ordained that there should be death, lest otherwise there be too many people and not enough food for them." Because death is inescapable, a biological and eco-logical necessity, its acceptance becomes a spiritual obligation, the only means of making life whole: "Whosoever shall seek to save his life shall lose it; and whosoever shall lose his life shall preserve it."

The opposing characteristics of the linear and cyclic visions might, then, be graphed something like this:

Linear	*Cyclic*
Progress. The conquest of nature.	Atonement with the creation.
The Promised Land motif in the Westward Movement.	Black Elk's sacred hoop, the community of creation.
Heavenly aspiration with-out earthly reconciliation or stewardship. The creation as commodity.	Reconciliation of heaven and earth in aspiration toward responsible life. The creation as source *and end*.
Training. Programming.	Education. Cultural process.
Possession.	Usufruct, relinquishment.
Quantity.	Quality.
Newness. The unique and "original."	Renewal. The recurring.
Life.	Life and death.

The linear vision flourishes in ignorance or contempt of the processes on which it depends. In the face of these processes our concepts and mechanisms are so unrealistic, so *impractical*, as to have the nature of fantasy. The processes are invariably cy-clic, rising and falling, taking and giving back, living and dying.

But the linear vision places its emphasis entirely on the rising phase of the cycle—on production, possession, life. It provides for no returns. Waste, for instance, is a concept that could have been derived only from the linear vision. According to the scheme of our present thinking, every human activity produces waste. This implies a profound contempt for correct discipline; it proposes, in the giddy faith of prodigals, that there can be production without fertility, abundance without thrift. We take and do not give back, and this causes waste. It is a hideous concept, and it is making the world hideous. It is consumption, a wasting disease. And this disease of our material economy becomes also the disease of our spiritual economy, and we have made a shoddy merchandise of our souls. We want the truth to be easy and spectacular, and so we waste our verities; we are always hastening from the essential to the novel. We want to have love without a return of devotion or fidelity; to us, Aphrodite is a peeping statistician, the seismographer of orgasms. We want a faith that demands no return of good work. And art—we want it to be instantaneous and effortless; we want it to involve no apprenticeship to a tradition or a discipline or a master, no devotion to an ideal of workmanship. We want our art to support the illusion that high achievement is within easy reach, for we want to believe that, though we are demeaned by our work and driven half crazy by our pleasures, we are all mute inglorious Miltons.

To take up again my theme of agriculture, it is obvious that the modern practice concentrates almost exclusively on the productive phase of the natural cycle. The means of production become more elaborate all the time, but the means of return —the building of health and fertility in the soil—are reduced more and more to the shorthand of chemicals. According to the industrial vision of it, the life of the farm does not rise and fall in the turning cycle of the year; it goes on in a straight line from one harvest to another. This, in the long run, may well be more productive of waste than of anything else. It wastes the soil. It wastes the animal manures and other organic residues that industrialized agriculture frequently fails to return to the soil. And what may be our largest agricultural waste is not usually recognized as such, but is thought to be both an urban product and an urban problem: the tons of garbage and sewage that are burned or buried or flushed into the rivers. This, like all waste, is the abuse of a resource. It was ecological stupidity of exactly this kind that destroyed Rome. The chemist Justus von Liebig wrote that "the sewers of the immense metropolis engulfed in the course of centuries the prosperity of Roman peasants. The Roman Campagna would no longer yield the means of feeding her population; these same sewers devoured the wealth of Sicily, Sardinia and the fertile lands of the coast of Africa."

To recognize the magnitude and the destructiveness of our "urban waste" is to recognize the shallowness of the notion that agriculture is only another form of technology to be turned over to a few specialists. The sewage and garbage problem of our cities suggests, rather, that a healthy agriculture is a cultural organism, not merely a universal necessity but a universal obligation as well. It suggests that, just as the cities exist within the environment, they also exist within agriculture. It suggests that, like farmers, city-dwellers have agricultural responsibilities: to use no more than necessary, to waste nothing, to return organic residues to the soil.

Our ecological or agricultural responsibilities, then, call for a corresponding set of disciplines that would be a part of the cultural common ground, and that each person would have an obligation to preserve in his behavior. Seeking his own ends by the correct means, he transcends selfishness and makes a just return to the ecological source; by his correct behavior, both the source and the means for its proper use are preserved. This is equally true of other cultural areas: it is the discipline, not the desire, that is the common ground. In politics, for example, it is only the personal career that can be advanced by "image making." *Politics* cannot be advanced except by honest, informed, open discourse—by determining and telling and implementing the truth, by assuming the truth's heavy responsibilities and great risks. The political careerist, by serving his "image" rather than the truth, becomes a consumer of the political disciplines. Similarly, in art, the common ground is workmanship, the artistic means, the technical possibility of art—not the insights or visions of particular artists. A person who practices an art without mastering its disciplines becomes his art's consumer; he obscures the means, and encumbers his successors. The art lives *for* its insights and visions, but it cannot live *upon* them. An art is inherited and handed down in its workmanly aspects. Workmanship is one of the means by which the artist prepares for—becomes worthy of, earns—his vision and his insights. "Art," A. R. Ammons says, "is the conscious preparation for the unconscious event. . . ."

Learning the correct and complete disciplines—the disciplines that take account of death as well as life, decay as well as growth, return as well as production—is an indispensable form of cultural generosity. It is the one effective way a person has of acknowledging and acting upon the fact of mortality: He will die, others will live after him.

One reason, then, for the disciplinary weakness of the linear vision is that it is incomplete. Another is that it sees history as always leading not to renewal but to the new: the road may climb hills and descend into valleys, but it is always going ahead; it never turns back on itself. "We have constructed a fate . . . that never turns aside," Thoreau said of the railroad; "its orbit does not look like a returning curve. . . ." But when the new is assumed to be a constant, discipline fails, for discipline is preparation, and the new cannot be prepared for; it cannot, in any very meaningful way, be expected. Here again we come upon one of the reasons for the generational disconnections that afflict us: all times, we assume, are different; we therefore have nothing to learn from our elders, nothing to teach our children. Civilization is thus reduced to a sequence of last-minute improvisations, desperately building today out of the wreckage of yesterday. There are two genres of writing that seem to me to be characteristic (or symptomatic) of the linear vision. The first is, not prophecy, though it is sometimes called that, but the most mundane and inquisitive taking of thought for the morrow. What will tomorrow be like? (We mean, what new machines or ideas will be invented by then?) What will the world be like in ten or fifty or a hundred years? Our preoccupation with these questions, besides being useless, is morbid and scared; mistaking appearance for substance, it assumes a condition of *absolute* change: The future will be *entirely* different from the past and the present, we think, because our vision of history and experience has not taught us to imagine persistence or recurrence or renewal. We disregard the necessary persistence of ancient needs and obligations, patterns and cycles, and assume that the human condition is entirely determined by human *devices*.

The other genre—complementary, obviously—is that of the death sentence. Because we see the human situation as perpetually changing, the new bearing down with annihilating force, we appear to ourselves to be living always at the end of possibilities. For the new to happen, the old must be destroyed. Our own lives, which are pleasant to us at least insofar as they represent a *kind* of life that we recognize, seem always on the verge of being replaced by a kind of life that is unrecognizable, or by a death that is equally so. Thus a common theme for the writers of feature articles and critical essays is the death or the impending death of something: of the fashions of dress and appearance, of the novel, of printing, of freedom, of Christianity, of Western civilization, of the human race, of the world, of God.

This genre is difficult to criticize, because there is always a certain justice, or likelihood of justice, in it. There is no denying that we fear the end of things because our way of life has brought so many things to an end. The sunlit road of progress never escapes a subterranean dread that threatens to undermine the pavement. Thoreau knew that the railroad was built upon the bodies of Irishmen, and not one of us but secretly wonders when *he* will be called upon to lie down and become a sleeper beneath the roadbed of progress. And so some of this death-sentence literature is faithfully reporting the destructiveness of the linear vision; it is the chorus of accusation and dread and mourning that accompanies Creon's defiance of the gods.

But at times it is no more than one of the sillier manifestations of the linear vision itself: the failure to see any pattern in experience, the failure to transform experience into useful memory. There is a sort of journalistic greenness in us that is continuously surprised by the seasons and the weather, as if these were no more than the inventions (or mistakes) of the meteorologists. History, likewise, is always a surprise to us; we read its recurring disasters as if they were the result merely of miscalculations of our intelligence—as if they could not have been foreseen in the flaws of our character. And the heralds of the push-button Eden of the future would be much put off to consider that those button-pushers will still have to deal with problems of food and sewage, with the picking up of scraps and the disposal of garbage, with building and maintenance and reclamation—with, that is, the fundamental work, much of it handwork, that is necessary to life; they would be even more put off to consider that the "quality of life" will not depend nearly so much on the distribution of push-buttons as on the manner and the quality of that fundamental and endlessly necessary work.

Some cycles revolve frequently enough to be well known in a man's lifetime. Some are complete only in the memory of several generations. And others are so vast that their motion can only be assumed: Like our galaxy, they appear to us to be remote and exalted, a Milky Way, when in fact they are near at hand, and we and all the humble motions of our days are their belongings. We are kept in touch with these cycles, not by technology or politics or any other strictly human device, but by our necessary biological relation to the world. It is only in the processes of the natural world, and in analogous and related processes of human culture, that the new may grow usefully old, and the old be made new.

The ameliorations of technology are largely illusory. They are always accompanied by penalties that are equal and opposite. Like the weather reports, they suggest the possibility of better solutions than they can provide; and by this suggestiveness—this glib and shallow optimism of gimcrackery—they have too often replaced older skills that were more serviceable to life in a mysterious universe. The farmer whose weather eye has been usurped by the radio has become less observant, has lost his old judicious fatalism with respect to the elements—and he is no more certain of the weather. It is by now obvious that these so-called blessings have not made us better or wiser. And their expense is growing rapidly out of proportion to their use. How many plastic innards can the human race afford? How many mountains can the world afford, to the strip miners, to light the whole outdoors and overheat our rooms? The limits, as Thoreau knew, have been in sight from the first: "To make a railroad round the world available to all mankind," he said in *Walden*, "is equivalent to grading the whole surface of the planet. Men have an indistinct notion that if they keep up this activity of joint stocks and spades long enough all will at length ride . . . but though a crowd rushes to the depot . . . when the smoke is blown away and the vapor condensed, it will be perceived that a few are riding, but the rest are run over. . . ."

We cannot look for happiness to any technological paradise or to any New Earth of outer space, but only to the world as it is, and as we have made it. The only life we may hope to live is here. It seems likely that if we are to reach the earthly paradise at all, we will reach it only when we have ceased to strive and hurry so to get there. There is no "there." We can only wait here, where we are, in the world, obedient to its processes, patient in its taking away, faithful to its returns. And as much as we may know, and all that we deserve, of earthly paradise will come to us.

8: KINDS OF DISCIPLINE

The disciplines we most readily think of are technical; they are the means by which we define and enact our relationship to the creation, and destroy or preserve the commonwealth of the living. There is no disputing the importance of these, and I have already said a good deal in their support. But alone, without the larger disciplines of community and faith, they fail of their meaning and their aim; they cannot even continue for very long. People who practice the disciplines of workmanship for their own sake—that is, as specializations—begin a process of degeneracy in those disciplines, for they remove from them the ultimate sense of use or effect by which their vitality and integrity are preserved. Without a proper sense of use a discipline declines from community responsibility to personal eccentricity; cut off from the common ground of experience and need, vision escapes into wishful or self-justifying fantasy, or into greed. Artists lose the awareness of an audience, craftsmen and merchants lose the awareness of customers and users.

Community disciplines, which can rise only out of the cyclic vision, are of two kinds. First there is the discipline of principle, the essence of the experience of the historical community. And second there is the discipline of fidelity to the living community, the community of family and neighbors and friends. As our society has become increasingly rootless and nomadic, it has become increasingly fashionable among the rhetoricians of dissatisfaction to advocate, or to seem to advocate, a strict and solitary adherence to principle in simple defiance of other people. "I don't care what they think" has become public currency with us; saying it, we always mean to imply that we are persons solemnly devoted to high principle—"rugged individuals" in the somewhat fictional sense Americans usually give to that term. In fact, this ready defiance of the opinions of others is a rhetorical fossil from our frontier experience. Once it meant that if our neighbors' opinions were repugnant to us, we were prepared either to kill our neighbors or to move west. Now it doesn't mean anything; it is adolescent bluster. For when there is no frontier to retreat to, the demands of one's community will be felt, and ways must be found to deal with them. The great moral labor of any age is probably not in the conflict of opposing principles, but in the tension between a living community and those principles that are the distillation of its experience. Thus the present anxiety and anguish in this country have very little to do with the much-heralded struggle between capitalism and communism, or left and right, and very much indeed to do with the rapidly building discord and tension between American principles and American behavior.

This sort of discord is the subject of tragedy. It is tragic because—outside the possibility of a renewal of harmony, which may depend upon the catharsis of tragedy—there are no possible resolutions that are not damaging. To choose community over principle is to accept in consequence a diminishment of the community's moral inheritance; it is to accept the great dangers and damages of life without principle. To choose principle over community is even worse, it seems to me, for that is to accept as the condition of being "right" a loneliness in which the right is ultimately meaningless; it is to destroy the only ground upon which principle can be enacted and renewed; it is to raise an ephemeral hope upon the ground of final despair.

Facing exactly this choice between principle and community, on April 20, 1861, Robert E. Lee resigned his commission in the army of the United States. Lee had clearly understood the evil of slavery. He disapproved and dreaded secession; almost alone among the Virginians, he foresaw the horrors that would follow. And yet he chose to go with his people. Having sent in his resignation, he wrote his sister: ". . . though I recognize no necessity for this state of things, and would have forborne and pleaded to the end for a redress of grievances, real or supposed, yet in my own person I had to meet the question whether I should take part against my native state. With all my devotion to the Union and the feeling of loyalty and duty of an American citizen, I have not been able to make up my mind to raise my hand against my relatives, my children, my home."

He was right. As a highly principled man, he could not bring himself to renounce the very ground of his principles. And devoted to that ground as he was, he held in himself much of his region's hope of the renewal of principle. His seems to me to have been an exemplary American choice, one that placed the precise Jeffersonian vision of a rooted devotion to community and homeland above the abstract "feeling of loyalty and duty of an American citizen." It was a tragic choice on the theme of Williams's maxim: "No ideas but in things."

If the profession of warfare has so declined in respectability since 1861 as to obscure my point, then change the terms. Say that a leader of our own time, in spite of his patriotism and his dependence on "the economy," nevertheless held his people and his place among them in such devotion that he would not lie to them or sell them shoddy merchandise or corrupt their language or degrade their environment. Say, in other words, that he would refuse to turn his abilities against his people. That is what Lee did, and there have been few public acts of as much integrity since.*

It is the intent of community disciplines, of course, to prevent such radical conflicts. If these disciplines are practiced at large among the members of the community, then the community holds together upon a basis of principle that is immediately clarified in feeling and behavior; and then destructive divisions, and the moral agonies of exceptional men, are averted.

---

*If loyalty to home and community allied one with a manifest evil, then it would be necessary to stand alone on principle. But it would be foolish to expect a choice so clean-cut.

In the Sermon on the Mount a major concern is with the community disciplines. The objective of this concern is a social ideal: ". . . all things whatsoever that men should do to you, do ye even so to them. . . ." But that everyone would do as he would be done by is hardly a realistic hope, and Jesus was speaking out of a moral tradition that was eminently realistic and tough-minded. It was a tradition that, in spite of its spiritual aspirations, was very worldly in its expectations: It would have spared Sodom for the sake of ten righteous men. And so the focus of the Sermon is not on the utopian social ideal of the Golden Rule, but on the *personal* ideal of nonresistance to evil: ". . . whosoever shall smite thee on thy right cheek, turn to him the other also. . . . Love your enemies, bless them that curse you, do good to them that hate you, and pray for them which despitefully use you, and persecute you. . . ." The point, I think, is that the anger of one man need not destroy the community, if it is contained by the peaceableness and long-suffering of another, but in the anger of *two* men, in anger repaid—"An eye for an eye, and a tooth for a tooth"—it is destroyed altogether.

Community, as a discipline, extends and enlarges the technical disciplines by looking at them within the perspective of their uses or effects. Community discipline imposes upon our personal behavior an ecological question: What is the effect, on our neighbors and on our place in the world, of what we do? It is aware that *all* behavior is social. It is aware, as the ecologists are aware, that there is a unity in the creation, and that the behavior and the fate of one creature must therefore affect the whole, though the exact relationships may not be known.

But essential as are the disciplines of technique and community, they are not sufficient in themselves. All such disciplines reach their limit of comprehensibility and at that point enter mystery. Thus an essential part of a discipline is that relinquishment or abandonment by which we acknowledge and accept its limits. We do not finally know what will be the result of our actions, however correct and excellent they may be. The good work we do today may be undone by some mere accident tomorrow. Our neighborly acts may be misunderstood and repaid with anger. With respect to what is to come, our real condition is that of abandonment; one of the primary functions of religion is to provide the ceremonial means of acknowledging this: We are in the hands of powers that we do not know.

The ultimate discipline, then, is faith: faith, if in nothing else, in the propriety of one's disciplines. We have obscured the question of faith by pretending that it is synonymous with the question of "belief," which is personal and not subject to scrutiny. But if one's faith is to have any public validity or force, then obviously it must meet some visible test. The test of faith is consistency—not the fanatic consistency by which one repudiates the influence of knowledge, but rather a consistency between principle and behavior. A man's behavior should be the creature of his principles, not the creature of his circumstances. The point has great practical bearing, because belief and the principles believed in, and whatever hope and promise are implied in them, are destroyed in contradictory behavior; hypocrisy salvages nothing but the hypocrite. If we put our faith in the truth, then we risk everything—the truth included —by telling lies. If we put our faith in peace, then we must see that violence makes us infidels. When we institute repressions to protect democracy from enemies abroad, we have already damaged it at home. The demands of faith are absolute: We must put all our eggs in one basket; we must burn our bridges.

An exemplary man of faith was Gideon, who reduced his army from thirty-two thousand to three hundred in earnest of his trust, and marched that remnant against the host of the Midianites, armed, not with weapons, but with "a trumpet in every man's hand, with empty pitchers, and lamps within the pitchers."

Beside this figure of Gideon, the hero as man of faith, let us place our own "defender," the Pentagon, which has faith in nothing except its own power. That, as the story of Gideon makes clear, is a dangerous faith for mere men; it places them in the most dangerous moral circumstance, that of *hubris*, in which one boasts that "mine own hand hath saved me." To be sure, the Pentagon is supposedly founded upon the best intentions and the highest principles, and there is a plea that justifies it in the names of Christianity, peace, liberty and democracy. But the Pentagon is an institution, not a person; and unless constrained by the moral vision of persons in them, institutions move in the direction of power and self-preservation, not high principle. Established, allegedly, in defense of "the free world," the Pentagon subsists complacently upon the involuntary servitude of millions of young men whose birthright, allegedly, is freedom. To wall our enemies out, it is walling us in.

Because its faith rests entirely in its own power, its mode of dealing with the rest of the world is not faith but suspicion. It recognizes no friends, for it knows that the face of friendship is the best disguise of an enemy. It has only enemies, or prospective enemies. It must therefore be prepared for *every possible* eventuality. It sees the future as a dark woods with a gunman behind every tree. It is passing through the valley of the shadow of death without a shepherd, and thus is never still. But as long as it can keep the public infected with its own state of mind, this spiritual dis-ease, it can survive without justification, and grow huge. Whereas the man of faith may go armed with only a trumpet and an empty pitcher and a lamp, the institution of suspicion arms with the death of the world; trusting nobody, it must stand ready to kill everybody.

The moral is that those who have no faith are apt to be much encumbered by their equipment, and overborne by their precautions. For the institution of suspicion there is no end of toiling and spinning. The Pentagon exists continually, not only on the brink of war, but on the brink of the exhaustion of its moral and material means. But the man of faith, even in the night, in the camp of his enemies, is at rest in the assurance of his trust and the correctness of his ways.* He has become the lily of the field.

9: THE LIKENESSES OF ATONEMENT (AT-ONE-MENT)

Living in our speech, though no longer in our consciousness, is an ancient system of analogies that clarifies a series of mutually defining and sustaining unities: of farmer and field, of husband and wife, of the world and God. The language both of our literature and of our everyday speech is full of references and

---

*That is not to suggest that faith may not be extremely dangerous, especially to the faithful.

allusions to this expansive metaphor of farming and marriage and worship. A man planting a crop is like a man making love to his wife, and vice versa: He is a husband or a husbandman. A man praying is like a lover, or he is like a plant in a field waiting for rain. As husbandman, a man is both the steward and the likeness of God, the greater husbandman. God is the lover of the world and its faithful husband. Jesus is a bridegroom. And he is a planter; his words are seeds. God is a shepherd and we are his sheep. And so on.

All the essential relationships are comprehended in this metaphor. A farmer's relation to his land is the basic and central connection in the relation of humanity to the creation; the agricultural relation *stands for* the larger relation. Similarly, marriage is the basic and central community tie; it begins and stands for the relation we have to family and to the larger circles of human association. And these relationships to the creation and to the human community are in turn basic to, and may stand for, our relationship to God—or to the sustaining mysteries and powers of the creation.

(These three relationships are dependent—and even intent —upon renewals of various sorts: of season, of fertility, of sexual energy, of love, of faith. And these concepts of renewal are always accompanied by concepts of loss or death; in order for the renewal to take place, the old must be not forgotten but relinquished; in order to become what we may be, we must cease to be as we are; in order to have life we must lose it. Our language bears abundant testimony to these deaths; the year's death that precedes spring; the burial of the seed before germination; sexual death, as in the Elizabethan metaphor; death as the definitive term of marriage; the spiritual death that must precede rebirth; the death of the body that must precede resurrection.)

As the metaphor comprehends all the essential relationships, so too it comprehends all the essential moralities. The moralities are ultimately emulative. For the metaphor does not merely perceive the likeness of these relationships. It perceives also that they are understandable only in terms of each other. They are the closed system of our experience; no instructions come from outside. A man finally cannot act upon the basis of absolute law, for the law is more fragmentary than his own experience; finally, he must emulate in one relationship what he knows of another. Thus, if the metaphor of atonement is alive in his consciousness, he will see that he should love and care for his land as for his wife, that his relation to his place in the world is as solemn and demanding, and as blessed, as marriage; and he will see that he should respect his marriage as he respects the mysteries and transcendent powers—that is, as a sacrament. Or—to move in the opposite direction through the changes of the metaphor—in order to care properly for his land he will see that he must emulate the Creator: to learn to use and preserve the open fields, as Sir Albert Howard said, he must look into the woods; he must study and follow natural process; he must understand the *husbanding* that, in nature, always accompanies providing.

Like any interlinking system, this one fails in the failure of any one of its parts. When we obscure or corrupt our understanding of any one of the basic unities, we begin to misunderstand all of them. The vital knowledge dies out of our consciousness and becomes fossilized in our speech and our culture. This is our condition now. We have severed the vital links of the atonement metaphor, and we did this initially, I think, by degrading and obscuring our connection to the land, by looking upon the land as merchandise and ourselves as its traveling salesmen.

I do not know how exact a case might be made, but it seems to me that there is an historical parallel, in white American history, between the treatment of the land and the treatment of women. The frontier, for instance, was notoriously exploitive of both, and I believe for largely the same reasons. Many of the early farmers seem to have worn out farms and wives with equal regardlessness, interested in both mainly for what they would produce, crops and dollars, labor and sons; they clambered upon their fields and upon their wives, struggling for an economic foothold, the having and holding that cannot come until both fields and wives are properly cherished. And today there seems to me a distinct connection between our nomadism (our "social mobility") and the nearly universal disintegration of marriages and families.

The prevalent assumption appears to be that marriage problems are problems strictly of "human relations": If the husband and wife will only assent to a number of truisms about "respect for the other person," "giving and taking," et cetera, and if they will only "understand" each other, then it is believed that their problems will be solved. The difficulty is that marriage is only partly a matter of "human relations," and only partly a circumstance of the emotions. It is also, and as much as anything, a practical circumstance. It is very much under the influence of things and people outside itself; that is, it must make a household, it must make a place for itself in the world and in the community. But with us, getting someplace always involves going somewhere. Every professional advance leads to a new place, a new house, a new neighborhood. Our marriages are always being cut off from what they have made; their substance is always disappearing into the thin air of human relations.

I think there is a limit to the portability of human relationships. Tribal nomads, when they moved, moved as a tribe; their personal and cultural identity—their household and community—accompanied them as they went. But our modern urban nomads are always moving away from the particulars by which they know themselves, and moving into abstraction (a house, a neighborhood, a job) in which they can only feel threatened by new particulars. The marriage becomes a sort of assembly-line product, made partly here and partly there, the whole of it never quite coming into view. Provided they stay married (which is unlikely) until the children leave (which is usually soon), the nomadic husband and wife who look to see what their marriage has been—that is to say, what it *is*—are apt to see only the lines in each other's face.

The carelessness of place that must accompany our sort of nomadism makes a vagueness in marriage that is its antithesis. And vagueness in marriage, the most sacred human bond and perhaps the basic metaphor of our moral and religious tradition, cannot help but produce a diminishment of reverence, and of the care for the earth that must accompany reverence.

When the metaphor of atonement ceases to live in our consciousness, we lose the means of relationship. We become isolated in ourselves, and our behavior becomes the erratic behavior of people who have no bonds and no limits.

10: THE PRACTICALITY OF MORALS

What I have been preparing at such length to say is that there is only one value: the life and health of the world. If there is only one value, it follows that conflicts of value are illusory, based upon perceptual error. Moral, practical, spiritual, esthetic, economic, and ecological values are all concerned ultimately with the same question of life and health. To the virtuous man, for example, practical and spiritual values are identical; it is only corruption that can see a difference. Esthetic value is always associated with sound values of other kinds. "Beauty is truth, truth beauty," Keats said, and I think we may take him at his word.* Or to say the same thing in a different way: Beauty is wholeness; it is health in the ecological sense of amplitude and balance. And ecology is long-term economics. If these identities are not apparent immediately, they are apparent *in time*. Time is the merciless, infallible critic of the specialized disciplines. In the ledgers that justify waste the ink is turning red.

Moral value, as should be obvious, is not separable from other values. An adequate morality would be ecologically sound; it would be esthetically pleasing. But the point I want to stress here is that it would be *practical*. Morality is long-term practicality.

Of all specialists the moralists are the worst, and the processes of disintegration and specialization that have characterized us for generations have made moralists of us all. We have obscured and weakened morality, first, by advocating it for its own sake—that is, by deifying it, as esthetes have deified art—and then, as our capacity for reverence has diminished, by allowing it to become merely decorative, a matter of etiquette.

What we have forgotten is the origin of morality in fact and circumstance; we have forgotten that the nature of morality is essentially practical. Moderation and restraint, for example, are necessary, not because of any religious commandment or any

---

*I now consider this reference to Keats to be misleading. The ideal beauty of the "Ode on a Grecian Urn" is very different from the common, earthly, mortal beauty I had in mind.

creed or code, but because they are among the assurances of good health and a sufficiency of goods. Likewise, discipline is necessary if the necessary work is to be done; also if we are to know transport, transcendence, joy. Loyalty, devotion, faith, self-denial are not ethereal virtues, but the concrete terms upon which the possibility of love is kept alive in this world. Morality is neither ethereal nor arbitrary; it is the definition of what is humanly possible, and it is the definition of the penalties for violating human possibility. A person who violates human limits is punished or he prepares a punishment for his successors, not necessarily because of any divine or human law, but because he has transgressed the order of things.* A live and adequate morality is an accurate perception of the order of things, and of humanity's place in it. By clarifying the human limits, morality tells us what we risk when we forsake the human to behave like false gods or like animals.

One would not wish to say—indeed, it is precisely my point that one *should* not say—that social *forms* will not change with changing conditions. They probably *will* do so, wholly regardless of whether or not they *should*. But I believe that it is erroneous to assume that a change of form implies a change of discipline. Under the influence of the rapid changes of modern life, it is persistently assumed that we are moving toward a justifiable relaxation of disciplines. This is wishful thinking, and it invites calamity, for the human place in the order of things, the human limits, the human tragedy remain the same. It seems altogether possible, as a final example, that for various reasons the forms of marriage will change.† But this does not promise a new age of benefit without obligation—which, I am afraid, is what many people mean by freedom. Though the forms of marriage may change, if it continues to exist in any form it will continue to rest upon the same sustaining disciplines, and to incorporate the same tragic awareness: that it is made "for better for worse, for richer for poorer, in sickness and in health, to love and to cherish, till death. . . ."

---

*The order of things, of course, *is* a law—and not a human one.
†But to change them quickly is simply to destroy them. It is, I think, foolish to think that ancient community forms can be satisfactorily altered or safely discarded by the intention of individuals.

II: THE SPRING OF HOPE

The most destructive of ideas is that extraordinary times justify extraordinary measures. This is the ultimate relativism, and we are hearing it from all sides. The young, the poor, the minority races, the Constitution, the nation, traditional values, sexual morality, religious faith, Western civilization, the economy, the environment, the world are all now threatened with destruction—so the arguments run—therefore let us deal with our enemies by whatever means are handiest and most direct; in view of our high aims history will justify and forgive. Thus the violent have always rationalized their violence.

But as wiser men have always known, all times are extraordinary in precisely this sense. In the condition of mortality all things are always threatened with destruction. The invention of atomic holocaust and the other man-made dooms renews for us the immediacy of the worldly circumstance as the religions have always defined it: We know "neither the day nor the hour. . . ."

Our bewilderment is not the time but our character. We have come to expect too much from outside ourselves. If we are in despair or unhappy or uncomfortable, our first impulse is to assume that this cannot be our fault; our second is to assume that some institution is not doing its duty. We are in the curious position of expecting from others what we can only supply ourselves. One of the Confucian ideals is that the "archer, when he misses the bullseye, turns and seeks the cause of the error in himself."

Goodness, wisdom, happiness, even physical comfort, are not institutional conditions. The real sources of hope are personal and spiritual, not public and political. A man is not happy by the dispensation of his government or by the fortune of his age. He is happy only in doing well what is in his power, and in being reconciled to what is not in his power. Thoreau, who knew such happiness, wrote in "Life Without Principle": "Of what consequence, though our planet explode, if there is no character involved in the explosion? In health we have not the least curiosity about such events. We do not live for idle amusement. I would not run round a corner to see the world blow up."

Asked why the Shakers, who expected the end of the world at any moment, were nevertheless consummate farmers and craftsmen, Thomas Merton replied: "When you expect the world to end at any moment, you know there is no need to hurry. You take your time, you do your work well."

In Defense of Literacy

IN A COUNTRY in which everybody goes to school, it may seem absurd to offer a defense of literacy, and yet I believe that such a defense is in order, and that the absurdity lies not in the defense, but in the necessity for it. The published illiteracies of the certified educated are on the increase. And the universities seem bent upon ratifying this state of things by declaring the acceptability in their graduates of adequate—that is to say, of mediocre—writing skills.

The schools, then, are following the general subservience to the "practical," as that term has been defined for us according to the benefit of corporations. By "practicality" most users of the term now mean whatever will most predictably and most quickly make a profit. Teachers of English and literature have either submitted, or are expected to submit, along with teachers of the more "practical" disciplines, to the doctrine that the purpose of education is the mass production of producers and consumers. This has forced our profession into a predicament that we will finally have to recognize as a perversion. As if awed by the ascendency of the "practical" in our society, many of us secretly fear, and some of us are apparently ready to say, that if a student is not going to become a teacher of his language, he has no need to master it.

In other words, to keep pace with the specialization—and the dignity accorded to specialization—in other disciplines, we have begun to look upon and to teach our language and literature as specialties. But whereas specialization is of the nature of the applied sciences, it is a perversion of the disciplines of language and literature. When we understand and teach these as specialties, we submit willy-nilly to the assumption of the "practical men" of business, and also apparently of education, that literacy is no more than an ornament: When one has become an efficient integer of the economy, *then* it is permissible, even desirable, to be able to talk about the latest novels. After all, the disciples of "practicality" may someday find themselves stuck in conversation with an English teacher.

I may have oversimplified that line of thinking, but not

much. There are two flaws in it. One is that, among the self-styled "practical men," the practical is synonymous with the immediate. The long-term effects of their values and their acts lie outside the boundaries of their interest. For such people a strip mine ceases to exist as soon as the coal has been extracted. Short-term practicality is long-term idiocy.

The other flaw is that language and literature are always *about* something else, and we have no way to predict or control what they may be about. They are about the world. We will understand the world, and preserve ourselves and our values in it, only insofar as we have a language that is alert and responsive to it, and careful of it. I mean that literally. When we give our plows such brand names as "Sod Blaster," we are imposing on their use conceptual limits that raise the likelihood that they will be used destructively. When we speak of man's "war against nature," or of a "peace offensive," we are accepting the limitations of a metaphor that suggests, and even proposes, violent solutions. When students ask for the right of "participatory input" at the meetings of a faculty organization, they are thinking of democratic process, but they are *speaking* of a convocation of robots, and are thus devaluing the very traditions that they invoke.

Ignorance of books and the lack of a critical consciousness of language were safe enough in primitive societies with coherent oral traditions. In our society, which exists in an atmosphere of prepared, public language—language that is either written or being read—illiteracy is both a personal and a public danger. Think how constantly "the average American" is surrounded by premeditated language, in newspapers and magazines, on signs and billboards, on TV and radio. He is forever being asked to buy or believe somebody else's line of goods. The line of goods is being sold, moreover, by men who are trained to make him buy it or believe it, whether or not he needs it or understands it or knows its value or wants it. This sort of selling is an honored profession among us. Parents who grow hysterical at the thought that their son might not cut his hair are *glad* to have him taught, and later employed, to lie about the quality of an automobile or the ability of a candidate.

What is our defense against this sort of language—this language-as-weapon? There is only one. We must know a better language. We must speak, and teach our children to speak, a language precise and articulate and lively enough to tell the truth about the world as we know it. And to do this we must know something of the roots and resources of our language; we must know its literature. The only defense against the worst is a knowledge of the best. By their ignorance people enfranchise their exploiters.

But to appreciate fully the necessity for the best sort of literacy we must consider not just the environment of prepared language in which most of us now pass most of our lives, but also the utter transience of most of this language, which is meant to be merely glanced at, or heard only once, or read once and thrown away. Such language is by definition, and often by calculation, not memorable; it is language meant to be replaced by what will immediately follow it, like that of shallow conversation between strangers. It cannot be pondered or effectively criticized. For those reasons an unmixed diet of it is destructive of the informed, resilient, critical intelligence that the best of our traditions have sought to create and to maintain —an intelligence that Jefferson held to be indispensable to the health and longevity of freedom. Such intelligence does not grow by bloating upon the ephemeral information and misinformation of the public media. It grows by returning again and again to the landmarks of its cultural birthright, the works that have proved worthy of devoted attention.

"Read not the Times. Read the Eternities," Thoreau said. Ezra Pound wrote that "literature is news that STAYS news." In his lovely poem, "The Island," Edwin Muir spoke of man's inescapable cultural boundaries and of his consequent responsibility for his own sources and renewals:

> Men are made of what is made,
> The meat, the drink, the life, the corn,
> Laid up by them, in them reborn.
> And self-begotten cycles close
> About our way; indigenous art
> And simple spells make unafraid
> The haunted labyrinths of the heart . . .

These men spoke of a truth that no society can afford to shirk for long: We are dependent, for understanding, and for consolation and hope, upon what we learn of ourselves from songs and stories. This has always been so, and it will not change.

I am saying, then, that literacy—the mastery of language and the knowledge of books—is not an ornament, but a necessity. It is impractical only by the standards of quick profit and easy power. Longer perspective will show that it alone can preserve in us the possibility of an accurate judgment of ourselves, and the possibilities of correction and renewal. Without it, we are adrift in the present, in the wreckage of yesterday, in the nightmare of tomorrow.

RECOLLECTED ESSAYS:
1965–1980
(1981)

The Making of a Marginal Farm

ONE DAY in the summer of 1956, leaving home for school, I stopped on the side of the road directly above the house where I now live. From there you could see a mile or so across the Kentucky River Valley, and perhaps six miles along the length of it. The valley was a green trough full of sunlight, blue in its distances. I often stopped here in my comings and goings, just to look, for it was all familiar to me from before the time my memory began: woodlands and pastures on the hillsides; fields and croplands, wooded slew-edges and hollows in the bottoms; and through the midst of it the tree-lined river passing down from its headwaters near the Virginia line toward its mouth at Carrollton on the Ohio.

Standing there, I was looking at land where one of my great-great-great-grandfathers settled in 1803, and at the scene of some of the happiest times of my own life, where in my growing-up years I camped, hunted, fished, boated, swam, and wandered—where, in short, I did whatever escaping I felt called upon to do. It was a place where I had happily been, and where I always wanted to be. And I remember gesturing toward the valley that day and saying to the friend who was with me: "That's all I need."

I meant it. It was an honest enough response to my recognition of its beauty, the abundance of its lives and possibilities, and of my own love for it and interest in it. And in the sense that I continue to recognize all that, and feel that what I most need is here, I can still say the same thing.

And yet I am aware that I must necessarily mean differently —or at least a great deal more—when I say it now. Then I was speaking mostly from affection, and did not know, by half, what I was talking about. I was speaking of a place that in some ways I knew and in some ways cared for, but did not live in. The differences between knowing a place and living in it, between cherishing a place and living responsibly in it, had not begun to occur to me. But they are critical differences, and understanding them has been perhaps the chief necessity of my experience since then.

I married in the following summer, and in the next seven years lived in a number of distant places. But, largely because I continued to feel that what I needed was here, I could never bring myself to want to live in any other place. And so we returned to live in Kentucky in the summer of 1964, and that autumn bought the house whose roof my friend and I had looked down on eight years before, and with it "twelve acres more or less." Thus I began a profound change in my life. Before, I had lived according to expectation rooted in ambition. Now I began to live according to a kind of destiny rooted in my origins and in my life. One should not speak too confidently of one's "destiny"; I use the word to refer to causes that lie deeper in history and character than mere intention or desire. In buying the little place known as Lanes Landing, it seems to me, I began to obey the deeper causes.

We had returned so that I could take a job at the University of Kentucky in Lexington. And we expected to live pretty much the usual academic life: I would teach and write; my "subject matter" would be, as it had been, the few square miles in Henry County where I grew up. We bought the tiny farm at Lanes Landing, thinking that we would use it as a "summer place," and on that understanding I began, with the help of two carpenter friends, to make some necessary repairs on the house. I no longer remember exactly how it was decided, but that work had hardly begun when it became a full-scale overhaul.

By so little our minds had been changed: this was not going to be a house to visit, but a house to live in. It was as though, having put our hand to the plow, we not only did not look back, but could not. We renewed the old house, equipped it with plumbing, bathroom, and oil furnace, and moved in on July 4, 1965.

Once the house was whole again, we came under the influence of the "twelve acres more or less." This acreage included a steep hillside pasture, two small pastures by the river, and a "garden spot" of less than half an acre. We had, besides the house, a small barn in bad shape, a good large building that once had been a general store, and a small garage also in usable condition. This was hardly a farm by modern standards, but it was land that could be used, and it was unthinkable that we would not use it. The land was not good enough to afford the possibility of a cash income, but it would allow us to grow our food—or most of it. And that is what we set out to do.

In the early spring of 1965 I had planted a small orchard; the next spring we planted our first garden. Within the following six or seven years we reclaimed the pastures, converted the garage into a henhouse, rebuilt the barn, greatly improved the garden soil, planted berry bushes, acquired a milk cow—and were producing, except for hay and grain for our animals, nearly everything that we ate: fruit, vegetables, eggs, meat, milk, cream, and butter. We built an outbuilding with a meat room and a food-storage cellar. Because we did not want to pollute our land and water with sewage, and in the process waste nutrients that should be returned to the soil, we built a composting privy. And so we began to attempt a life that, in addition to whatever else it was, would be responsibly agricultural. We used no chemical fertilizers. Except for a little rotenone, we used no insecticides. As our land and our food became healthier, so did we. And our food was of better quality than any that we could have bought.

We were not, of course, living an idyll. What we had done could not have been accomplished without difficulty and a great deal of work. And we had made some mistakes and false starts. But there was great satisfaction, too, in restoring the neglected land, and in feeding ourselves from it.

Meanwhile, the forty-acre place adjoining ours on the down-river side had been sold to a "developer," who planned to divide it into lots for "second homes." This project was probably doomed by the steepness of the ground and the difficulty of access, but a lot of bulldozing—and a lot of damage—was done before it was given up. In the fall of 1972, the place was offered for sale and we were able to buy it.

We now began to deal with larger agricultural problems. Some of this new land was usable; some would have to be left in trees. There were perhaps fifteen acres of hillside that could be reclaimed for pasture, and about two and a half acres of excellent bottomland on which we would grow alfalfa for hay. But it was a mess, all of it badly neglected, and a considerable portion of it badly abused by the developer's bulldozers. The hillsides were covered with thicket growth; the bottom was shoulder high in weeds; the diversion ditches had to be restored; a bulldozed gash meant for "building sites" had to be mended; the barn needed a new foundation, and the cistern a new top; there were no fences. What we had bought was less a farm than a reclamation project—which has now, with a later purchase, grown to seventy-five acres.

While we had only the small place, I had got along very well with a Gravely "walking tractor" that I owned, and an old Farmall A that I occasionally borrowed from my Uncle Jimmy. But now that we had increased our acreage, it was clear that I could not continue to depend on a borrowed tractor. For a while I assumed that I would buy a tractor of my own. But because our land was steep, and there was already talk of a fuel shortage—and because I liked the idea—I finally decided to buy a team of horses instead. By the spring of 1973, after a lot of inquiring and looking, I had found and bought a team of five-year-old sorrel mares. And—again by the generosity of my Uncle Jimmy, who has never thrown any good thing away—I had enough equipment to make a start.

Though I had worked horses and mules during the time I was growing up, I had never worked over ground so steep and problematical as this, and it had been twenty years since I had worked a team over ground of any kind. Getting started again, I anticipated every new task with uneasiness, and sometimes with dread. But to my relief and delight, the team and I did all that needed to be done that year, getting better as we went along. And over the years since then, with that team and others, my son and I have carried on our farming the way it was carried on in my boyhood, doing everything with our horses except baling the hay. And we have done work in places and in weather in which a tractor would have been useless. Experience has shown us—or re-shown us—that horses are not only a satisfactory and economical means of power, especially on such small places as ours, but are probably *necessary* to the most conservative use of steep land. Our farm, in fact, is surrounded by potentially excellent hillsides that were maintained in pasture until tractors replaced the teams.

Another change in our economy (and our lives) was accomplished in the fall of 1973 with the purchase of our first wood-burning stove. Again the petroleum shortage was on our minds, but we also knew that from the pasture-clearing we had ahead of us we would have an abundance of wood that otherwise would go to waste—and when that was gone we would still have our permanent wood lots. We thus expanded our subsistence income to include heating fuel, and since then have used our furnace only as a "backup system" in the coldest weather and in our absences from home. The horses also contribute significantly to the work of fuel-gathering; they will go easily into difficult places and over soft ground or snow where a truck or a tractor could not move.

As we have continued to live on and from our place, we have slowly begun its restoration and healing. Most of the scars have now been mended and grassed over, most of the washes stopped, most of the buildings made sound; many loads of rocks have been hauled out of the fields and used to pave entrances or fill hollows; we have done perhaps half of the necessary fencing. A great deal of work is still left to do, and some of it—the rebuilding of fertility in the depleted hillsides—will take longer than we will live. But in doing these things we have begun a restoration and a healing in ourselves.

I should say plainly that this has not been a "paying proposition." As a reclamation project, it has been costly both in money and in effort. It seems at least possible that, in any other place, I might have had little interest in doing any such thing. The reason I have been interested in doing it here, I think, is that I have felt implicated in the history, the uses, and the attitudes that have depleted such places as ours and made them "marginal."

I had not worked long on our "twelve acres more or less" before I saw that such places were explained almost as much by their human history as by their nature. I saw that they were not "marginal" because they ever were unfit for human use, but because in both culture and character *we* had been unfit to use them. Originally, even such steep slopes as these along the lower Kentucky River Valley were deep-soiled and abundantly fertile; "jumper" plows and generations of carelessness impoverished them. Where yellow clay is at the surface now, five feet of good soil may be gone. I once wrote that on some of the nearby uplands one walks as if "knee-deep" in the absence of the original soil. On these steeper slopes, I now know, that absence is shoulder-deep.

That is a loss that is horrifying as soon as it is imagined. It happened easily, by ignorance, indifference, "a little folding of the hands to sleep." It cannot be remedied in human time; to build five feet of soil takes perhaps fifty or sixty thousand years. This loss, once imagined, is potent with despair. If a people in adding a hundred and fifty years to itself subtracts fifty thousand from its land, what is there to hope?

And so our reclamation project has been, for me, less a matter of idealism or morality than a kind of self-preservation. A destructive history, once it is understood as such, is a nearly insupportable burden. Understanding it is a disease of understanding, depleting the sense of efficacy and paralyzing effort, unless it finds healing work. For me that work has been partly of the mind, in what I have written, but that seems to have depended inescapably on work of the body and of the ground. In order to affirm the values most native and necessary to me —indeed, to affirm my own life as a thing decent in possibility —I needed to know in my own experience that this place did not have to be abused in the past, and that it can be kindly and conservingly used now.

With certain reservations that must be strictly borne in mind, our work here has begun to offer some of the needed proofs.

Bountiful as the vanished original soil of the hillsides may have been, what remains is good. It responds well—sometimes astonishingly well—to good treatment. It never should have been plowed (some of it never should have been cleared), and it never should be plowed again. But it can be put in pasture without plowing, and it will support an excellent grass sod that will in turn protect it from erosion, if properly managed and not overgrazed.

Land so steep as this cannot be preserved in row crop cultivation. To subject it to such an expectation is simply to ruin it, as its history shows. Our rule, generally, has been to plow no steep ground, to maintain in pasture only such slopes as can be safely mowed with a horse-drawn mower, and to leave the rest in trees. We have increased the numbers of livestock on our pastures gradually, and have carefully rotated the animals from field to field, in order to avoid overgrazing. Under this use and care, our hillsides have mended and they produce more and better pasturage every year.

As a child I always intended to be a farmer. As a young man, I gave up that intention, assuming that I could not farm and do the other things I wanted to do. And then I became a farmer almost unintentionally and by a kind of necessity. That wayward and necessary becoming—along with my marriage, which has been intimately a part of it—is the major event of my life. It has changed me profoundly from the man and the writer I would otherwise have been.

There was a time, after I had left home and before I came back, when this place was my "subject matter." I meant that too, I think, on the day in 1956 when I told my friend, "That's all I need." I was regarding it, in a way too easy for a writer, as a mirror in which I saw myself. There was obviously a sort of narcissism in that—and an inevitable superficiality, for only the surface can reflect.

In coming home and settling on this place, I began to *live* in my subject, and to learn that living in one's subject is not at all the same as "having" a subject. To live in the place that is one's subject is to pass through the surface. The simplifications of distance and mere observation are thus destroyed. The obsessively regarded reflection is broken and dissolved. One sees that the mirror was a blinder; one can now begin to see where one is. One's relation to one's subject ceases to be merely emotional or aesthetical, or even merely critical, and becomes problematical, practical, and responsible as well. Because it must. It is like marrying your sweetheart.

Though our farm has not been an economic success, as such success is usually reckoned, it is nevertheless beginning to make a kind of economic sense that is consoling and hopeful. Now that the largest expenses of purchase and repair are behind us, our income from the place is beginning to run ahead of expenses. As income I am counting the value of shelter, subsistence, heating fuel, and money earned by the sale of livestock. As expenses I am counting maintenance, newly purchased equipment, extra livestock feed, newly purchased animals, reclamation work, fencing materials, taxes, and insurance.

If our land had been in better shape when we bought it, our expenses would obviously be much smaller. As it is, once we have completed its restoration, our farm will provide us a home, produce our subsistence, keep us warm in winter, and earn a modest cash income. The significance of this becomes apparent when one considers that most of this land is "unfarmable" by the standards of conventional agriculture, and that most of it was producing nothing at the time we bought it.

And so, contrary to some people's opinion, it *is* possible for a family to live on such "marginal" land, to take a bountiful subsistence and some cash income from it, and, in doing so, to improve both the land and themselves. (I believe, however, that, at least in the present economy, this should not be attempted without a source of income other than the farm. It is now extremely difficult to pay for the best of farmland by farming it, and even "marginal" land has become unreasonably expensive. To attempt to make a living from such land is to impose a severe strain on land and people alike.)

I said earlier that the success of our work here is subject to reservations. There are only two of these, but both are serious.

The first is that land like ours—and there are many acres of such land in this country—can be conserved in use only by competent knowledge, by a great deal more work than is required by leveler land, by a devotion more particular and disciplined than patriotism, and by ceaseless watchfulness and care. All these are cultural values and resources, never sufficiently abundant in this country, and now almost obliterated by the contrary values of the so-called "affluent society."

One of my own mistakes will suggest the difficulty. In 1974 I dug a small pond on a wooded hillside that I wanted to pasture occasionally. The excavation for that pond—as I should have anticipated, for I had better reason than I used—caused the hillside to slump both above and below. After six years the slope has not stabilized, and more expense and trouble will be required to stabilize it. A small hillside farm will not survive many mistakes of that order. Nor will a modest income.

The true remedy for mistakes is to keep from making them. It is not in the piecemeal technological solutions that our society now offers, but in a change of cultural (and economic) values that will encourage in the whole population the necessary respect, restraint, and care. Even more important, it is in the possibility of settled families and local communities, in which the knowledge of proper means and methods, proper moderations and restraints, can be handed down, and so accumulate in place and stay alive; the experience of one generation is not adequate to inform and control its actions. Such possibilities are not now in sight in this country.

The second reservation is that we live at the lower end of the Kentucky River watershed, which has long been intensively used, and is increasingly abused. Strip mining, logging, extractive farming, and the digging, draining, roofing, and paving that go with industrial and urban "development," all have seriously depleted the capacity of the watershed to retain water. This means not only that floods are higher and more frequent than they would be if the watershed were healthy, but that the floods subside too quickly, the watershed being far less a sponge, now, than it is a roof. The floodwater drops suddenly out of the river, leaving the steep banks soggy, heavy, and soft. As a result, great strips and blocks of land crack loose and slump, or they give way entirely and disappear into the river in what people here call "slips."

The flood of December 1978, which was unusually high, also went down extremely fast, falling from banktop almost to pool stage within a couple of days. In the aftermath of this rapid "drawdown," we lost a block of bottomland an acre square. This slip, which is still crumbling, severely damaged our place, and may eventually undermine two buildings. The same flood started a slip in another place, which threatens a third building. We have yet another building situated on a huge (but, so far, very gradual) slide that starts at the river and, aggravated by two state highway cuts, goes almost to the hilltop. And we have serious river bank erosion the whole length of our place.

What this means is that, no matter how successfully we may control erosion on our hillsides, our land remains susceptible to a more serious cause of erosion that we cannot control. Our river bank stands literally at the cutting edge of our nation's consumptive economy. This, I think, is true of many "marginal" places—it is true, in fact, of many places that are not marginal. In its consciousness, ours is an upland society; the ruin of watersheds, and what that involves and means, is little considered. And so the land is heavily taxed to subsidize an "affluence" that consists, in reality, of health and goods stolen from the unborn.

Living at the lower end of the Kentucky River watershed is what is now known as "an educational experience"—and not an easy one. A lot of information comes with it that is severely damaging to the reputation of our people and our time. From where I live and work, I never have to look far to see that the earth does indeed pass away. But however that is taught, and however bitterly learned, it is something that should be known, and there is a certain good strength in knowing it. To spend one's life farming a piece of the earth so passing is, as many would say, a hard lot. But it is, in an ancient sense, the human lot. What saves it is to love the farming.

THE UNSETTLING
OF AMERICA
(1977)

For Maurice Telleen

Contents

Preface

This book was meant to be a criticism of what I have called modern or orthodox agriculture. As I now realize, it is more a review than a criticism. Criticism requires a subject that is "finished." When agriculture is "finished," no would-be critic will be available. I am therefore constrained to accept my demotion as a privilege.

Nevertheless, there is a difficulty in writing a book on so inherently topical a subject as agricultural policy, and this difficulty is time: events that were the immediate cause of the book may be "finished" before the writing is. No reader of this book can fail to observe that it deals at length with the assumptions and policies of former Secretary of Agriculture Earl L. Butz, though Mr. Butz and the administration he served are now out of office.

I can only insist that my book is not for that reason out-of-date. Secretary Butz's tenure in the Department of Agriculture, and even his influence, are matters far more transient than the power and the values of those whose interests he represented. Moreover, the cultural issues that I attempt to deal with have been with us since our history began, and, barring miracle or catastrophe, they will be with us for a long time to come.

As a matter of fact, this book's origins go back farther than the secretaryship of Mr. Butz. The first notes I made for it were incited by a news story in the summer of 1967 on the report of President Johnson's "special commission on federal food and fiber policies."

The commission said, according to an article in the *Louisville Courier-Journal*, that the country's biggest farm problem was a surplus of farmers: ". . . the technological advances in agriculture have so greatly reduced the need for manpower that too many people are trying to live on a national farm income wholly inadequate for them." The proposed solutions were to find "better opportunities for the farm people," "a more comprehensive national employment policy," "retraining programs," "improved general educational facilities," etc.

Both the commission and the writer of the article had obviously *taken for granted* that the lives and communities of small farmers then still on the farm—and those of the 25 million who had left the farm since 1940—were of less value than "technological advances in agriculture." There seemed also to be no official doubt that adequate solutions were to be found in government-supplied "opportunities," facilities, and programs. Reading that article, I realized that my values were not only out of fashion, but under powerful attack. I saw that I was a member of a threatened minority. That is what set me off.

W.B.

Preface to the Second Edition

When I was working on this book—from 1974 to 1977—the long agricultural decline that it deals with was momentarily disguised as a "boom." The big farmers were getting bigger with the help of inflated land prices and borrowed money, and the foreign demand for American farm products was strong, so from the official point of view the situation looked good. The big were *supposed* to get bigger. Foreigners were *supposed* to be in need of our products. The official point of view, foreshortened as usual by statistics, superstitious theory, and wishful prediction, was utterly complacent. Then Secretary of Agriculture Earl L. Butz issued the most optimistic, the most widely obeyed, and the worst advice ever given to farmers: that they should plow "fencerow to fencerow."

That the situation was *not* good—for farms or farmers or rural communities or nature or the general public—was even then evident to any experienced observer who would turn aside from the preconceptions of "agribusiness" and look at the marks of deterioration that were plainly visible. And now, almost a decade later, it is evident to everyone that, at least for farmers and rural communities, the situation is catastrophic: Farmers are losing their farms, some are killing themselves, some in the madness of despair are killing other people, and rural economy and rural life are gravely stricken. The agricultural economists chart the "liquidations of assets," the "shake-outs," and the "downturns," apparently amazed that now even the large "progressive" and "efficient" farmers are in trouble.

But this is not just a financial crisis for country people. Critical questions are being asked of our whole society: Are we, or are we not, going to take proper care of our land, our country? And do we, or do we not, believe in a democratic distribution of usable property? At present, these questions are being answered in the negative. Our soil erosion rates are worse now than during the years of the Dust Bowl. In the arid lands of the West, we are overusing and wasting the supplies of water. Toxic pollution from agricultural chemicals is a growing problem. We are closer every day to the final destruction of private ownership not only of small family farms, but of small usable properties of all kinds. Every problem I dealt with in this book, in fact, has grown worse since the book was written.

The one improvement has been in public concern about the problems. Among farmers there is growing distrust of the "agribusiness" line of talk and growing interest in agricultural health and sanity. Among city people there is a growing awareness that sane and healthy agriculture requires an informed urban constituency. There is hope in these developments and in the continued existence of a remnant of excellent small farms and farmers.

Some prominent agricultural economists are still finding it possible to pretend that the only issues involved are economic, but that possibility is diminishing. I recently attended a meeting at which an agricultural economist argued that there is no essential difference between owning and renting a farm. A farmer stood up in the audience and replied: "Professor, I don't think our ancestors came to America in order to *rent* a farm."

'Nough said.

W.B.
March 1986

Acknowledgments

Anything that I will ever have to say on the subject of agriculture can be little more than a continuation of talk begun in childhood with my father and with my late friend Owen Flood. Their conversation, first listened to and then joined, was my first and longest and finest instruction. From them, before I knew I was being taught, I learned to think of the meanings, the responsibilities, and the pleasures of farming.

But this book's greatest immediate debt is acknowledged in the dedication. Maury Telleen has been an indefatigable friend both to my book and to me. He has arranged indispensable meetings, written me letters, sent me clippings and books, talked to me on the phone, read my manuscript, borne up through hours of fervent conversation in his house, in my house, and over many hundreds of miles of happy agricultural travels. To his wife, Jeannine, I am indebted for understanding and for hospitality.

I am, of course, hopelessly in debt to my own wife, Tanya, for keeping farm and household together during my absences, and for enduring both my travels and my travails. But she has also participated in the thinking-out of this book, has been its critic, and, finally, its typist.

My neighbor, Tom Grissom, gave me invaluable help with the preliminary reading, and in many talks helped me to clarify my ideas.

For reading the manuscript and other kindnesses, I am grateful to Robert Rodale, Jerry Goldstein, Cia and Ed McClanahan, James Baker Hall, Joan Hall, Gary Snyder, Jim and Barbara Foote, and my father.

Stephen B. Brush, Clarence Van Sant, Gurney Norman, Ben Webb, and David Budbill generously gave me permission to quote from their letters, and Stephen Brush allowed me to make use of his excellent paper on Andean agriculture.

For hospitality and other kindnesses on my travels, I thank Tommy Shoup, Everett Hildebrandt, Roger Blobaum, Mr. and Mrs. Ilo Kusserow, Mr. and Mrs. Mike Jessen, Mr. and Mrs.

Monroe J. Miller, Arnold Hockett, Clarence Van Sant, Mr. and Mrs. John Heinrich, Patty Kaminsky.

This list of my debts will make clear that much of the good of my book has been the gift of other people. Its flaws, on the other hand, have been furnished exclusively by me.

Who so hath his minde on taking,
hath it no more on what he hath taken.

MONTAIGNE, III. VI

So many goodly citties ransacked and razed; so many nations destroyed and made desolate; so infinite millions of harmelesse people of all sexes, states and ages, massacred, ravaged and put to the sword; and the richest, the fairest and the best part of the world topsiturvied, ruined and defaced for the traffick of Pearles and Pepper: Oh mechanicall victories, oh base conquest.

MONTAIGNE

CHAPTER ONE

The Unsettling of America

ONE OF THE peculiarities of the white race's presence in America is how little intention has been applied to it. As a people, wherever we have been, we have never really intended to be. The continent is said to have been discovered by an Italian who was on his way to India. The earliest explorers were looking for gold, which was, after an early streak of luck in Mexico, always somewhere farther on. Conquests and foundings were incidental to this search—which did not, and could not, end until the continent was finally laid open in an orgy of goldseeking in the middle of the 19th century. Once the unknown of geography was mapped, the industrial marketplace became the new frontier, and we continued, with largely the same motives and with increasing haste and anxiety, to displace ourselves—no longer with unity of direction, like a migrant flock, but like the refugees from a broken ant hill. In our own time we have invaded foreign lands and the moon with the high-toned patriotism of the conquistadors, and with the same mixture of fantasy and avarice.

That is too simply put. It is substantially true, however, as a description of the dominant tendency in American history. The temptation, once that has been said, is to ascend altogether into rhetoric and inveigh equally against all our forebears and all present holders of office. To be just, however, it is necessary to remember that there has been another tendency: the tendency to stay put, to say, "No farther. This is the place." So far, this has been the weaker tendency, less glamorous, certainly less successful. It is also the older of these tendencies, having been the dominant one among the Indians.

The Indians did, of course, experience movements of population, but in general their relation to place was based upon old usage and association, upon inherited memory, tradition, veneration. The land was their homeland. The first and greatest American revolution, which has never been superseded, was the coming of people who did *not* look upon the land as a homeland. But there were always those among the newcomers who saw that they had come to a good place and who saw its domestic possibilities. Very early, for instance, there were men who wished to establish agricultural settlements rather than quest for gold or exploit the Indian trade. Later, we know that every advance of the frontier left behind families and communities who intended to remain and prosper where they were.

But we know also that these intentions have been almost systematically overthrown. Generation after generation, those who intended to remain and prosper where they were have been dispossessed and driven out, or subverted and exploited where they were, by those who were carrying out some version of the search for El Dorado. Time after time, in place after place, these conquerors have fragmented and demolished traditional communities, the beginnings of domestic cultures. They have always said that what they destroyed was outdated, provincial, and contemptible. And with alarming frequency they have been believed and trusted by their victims, especially when their victims were other white people.

If there is any law that has been consistently operative in American history, it is that the members of any *established* people or group or community sooner or later become "redskins"—that is, they become the designated victims of an utterly ruthless, officially sanctioned and subsidized exploitation. The colonists who drove off the Indians came to be intolerably exploited by their imperial governments. And that alien imperialism was thrown off only to be succeeded by a domestic version of the same thing; the class of independent small farmers who fought the war of independence has been exploited by, and recruited into, the industrial society until by now it is almost extinct. Today, the most numerous heirs of the farmers of Lexington and Concord are the little groups scattered all over the country whose names begin with "Save": Save Our Land, Save the Valley, Save Our Mountains, Save Our Streams, Save Our Farmland. As so often before, these are *designated* victims—people without official sanction, often without official friends, who are struggling to preserve their places, their values, and their lives as they know them and prefer to live them against the agencies of their own government which are using their own tax moneys against them.

The only escape from this destiny of victimization has been to "succeed"—that is, to "make it" into the class of exploiters, and then to remain so specialized and so "mobile" as to be unconscious of the effects of one's livelihood. This escape is, of course, illusory, for one man's producer is another's consumer, and even the richest and most mobile will soon find it hard to escape the noxious effluents and fumes of the various public services.

Let me emphasize that I am not talking about an evil that is merely contemporary or "modern," but one that is as old in America as the white man's presence here. It is an intention that was *organized* here almost from the start. "The New World," Bernard DeVoto wrote in *The Course of Empire*, "was a constantly expanding market . . . Its value in gold was enormous but it had still greater value in that it expanded and integrated the industrial systems of Europe."

And he continues: "The first belt-knife given by a European to an Indian was a portent as great as the cloud that mushroomed over Hiroshima . . . Instantly the man of 6000 B.C. was bound fast to a way of life that had developed seven and a half millennia beyond his own. He began to live better and he began to die."

The principal European trade goods were tools, cloth, weapons, ornaments, novelties, and alcohol. The sudden availability of these things produced a revolution that "affected every aspect of Indian life. The struggle for existence . . . became easier. Immemorial handicrafts grew obsolescent, then obsolete. Methods of hunting were transformed. So were methods—and the purposes—of war. As war became deadlier in purpose and armament a surplus of women developed, so that marriage customs changed and polygamy became common. The increased usefulness of women in the preparation of pelts worked to the same end . . . Standards of wealth, prestige, and honor changed. The Indians acquired commercial values and developed business cults. They became more mobile. . . .

"In the sum it was cataclysmic. A culture was forced to change much faster than change could be adjusted to. All corruptions of culture produce breakdowns of morale, of communal integrity, and of personality, and this force was as strong as any other in the white man's subjugation of the red man."

I have quoted these sentences from DeVoto because, the obvious differences aside, he is so clearly describing a revolution that did not stop with the subjugation of the Indians, but went on to impose substantially the same catastrophe upon the small farms and the farm communities, upon the shops of small local tradesmen of all sorts, upon the workshops of independent craftsmen, and upon the households of citizens. It is a revolution that is still going on. The economy is still substantially that of the fur trade, still based on the same general kinds of commercial items: technology, weapons, ornaments, novelties, and drugs. The one great difference is that by now the revolution has deprived the mass of consumers of any independent access to the staples of life: clothing, shelter, food, even water. Air remains the only necessity that the average user can still get for himself, and the revolution has imposed a heavy tax on that by way of pollution. Commercial conquest is far more thorough and final than military defeat. The Indian became a redskin, not by loss in battle, but by accepting a dependence on traders that made *necessities* of industrial goods. This is not merely history. It is a parable.

DeVoto makes it clear that the imperial powers, having made themselves willing to impose this exploitive industrial economy upon the Indians, could not then keep it from contaminating their own best intentions: "More than four-fifths of the wealth of New France was furs, the rest was fish, and it had no agricultural wealth. One trouble was that whereas the crown's imperial policy required it to develop the country's agriculture, the crown's economy required the colony's furs, an adverse interest." And La Salle's dream of developing Louisiana (agriculturally and otherwise) was frustrated because "The interest of the court in Louisiana colonization was to secure a bridgehead for an attack on the silver mines of northern Mexico. . . ."

One cannot help but see the similarity between this foreign colonialism and the domestic colonialism that, by policy, converts productive farm, forest, and grazing lands into strip mines. Now, as then, we see the abstract values of an industrial economy preying upon the native productivity of land and people. The fur trade was only the first establishment on this continent of a mentality whose triumph is its catastrophe.

My purposes in beginning with this survey of history are (1) to show how deeply rooted in our past is the mentality of exploitation; (2) to show how fundamentally revolutionary it is; and (3) to show how crucial to our history—hence, to our own minds—is the question of how we will relate to our land. This question, now that the corporate revolution has so determinedly invaded the farmland, returns us to our oldest crisis.

We can understand a great deal of our history—from Cortés' destruction of Tenochtitlán in 1521 to the bulldozer attack on the coalfields four-and-a-half centuries later—by thinking of ourselves as divided into conquerors and victims. In order to understand our own time and predicament and the work that is to be done, we would do well to shift the terms and say that we are divided between exploitation and nurture. The first set of terms is too simple for the purpose because, in any given situation, it proposes to divide people into two mutually exclusive groups; it becomes complicated only when we are dealing with situations in succession—as when a colonist who persecuted the Indians then resisted persecution by the crown. The terms exploitation and nurture, on the other hand, describe a division not only between persons but also within persons. We are all to some extent the products of an exploitive society, and it would be foolish and self-defeating to pretend that we do not bear its stamp.

Let me outline as briefly as I can what seem to me the characteristics of these opposite kinds of mind. I conceive a strip-miner to be a model exploiter, and as a model nurturer I take the old-fashioned idea or ideal of a farmer. The exploiter is a specialist, an expert; the nurturer is not. The standard of the exploiter is efficiency; the standard of the nurturer is care. The exploiter's goal is money, profit; the nurturer's goal is health—his land's health, his own, his family's, his community's, his country's. Whereas the exploiter asks of a piece of land only how much and how quickly it can be made to produce, the nurturer asks a question that is much more complex and difficult: What is its carrying capacity? (That is: How much can be taken from it without diminishing it? What can it produce *dependably* for an indefinite time?) The exploiter wishes to earn as much as possible by as little work as possible; the nurturer expects, certainly, to have a decent living from his work, but his characteristic wish is to work *as well* as possible. The competence of the exploiter is in organization; that of the nurturer is in order—a human order, that is, that accommodates itself both to other order and to mystery. The exploiter typically serves an institution or organization; the nurturer serves land, household, community, place. The exploiter thinks in terms of numbers, quantities, "hard facts"; the nurturer in terms of character, condition, quality, kind.

It seems likely that all the "movements" of recent years have been representing various claims that nurture has to make against exploitation. The women's movement, for example, when its energies are most accurately placed, is arguing the cause of nurture; other times it is arguing the right of women to be exploiters—which men have no *right* to be. The exploiter is clearly the prototype of the "masculine" man—the wheeler-dealer whose "practical" goals require the sacrifice of flesh, feeling, and principle. The nurturer, on the other hand, has always passed with ease across the boundaries of the so-called sexual roles. Of necessity and without apology, the preserver of seed, the planter, becomes midwife and nurse. Breeder is always metamorphosing into brooder and back again. Over and over again, spring after spring, the questing mind, idealist and visionary, must pass through the planting to become nurturer of the real. The farmer, sometimes known as husbandman, is by definition half mother; the only question is how good a mother he or she is. And the land itself is not mother or father only, but both. Depending on crop and season, it is at one time receiver of seed, bearer and nurturer of young; at another, raiser of seed-stalk, bearer and shedder of seed. And in response to these changes, the farmer crosses back and forth from one zone of spousehood to another, first as planter and then as gatherer. Farmer and land are thus involved in a sort of dance in which the partners are always at opposite sexual poles, and the lead keeps changing; the farmer, as seed-bearer, causes growth; the land, as seed-bearer, causes the harvest.

The exploitive always involves the abuse or the perversion of nurture and ultimately its destruction. Thus, we saw how far the exploitive revolution had penetrated the official character when our recent secretary of agriculture remarked that "Food is a weapon." This was given a fearful symmetry indeed when, in discussing the possible use of nuclear weapons, a secretary of defense spoke of "palatable" levels of devastation. Consider the associations that have since ancient times clustered around the idea of food—associations of mutual care, generosity, neighborliness, festivity, communal joy, religious ceremony—and you will see that these two secretaries represent a cultural catastrophe. The concerns of farming and those of war, once thought to be diametrically opposed, have become identical. Here we have an example of men who have been made vicious, not presumably by nature or circumstance, but by their *values*.

Food is *not* a weapon. To use it as such—to foster a mentality willing to use it as such—is to prepare, in the human character and community, the destruction of the sources of food. The first casualties of the exploitive revolution are character and community. When those fundamental integrities are devalued and broken, then perhaps it is inevitable that food will be looked upon as a weapon, just as it is inevitable that the earth will be looked upon as fuel and people as numbers or machines. But character and community—that is, culture in the broadest, richest sense—constitute, just as much as nature, the source of food. Neither nature nor people alone can produce human sustenance, but only the two together, culturally wedded. The poet Edwin Muir said it unforgettably:

> Men are made of what is made,
> The meat, the drink, the life, the corn,
> Laid up by them, in them reborn.
> And self-begotten cycles close
> About our way; indigenous art
> And simple spells make unafraid
> The haunted labyrinths of the heart
> And with our wild succession braid
> The resurrection of the rose.

To think of food as a weapon, or of a weapon as food, may give an illusory security and wealth to a few, but it strikes directly at the life of all.

The concept of food-as-weapon is not surprisingly the doctrine of a Department of Agriculture that is being used as an instrument of foreign political and economic speculation. This militarizing of food is the greatest threat so far raised against the farmland and the farm communities of this country. If present attitudes continue, we may expect government policies that will encourage the destruction, by overuse, of farmland. This, of course, has already begun. To answer the official call for more production—evidently to be used to bait or bribe foreign countries—farmers are plowing their waterways and permanent pastures; lands that ought to remain in grass are being planted in row crops. Contour plowing, crop rotation, and other conservation measures seem to have gone out of favor or fashion in official circles and are practiced less and less on the farm. This exclusive emphasis on production will accelerate the mechanization and chemicalization of farming, increase the price of land, increase overhead and operating costs, and thereby further diminish the farm population. Thus the tendency, if not the intention, of Mr. Butz's confusion of farming and war, is to complete the deliverance of American agriculture into the hands of corporations.

The cost of this corporate totalitarianism in energy, land, and social disruption will be enormous. It will lead to the exhaustion of farmland and farm culture. Husbandry will become an extractive industry; because maintenance will entirely give way to production, the fertility of the soil will become a limited, unrenewable resource like coal or oil.

This may not happen. It *need* not happen. But it is necessary to recognize that it *can* happen. That it can happen is made evident not only by the words of such men as Mr. Butz, but more clearly by the large-scale industrial destruction of farmland already in progress. If it does happen, we are familiar enough with the nature of American salesmanship to know that it will be done in the name of the starving millions, in the name of liberty, justice, democracy, and brotherhood, and to free the world from communism. We must, I think, be prepared to see, and to stand by, the truth: that the land should not be destroyed for *any* reason, not even for any apparently good reason. We must be prepared to say that enough food, year after year, is possible only for a limited number of people, and that this possibility can be preserved only by the steadfast, knowledgeable *care* of those people. Such "crash programs" as apparently have been contemplated by the Department of Agriculture in recent years will, in the long run, cause more starvation than they can remedy.

Meanwhile, the dust clouds rise again over Texas and Oklahoma. "Snirt" is falling in Kansas. Snow drifts in Iowa and the Dakotas are black with blown soil. The fields lose their humus and porosity, become less retentive of water, depend more on pesticides, herbicides, chemical fertilizers. Bigger tractors become necessary because the compacted soils are harder to work—and their greater weight further compacts the soil. More and bigger machines, more chemical and methodological shortcuts are needed because of the shortage of manpower on the farm—and the problems of overcrowding and unemployment increase in the cities. It is estimated that it now costs (by erosion) two bushels of Iowa topsoil to grow one bushel of corn. It is variously estimated that from five to twelve calories of fossil fuel energy are required to produce one calorie of hybrid corn energy. An official of the National Farmers Union says that "a farmer who earns $10,000 to $12,000 a year typically leaves an estate valued at about $320,000"—which means that when that farm is financed again, either by a purchaser or by an heir (to pay the inheritance taxes), it simply cannot support its new owner and pay for itself. And the *Progressive Farmer* predicts the disappearance of 200,000 to 400,000 farms each year during the next twenty years if the present trend continues.

The first principle of the exploitive mind is to divide and conquer. And surely there has never been a people more ominously and painfully divided than we are—both against each other and within ourselves. Once the revolution of exploitation is under way, statesmanship and craftsmanship are gradually replaced by salesmanship.* Its stock in trade in politics is to sell despotism and avarice as freedom and democracy. In business it sells sham and frustration as luxury and satisfaction. The "constantly expanding market" first opened in the New World by the fur traders is still expanding—no longer so much by expansions of territory or population, but by the calculated outdating, outmoding, and degradation of goods and by the hysterical self-dissatisfaction of consumers that is indigenous to an exploitive economy.

---

*The *craft* of persuading people to buy what they do not need, and do not want, for more than it is worth.

This gluttonous enterprise of ugliness, waste, and fraud thrives in the disastrous breach it has helped to make between our bodies and our souls. As a people, we have lost sight of the profound communion—even the union—of the inner with the outer life. Confucius said: "If a man have not order within him / He can not spread order about him. . . ." Surrounded as we are by evidence of the disorders of our souls and our world, we feel the strong truth in those words as well as the possibility of healing that is in them. We see the likelihood that our surroundings, from our clothes to our countryside, are the products of our inward life—our spirit, our vision—as much as they are products of nature and work. If this is true, then we cannot live as we do and be as we would like to be. There is nothing more absurd, to give an example that is only apparently trivial, than the millions who wish to live in luxury and idleness and yet be slender and good-looking. We have millions, too, whose livelihoods, amusements, and comforts are all destructive, who nevertheless wish to live in a healthy environment; they want to run their recreational engines in clean, fresh air. There is now, in fact, no "benefit" that is not associated with disaster. That is because power can be disposed morally or harmlessly only by thoroughly unified characters and communities.

What caused these divisions? There are no doubt many causes, complex both in themselves and in their interaction. But pertinent to all of them, I think, is our attitude toward work. The growth of the exploiters' revolution on this continent has been accompanied by the growth of the idea that work is beneath human dignity, particularly any form of hand work. We have made it our overriding ambition to escape work, and as a consequence have debased work until it is only fit to escape from. We have debased the products of work and have been, in turn, debased by them. Out of this contempt for work arose the idea of a nigger: at first some person, and later some thing, to be used to relieve us of the burden of work. If we began by making niggers of people, we have ended by making a nigger of the world. We have taken the irreplaceable energies and materials of the world and turned them into jimcrack "labor-saving devices." We have made of the rivers and oceans and winds niggers to carry away our refuse, which we think we are too good to dispose of decently ourselves. And in doing this to the world that is our common heritage and bond, we have returned to making niggers of people: we have become each other's niggers.

But is work something that we have a right to escape? And can we escape it with impunity? We are probably the first entire people ever to think so. All the ancient wisdom that has come down to us counsels otherwise. It tells us that work is necessary to us, as much a part of our condition as mortality; that good work is our salvation and our joy; that shoddy or dishonest or self-serving work is our curse and our doom. We have tried to escape the sweat and sorrow promised in Genesis—only to find that, in order to do so, we must forswear love and excellence, health and joy.

Thus we can see growing out of our history a condition that is physically dangerous, morally repugnant, ugly. Contrary to the blandishments of the salesmen, it is not particularly comfortable or happy. It is not even affluent in any meaningful sense, because its abundance is dependent on sources that are being rapidly exhausted by its methods. To see these things is to come up against the question: Then what *is* desirable?

One possibility is just to tag along with the fantasists in government and industry who would have us believe that we can pursue our ideals of affluence, comfort, mobility, and leisure indefinitely. This curious faith is predicated on the notion that we will soon develop unlimited new sources of energy: domestic oil fields, shale oil, gasified coal, nuclear power, solar energy, and so on. This is fantastical because the basic cause of the energy crisis is not scarcity; it is moral ignorance and weakness of character. We don't know *how* to use energy, or what to use it *for*. And we cannot restrain ourselves. Our time is characterized as much by the abuse and waste of human energy as it is by the abuse and waste of fossil fuel energy. Nuclear power, if we are to believe its advocates, is presumably going to be well used by the same mentality that has egregiously devalued and misapplied man- and womanpower. If we had an unlimited supply of solar or wind power, we would use that destructively, too, for the same reasons.

Perhaps all of those sources of energy are going to be developed. Perhaps all of them can sooner or later be developed without threatening our survival. But not all of them together can guarantee our survival, and they cannot define what is desirable. We will not find those answers in Washington, D.C., or in the laboratories of oil companies. In order to find them, we will have to look closer to ourselves.

I believe that the answers are to be found in our history: in its until now subordinate tendency of settlement, of domestic permanence. This was the ambition of thousands of immigrants; it is formulated eloquently in some of the letters of Thomas Jefferson; it was the dream of the freed slaves; it was written into law in the Homestead Act of 1862. There are few of us whose families have not at some time been moved to see its vision and to attempt to enact its possibility. I am talking about the idea that as many as possible should share in the ownership of the land and thus be bound to it by economic interest, by the investment of love and work, by family loyalty, by memory and tradition. How much land this should be is a question, and the answer will vary with geography. The Homestead Act said 160 acres. The freedmen of the 1860s hoped for forty. We know that, particularly in other countries, families have lived decently on far fewer acres than that.

The old idea is still full of promise. It is potent with healing and with health. It has the power to turn each person away from the big-time promising and planning of the government, to confront in himself, in the immediacy of his own circumstances and whereabouts, the question of what methods and ways are best. It proposes an economy of necessities rather than an economy based upon anxiety, fantasy, luxury, and idle wishing. It proposes the independent, free-standing citizenry that Jefferson thought to be the surest safeguard of democratic liberty. And perhaps most important of all, it proposes an agriculture based upon intensive work, local energies, care, and long-living communities—that is, to state the matter from a consumer's point of view: a dependable, long-term food supply.

This is a possibility that is obviously imperiled—by antipathy in high places, by adverse public fashions and attitudes, by the deterioration of our present farm communities and traditions, by the flawed education and the inexperience of our young people. Yet it alone can promise us the continuity of attention and devotion without which the human life of the earth is impossible.

Sixty years ago, in another time of crisis, Thomas Hardy wrote these stanzas:

> Only a man harrowing clods
>   In a slow silent walk
> With an old horse that stumbles and nods
>   Half asleep as they stalk.

> Only thin smoke without flame
>   From the heaps of couch-grass;
> Yet this will go onward the same
>   Though Dynasties pass.

Today most of our people are so conditioned that they do not wish to harrow clods either with an old horse or with a new tractor. Yet Hardy's vision has come to be more urgently true than ever. The great difference these sixty years have made is that, though we feel that this work must go onward, we are not so certain that it will. But the care of the earth is our most ancient and most worthy and, after all, our most pleasing responsibility. To cherish what remains of it, and to foster its renewal, is our only legitimate hope.

. . . wanting good government in their states, they first established order in their own families; wanting order in the home, they first disciplined themselves . . .

CONFUCIUS, *The Great Digest*

CHAPTER TWO

The Ecological Crisis as a Crisis of Character

IN JULY OF 1975 it was revealed by William Rood in the *Los Angeles Times* that some of our largest and most respected conservation organizations owned stock in the very corporations and industries that have been notorious for their destructiveness and for their indifference to the concerns of conservationists. The Sierra Club, for example, had owned stocks and bonds in Exxon, General Motors, Tenneco, steel companies "having the worst pollution records in the industry," Public Service Company of Colorado, "strip-mining firms with 53 leases covering nearly 180,000 acres and pulp-mill operators cited by environmentalists for their poor water pollution controls."

These investments proved deeply embarrassing once they were made public, but the Club's officers responded as quickly as possible by making appropriate changes in its investment policy. And so if it were only a question of policy, these investments could easily be forgotten, dismissed as aberrations of the sort that inevitably turn up now and again in the workings of organizations. The difficulty is that, although the investments were absurd, they were *not* aberrant; they were perfectly representative of the modern character. These conservation groups were behaving with a very ordinary consistency; they were only doing as organizations what many of their members were, and are, doing as individuals. They were making convenience of enterprises that they knew to be morally, and even practically, indefensible.

We are dealing, then, with an absurdity that is not a quirk or an accident, but is fundamental to our character as a people. The split between what we think and what we do is profound. It is not just possible, it is altogether to be expected, that our society would produce conservationists who invest in strip-mining companies, just as it must inevitably produce asthmatic executives whose industries pollute the air and vice-presidents of pesticide corporations whose children are dying of cancer.

245

And these people will tell you that this is the way the "real world" works. They will pride themselves on their sacrifices for "our standard of living." They will call themselves "practical men" and "hardheaded realists." And they will have their justifications in abundance from intellectuals, college professors, clergymen, politicians. The viciousness of a mentality that can look complacently upon disease as "part of the cost" would be obvious to any child. But this is the "realism" of millions of modern adults.

There is no use pretending that the contradiction between what we think or say and what we do is a limited phenomenon. There is no group of the extra-intelligent or extra-concerned or extra-virtuous that is exempt. I cannot think of any American whom I know or have heard of, who is not contributing in some way to destruction. The reason is simple: to live undestructively in an economy that is overwhelmingly destructive would require of any one of us, or of any small group of us, a great deal more work than we have yet been able to do. How could we divorce ourselves completely and yet responsibly from the technologies and powers that are destroying our planet? The answer is not yet thinkable, and it will not be thinkable for some time—even though there are now groups and families and persons everywhere in the country who have begun the labor of thinking it.

And so we are by no means divided, or readily divisible, into environmental saints and sinners. But there *are* legitimate distinctions that need to be made. These are distinctions of degree and of consciousness. Some people are less destructive than others, and some are more conscious of their destructiveness than others. For some, their involvement in pollution, soil depletion, strip-mining, deforestation, industrial and commercial waste is simply a "practical" compromise, a necessary "reality," the price of modern comfort and convenience. For others, this list of involvements is an agenda for thought and work that will produce remedies.

People who thus set their lives against destruction have necessarily confronted in themselves the absurdity that they have recognized in their society. They have first observed the tendency of modern organizations to perform in opposition to their stated purposes. They have seen governments that exploit and oppress the people they are sworn to serve and protect, medical procedures that produce ill health, schools that pre-serve ignorance, methods of transportation that, as Ivan Illich says, have "created more distances than they . . . bridge." And they have seen that these public absurdities are, and can be, no more than the aggregate result of private absurdities; the corruption of community has its source in the corruption of character. This realization has become the typical moral cri-sis of our time. Once our personal connection to what is wrong becomes clear, then we have to choose: we can go on as before, recognizing our dishonesty and living with it the best we can, or we can begin the effort to change the way we think and live.

The disease of the modern character is specialization. Looked at from the standpoint of the social *system*, the aim of specialization may seem desirable enough. The aim is to see that the responsibilities of government, law, medicine, engi-neering, agriculture, education, etc., are given into the hands of the most skilled, best prepared people. The difficulties do not appear until we look at specialization from the opposite standpoint—that of individual persons. We then begin to see the grotesquery—indeed, the impossibility—of an idea of community wholeness that divorces itself from any idea of per-sonal wholeness.

The first, and best known, hazard of the specialist system is that it produces specialists—people who are elaborately and expensively trained *to do one thing*. We get into absurdity very quickly here. There are, for instance, educators who have noth-ing to teach, communicators who have nothing to say, medical doctors skilled at expensive cures for diseases that they have no skill, and no interest, in preventing. More common, and more damaging, are the inventors, manufacturers, and salesmen of devices who have no concern for the possible effects of those devices. Specialization is thus seen to be a way of institutional-izing, justifying, and paying highly for a calamitous disintegra-tion and scattering-out of the various functions of character: workmanship, care, conscience, responsibility.

Even worse, a system of specialization requires the abdica-tion to specialists of various competences and responsibilities that were once personal and universal. Thus, the average—one is tempted to say, the ideal—American citizen now consigns the problem of food production to agriculturists and "agri-businessmen," the problems of health to doctors and sanitation experts, the problems of education to school teachers and educators, the problems of conservation to conservationists, and so on. This supposedly fortunate citizen is therefore left with only two concerns: making money and entertaining himself. He earns money, typically, as a specialist, working an eight-hour day at a job for the quality or consequences of which somebody else—or, perhaps more typically, nobody else—will be responsible. And not surprisingly, since he can do so little else for himself, he is even unable to entertain himself, for there exists an enormous industry of exorbitantly expensive specialists whose purpose is to entertain him.

The beneficiary of this regime of specialists ought to be the happiest of mortals—or so we are expected to believe. *All* of his vital concerns are in the hands of certified experts. He is a certified expert himself and as such he earns more money in a year than all his great-grandparents put together. Between stints at his job he has nothing to do but mow his lawn with a sit-down lawn mower, or watch other certified experts on television. At suppertime he may eat a tray of ready-prepared food, which he and his wife (also a certified expert) procure at the cost only of money, transportation, and the pushing of a button. For a few minutes between supper and sleep he may catch a glimpse of his children, who since breakfast have been in the care of education experts, basketball or marching-band experts, or perhaps legal experts.

The fact is, however, that this is probably the most unhappy average citizen in the history of the world. He has not the power to provide himself with anything but money, and his money is inflating like a balloon and drifting away, subject to historical circumstances and the power of other people. From morning to night he does not touch anything that he has produced himself, in which he can take pride. For all his leisure and recreation, he feels bad, he looks bad, he is overweight, his health is poor. His air, water, and food are all known to contain poisons. There is a fair chance that he will die of suffocation. He suspects that his love life is not as fulfilling as other people's. He wishes that he had been born sooner, or later. He does not know why his children are the way they are. He does not understand what they say. He does not care much and does not know why he does not care. He does not know what his wife wants or what he wants. Certain advertisements and pictures in magazines make him suspect that he is basically unattractive. He feels that all his possessions are under threat of pillage. He does not know what he would do if he lost his job, if the economy failed, if the utility companies failed, if the police went on strike, if the truckers went on strike, if his wife left him, if his children ran away, if he should be found to be incurably ill. And for these anxieties, of course, he consults certified experts, who in turn consult certified experts about *their* anxieties.

It is rarely considered that this average citizen is anxious because he *ought* to be—because he still has some gumption that he has not yet given up in deference to the experts. He ought to be anxious, because he is helpless. That he is dependent upon so many specialists, the beneficiary of so much expert help, can only mean that he is a captive, a potential victim. If he lives by the competence of so many other people, then he lives also by their indulgence; his own will and his own reasons to live are made subordinate to the mere tolerance of everybody else. He has *one* chance to live what he conceives to be his life: his own small specialty within a delicate, tense, everywhere-strained system of specialties.

From a public point of view, the specialist system is a failure because, though everything is done by an expert, very little is done well. Our typical industrial or professional product is both ingenious and shoddy. The specialist system fails from a personal point of view because a person who can do only one thing can do virtually nothing for himself. In living in the world by his own will and skill, the stupidest peasant or tribesman is more competent than the most intelligent worker or technician or intellectual in a society of specialists.

What happens under the rule of specialization is that, though society becomes more and more intricate, it has less and less structure. It becomes more and more organized, but less and less orderly. The community disintegrates because it loses the necessary understandings, forms, and enactments of the relations among materials and processes, principles and actions, ideals and realities, past and present, present and future, men and women, body and spirit, city and country, civilization and wilderness, growth and decay, life and death—just as the individual character loses the sense of a responsible involvement in these relations. No longer does human life rise from the earth like a pyramid, broadly and considerately founded upon its sources. Now it scatters itself out in a reckless horizontal sprawl, like a disorderly city whose suburbs and pavements destroy the fields.

The concept of country, homeland, dwelling place becomes simplified as "the environment"—that is, what surrounds us. Once we see our place, our part of the world, as *surrounding* us, we have already made a profound division between it and ourselves. We have given up the understanding—dropped it out of our language and so out of our thought—that we and our country create one another, depend on one another, are literally part of one another; that our land passes in and out of our bodies just as our bodies pass in and out of our land; that as we and our land are part of one another, so all who are living as neighbors here, human and plant and animal, are part of one another, and so cannot possibly flourish alone; that, therefore, our culture must be our response to our place, our culture and our place are images of each other and inseparable from each other, and so neither can be better than the other.

Because by definition they lack any such sense of mutuality or wholeness, our specializations subsist on conflict with one another. The rule is never to cooperate, but rather to follow one's own interest as far as possible. Checks and balances are all applied externally, by opposition, never by self-restraint. Labor, management, the military, the government, etc., never forbear until their excesses arouse enough opposition to *force* them to do so. The good of the whole of Creation, the world and all its creatures together, is never a consideration because it is never thought of; our culture now simply lacks the means for thinking of it.

It is for this reason that none of our basic problems is ever solved. Indeed, it is for this reason that our basic problems are getting worse. The specialists are profiting too well from the symptoms, evidently, to be concerned about cures—just as the myth of imminent cure (by some "breakthrough" of science or technology) is so lucrative and all-justifying as to foreclose any possibility of an interest in prevention. The problems thus become the stock in trade of specialists. The so-called professions survive by endlessly "processing" and talking about problems that they have neither the will nor the competence to solve. The doctor who is interested in disease but not in health is clearly in the same category with the conservationist who invests in the destruction of what he otherwise intends to preserve. They both have the comfort of "job security," but at the cost of ultimate futility.

One of the most troubling characteristics of the specialist mentality is its use of money as a kind of proxy, its willingness to transmute the powers and functions of life into money. "Time is money" is one of its axioms and the source of many evils—among them the waste of both time and money. Akin to the idea that time is money is the concept, less spoken but as commonly assumed, that we may be adequately represented by money. The giving of money has thus become our characteristic virtue.

But to give is not to do. The money is given *in lieu* of action, thought, care, time. And it is no remedy for the fragmentation of character and consciousness that is the consequence of specialization. At the simplest, most practical level, it would be difficult for most of us to give enough in donations to good causes to compensate for, much less remedy, the damage done by the money that is taken from us and used destructively by various agencies of the government and by the corporations that hold us in captive dependence on their products. More important, even if we *could* give enough to overbalance the official and corporate misuse of our money, we would still not solve the problem: the willingness to be represented by money involves a submission to the modern divisions of character and community. The remedy safeguards the disease.

This has become, to some extent at least, an argument against institutional solutions. Such solutions necessarily fail to solve the problems to which they are addressed because, by definition, they cannot consider the real causes. The only real, practical, hope-giving way to remedy the fragmentation that is the disease of the modern spirit is a small and humble way—a way that a government or agency or organization or institution will never think of, though a *person* may think of it: one must begin in one's own life the private solutions that can only *in turn* become public solutions.

If, for instance, one is aware of the abuses and extortions to which one is subjected as a modern consumer, then one may join an organization of consumers to lobby for consumer-protection legislation. But in joining a consumer organization, one defines oneself as a consumer *merely*, and a mere consumer is by definition a dependent, at the mercy of the manufacturer and the salesman. If the organization secures the desired legislation, then the consumer becomes the dependent not only of the manufacturer and salesman, but of the agency that enforces the law, and is at its mercy as well. The law enacted may be a good one, and the enforcers all honest and effective; even so, the consumer will understand that one result of his effort has been to increase the number of people of whom he must beware.

The consumer may proceed to organization and even to legislation by considering only his "rights." And most of the recent talk about consumer protection has had to do with the consumer's rights. Very little indeed has been said about the consumer's responsibilities. It may be that whereas one's rights may be advocated and even "served" by an organization, one's responsibilities cannot. It may be that when one hands one's responsibilities to an organization, one becomes by that divestiture irresponsible. It may be that responsibility is intransigently a personal matter—that a responsibility can be fulfilled or failed, but cannot be got rid of.

If a consumer begins to think and act in consideration of his responsibilities, then he vastly increases his capacities as a person. And he begins to be effective in a different way—a way that is smaller perhaps, and certainly less dramatic, but sounder, and able sooner or later to assume the force of example.

A responsible consumer would be a critical consumer, would refuse to purchase the less good. And he would be a moderate consumer; he would know his needs and would not purchase what he did not need; he would sort among his needs and study to reduce them. These things, of course, have been often said, though in our time they have not been said very loudly and have not been much heeded. In our time the rule among consumers has been to spend money recklessly. People whose governing habit is the relinquishment of power, competence, and responsibility, and whose characteristic suffering is the anxiety of futility, make excellent spenders. They are the ideal consumers. By inducing in them little panics of boredom, powerlessness, sexual failure, mortality, paranoia, they can be made to buy (or vote for) virtually anything that is "attractively packaged." The advertising industry is founded upon this principle.

What has not been often said, because it did not need to be said until fairly recent times, is that the responsible consumer must also be in some way a producer. Out of his own resources and skills, he must be equal to some of his own needs. The household that prepares its own meals in its own kitchen with some intelligent regard for nutritional value, and thus depends on the grocer only for selected raw materials, exercises an influence on the food industry that reaches from the store all the way back to the seedsman. The household that produces some or all of its own food will have a proportionately greater influence. The household that can provide some of its own pleasures will not be helplessly dependent on the entertainment industry, will influence it by not being helplessly dependent on it, and will not support it thoughtlessly out of boredom.

The responsible consumer thus escapes the limits of his own dissatisfaction. He can choose, and exert the influence of his choosing, because he has given himself choices. He is not confined to the negativity of his complaint. He influences the market by his freedom. This is no specialized act, but an act that is substantial and complex, both practically and morally. By making himself responsibly free, a person changes both his life and his surroundings.

It is possible, then, to perceive a critical difference between responsible consumers and consumers who are merely organized. The responsible consumer slips out of the consumer category altogether. He is a responsible consumer incidentally, almost inadvertently; he is a responsible consumer because he lives a responsible life.

The same distinction is to be perceived between organized conservationists and responsible conservationists. (A responsible consumer *is*, of course, a responsible conservationist.) The conservationists who are merely organized function as

specialists who have lost sight of basic connections. Conservation organizations hold stock in exploitive industries because
they have no clear perception of, and therefore fail to be responsible for, the connections between what they say and what
they do, what they desire and how they live.

The Sierra Club, for instance, defines itself by a slogan
which it prints on the flaps of its envelopes. Its aim, according to the slogan, is ". . . to explore, enjoy, and protect the
nation's scenic resources . . ." To some extent, the Club's
current concerns and attitudes belie this slogan. But there is
also a sense in which the slogan defines the limits of organized
conservation—some that have been self-imposed, others that
are implicit in the nature of organization.

The key word in the slogan is "scenic." As used here, the
word is a fossil. It is left over from a time when our comforts
and luxuries were accepted simply as the rewards of progress
to an ingenious, forward-looking people, when no threat was
perceived in urbanization and industrialization, and when conservation was therefore an activity oriented toward vacations.
It was "good to get out of the city" for a few weeks or weekends a year, and there was understandable concern that there
should remain pleasant places to go. Some of the more adventurous vacationers were even aware of places of unique beauty
that would be defaced if they were not set aside and protected.
These people were effective in their way and within their limits,
and they started the era of wilderness conservation. The results
will give us abundant reasons for gratitude as long as we have
sense enough to preserve them. But wilderness conservation
did little to prepare us either to understand or to oppose the
general mayhem of the all-outdoors that the industrial revolution has finally imposed upon us.

Wilderness conservation, we can now see, is specialized
conservation. Its specialization is memorialized, in the Sierra
Club's slogan, in the word "scenic." A scene is a place "as seen
by a viewer." It is a "view." The appreciator of a place perceived
as scenic is merely its observer, by implication both different
and distant or detached from it. The connoisseur of the scenic
has thus placed strict limitations both upon the sort of place he
is interested in and upon his relation to it.

But even if the slogan were made to read ". . . to explore, enjoy, and protect the nation's resources . . . ," the most critical concern would still be left out. For while conservationists are exploring, enjoying, and protecting the nation's resources, they are also *using* them. They are drawing their lives from the nation's resources, scenic and unscenic. If the resolve to explore, enjoy, and protect does not create a moral energy that will define and enforce responsible use, then organized conservation will prove ultimately futile. And this, again, will be a failure of character.

Although responsible use may be defined, advocated, and to some extent required by organizations, it cannot be implemented or enacted by them. It cannot be effectively enforced by them. The use of the world is finally a personal matter, and the world can be preserved in health only by the forbearance and care of a multitude of persons. That is, the possibility of the world's health will have to be defined in the characters of persons as clearly and as urgently as the possibility of personal "success" is now so defined. Organizations may promote this sort of forbearance and care, but they cannot provide it.

CHAPTER THREE

The Ecological Crisis
as a Crisis of Agriculture

ONE REASON THAT an organization cannot properly enact our relationship to the world is that an organization
cannot define that relationship except in general terms, and
no matter how general may be a person's attitude toward the
world, his impact upon it must become specific and tangible
at some point. Sooner or later in his behalf—whether he approves or understands or not—a strip-miner's bulldozer tears
into a mountainside, a stand of trees is clear-cut, a gully washes
through a cornfield.

The conservation movement has never resolved this dilemma. It has never faced it. Until very recently—until pollution and strip-mining became critical issues—conservationists
divided the country into that land which they wished to preserve and enjoy (the wilderness areas) and that which they con-
signed to use by *other* people. With the increase of pollution
and mining, their interest has become two-branched, to include, along with the pristine, the critically abused. At present
the issue of *use* is still in its beginning.

Because of this, the mentality of conservation is divided, and
disaster is implicit in its division. It is divided between its intentional protection of some places and some aspects of "the
environment" and its inadvertent destruction of others. It is
variously either vacation-oriented or crisis-oriented. For the
most part, it is not yet sensitive to the impact of daily living
upon the sources of daily life. The typical present-day conservationist will fight to preserve what he enjoys; he will fight
whatever directly threatens his health; he will oppose any ecological violence large or dramatic enough to attract his attention. But he has not yet worried much about the impact of his
own livelihood, habits, pleasures, or appetites. He has not, in
short, addressed himself to the problem of use. He does not
have a definition of his relationship to the world that is sufficiently elaborate and exact.

The problem is well defined in a letter I received from David Budbill of Wolcott, Vermont:

"What I've noticed around here with the militant ecology people (don't get me wrong, I, like you, consider myself one of them) is a syndrome I call the Terrarium View of the World: nature always at a distance, under glass.

"Down-country people come up here, buy a 30-acre meadow, then when you ask them what they plan to *do* with it, they look at you like you're some kind of war criminal and say, 'Why, nothing! We want to leave it *just* the way it is!' They think they're protecting the environment, even though they've forgotten, or never knew, that nature abhors a vacuum . . . and in a couple of years their meadow is full of hardhack and berries and young gray birch and red maple. Pretty soon they can't even walk through the brush it's so thick. They treat the land like any other possession, object, they own, set it aside, watch it, passively, not wanting to, nay! thinking it abhorrent to engage in a living relationship with it. . . .

"Another thing folks like this do is buy land and immediately post it (to protect the animals, or their investment, I guess) then go back home. . . . The old guy or the young guy who has always hunted deer on that piece is mad. The excuse for posting (protection) is a thinly disguised cover for the real notion which has to do with the possessive, capitalist ideas about property. I'm not opposed to private property, like it even, but the folks I'm talking about, in their posting, violate . . . a strong local tradition of free trespass. There are disadvantages to free trespass, abuses, we've suffered them, but what's good about it is it understands something about use and sharing. The upper-class eco-folks lack this understanding . . .

". . . we always, with our neighbor, pick apples in the fall off trees on a down-country owner's land. There is a feeling we have the *right* to do that, a feeling that the sin is not trespass, the sin is letting the apples go to waste.

"What I'm trying to get at is that in the environmental movement there are some ugly, elitist, class-struggle type things operating. The best example of this around here is the controversy over trailers. The Audubon types (I'm a member of Audubon) are fighting . . . terribly hard to zone trailers out of areas like this, put them in trailer parks or eliminate them altogether. Well, a trailer is the only living space a working man around here can afford. And if he, say, inherits 3 acres from a parent and wants to put a trailer on it, the eco-folks would like to say no, which is a dandy way to ghettoize the poor. There are so many elements of class struggle lying under the attitudes of a lot of environmentalists; it's scary. . . . Their view of the natural world is so delicate and precious, terrarium-like, picture-windowish. I know nature is precious and delicate. I also know it is incredibly tough and resilient, has unbelievable power to respond to and flourish with kindly use.

". . . I don't care about the landscape if I am to be excluded from it. Why should I? In Audubon magazine almost always the beautiful pictures are without man; the ugly ones with him. Such self hatred! I keep wanting to write to them and say, 'Look! my name is David Budbill and I belong to the chain of being too, as a participant not an observer (nature is not television!) and the question isn't to use or not to use but rather *how* to use.'"

The conservationist congratulates himself, on the one hand, for his awareness of the severity of human influence on the natural world. On the other hand, in his own contact with that world, he can think of nothing but to efface himself—to leave it *just* the way it is.

This is an important issue, and I want to be careful not to oversimplify it. What has to be acknowledged at the outset is that wilderness conservation is important and that it has its place in any conservation program, just as the wilderness has its place in human memory and culture. It seems likely to me that the concern for wilderness must stand at the apex of the conservation effort, just as it probably must stand at the apex of consciousness in any decent culture. There are several reasons for this:

1. Our biological roots as well as our cultural roots are in nature. We began in a world that was pristine, undiminished by anything we had done, and at various times in our history the unspoiled wilderness has again imposed itself, its charming and forbidding *invitation*, upon our consciousness. It is important that we should preserve this memory. We need places in reach of every community where children can imagine the prehistoric and the beginning of history: the unknown, the trackless, the first comers.

2. If we are to be properly humble in our use of the world, we need places that we do not use at all. We need the experience of leaving something alone. We need places that we forbear to change, or influence by our presence, or impose on even by our understanding; places that we accept as influences upon us, not the other way around, that we enter with the sense, the pleasure, of having nothing to do there; places that we must enter in a kind of cultural nakedness, without comforts or tools, to submit rather than to conquer. We need what other ages would have called sacred groves. We need groves, anyhow, that we would treat as if they were sacred—in order, perhaps, to perceive their sanctity.

3. We need wilderness as a standard of civilization and as a cultural model. Only by preserving areas where nature's processes are undisturbed can we preserve an accurate sense of the impact of civilization upon its natural sources. Only if we know how the land *was* can we tell how it *is*. Records, figures, statistics will not suffice; to know, in the true sense, is to see. We must see the difference—in rates of erosion, for instance, or in soil structure or fertility—in order to keep it as small as possible. As a cultural model, the wilderness is probably indispensable. Sir Albert Howard suggests that it is when he says that farmers should pattern the maintenance of their fields after the forest floor, for the forces of growth and the forces of decay are in balance there.

But we cannot hope—for reasons practical and humane, we cannot even wish—to preserve more than a small portion of the land in wilderness. Most of it we will have to use. The conservation mentality swings from self-righteous outrage to self-deprecation because it has neglected this issue. Its self-contradictions can only be reconciled—and the conservation impulse made to function as ubiquitously and variously as it needs to—by understanding, imagining, and living out the possibility of "kindly use." Only that can dissolve the boundaries that divide people from the land and its care, which together are the source of human life. There are many kinds of land use, but the one that is most widespread and in need of consideration is that of agriculture.

For us, the possibility of kindly use is weighted with problems. In the first place, this is not ultimately an organizational or institutional solution. Institutional solutions tend to narrow and simplify as they approach action. A large number of people can act together only by defining the point or the line on which their various interests converge. Organizations tend to move toward single objectives—a ruling, a vote, a law—and they find it relatively simple to cohere under acronyms and slogans.

But kindly use is a concept that of necessity broadens, becoming more complex and diverse, as it approaches action. The land is too various in its kinds, climates, conditions, declivities, aspects, and histories to conform to any generalized understanding or to prosper under generalized treatment. The use of land cannot be both general and kindly—just as the forms of good manners, generally applied (applied, that is, without consideration of differences), are experienced as indifference, bad manners. To treat every field, or every part of every field, with the same consideration is not farming but industry. Kindly use depends upon intimate knowledge, the most sensitive responsiveness and responsibility. As knowledge (hence, use) is generalized, essential values are destroyed. As the householder evolves into a consumer, the farm evolves into a factory—with results that are potentially calamitous for both.

The understanding of kindly use in agriculture must encompass both farm and household, for the mutuality of influence between them is profound. Once, of course, the idea of a farm included the idea of a household: an integral and major part of a farm's economy was the economy of its own household; the family that owned and worked the farm lived from it. But the farm also helped to feed other households in towns and cities. These households were dependent on the farms, but not passively so, for their dependence was limited in two ways. For one thing, the town or city household was itself often a producer of food: at one time town and city lots routinely included garden space and often included pens and buildings to accommodate milk cows, fattening hogs, and flocks of poultry. For another thing, the urban household carefully selected and prepared the food that it bought; the neighborhood shops were suppliers of kitchen raw materials to local households, of whose needs and tastes the shopkeepers had personal knowledge. The shopkeepers were under the direct influence and discipline of their customers' wants, which they had to supply honestly if they hoped to prosper. The household was therefore not merely a unit in the economy of food production; its members practiced essential productive skills. The consumers of food were also producers or processors of food, or both.

This collaboration of household and farm was never, in America, sufficiently thrifty or sufficiently careful of soil fertility. It is tempting to suppose that, given certain critical historical and cultural differences, they might have developed sufficient thrift and care. As it happened, however, the development went in the opposite direction. The collaborators purified their roles—the household became simply a house or residence, purely consumptive in its function; the farm ceased to be a place to live and a way of life and became a unit of production—and their once collaborative relationship became competitive. Between them the merchant, who had been only a supplier of raw materials, began to usurp the previous functions of both household and farm, becoming increasingly both a processor and producer. And so an enterprise that once had some susceptibility to qualitative standards—standards of personal taste and preference at one end and of good husbandry at the other—has come more and more under the influence of standards that are merely economic or quantitative. The consumer wants food to be as cheap as possible. The producer wants it to be as expensive as possible. Both want it to involve as little labor as possible. And so the standards of cheapness and convenience, which are irresistibly simplifying and therefore inevitably exploitive, have been substituted for the standard of health (of both people and land), which would enforce consideration of essential complexities.

Social fashion, delusion, and propaganda have combined to persuade the public that our agriculture is for the best of reasons the envy of the Modern World. American citizens are now ready to believe without question that it is entirely good, a grand accomplishment, that each American farmer now "feeds himself and 56 others." They are willing to hear that "96 percent of America's manpower is freed from food production" —without asking what it may have been "freed" *for*, or how many as a consequence have been "freed" from employment of any kind. The "climate of opinion" is now such that a recent assistant secretary of agriculture could condemn the principle of crop rotation without even an acknowledgment of the probable costs in soil depletion and erosion, and former Secretary of Agriculture Butz could say with approval that in 1974 "only 4 percent of all U.S. farms . . . produced almost 50 percent of all farm goods," without acknowledging the human—and, indeed, the agricultural—penalties.

What these men were praising—what such men have been praising for so long that the praise can be uttered without thought—is a disaster that is both agricultural and cultural: the generalization of the relationship between people and land. That one American farmer can now feed himself and fifty-six other people may be, within the narrow view of the specialist, a triumph of technology; by no stretch of reason can it be considered a triumph of agriculture or of culture. It has been made possible by the substitution of energy for knowledge, of methodology for care, of technology for morality. This "accomplishment" is not primarily the work of farmers—who have been, by and large, its victims—but of a collaboration of corporations, university specialists, and government agencies. It is therefore an agricultural development not motivated by agricultural aims or disciplines, but by the ambitions of merchants, industrialists, bureaucrats, and academic careerists. We should not be surprised to find that its effect on both the farmland and the farm people has been ruinous. It has divided all land into two kinds—that which permits the use of large equipment and that which does not. And it has divided all farmers into two kinds—those who have sufficient "business sense" and managerial ability to handle the large acreages necessary to finance large machines and those who do not.

Those lands that are too steep or stony or small-featured to be farmed with big equipment are increasingly not farmed at all, but are abandoned to weeds and bushes, often with the gullies of previous bad use unrepaired. That these lands can often be made highly productive with kindly use is simply of no interest; we now have neither the small technology nor the small economics nor the available work force necessary to make use of them. What might be the importance of these "marginal" lands, and of an agricultural technology and economy appropriate to them, in light of population growth is a question that the agriculture experts apparently would be embarrassed to consider, so entranced are they by the glamor of bigness.

As for the farm families who cannot "get bigger" and therefore have to "get out," they are apparently written off as a reasonable, quite ordinary, and altogether bearable expense. Former Secretary Butz could praise the business acumen of the new big-time American farmer ("In all likelihood he knows as much about financing and business accountability as his banker"), evidently without wondering what may be the *agricultural* import or effect of such knowledge, or if somewhere there might not be an excellent farmer who is *not* more acute, in a business way, than his banker. But this is the catch in our almost religious dependence on experts: Mr. Butz is a farm expert, and a farm expert is by definition not a farmer; he has changed sides. I have at hand fifteen speeches by Mr. Butz and his assistant secretaries, all of which praise the productivity—that is, the business success—of the American farmer, and none of which mentions any problem of land maintenance or any problem of the small farmer.

A sampling of quotations from one of these speeches—one made by former Assistant Secretary Richard E. Bell—will give the gist and the manner of official agricultural thinking:

". . . true agripower . . . generates agridollars through agricultural exports."

"True agripower is the capacity of less than 5 percent of America's population to feed itself and the remaining 95 percent with enough food left over to meet market demands of other nations and still provide food assistance for poor people throughout the world."

"Agripower should not be a political tool. Feeding people . . . is too serious a matter to be left to political manipulation."

"Once again growth in U.S. farm productivity . . . is on the rise. . . . We no longer have the acreage limitations which for so many years served to restrict grain and cotton production. . . ."

". . . the real measure [of agricultural strength] is productivity, combined with processing and marketing efficiency."

"Years ago, farm operations were highly diversified, but today, farmers are concentrating on fewer and much larger crop or livestock enterprises. Now, many one- or two-enterprise farms exist where there were formerly three to five enterprises.

"And with the spread of sophisticated machinery, farm sizes have expanded as their numbers have declined—stretching from an average 195 acres in the 1940s to about 390 in the 1970s.

"Specialization and growth are aided by the ready availability of purchased inputs and custom services."

"With additional income earned from exports, U.S. farmers are able to purchase more household appliances, farm equipment, building supplies, and other capital and consumer goods."

"Agridollars have gone a long way toward offsetting our petro-dollar drain."

"Less than 5 percent . . . of all grain moving between countries goes for food assistance."

"It is evident that U.S. agripower is a major force in the world's exchange of goods and services. Agripower is, unquestionably, an even greater force than petropower in man's survival in the future. Man can and has survived without petroleum, but he cannot live without food."

And that was the official line on agriculture during the Butz years. There is nothing in it that was not representative: the self-congratulation, the confusions of purpose, the complacency, the jargon, the sprains and ruptures of sense, the ignorance or ignoring of consequence, the social and economic prejudices ritualized in progressivist clichés. And nowhere that I have seen was the official line more complicated than this, more aware of costs or inequities or conflicts or problems.

We would do well to examine these statements in more detail, for they are not just the political policies of ex-officials. They represent very well the prevalent assumptions of agricultural bureaucrats, academicians, and businessmen.

"Agripower," it will be noted, is not measured by the fertility or health of the soil, or the health, wisdom, thrift, or stewardship of the farming community. It is measured by its ability to produce a marketable surplus, which "generates agridollars." It is to be measured by "productivity, combined with processing and marketing efficiency." The income from this increased production, we are told, is spent by farmers not for soil maintenance or improvement, water conservation, or erosion control, but for "purchased inputs": "household appliances, farm equipment, building supplies, and other capital and consumer goods." I do not mean that we should necessarily begrudge the farmer these purchases; I am only noticing that, to Mr. Bell, the farmer does not prosper to become a better farmer, but to become a bigger spender. The assistant secretary was applying to farming a standard of judgment that is economic, not agricultural. Farming is defined here purely to suit the purposes of a businessman.

Mr. Bell makes the benign assertion that this "agripower" feeds people, including the poor of the world, and is therefore too important to be put to political *use*. But when this subject is reverted to at the end of the speech, we find that "U.S. agripower" is a major force in world trade, a force intended to offset the "petropower" of other countries. And we have the assurance that, after all, "Less than 5 percent . . . of all grain moving between countries goes for food assistance" to the poor. (And, of course, all of this must be weighed against former Secretary Butz's avowal that "Food is a weapon.")*

Next we hear the routine self-congratulation of the department on the increase of productivity following the removal of production controls (the only agricultural problem acknowledged in any of these speeches). Our agriculture policy is now based on the principle of "full production"—an obscure notion that former Secretary Butz and his colleagues paraded

*A friend has pointed out the "incredible cheek" of calling food "agripower" and then warning against its use as "a political tool."

before their audiences like the True Cross. As businessmen and politicians, perhaps they did not know how strenuously agricultural production must be qualified by the restraints and disciplines of soil maintenance and conservation. Perhaps they did not know what "full production" means in present practice— present technology, methods, and economic urgencies having replaced those restraints and disciplines. In practice, however, "full production" means that on farm after farm fence rows, windbreaks, and waterways have been plowed, steep slopes put under cultivation, and soil stewardship generally neglected. It means that production is being paid for, not just with labor, money, and fuel, *but with land*.

But the most remarkable and significant part of Mr. Bell's speech is the one in which he applauds the most degenerative, dangerous, costly, and socially disruptive "achievements" of American agriculture: (1) "economy of size," which means the gathering of farmland into the ownership of fewer and fewer people—not farmers necessarily but an "agribusiness elite"— and the consequent dispossession of millions of small farmers and farm families; and (2) specialization, which means the abandonment of the ancient, proven principle of agricultural diversity—agricultural stability through diversity—with its attendant principles of mixed husbandry of plants and animals and crop rotation. It is now, for the first time, deemed provident and wise to put all the eggs in one basket.

The giveaway is in the curiously pleased-sounding statement that "specialization and growth are aided by the ready availability of purchased inputs. . . ." This betrays, for one thing, how far we have abandoned the old ideal that the farm should aim at economic independence; that is, it should be far more productive than consumptive, more a source than a consumer of material goods. This old ideal sought to preserve the farmer on the farm; that was of necessity its first objective. But it also sought to keep the source of food independent of any but agricultural means—an aim that ought to recommend itself, it would seem, to a fairly ordinary intelligence. Its desirability becomes altogether clear when one considers that a farm—given the *appropriate* technology, the recovery and return of organic wastes to the soil, an economy that is not exploitive, and a sufficient human work force—can achieve a high measure of economic independence.

None of this was clear to the intellectuals of the Department of Agriculture, and no doubt they were thereby saved a good deal of worry. For one of the "purchased inputs," on the "ready availability" of which our agriculture now absolutely depends, is petroleum—for which we are not only dependent on non-agricultural sources, but on other nations. That we should have an agriculture based as much on petroleum as on the soil—that we need petroleum exactly as much as we need food and must have it *before* we can eat—may seem absurd. It *is* absurd. It is nevertheless true. And it exposes the hollowness of Mr. Bell's contention that "Agripower is, unquestionably, an even greater force than petropower in man's survival in the future. Man can and has survived without petroleum, but he cannot live without food." The two powers are now clearly the same. That the two are not only interdependent, but competitive as well, suggests more forcibly than Mr. Butz's words that "Food is a weapon."

And so, far from the concerns of "kindly use" that alone can assure a permanent agriculture and a permanent food supply, the Department of Agriculture is lost in the paper clouds of "agribusiness," propagating statistical proofs of visibly ruinous agricultural practices. One can imagine the average American nodding over these "expert" reports and projections. Whether he is nodding because he agrees or because he is asleep does not matter; there is no difference.

Thus the estrangement of consumer and producer, their evolution from collaborators in food production to competitors in the food market, involves a process of oversimplification on both sides. The consumer withdraws from the problems of food production, hence becomes ignorant of them and often scornful of them; the producer no longer sees himself as intermediary between people and land—the people's representative on the land—and becomes interested only in production. The consumer eats worse, and the producer farms worse. And, in their estrangement, waste is institutionalized. Without regret, with less and less interest in the disciplines of thrift and conservation, with, in fact, the assumption that this is the way of the world, our present agriculture wastes topsoil, water, fossil fuel, and human energy—to name only the most noticeable things. Consumers participate "innocently" or ignorantly in all these farm wastes and add to them wastes that are urban or consumptive in nature: mainly all the materials and energy that go into unnecessary processing and packaging, as well as tons of organic matter (highly valuable—and certainly, in the long run, necessary—as fertilizer) that they flush down their drains or throw out as garbage.

What this means for conservationists is that, as consumers, they may be using—and abusing—more land by proxy than they are conserving by the intervention of their organizations. We now have more people using the land (that is, living from it) and fewer thinking about it than ever before. We are eating thoughtlessly, as no other entire society ever has been able to do. We are eating—drawing our lives out of our land—thoughtlessly. If we study carefully the implications of that, we will see that the agricultural crisis is not merely a matter of supply and demand to be remedied by some change of government policy or some technological "breakthrough." It is a crisis of culture.

CHAPTER FOUR

The Agricultural Crisis as a Crisis of Culture

IN MY BOYHOOD, Henry County, Kentucky, was not just a rural county, as it still is—it was a *farming* county. The farms were generally small. They were farmed by families who lived not only upon them, but within and *from* them. These families grew gardens. They produced their own meat, milk, and eggs. The farms were highly diversified. The main money crop was tobacco. But the farmers also grew corn, wheat, barley, oats, hay, and sorghum. Cattle, hogs, and sheep were all characteristically raised on the same farms. There were small dairies, the milking more often than not done by hand. Those were the farm products that might have been considered major. But there were also minor products, and one of the most important characteristics of that old economy was the existence of markets for minor products. In those days a farm family could easily market its surplus cream, eggs, old hens, and frying chickens. The power for field work was still furnished mainly by horses and mules. There was still a prevalent pride in workmanship, and thrift was still a forceful social ideal. The pride of most people was still in their homes, and their homes looked like it.

This was by no means a perfect society. Its people had often been violent and wasteful in their use of the land and of each other. Its present ills had already taken root in it. But I have spoken of its agricultural economy of a generation ago to suggest that there were also good qualities indigenous to it that might have been cultivated and built upon.

That they were not cultivated and built upon—that they were repudiated as the stuff of a hopelessly outmoded, unscientific way of life—is a tragic error on the part of the people themselves; and it is a work of monstrous ignorance and irresponsibility on the part of the experts and politicians, who have prescribed, encouraged, and applauded the disintegration of such farming communities all over the country.

In the decades since World War II the farms of Henry County have become increasingly mechanized. Though they are still comparatively diversified, they are less diversified than they used to be. The holdings are larger, the owners are fewer. The land is falling more and more into the hands of speculators and professional people from the cities, who—in spite of all the scientific agricultural miracles—still have much more money than farmers. Because of big technology and big economics, there is more abandoned land in the county than ever before. Many of the better farms are visibly deteriorating, for want of manpower and time and money to maintain them properly. The number of part-time farmers and ex-farmers increases every year. Our harvests depend more and more on the labor of old people and young children. The farm people live less and less from their own produce, more and more from what they buy. The best of them are more worried about money and more overworked than ever before. Among the people as a whole, the focus of interest has largely shifted from the household to the automobile; the ideals of workmanship and thrift have been replaced by the goals of leisure, comfort, and entertainment. For Henry County plays its full part in what Maurice Telleen calls "the world's first broad-based hedonism." The young people expect to leave as soon as they finish high school, and so they are without permanent interest; they are generally not interested in anything that cannot be reached by automobile on a good road. Few of the farmers' children will be able to afford to stay on the farm—perhaps even fewer will wish to do so, for it will cost too much, require too much work and worry, and it is hardly a fashionable ambition.

And nowhere now is there a market for minor produce: a bucket of cream, a hen, a few dozen eggs. One cannot sell milk from a few cows anymore; the law-required equipment is too expensive. Those markets were done away with in the name of sanitation—but, of course, to the enrichment of the large producers. We have always had to have "a good reason" for doing away with small operators, and in modern times the good reason has often been sanitation, for which there is apparently no small or cheap technology. Future historians will no doubt remark upon the inevitable association, with us, between sanitation and filthy lucre. And it is one of the miracles of science and hygiene that the germs that used to be in our food have been replaced by poisons.

In all this, few people whose testimony would have mattered have seen the connection between the "modernization" of agricultural techniques and the disintegration of the culture and the communities of farming—and the consequent disintegration of the structures of urban life. What we have called agricultural progress has, in fact, involved the forcible displacement of millions of people.

I remember, during the fifties, the outrage with which our political leaders spoke of the forced removal of the populations of villages in communist countries. I also remember that at the same time, in Washington, the word on farming was "Get big or get out"—a policy which is still in effect and which has taken an enormous toll. The only difference is that of method: the force used by the communists was military; with us, it has been economic—a "free market" in which the freest were the richest. The attitudes are equally cruel, and I believe that the results will prove equally damaging, not just to the concerns and values of the human spirit, but to the practicalities of survival.

And so those who could not get big have got out—not just in my community, but in farm communities all over the country. But as a social or economic goal, bigness is totalitarian; it establishes an inevitable tendency toward the *one* that will be the biggest of all. Many who got big to stay in are now being driven out by those who got bigger. The aim of bigness implies not one aim that is not socially and culturally destructive.

And this community-killing agriculture, with its monomania of bigness, is not primarily the work of farmers, though it has burgeoned on their weaknesses. It is the work of the institutions of agriculture: the university experts, the bureaucrats, and the "agribusinessmen," who have promoted so-called efficiency at the expense of community (and of real efficiency), and quantity at the expense of quality.

In 1973, 1,000 Kentucky dairies went out of business. They were the victims of policies by which we imported dairy products to compete with our own and exported so much grain as to cause a drastic rise in the price of feed. And, typically, an agriculture expert at the University of Kentucky, Dr. John Nicolai, was optimistic about this failure of 1,000 dairymen, whose cause he is supposedly being paid—partly with *their* tax money—to serve. They were inefficient producers, he said, and they needed to be eliminated.

He did not say—indeed, there was no indication that he had ever considered—what might be the limits of his criterion or his logic. Did he propose to applaud this process year after year until "biggest" and "most efficient" become synonymous with "only"? Did these dairymen have any value not subsumed under the heading of "efficiency"? And who benefited by their failure? Assuming that the benefit reached beyond the more "efficient" (that is, the bigger) producers to lower the cost of milk to consumers, do we then have a formula by which to determine how many consumer dollars are equal to the livelihood of one dairyman? Or is *any* degree of "efficiency" worth any cost? I do not think that this expert knows the answers. I do not think that he is under any pressure—scholarly, professional, moral, or otherwise—to ask the questions. This sort of regardlessness is invariably justified by pointing to the enormous productivity of American agriculture. But any abundance, in any amount, is illusory if it does not safeguard its producers, and in American agriculture it is now virtually the accepted rule that abundance will destroy its producers.

And along with the rest of society, the established agriculture has shifted its emphasis, and its interest, from quality to quantity, having failed to see that in the long run the two ideas are inseparable. To pursue quantity alone is to destroy those disciplines in the producer that are the only assurance of quantity. What is the effect on quantity of persuading a producer to produce an inferior product? What, in other words, is the relation of pride or craftsmanship to abundance? That is another question the "agribusinessmen" and their academic collaborators do not ask. They do not ask it because they are afraid of the answer: The preserver of abundance is excellence.

My point is that food is a cultural product; it cannot be produced by technology alone. Those agriculturists who think of the problems of food production solely in terms of technological innovation are oversimplifying both the practicalities of production and the network of meanings and values necessary to define, nurture, and preserve the practical motivations. That the discipline of agriculture should have been so divorced from other disciplines has its immediate cause in the compartmental structure of the universities, in which complementary, mutually sustaining and enriching disciplines are divided, according to "professions," into fragmented, one-eyed specialties. It is suggested, both by the organization of the universities and by the kind of thinking they foster, that farming shall be the responsibility only of the college of agriculture, that law shall be in the sole charge of the professors of law, that morality shall be taken care of by the philosophy department, reading by the English department, and so on. The same, of course, is true of government, which has become another way of institutionalizing the same fragmentation.

However, if we conceive of a culture as one body, which it is, we see that all of its disciplines are everybody's business, and that the proper university product is therefore not the whittled-down, isolated mentality of expertise, but a mind competent in all its concerns. To such a mind it would be clear that there are agricultural disciplines that have nothing to do with crop production, just as there are agricultural obligations that belong to people who are not farmers.

A culture is not a collection of relics or ornaments, but a practical necessity, and its corruption invokes calamity. A healthy culture is a communal order of memory, insight, value, work, conviviality, reverence, aspiration. It reveals the human necessities and the human limits. It clarifies our inescapable bonds to the earth and to each other. It assures that the necessary restraints are observed, that the necessary work is done, and that it is done well. A healthy *farm* culture can be based only upon familiarity and can grow only among a people soundly established upon the land; it nourishes and safe-guards a human intelligence of the earth that no amount of technology can satisfactorily replace. The growth of such a culture was once a strong possibility in the farm communities of this country. We now have only the sad remnants of those communities. If we allow another generation to pass without doing what is necessary to enhance and embolden the possibility now perishing with them, we will lose it altogether. And then we will not only invoke calamity—we will deserve it.

Several years ago I argued with a friend of mine that we might make money by marketing some inferior lambs. My friend thought for a minute and then he said, "I'm in the business of producing *good* lambs, and I'm not going to sell any other kind." He also said that he kept the weeds out of his crops for the same reason that he washed his face. The human race has survived by that attitude. It can survive *only* by that attitude—though the farmers who have it have not been much acknowledged or much rewarded.

Such an attitude does not come from technique or technology. It does not come from education; in more than two decades in universities I have rarely seen it. It does not come even from principle. It comes from a passion that is culturally prepared—a passion for excellence and order that is handed down to young people by older people whom they respect and love. When we destroy the possibility of that succession, we will have gone far toward destroying ourselves.

It is by the measure of culture, rather than economics or technology, that we can begin to reckon the nature and the cost of the country-to-city migration that has left our farmland in the hands of only five percent of the people. From a cultural point of view, the movement from the farm to the city involves a radical simplification of mind and of character.

A competent farmer is his own boss. He has learned the disciplines necessary to go ahead on his own, as required by economic obligation, loyalty to his place, pride in his work. His workdays require the use of long experience and practiced judgment, for the failures of which he knows that he will suffer. His days do not begin and end by rule, but in response to necessity, interest, and obligation. They are not measured by the clock, but by the task and his endurance; they last as long as necessary or as long as he can work. He has mastered intricate formal patterns in ordering his work within the overlapping cycles—human and natural, controllable and uncontrollable—of the life of a farm.

Such a man, upon moving to the city and taking a job in industry, becomes a specialized subordinate, dependent upon the authority and judgment of other people. His disciplines are no longer implicit in his own experience, assumptions, and values, but are imposed on him from the outside. For a complex responsibility he has substituted a simple dutifulness. The strict competences of independence, the formal mastery, the complexities of attitude and know-how necessary to life on the farm, which have been in the making in the race of farmers since before history, all are replaced by the knowledge of some fragmentary task that may be learned by rote in a little while.

Such a simplification of mind is easy. Given the pressure of economics and social fashion that has been behind it and the decline of values that has accompanied it, it may be said to have been gravity-powered. The reverse movement—a reverse movement *is* necessary, and some have undertaken it—is uphill, and it is difficult. It cannot be fully accomplished in a generation. It will probably require several generations—enough to establish complex local cultures with strong communal memories and traditions of care.

There seems to be a rule that we can simplify our minds and our culture only at the cost of an oppressive social and mechanical complexity. We can simplify our society—that is, make ourselves free—only by undertaking tasks of great mental and cultural complexity. Farming, the *best* farming, is a task that calls for this sort of complexity, both in the character of the farmer and in his culture. To simplify either one is to destroy it.

That is because the best farming requires a farmer—a husbandman, a nurturer—not a technician or businessman. A technician or a businessman, given the necessary abilities and ambitions, can be made in a little while, by training. A good farmer, on the other hand, is a cultural product; he is made by a sort of training, certainly, in what his time imposes or demands, but he is also made by generations of experience. This essential experience can only be accumulated, tested, preserved, handed down in settled households, friendships, and communities that are deliberately and carefully native to their own ground, in which the past has prepared the present and the present safeguards the future.

The concentration of the farmland into larger and larger holdings and fewer and fewer hands—with the consequent increase of overhead, debt, and dependence on machines—is thus a matter of complex significance, and its agricultural significance cannot be disentangled from its cultural significance. It *forces* a profound revolution in the farmer's mind: once his investment in land and machines is large enough, he must forsake the values of husbandry and assume those of finance and technology. Thenceforth his thinking is not determined by agricultural responsibility, but by financial accountability and the capacities of his machines. Where his money comes from becomes less important to him than where it is going. He is caught up in the drift of energy and interest away from the land. Production begins to override maintenance. The economy of money has infiltrated and subverted the economies of nature, energy, and the human spirit. The man himself has become a consumptive machine.

For some time now ecologists have been documenting the principle that "you can't do one thing"—which means that in a natural system whatever affects one thing ultimately affects everything. Everything in the Creation is related to everything else and dependent on everything else. The Creation is one. It is a uni-verse, a whole, the parts of which are all "turned into one."

A good agricultural system, which is to say a durable one, is similarly unified. In the 1940s, the great British agricultural scientist, Sir Albert Howard, published *An Agricultural Testament* and *The Soil and Health*, in which he argued against the influence in agriculture of "the laboratory hermit" who had substituted "that dreary precept [official organization] for the soul-shaking principle of that essential freedom needed by the seeker after truth." And Howard goes on to speak of the disruptiveness of official organization: "The natural universe, which is one, has been halved, quartered, fractioned. . . . Real organization always involves real responsibility: the official organization of research tries to retain power and avoid responsibility by sheltering behind groups of experts." Howard himself began as a laboratory hermit: "I could not take my own advice before offering it to other people." But he saw the significance of the "wide chasm between science in the laboratory and practice in the field." He devoted his life to bridging that chasm. His is the story of a fragmentary intelligence seeking both its own wholeness and that of the world. The aim that he finally realized in his books was to prepare the way "for treating the whole problem of health in soil, plant, animal, and man as one great subject." He unspecialized his vision, in other words, so as to see the necessary unity of the concerns of agriculture, as well as the convergence of these concerns with concerns of other kinds: biological, historical, medical, moral, and so on. He sought to establish agriculture upon the same unifying cycle that preserves health, fertility, and renewal in nature: the Wheel of Life (as he called it, borrowing the term from religion), by which "Death supersedes life and life rises again from what is dead and decayed."

It remains only to say what has often been said before—that the best human cultures also have this unity. Their concerns and enterprises are not fragmented, scattered out, at variance or in contention with one another. The people and their work and their country are members of each other and of the culture. If a culture is to hope for any considerable longevity, then the relationships within it must, in recognition of their interdependence, be predominantly cooperative rather than competitive. A people cannot live long at each other's expense or at the expense of their cultural birthright—just as an agriculture cannot live long at the expense of its soil or its work force, and just as in a natural system the competitions among species must be limited if all are to survive.

In any of these systems, cultural or agricultural or natural, when a species or group exceeds the principle of usufruct (literally, the "use of the fruit"), it puts itself in danger. Then, to use an economic metaphor, it is living off the principal rather than the interest. It has broken out of the system of nurture and has become exploitive; it is destroying what gave it life and what it depends upon to live. In all of these systems a fundamental principle must be the protection of the source: the seed, the food species, the soil, the breeding stock, the old and the wise, the keepers of memories, the records.

And just as competition must be strictly curbed within these systems, it must be strictly curbed *among* them. An agriculture cannot survive long at the expense of the natural systems that support it and that provide it with models. A culture cannot survive long at the expense either of its agricultural or of its natural sources. To live at the expense of the source of life is obviously suicidal. Though we have no choice but to live at the expense of other life, it is necessary to recognize the limits and dangers involved: past a certain point in a unified system, "other life" is our own.

The definitive relationships in the universe are thus not

competitive but interdependent. And from a human point of view they are analogical. We can build one system only within another. We can have agriculture only within nature, and culture only within agriculture. At certain critical points these systems have to conform with one another or destroy one another.

Under the discipline of unity, knowledge and morality come together. No longer can we have that paltry "objective" knowledge so prized by the academic specialists. To know anything at all becomes a moral predicament. Aware that there is no such thing as a specialized—or even an entirely limitable or controllable—effect, one becomes responsible for judgments as well as facts. Aware that as an agricultural scientist he had "one great subject," Sir Albert Howard could no longer ask, What can I do with what I know? without at the same time asking, How can I be responsible for what I know?

And it is within unity that we see the hideousness and destructiveness of the fragmentary—the kind of mind, for example, that can introduce a production machine to increase "efficiency" without troubling about its effect on workers, on the product, and on consumers; that can accept and even applaud the "obsolescence" of the small farm and not hesitate over the possible political and cultural effects; that can recommend continuous tillage of huge monocultures, with massive use of chemicals and no animal manure or humus, and worry not at all about the deterioration or loss of soil. For cultural patterns of responsible cooperation we have substituted this moral ignorance, which is the etiquette of agricultural "progress."

Dreams of the far future destiny of man were dragging up from its shallow and unquiet grave the old dream of Man as God. The very experience of the dissecting room and the pathological laboratory were breeding a conviction that the stifling of all deep-set repugnances was the first essential for progress.

C. S. LEWIS, *That Hideous Strength*

CHAPTER FIVE

Living in the Future:
The "Modern" Agricultural Ideal

THE DOMESTICATION OF ABSENCE

IT is impossible to divorce the question of what we do from the question of where we are—or, rather, where we think we are. That no sane creature befouls its own nest is accepted as generally true. What we conceive to be our nest, and where we think it is, are therefore questions of the greatest importance. Do we, for instance, carry on our work in our nest or do we only reside and get our mail there? Is our nest a place of consumption only or is it also a place of production? Is it the source of necessary goods, energies, and "services," or only their destination?

I have already spoken of the highly simplified role of the modern household with respect to the production and preparation of food: it has set itself increasingly aside from production and preparation and become more and more a place for the consumption of food produced and prepared elsewhere. But this setting aside of the nest or residence from the sources of life is more general and even more serious than that would indicate. The modern home, even more than the government and universities, has institutionalized the divisions and fragmentations of modern life.

With its array of gadgets and machines, all powered by energies that are destructive of land or air or water, and connected to work, market, school, recreation, etc., by gasoline engines, the modern home is a veritable factory of waste and destruction. It is the mainstay of the economy of money. But within the economies of energy and nature, it is a catastrophe. It takes in the world's goods and converts them into garbage, sewage, and noxious fumes—for none of which we have found a use.

And the modern household's direct destructiveness of the world bears a profound relation—as cause or effect or both—to the fundamental moral disconnections for which it also stands. It divorces us from the sources of our bodily life; as a people, we no longer know the earth we come from, have no respect for it, keep no responsibilities to it. And few who are acquainted with the young can doubt that the modern home has also failed as a place of instruction and that the schools are failing under the burden of that deeper failure.

But nowhere is the destructive influence of the modern home so great as in its remoteness from work. When people do not live where they work, they do not feel the effects of what they do. The people who make wars do not fight them. The people responsible for strip-mining, clear-cutting of forests, and other ruinations do not live where their senses will be offended or their homes or livelihoods or lives immediately threatened by the consequences. The people responsible for the various depredations of "agribusiness" do not live on farms. They—like many others of less wealth and power—live in ghettos of their own kind in homes full of "conveniences" which signify that all is well. In an automated kitchen, in a gleaming, odorless bathroom, in year-round air-conditioning, in color TV, in an easy chair, the world is redeemed. If what God made can be made by humans into *this*, then what can be wrong?

The modern home is so destructive, I think, because it is a generalization, a product of factory and fashion, an everyplace or a noplace. Modern houses, like airports, are extensions of each other; they do not vary much from one place to another. A person standing in a modern room anywhere might imagine himself anywhere else—much as he could if he shut his eyes. The modern house is not a response to its place, but rather to the affluence and social status of its owner. It is the first means by which the modern mentality imposes itself upon the world. The industrial conquistador, seated in his living room in the evening in front of his TV set, many miles from his work, can easily forget where he is and what he has done. He is everywhere or nowhere. Everything around him, everything on TV, tells him of his success: *his* comfort is the redemption of the world. His home is the emblem of his status, but it is not the center of his interest or of his consciousness. The history of our time has been to a considerable extent the movement of the center of consciousness away from home.

Once, some farmers, particularly in Europe, lived in their barns—and so were both at work and at home. Work and rest, work and pleasure, were continuous with each other, often not distinct from each other at all. Once, shopkeepers lived in, above, or behind their shops. Once, many people lived by "cottage industries"—home production. Once, households were producers and processors of food, centers of their own maintenance, adornment, and repair, places of instruction and amusement. People were born in these houses, and lived and worked and died in them. Such houses were not generalizations. Similar to each other in materials and design as they might have been, they nevertheless looked and felt and smelled different from each other because they were articulations of particular responses to their places and circumstances.

THE VAGRANT SOVEREIGN

The modern specialist and/or industrialist in his modern house can probably have no very clear sense of where he is. His sense of his whereabouts is abstract: he is in a certain "line" as signified by his profession, in a certain "bracket" as signified by his income, and in a certain "crowd" as signified by his house and his amusements. Where he is matters only in proportion to the number of other people's effects he has to put up with. Geography is defined for him by his house, his office, his commuting route, and the interiors of shopping centers, restaurants, and places of amusement—which is to say that his geography is artificial; he could be anywhere, and he usually is.

This generalized sense of worldly whereabouts is a reflection of another kind of bewilderment: this modern person does not know where he is morally either. He assumes, as he has clearly been taught to assume, that as a member of the human race he is sovereign in the universe. He assumes that there is nothing that he *can* do that he should not do, nothing that he *can* use that he should not use. His "success"—which at present is indisputable—is that he has escaped any order that might imply restraints or impose limits. He has, like the heroes of fantasy, left home—left behind all domestic ties and restraints —and gone out into the world to seek his fortune.

This mentality has been long in the making, and its rise evidently parallels the exploitation of the New World. Carl Sauer wrote: "The Modern Age began with the extension of royal absolution overseas. The crowns gave patents to individuals to discover, take possession, and govern islands or mainland, inhabited or uninhabited. The crown took to itself the title to land and people, first claimed for it by formal act. Thus Columbus planted the flag as he landed, the natives being bemused spectators. Thus Cabot without having sight of a native. Thus Juan de la Cosa entered on his map the flags of three nations. *The course of colonial empire began with disregard of native rights and persons* [my emphasis]. The Portuguese loaded the first cargo of black slaves when they reached the Bay of Arguin, and they did the same with Indians in New Foundland. Columbus estimated the prospects of slave trade when he landed in the West Indies. The Colonial idea as it took shape in the fifteenth century was untroubled by any concern other than to establish priority over other European nations."

Economic exploitation and competition as we now know them were thus established at the beginning of American history. Or perhaps it would be truer to say that they were established by the beginning of American history—for they do not seem to have risen so much out of theory or vision or desire or decree as out of newly opened distance and space. The new reaches of oceanic navigation, the discovery of new lands across what shortly before had been inconceivable distances, seem to have forced the European mind out of its old moral order. Those first discoverers carried the patents of their sovereigns, but they carried them into places altogether new to them, beyond what had been imagined, much less what had been culturally ordered. And so no matter the flags and pronouncements and the other trappings of fealty—the sovereignty that crossed the surf onto the shore of the New World was a new sovereignty of the human mind. What appeared to the eyes of the discoverers was not one of the orders of Creation that required respect or deference for its own sake. What they saw was a great concentration of "natural resources"—to be used according to purposes exterior to them. That some of those resources were human beings mattered not at all.

And so at the same time that they "discovered" America, these men invented the modern condition of being away from home. On the new shores the old orders of domesticity, respect, deference, humility fell away from them; they arrived contemptuous of whatever existed before their own coming, disdainful beyond contempt of native creatures or values or orders, ravenous for their own success. They began the era of absolute human sovereignty—which is to say the era of absolute human presumption. They invented us: the flag of Ferdinand and Isabella in the hand of Columbus on the shores of the Indies becomes Old Glory in the hand of Neil Armstrong on the moon. An infinitely greedy sovereign is afoot in the universe, staking his claims.

THE MANUFACTURED PARADISE

But our experience of sovereignty suggests that it becomes dangerous when it defines itself exclusively in terms of what is inferior to it, neglecting or ignoring what is superior to it. That is to say that sovereignty is a safe concept only when its place is symmetrically defined. Thus, once, the place of humans was thought to be above the animals and below the angels— between the natural and the divine. Then, by understanding and accepting that human place in the order of things, people could see that their privileges were limited and safeguarded by certain responsibilities. They could see, moreover, that only evil could be the result of the transgression of these limits: one could not escape the human condition except sinfully, by pride or by degradation.

The growth of what is called the Modern World has been, by turns, both the cause and the effect of the destruction of that old sense of universal order. The most characteristically modern behavior, or misbehavior, was made possible by a redefinition of humanity which allowed it to claim, not the sovereignty of its place, neither godly nor beastly, in the order of things, but rather an absolute sovereignty, placing the human will in charge of itself and of the universe.

And having thus usurped the whole Chain of Being, conceiving itself, in effect, both creature and creator, humanity set itself a goal that in those circumstances was fairly predictable: it would make an Earthly Paradise. This projected Paradise was no longer that of legend: the lost garden that might be rediscovered by some explorer or navigator. This new Paradise was to be invented and built by human intelligence and industry.

And by machines. For the agent of our escape from our place in the order of Creation, and of our godlike ambition to make a Paradise, was the machine—not only as instrument, but even more powerfully as metaphor. Once, the governing human metaphor was pastoral or agricultural, and it clarified, and so preserved in human care, the natural cycles of birth, growth, death, and decay. But modern humanity's governing metaphor is that of the machine. Having placed ourselves in charge of Creation, we began to mechanize both the Creation itself and our conception of it. We began to see the whole Creation merely as raw material, to be transformed by machines into a manufactured Paradise.

And so the machine did away with mystery on the one hand and multiplicity on the other. The Modern World would respect the Creation only insofar as it could be *used* by humans. Henceforth, by definition, by principle, we would be unable to leave anything as it was. The usable would be used; the useless would be sacrificed in the use of something else. By means of the machine metaphor we have eliminated any fear or awe or reverence or humility or delight or joy that might have restrained us in our use of the world. We have indeed learned to act as if our sovereignty were unlimited and as if our intelligence were equal to the universe. Our "success" is a catastrophic demonstration of our failure. The industrial Paradise is a fantasy in the minds of the privileged and the powerful; the reality is a shambles.

THE COLONIZATION OF THE FUTURE

The generalization of vital connections and the assumption of unlimited human sovereignty go a long way toward explaining the displacement of the modern mind. But they do not explain *how* it happened. It can be said that the motive has often been greed, but even that does not satisfy, for greed has always existed. It is necessary to account for a new intensity of greed—a greed newly empowered, under no constraint to see itself as evil, allied (so it believes) with a manifest destiny and the way of the world. There must have been, not just a shift of basic assumptions, not just a motive, but also some kind of vision or dream or psychic lure.

It has been, I think, the future. What has drawn the Modern World into being is a strange, almost occult yearning for the future. The modern mind longs for the future as the medieval mind longed for Heaven. The great aim of modern life has been to improve the future—or even just to *reach* the future, assuming that the future will inevitably be "better." One of the oddest terms of praise in our language is "futuristic." "Far out," as a term of universal approbation, is perhaps a lineal descendant. Such terms are used to identify the signs and landmarks that confirm that we are indeed on the right road to the future, that we are getting there, that at any moment we may at last arrive. And this is no elitist obsession; it is commonplace. Politicians understand very well the power of the promise to build a better or more prosperous or more secure future. Parents characteristically strive and sacrifice to make a better or more secure future for their children. Workers work toward a secure future in which they will retire and enjoy themselves. Our obsession with security is a measure of the power we have granted the future to hold over us.*

The future has been envisioned, dreamed, projected, painted for us by prophets of every kind: scientists, comic-book writers, novelists, philosophers, politicians, industrialists, professors. And, of course, by ourselves; the cult of the future has turned us all into prophets. The future is the time when science will have solved all our problems, gratified all our desires; when we will all live in perfect ease in an air-conditioned, fully automated womb; when all the work will be done by machines so sophisticated that they will not only clothe, house, and feed us, but think for us, play our games, paint our pictures, write our poems. It is the Earthly Paradise, the Other Shore, where all will be well. And if we are living for the future, then history is on our side—or so we are at liberty to think, for the needed proofs are never at hand. That there has for some time been growing a cult of dread of the future testifies not only to the

*The following sentences are from a recent oil company advertisement:

"We have always been a nation more interested in the promise of the future than in the events of the past.

"Here at Atlantic Richfield we see the future as an exciting time. The best of times."

innate silliness and frivolity of this vision, but to its power. The adoration of the future may be beginning to falter, but it is still dominant, still available and useful to the exploitive mind.

There is no aspect of our life as a people that is not now under the dominance of this industrial dream of the future-as-Paradise. All our implements—automobiles, tractors, kitchen utensils, etc.—have always been conceived by the modern mind as in a kind of progress or pilgrimage toward their future forms. The automobile-of-the-future, the kitchen-of-the-future, the classroom-of-the-future have long figured more actively in our imaginations, plans, and desires than whatever versions of these things we may currently have. We long ago gave up the wish to have things that were adequate or even excellent; we have preferred instead to have things that were up-to-date. But to be up-to-date is an ambition with built-in panic: our possessions cannot be up-to-date more than momentarily unless we can stop time—or somehow get ahead of it. The only possibility of satisfaction is to be driving *now* in one's *future* automobile.

It is no doubt impossible to live without thought of the future; hope and vision can live nowhere else. But the only possible guarantee of the future is responsible behavior in the present. When supposed future needs are used to justify misbehavior in the present, as is the tendency with us, then we are both perverting the present and diminishing the future. But the most prolific source of justifications for exploitive behavior has been the future. The exploitive mind characteristically puts itself in charge of the future. The future is a time that cannot conceivably be reached except by industrial progress and economic growth. The future, so full of material blessings, is nevertheless threatened with dire shortages of food, energy, and security unless we exploit the earth even more "freely," with greater speed and less caution. The obvious paradoxes involved in this—that we are using up future necessities in order to make a more abundant future; that final loss has been made a calculated strategy of annual gain—have so far been understood to no great effect. The great convenience of the future as a context of behavior is that nobody knows anything about it. No rational person can *see* how using up the topsoil or the fossil fuels as quickly as possible can provide greater security for the future; but if enough wealth and power can conjure up the audacity to *say* that it can, then sheer fantasy is given the force of truth; the future becomes reckonable as even the past has never been. It is as if the future is a newly discovered continent which the corporations are colonizing. They have made "redskins" of our descendants, holding them subject to alien values, while their land is plundered of anything that can be shipped home and sold.

Nowhere is the cult of the future stronger than in agriculture. One reason for this is that farming has been harder to industrialize than manufacturing, and when industrialization has come, it has not brought shorter hours or greater ease or less worry. A great deal of the strain of the industrial revolution has been borne by farmers, and so it has been fairly easy to secure their allegiance to the future, when more industrialization will supposedly bring a better farm economy. The industrialization of farming as we now have it is not something that farmers would have bought all in a piece; as a group they have been too traditional or conservative for that. Instead, it has been sold to them in stages, one implement at a time. The reduction of available manpower by each new machine created the need for a better machine or a different one. In the practical circumstances of the modern farm, the popular yearning for the future is directly felt as a yearning for relief from weariness and worry.

Another reason for the dominance of the future over agriculture is that projected rates of population growth have become the all-purpose threat and justifier of the apologists of the agricultural establishment. Millions are threatened with starvation —so the argument runs—therefore we must continue to farm in larger monocultures on larger holdings with fewer farmers, larger and more expensive machines, more chemicals. The hunger of these future millions is now the foundation of policy in the Department of Agriculture. Hunger supports the department's charitable rhetoric ("Feeding people . . . is too serious a matter to be left to political manipulation"), its realpolitik ("Food is a weapon"), and its self-justification ("true agripower . . . generates agridollars through agricultural exports"). How the future might be served by careless and destructive practices in the present is a question that is simply overridden by the brazen glibness of official optimism. If there is a food crisis, then, according to specialist logic, we must produce more food more carelessly than ever before. The energy crisis has been used, by the same logic, to justify the squandering of fuels.

LET THEM EAT THE FUTURE

As a sampler both of prevalent agricultural trends and official attitudes, as well as of the popular gullibility with which they have been received, one could not do better than an article entitled "The Revolution in American Agriculture" in the *National Geographic* of February 1970.

We should remember that in 1970 revolution was a controversial subject in America. We had spent the past half century in various stages of panic over the fact or the alleged possibility of communist revolution. And, during the decade just past, a good many of our people, mostly young, had begun to think of themselves as revolutionaries; some of them had even begun to *act* like revolutionaries. That the *National Geographic* could speak at such a time of an agricultural revolution could only indicate that a revolution of this kind, as opposed to a political revolution, was entirely acceptable to most Americans; it was simply part of the industrial revolution, which, after all, had become their way of life. That the industrial revolution, and the agricultural revolution along with it, had been real revolutions, surely the most powerful ever experienced, with real consequences, some of them political, and by no means all good—none of that mattered. The agricultural revolution, so far as the *National Geographic* and its readers were concerned, was a "good" revolution.

The author of the article, Mr. Jules B. Billard, is identified as a member of the magazine's senior editorial staff. But nowhere does he display the independence of judgment that one would expect either of a geographer or an editor. During most of the article he is in the grip of the ignorant awe, the greenhorn's ecstasy, that has been as necessary to this revolution as the ball bearing. During his encounters with the various manifestations of agricultural progress, Mr. Billard "marveled" twice; he was "staggered," "fascinated," "astounded," and "jolted" once each; he experienced two "jolting awakenings," the second more jolting than the first; and once his "mind churned."

The following is an inventory of Mr. Billard's revolutionary wonders:

"You can have strawberries in January, fresh oranges and lettuce the year round."

"Of the 6,000 to 8,000 items in the typical supermarket, 40 percent were not there a dozen years ago."

". . . in a single lifetime United States agriculture has advanced more than in all the preceding millenniums of man's labor on the land."

". . . I watched a factory-on-wheels move down celery rows . . . doing the work of forty men."

"I handled tomatoes bred for machine harvesting."

"I learned about heating cables buried underground to warm the soil so asparagus can grow in December . . ."

"Because only one person in 43 is needed to produce food, others can become doctors, teachers, shoemakers, janitors . . ." [This is a quote from former Agriculture Secretary Clifford Hardin.]

"Today 90 percent of the [California] tomato crop is picked mechanically."

". . . an incredible parade of machines are at work today on U.S. farms: acre-eaters . . . self-propelled combines that permit a man to ride in an air-conditioned cab to harvest a crop of corn that used to take a crew of 80 hands. Monster road-building machinery to level terraces or shape rice fields. Helicopters to spray cucumber fields. In all such a host of devices that today U.S. farmers are investing eight times as much capital as they did thirty years ago."

"Automated feeders, waterers, ventilators, and other labor savers make it possible for one man to take care of 100,000 broilers . . ."

"Block-long buildings, each housing 90,000 White Leghorns, cooped five birds to a 16-by-18 inch cage . . ."

". . . meatless dishes tasting like chicken, beef, or ham."

These accomplishments sort themselves readily into two categories: the frivolous and the problematic. The frivolity of strawberries in January, asparagus in December, and wheat or soybean products that taste like chicken is simply never acknowledged. Nor are the implications of the enormous increase of "items" in the supermarkets. By the values of gee-whiz journalism *any* increase is marvelous.

Nor is there any acknowledgment of the influence of "monster" technology ("acre-eaters") on the soil, the produce, the farm communities, and the lives and characters of farmers.

It is harder to ignore the enormous increase of indebtedness and overhead that has accompanied the enlargement of farm technology. Mr. Billard quotes an Iowa banker: "In 1920 . . . $5,000 was a big loan, and people hesitated to borrow. Now a $40,000 loan is commonplace, and having mortgage after mortgage is an accepted thing. I occasionally wonder whether the average farmer will ever get out of debt." The article gives examples of the enormous acreages and costs involved in several up-to-date operations. But these figures are simply left lying; in Mr. Billard's mind they evidently stand for nothing except the bigness of modern agriculture—which he approves of, so far as one can tell, because it amazes him. The Iowa banker's statement, doubtful as it may seem out of context, is made *in praise* of credit. Nowhere is there a question of the advisability of basing so large an enterprise on credit, or of the influence of routine indebtedness on a people's character. Nowhere is there a suspicion that there might be any worth in the old rural virtues of solvency and thrift.

The economic and moral uncertainty of living on credit is evidently—and typically—thought to be compensated by an improved standard of living: "Today [the farm wife is] as likely to be mini-skirted as her city sister, and as likely to own a dishwasher or self-cleaning oven or color television set. And her husband, who drives a tractor with an automatic transmission and uses power tools to eliminate back-straining labor, is as likely to have gone to college as his town cousin." That this standard of living is entirely material and entirely urban is characteristic of the prejudices that underlie the article.

The industrialization of animal husbandry is likewise seriously oversimplified. In addition to the ethical questions involved, the use of animals as machines—penning them in feed lots and cages—creates an enormous pollution problem. Mr. Billard acknowledges that this problem exists. He even cites a dubious solution: spreading the manure of 20,000 cattle on the pastures of a 320-acre farm which also contains the feed lots, a drainage pond, and a feed mill. But he also notes that in 1968 American farmers spread "nearly forty million tons" of chemical fertilizers, or "260 pounds for each acre under cultivation." The manure problem is separated from these figures on fertilizer consumption by fourteen pages. The dependence of our farmers on chemical fertilizers is not seen as a problem, and so the connection is missed. Mr. Billard forgot, or he never knew, that once plants and animals were raised together on the same farms—which therefore neither produced unmanageable surpluses of manure, to be wasted and to pollute the water supply, nor depended on such quantities of commercial fertilizer. The genius of American farm experts is very well demonstrated here: they can take a solution and divide it neatly into two problems.

That the agricultural revolution has displaced large numbers of people and put large numbers out of work is also acknowledged by Mr. Billard. But like the society as a whole, he has no trouble accepting this as part of the inevitable cost of progress. Lest anyone should become concerned about it, he includes early in his article a formula that makes it all right: "'Machines do replace labor,' G. E. VandenBerg told me . . . in his office at the USDA's Agricultural Research Center in Beltsville, Maryland. 'However, it is the scarcity of labor that really spurs adoption of machines.'"

Nevertheless, twenty-four pages later, Mr. Billard is saying:

"Squeezed between higher operating costs and what he gets for his produce, the man on the farm must become more efficient or give up." So apparently there is a problem after all. But another "agribusiness" formula is immediately invoked to assuage the moral discomfort: when all else fails to disguise the indifference of official agriculture to all human concerns, one can always fall back on efficiency. And so it appears that the failure of so many small farmers over so many years is really a kind of justice: it is their own fault; they ought to have been more efficient; if they had to get bigger in order to be more efficient, then they ought to have got bigger.

But suppose there is no room to get bigger unless somebody is driven out. In that case, one must have recourse to the law of compensation. This is the favorite law of the exploiter. It holds that for every loss there is a gain that is opposite and at least equal. This law is good fortune itself, for it means that you can do no wrong. Mr. Billard is an ardent observer of the law of compensation. "How many have given up," he writes, "can be seen in such figures as these: In 1910 our farm population accounted for a third of the U.S. total. By 1969 it was a mere twentieth. People leave rural areas at an average rate of 650,000 a year; many drift into cities where they join past migrants in the ghettos—to become added tinder for the riots that can be labeled one of the social consequences of the agricultural revolution." And he goes on: "When people leave the farm, rural communities . . . likewise wither away."

Here surely is cause for mourning: a forced migration of people greater than any in history, the foretelling of riots in the cities and the failure of human community in the country. But no. On the contrary: "Not all small towns are dying. The smog and the traffic and the social unrest of megalopolis prompts a second look at advantages of living in smaller communities. Industry, freed by jet planes and superhighways from dependence on nearby markets, shifts its plants away from cities. Employees are drawn by such appeals as being able, ten minutes after leaving work, to be out on the golf course, or roaming the woods with gun and dog, or watching kids and crops grow in a handful of acres a man can call his own."

Thus, if country people are forced to move into the city, that is made up for, according to Mr. Billard, by the movement of city people, and the city itself, into the country. But that only *looks* like a balanced equation. The people who move into the city and those who move out into the country are hardly the same people. The country community (of "inefficient" and therefore socially negligible people) is broken up, to be replaced by an influx of urban people who (however "efficient") have no economic or cultural ties to the land and are not a community. In this exchange we lose country people, we lose community, and we lose land. And we also lose the "inner city," which is abandoned to those who cannot perform "efficiently" either in the city or in the country.

But probably the most interesting feature of Mr. Billard's account of this exchange is the importance he attaches to "watching kids and crops grow on a handful of acres a man can call his own." Why, one wonders, does this feeling assert itself when the handful of acres is owned by an urban migrant, but not when they are owned by a farmer? How, rationally, can one hold the small farm in contempt as the living of a farm family and then sentimentalize over it as the "country place" or hobby of an executive? It cannot be done unless it is assumed that an executive is more deserving of a small farm because, as an urban or a professional person, he is superior to a farmer.

The callousness and smugness of this attitude is fully displayed in the caption to two pictures, one showing several members of a black family in their house and the other showing a modern cotton-picking machine at work. The caption is headlined dramatically: "When machines displace people"—and it reads: "Through the years Ruth Anderson's husband had worked the sweltering cotton fields around Isola, Mississippi. In late spring Ed Anderson chopped cotton . . . Summers he picked the cotton at $2.50 a hundred pounds. Between having her nine children, four of whom she tends above in the family's one-room shanty, Mrs. Anderson worked beside her husband. During picking season they brought home as much as $10 a day, and they got by.

"Then onto the fields rolled machines . . . that harvested as much in a day as could 80 men. Picking jobs vanished. Herbicides came on the market to kill weeds; they killed the chopping, too.

"Lacking a skill for steady work, the Andersons joined the

hapless millions of rural refugees who, uprooted by mechanized farming, often drift to big cities seeking jobs.

"To help stem this flow, civil-rights groups, foundations, and
the National Council of Churches support a self-help community called Freedom City . . ."

So much for the Andersons. We are evidently expected to
assume that their plight, and the plight of millions like them,
is exactly offset by Freedom City, which is trying "to help stem
this flow"—as if the flow can be stemmed until every last "inefficient" field worker has entered a ghetto and gone on welfare.
And then we are asked to turn away and marvel at the big
machine that can do the work of eighty men, whose working
conditions were, after all, "sweltering," and whose getting-by
economy was out of fashion, if not slightly contemptible.

We are shown another farm family—three generations of
them—at the dinner table on their 130-acre Long Island farm,
part of which they have owned since 1737. These people, too,
"stand at a crossroads: Either they mechanize and expand, or
rising costs, high taxes, and big farm competition will drive
them from the land." There is not a word or implication of so
much as a doubt about the economic conditions that constitute this "crossroads," not the smallest curiosity as to what may
be the cost. Nearly two-and-a-half centuries of family history
on the same farm amounts simply to nothing if it can't pay the
taxes. The fate of this family is offered as merely interesting, a
kind of journal-fodder.

What excuses this human waste, this destruction of preserving traditions and associations, this moral indifference? It is the
future—the future as both threat and lure—the secular Hell
and Heaven of the enraptured booster.

Early in his article Mr. Billard refers to a possibility that is certainly grave, certainly to be taken seriously, but which has nevertheless become the routine curtain-raiser of "agribusiness"
propagandists, who use it not for the rigorous self-evaluation
that it requires, but shamelessly and tirelessly to justify themselves. They are fanatical believers in themselves, and like all
fanatics they need an apocalypse—some ultimate bugaboo to
shove into the face of doubt. What we have here is everybody's
worry, but the farm experts and agribusinessmen would like us

to leave it to them: "Earth's numbers now stand at 3.6 billion, and could double in 35 years. This . . . raises the specter of a famine more catastrophic than the world has ever seen."

Of course it does. And that means that we should be at work overhauling all our assumptions about ourselves and what we have done and what we are capable of doing, all our attitudes toward life and its complex sources, all our resources of technique and technology. If we are heading toward apocalypse, then obviously we must undertake an ordeal of preparation. We must cleanse ourselves of slovenliness, laziness, and waste. We must learn to discipline ourselves, to restrain ourselves, to need less, to care more for the needs of others. We must understand what the health of the earth requires, and we must put that before all other needs. If a catastrophic famine is possible, then let us undertake the labors of wisdom and make the necessary sacrifices of luxury and comfort.

But, according to Mr. Billard, this is not for ordinary people to worry about. The agriculture experts, industrialists, and scientists are going to take care of it: "The spread of modern agriculture can help assure the underdeveloped two-thirds of the world the freedom from hunger it gives the economically advanced one-third."

And by the end of his article Mr. Billard has entered into the glory of the true future-rapture. He talked to "Dr. Irving, of the Department of Agriculture," who said of the future: "Agriculture will be highly specialized. . . . Farms in one area will concentrate on growing oranges, those in another area tomatoes, in another potatoes—capitalizing on the competitive advantage soil or climate gives for a particular crop.

"Fields will be larger, with fewer trees, hedges, and roadways. Machines will be bigger and more powerful. . . . They'll be automated, even radio-controlled, with closed circuit TV to let an operator sitting on a front porch monitor what is going on. . . .

"Weather control may tame hailstorm and tornado dangers. . . . Atomic energy may supply power to level hills or provide irrigation water from the sea."

It was at this point that Mr. Billard's "mind churned with the implications of such developments building on the progress of the past." Gone are the fears of famine. Gone are any thoughts of displaced small farmers and farm workers, or of the threat of riots in the cities. Mr. Billard has risen right over apocalypse into Heaven itself. He ends by quoting triumphantly a remark by a Brazilian official: "'We are concerned about the future of agriculture in Brazil. . . . In your country, you *are* in the future.'"

And so, of course, are the Andersons, with their nine children, their "one-room shanty," and no job—which ought to be reassuring to the people in the "underdeveloped two-thirds of the world," who are still trapped back there in the present.

The final two pages of Mr. Billard's article carry an "artist's conception" of the agricultural future, which is a veritable paradigm of the agribusiness ambition. The caption reads as follows:

"Farm of the future: Grainfields stretch like fairways and cattle pens resemble high-rise apartments in a farm of the early 21st century, as portrayed by artist David Meltzer *with the guidance of U.S. Department of Agriculture specialists* [my emphasis].

"Attached to a modernistic farm house, a bubble-topped control tower hums with a computer, weather reports, and a farm-price ticker tape. A remote-controlled tiller-combine glides across a 10-mile-long wheat field on tracks that keep the heavy machine from compacting the soil. Threshed grain, funneled into a pneumatic tube beside the field, flows into storage elevators rising close to a distant city. The same machine that cuts the grain prepares the land for another crop. A similar device waters neighboring strips of soybeans as a jet-powered helicopter sprays insecticides.

"Across a service road, conical mills blend feed for beef cattle, fattening in multilevel pens that conserve ground space. Tubes carry the feed to be mechanically distributed. A central elevator transports the cattle up and down, while a tubular side drain flushes wastes to be broken down for fertilizer. Beside the farther pen, a processing plant packs beef into cylinders for shipment to market by helicopter and monorail. Illuminated plastic domes provide controlled environments for growing high-value crops such as strawberries, tomatoes, and celery. Near a distant lake and recreation area, a pumping plant supplies water for the vast operation."

THE ORGANIZATION OF DISORDER

The cooperation of Department of Agriculture specialists in this visualization of a completely industrialized agriculture-of-the-future makes it as official, it would seem, as any vision of the future could be. And that this sort of thing is not an isolated aberration of overexcited journalism, but a confirmed habit—even the theoretical context—of agricultural expertise, is suggested by an article in the October 1974 issue of the *American Farmer* (voice of the American Farm Bureau Federation). This article is about a "dream farm" of 2076 A.D.—a model constructed by a group of South Dakota State University agricultural engineering students. This farm of the future is described as follows:

"The farm of 9 square miles will use only about 1,800 acres, less than one-fourth of which is for production.* The remainder will be a buffer or 'relaxed' zone for recreation, wildlife, and living under the 'blending with human values' aspect of the overall planning.

"Livestock will be housed (and products processed) in a 15-story, 150′ × 200′ building. It will also contain power facilities, administrative headquarters, veterinary facilities, repair shops, refrigeration and packaging units, storage, research labs, water and waste treatment facilities. At capacity, the high-rise building will house 2,500 feeder cattle, 600 cow-calf units, 500 dairy cattle, 2,500 sheep, 6,750 finishing hogs, space for 150 sows and litters, 1,000 turkeys, and 15,000 chickens.

"Crops will be grown year around under plastic covers that provide precise climate control in three circular fields each a mile in diameter. At any given time, regardless of weather, one field or crop will be in the planting stage, another in the growing stage, and the third in the harvesting stage. Exceptionally high yields mean that only a fourth of the total 5,760-acre farm area would be needed for agricultural production.

"Only a half-inch of water will be needed for each crop. That's because evapotranspiration from growing plants would be recycled under massive, permanent plastic enclosures . . .

---

*This sentence is hard to understand. The acreage to be used for production is evidently one-fourth of the nine square miles, not of the 1800 acres. But 1800 acres is *more* than one-fourth of nine square miles, not less.

"Underground magnetic patterns, arranged to fit crop or machine, will attract specially-treated seed blasted from overhead tubes in the enclosures.

"If tillage is needed, it will be done by electromagnetic waves. Air-supported, remotely controlled machines will harvest entire plants because by 2076 A.D. the students believe multiple uses will be needed and found for most crops.

"'Trickle' irrigation is to be electronically monitored to provide subsurface moisture automatically whenever needed.

"Recycling human, animal and crop wastes will be a key to the operation of the farm. Carbon dioxide from the respiration of livestock is to be piped into the circular enclosures for use by crops in exchange for the oxygen transpired by crops for use by livestock.

"Weed control is not anticipated as a problem because weeds would be eradicated under the field covers."

Soon after reading this article I wrote to Dr. Milo A. Hellickson, Associate Professor in Agricultural Engineering at South Dakota State University. My letter asked the following questions:

"1. Was any attention given to the possible social and economic effects of the projected innovations? Was it envisioned that this sort of farm would entirely replace the relatively small owner-operated farm? What would be its effect upon population patterns? Would it make food more or less expensive? What would be the energy requirements of such an operation, and what would be the sources of the required energy?

"2. What political consequences were anticipated? What, for instance, would be the impact . . . upon the doctrines of personal liberty and private property?

"3. What would be the effect upon the consumer? Would there be more or less choice of variety and quality?

"4. What would be the effect on the environment? For instance, roofing so large an acreage would present an unprecedented drainage problem. What did your students propose to do with the runoff? Would such a farm be built only in a desert area, or would it be feasible in an area of abundant rain fall?"

I received a prompt and very cordial reply from Dr. Hellickson, who responded to my questions as follows:

"Attention was given to the social and economic effect of the innovations. Essentially we feel that these developments would most likely fit individually into various farming operations and would not necessarily be all concentrated into one farmstead. As a matter of convenience in construction and so as not to alienate any particular phase of the agricultural industry, the model is constructed incorporating all the areas. Therefore, it would be equally possible in the future to maintain the smaller owner-operated farm and this then would cause little change in the distribution of the population. Specifically, we are thinking of energy captured from the sun, solar energy, as the sole energy source. I wouldn't even attempt to make a guess concerning expense, since this is such an area of dynamic change.

"Hopefully the above paragraph also answers question two. We are in no way advocating the elimination of the free enterprise system or the reduction of privately owned land.

"As per question three, I would see little change in the variety or quality of products available. . . . If anything, quality might be improved through the reduction or elimination of disease and through better handling systems.

"Hopefully this system would improve the environment by eliminating air pollution from the livestock building and also eliminating erosion from the cropped area. Runoff from the roof areas is proposed to be used as the water source for the irrigation system and for live-stock and humans. Naturally, adequate facilities must be included to handle unusually large rainfalls."

There is no quarreling with the professed aim of either of these farms-of-the-future, which is an abundance of food. And they have other aspects that are praiseworthy: the conversion of wastes into fertilizer and the reliance of the South Dakota model on solar energy. But we are still left with the question of what will be the costs, not just of construction and materials, which would be passed on to consumers in the price of food, but costs of other kinds: social, cultural, political, nutritional, etc. And we still must ask if there may not be less costly ways to achieve the same ends.

The issue that is raised most directly by these farms-of-the-future is that of control. The ambition underlying these model farms is that of total control—a totally controlled agricultural environment. Nowhere is the essential totalitarianism and the essential weakness of the specialist mind more clearly displayed than in this ambition. Confronted with the living substance of farming—the complexly, even mysteriously interrelated lives on which it depends, from the microorganisms in the soil to the human consumers—the agriculture specialist can think only of subjecting it to total control, of turning it into a machine.

But total human control is just as impossible now as it ever was—or so the available evidence constrains one to believe. Nothing, for instance, could be more organized than one of our large cities, with its geometric streets, its numbered houses, its numbered citizens, its charted routes and zones, its great numbers of police and other functionaries charged to keep order—and yet nothing could be more chaotic than one of these same cities during rush hour or after dark or during a riot or a garbage collectors' strike. In the modern city unprecedented organization and unprecedented disorder exist side by side; one could argue that they have a symbiotic relationship, that they feed and thrive upon each other. It is not difficult to think of any number of such examples in government, education, industry, medicine, agriculture—wherever the specialist has come with his controls.

The reason would seem to be that the specialist and the idea of total control also have a symbiotic relationship, that neither can exist without the other. The specialist puts himself in charge of *one* possibility. By leaving out all other possibilities, he enfranchises his little fiction of total control. Leaving out all the "non-functional" or otherwise undesirable possibilities, he makes a rigid, exclusive boundary within which absolute control becomes, if not possible, at least conceivable.

But what the specialist never considers is that such a boundary is, in itself, profoundly disruptive. Its first disruption is in his mind, for having enclosed the possibility of control that is within his competence to imagine and desire, he becomes the enemy of all other possibilities. And, secondly, having chosen the possibility of total control within a small and highly simplified enclosure, he simply abandons the rest, leaves it totally *out* of control; that is, he forsakes or even repudiates the complex, partly mysterious patterns of interdependence and cooperation, controllable only within limits, by which human culture joins itself to its sources in the natural world.

This attempt at total control is an invitation to disorder. And the rule seems to be that the more rigid and exclusive is the specialist's boundary, and the stricter the control within it, the more disorder rages around it. One can make a greenhouse and grow summer vegetables in the wintertime, but in doing so one creates a vulnerability to the weather and a possibility of failure where none existed before. The control by which a tomato plant lives through January is much more problematic than the natural order by which an oak tree or a titmouse lives through January. The patterns of cooperation are safer than the mechanisms of exclusion, even though they lack the illusory safety of "control."

Because of his dependence on boundaries and controls, the genre or mode of the specialist is the "model." The necessary context of a model is the future. The qualifications of the present, of *living*, do not affect it, nor do the non-functional or the undesirable. It is remote even from probable difficulties of the future. Thus the language of the article in the *American Farmer*, having to do with the inward workings of the South Dakota State model, is confident and for the most part it is exact. But Dr. Hellickson's responses to my questions about its *influence* are tentative, conjectural, and hopeful. The model perfectly empowers the machine metaphor: only the "working parts" need be admitted. Therefore, if one is going to make a "model farm," one must give it a boundary, if possible a roof, that will keep out whatever does not "work." Weeds, insects, diseases do not work; leave them out. The weather works only sometimes, or on the average; leave the weather out. The work can be done by machines; leave the people out. But chemicals and drugs, no matter how dangerous, *do* work; they are part of the boundary, so they can be let in.

It may be a bit startling at this point to realize that what has been left out of this enclosure is health. As soon as pests, parasites, diseases, climatic fluctuations and extremes are left out, resistance to these things is also left out; and this resistance, in the soil and in the lives that come from the soil, is what we call health. And so for total control we have given up health—which is also a kind of control, safer by far than a plastic roof, but never total.

The model is an ideal and is surely meant to function as an ideal. But it is a *mechanical* ideal, and an exclusive one. Furthermore, its connections with the past and the present are severed; always implicit in a model is the idea of *replacing* what has survived of the past, what exists in the present. These characteristics divide the model radically from ideals of the more usual sort. Such ideals as honesty or generosity or gentleness or symmetry do indeed have an influence on the future, but we recognize them from what we have known of them in the past and from what they require of us in the present. Like health, they are required to survive among us in the presence of what they must resist; they survive in culture, in community, and in the characters of people. They are known and valued not because they have been modeled, but because they have been *exemplified*. The specialist, on the other hand, is interested only in the model, never in the example. He is interested in the future of farming, not in its history.

That is why the influence of his work does not interest him; if he puts a machine into the field to "save labor," he does not ask the fate of the replaced people.* He is working "in the future," which puts him at liberty simply to leave out whatever is displaced or whatever does not work. That is why there are no more people in these scenes of future farms than in the landscape photographs in conservation magazines; neither the agriculture specialist nor the conservation specialist has any idea where people belong in the order of things. Neither can conceive of a domesticated or a humane landscape. People are complex, contradictory, unpredictable; they are perceived by the specialist as a kind of litter, pollutants of pure nature on the one hand and of pure technology, total control, on the other.

WHERE ARE THE PEOPLE?

By the power of a model, the specialist turns the future into a greenhouse of fantasies. But the model also empowers the fantasies with influence over present life—and, of course, over future life. And so, considering these model farms, one asks,

---

*This is the flaw in the doctrine of labor-saving. Labor-saving machines are *supposed* to make jobs easier. In fact, they destroy jobs. Instead of ameliorating work, they replace workers. What makes work easier and more pleasant without reducing employment is collaboration, neighbors helping each other. "Many hands make light work."

Where are the people? out of self-interest and with some trep-
idation. The *National Geographic* model shows, as far as I can
tell, only one "farmer." He is standing in the "bubble-topped
control tower," presumably operating the whole farm by remote control. The article on the South Dakota State model
mentions that "The hired hand gets the imposing new title
of 'manager.'" Allowing for shifts, vacations, etc., these model
farms evidently require a staff of only half a dozen or so to do
the actual "work." Most of the jobs for people would evidently
be non-agricultural: jobs of construction, maintenance, transport, etc.

And where are the *other* people—the ones who are not doing the computer-work of future farming? Well, the *National
Geographic* picture shows some highway traffic that may or
may not be remote-controlled. It shows "a distant city" and
"a distant lake and recreation area." The South Dakota State
model includes a zone for "recreation, wildlife, and living."

The agriculture prophets evidently think that they have left
people pretty much to their own devices: they will have places
to live and places to work and places for recreation, and, thanks
to the completely controlled farms-of-the-future, they will
have plenty to eat.

The specialists who conceived these models are American
citizens. They undoubtedly believe in the doctrines of personal
liberty and dignity, equality of opportunity, etc. If asked, they
would undoubtedly say that the people outside the boundaries
of these farms would benefit from them in every way: they
would not only have areas especially allotted to them for living,
working, and recreation, they would also have more freedom,
dignity, and equality of opportunity than ever before.

But one must ask if they would not say these things
thoughtlessly—because they are the right things to say in a democracy, or the most persuasive things to say, or because they
are in the habit of saying them. It is clear, at least, that official
policies—and these model farms represent official policy—have
come to be *routinely* justified in this country on the grounds
that they will uphold freedom, dignity, and equality of opportunity. There is no official depredation that one can think of
that has not been initially so justified. The skids are greased
with unctions of democracy.

But these assurances are always incidental, outside the boundary of whatever allegedly benign (and profitable) innovation is at hand. People are not going to be free or dignified or even well fed just because some specialist *says* that they will be. Or says that they will be *allowed to be, in certain areas*—for that is what these "agribusiness" visionaries are in fact saying. People will be *allowed* to be free to do *certain* things in *certain* places prescribed by *other* people. They will be free to work in the places set aside for work, free to play or relax in places set aside for recreation, free to live (whatever that may mean) in places set aside for living.

Thus there are several things that people will *not* be free to do in the nation-of-the-future that will be fed by these farms-of-the-future. They will not live where they work or work where they live. They will not work where they play. And they will not, above all, play where they work. There will be no singing in those fields. There will be no crews of workers or neighbors laughing and joking, telling stories, or competing at tests of speed or strength or skill. There will be no holiday walks or picnics in those fields because, in the first place, the fields will be ugly, all graces of nature having been ruled out, and, in the second place, they will be dangerous.

Very few people, more likely none of them, will own those farms. Very few will work on them. Most of them, more even than the ninety-five percent that *now* live in urban situations, will live remote from the farmland, divided from it by distance, by "buffer zones," by economics, by official structure. They will have nothing to say about how the land is used or the kind or quality of its produce. For these farms are obviously designed for the ownership and management of huge "agribusiness" corporations that will control them "privately" and control the market as well. The people will eat what the corporations decide for them to eat. They will be detached and remote from the sources of their life, joined to them only by corporate tolerance. They will have become consumers purely —consumptive machines—which is to say, the slaves of producers. What these model farms very powerfully suggest, then, is that the concept of total control may be impossible to confine within the boundaries of the specialist enterprise—that it is impossible to mechanize production without mechanizing consumption, impossible to make machines of soil, plants, and animals without making machines also of people.

It is important to recognize that in the minds both of the agribusiness specialists and of their believers and supporters among the public these representations of technological totalitarianism rest side by side with conventional good intentions. Mr. Billard, in his *National Geographic* article, is careful to write a paragraph of reassurance about the future of the family farm and the welfare of consumers: ". . . farms grossing more than $10,000 a year expand in number, with those in the more-than-$40,000 category increasing rapidly. The family farm figures largest in this growth. It accounts for 95 percent of all farms and 64 percent of total marketings. Corporate behemoths play no greater role today than 20 years ago; the specter of their progressively gobbling up all the farmland and in the end holding consumers at their mercy seems farfetched."

That, of course, depends on how you define "family farm" and "corporate behemoth." And beside the Department of Agriculture figures quoted earlier, and the testimony of his own article, Mr. Billard's reassurance is a mere hopeful assertion, not very reassuring.

And Dr. Hellickson, replying to my question about the possible influence of his students' work on personal liberty and private property, said: "We are in no way advocating the elimination of the free enterprise system or the reduction of privately owned land."

Sometimes I ask myself if it may not be that these reassurances are given cynically by people who know very well that they are turned against what they wish to appear to uphold. Though I leave open the possibility that this occasionally may be so, I have concluded so far that most often it is not. I believe that both Mr. Billard and Dr. Hellickson are sincere in their belief that the innovations they praise or advocate will not adversely affect traditional values, the supply or the quality of food, or the life of farming. I believe this of Dr. Hellickson in spite of his substitution of "the free enterprise system" for my phrase "personal liberty."

It is nevertheless a part of the significance of the statements of both men that they embody a large, if unconscious, moral contradiction. In this, it seems to me, they represent very accurately the flawed consciousness of our society, which is everywhere eagerly conniving in the destruction of what it says, and thinks, it wants to preserve.

For no matter what these gentlemen say, the private ownership of farmland and public concern for the health of farming are both diminishing at an alarming rate, and they are diminishing because of the big economics and big technology represented by these visions of future agriculture.

They are diminishing because as a society we have abandoned any interest in the survival of anything small. We seem to have adopted a moral rule of thumb according to which anything big is better than anything small. As a result, the agricultural establishment has simply looked away from the possibility of an economics and a technology suited to the needs and aims of the small farmer.

Some time ago I took part in a conference on agriculture, at which one of the speakers was an executive of Deere and Company. This man was asked by someone in the audience if he and his company were interested in small-farm technology. He replied that indeed they were. But as I remember that was all he said; he spoke of none of the particulars of such technology, in which he evidently had at least no personal interest. What did interest him, as I learned later in conversation, was the impending development of a 600-horsepower tractor, which eventually would be operated by remote control. In the face of such an interest, empowered as it is by official sanction, tax-supported research, and vast sums of money, the small farmer is not so much condemned as written off as a necessary expenditure. A price is put on his way of life which he is less and less able to meet.

DESERTS OF VAST TECHNOLOGY

As specialists, the agricultural scientists and "agribusinessmen" find it easy to talk as if the influence of big agricultural technology can be confined neatly to the "field" of agriculture. That it cannot be is already proven by the powerful *urban* influence that such technology has already had. And to the big-thinking, non-agricultural mind, food is merely a resource, like energy and raw materials, and so agricultural technology is not different from any other. About grain, fuel, and ore the only questions are: How much? and How fast?

Because big technology is so simplifying, the future looks, not bright, but absolutely perfect to F. M. Esfandiary, who teaches "long-range planning" in New York City's New School for Social Research. In an article entitled "Homo Sapiens, the Manna Maker," Mr. Esfandiary sees the future as an earthly Heaven in which, by the miracles of technology, humans will usurp the role of God—who, it may be recalled, was once thought to be the only maker of manna. The following quotations will give the gist of his argument:

"The world is moving toward an age of limitless abundance —abundant energy, food, raw materials."

"Solar power, nuclear fusion, geothermal energy, recycled energy, wind energy, hydrogen fuel—these sources will soon provide cheap, nonpolluting, limitless energy, enough to last for millions of years."

"Agriculture is undergoing an epochal revolution. We are evolving from feudal and industrial agriculture to cybernated food production. Computers, remote control cultivators, television monitors, sensors, data banks can now automatically run thousands of acres of cultivated land. A couple of telefarm operators can feed a million people."

"We now have the capability to extract limitless raw materials from recycled wastes, rocks, the earth's interiors, the ocean floors, space."

That is the "objective" part of the argument. There follows a series of paragraphs that must hold the world record for rhetorical passion. It appears that Mr. Esfandiary is mad at us because we do not duck our heads and hurry right on into the future.

"How absurd the American panic over scarcity when we are entering an age of abundance. How absurd to focus on 'finiteness' at the period in evolution when our world is transcending finiteness, opening up the infinite resources of an infinite universe.

"How outrageous that after centuries of privation and sacrifice leaders can come up with nothing more than yet more sacrifice. How short-sighted the exhortation to no-growth at precisely the time when we urgently need more and more growth—growth not *within* but *beyond* industrialism.

"How retrogressive the preachings to lower living standards of the relatively rich to raise conditions of the poor, at a time when we can raise *everybody's* living conditions by vigorously developing and spreading abundance, not sharing scarcity."

The common assumption is that mechanization involves the giving over of certain tasks or functions to machines. In these paragraphs by Mr. Esfandiary a very different assumption, that may always have been implicit in the advocacy of industrial revolution, comes to the surface: he is proposing that we give over *everything* to machines. He is berating us, with the fervor of an evangelist, because we do not abandon ourselves to machines as people of faith abandon themselves to God. He is berating us, in fact, for not *being* gods or at least acting as if we were gods.

The crucial concept here is that of "limitless" or "infinite" quantity. By "limitless" and "infinite" Mr. Esfandiary undoubtedly means only "inconceivable." At any rate, people who have desired material *quantities* on such a scale have always been recognized as evil, and their stories have always involved a sort of ecological justice: godly appetite very quickly led beyond human competence, invariably with disastrous consequences. Mr. Esfandiary's unlimited, if theoretical, gluttony is licensed and given an illusory respectability because of its claim to be "scientific"—godly appetite may be within the competence of a computer—and because, as a "long-range planner," he does his theorizing in the future, where it cannot very handily be called to account.

It is nevertheless clear that Mr. Esfandiary's "future" calls for unprecedented violence. It would require the sacrifice of every value that is not quantitative. The technology of infinity (however that might be defined) would be vast and exclusive. It would be completely totalitarian, whether "publicly" or "privately" owned. It would overthrow the whole issue of control, for it would *be* the control. Since everyone would be totally dependent upon it, it would necessarily be everyone's first consideration. It might at first seem that enormous power would lie in the hands of the "couple of telefarm operators" who would be feeding a million people; but it seems more likely that they, too, would be the absolute slaves of their machinery, no less dependent on it than the million. The machine would become an anti-god—if not infinite, at least absolute. To have even the illusion of infinite quantity, we would have to debase both the finite and the infinite; we would have to sacrifice both flesh and spirit. It is an old story. Evil is offering us the world: "All these things will I give thee, if thou wilt fall down and worship me." And we have only the old paradox for an answer: If we accept all on that condition, we lose all.

What is new is the *guise* of the evil: a limitless technology, dependent upon a limitless morality, which is to say upon no morality at all. How did such a possibility become thinkable? It seems to me that it is implicit in the modern separation of life and work. It is implicit in the assumption that we can live entirely apart from our way of making a living. It is implicit in the idea of the agricultural engineering students at South Dakota State University that their farm-of-the-future would require "blending with human values." To propose to blend such a farm with human values is simply to acknowledge that it *has* no human values, that human values have been removed from it. (The analogy is not accidental, I think, between this "blending with human values" and the "enrichment" of bread after the nutrients have been removed from the wheat.) If human values are removed from production, how can they be preserved in consumption? How can we value our lives if we devalue them in making a living?

If we do not live where we work, and when we work, we are wasting our lives, and our work too.

CHAPTER SIX

The Use of Energy

"ENERGY," SAID William Blake, "is Eternal Delight." And the scientific prognosticators of our time have begun to speak of the eventual opening, for human use, of "infinite" sources of energy. In speaking of the use of energy, then, we are speaking of an issue of religion, whether we like it or not.

Religion, in the root sense of the word, is what binds us back to the source of life. Blake also said that "Energy is the only life . . ." And it is superhuman in the sense that humans cannot create it. They can only refine or convert it. And they are bound to it by one of the paradoxes of religion: they cannot have it except by losing it; they cannot use it except by destroying it. The lives that feed us have to be killed before they enter our mouths; we can only use the fossil fuels by burning them up. We speak of electrical energy as "current": it exists only while it runs away; we use it only by delaying its escape. To receive energy is at once to live and to die.

Perhaps from an "objective" point of view it is incorrect to say that we can destroy energy; we can only change it. Or we can destroy it only in its current form. But from a human point of view, we can destroy it also by wasting it—that is, by changing it into a form in which we cannot use it again. As users, we can preserve energy in cycles of use, passing it again and again through the same series of forms; or we can waste it by using it once in a way that makes it irrecoverable. The human pattern of cyclic use is exemplified in the small Oriental peasant farms described in F. H. King's *Farmers of Forty Centuries*, in which all organic residues, plant and animal and human, were returned to the soil, thus keeping intact the natural cycle of "birth, growth, maturity, death, and decay" that Sir Albert Howard identified as the "Wheel of Life." The pattern of wasteful use is exemplified in the modern sewage system and the internal combustion engine. With us, the wastes that escape use typically become pollutants. This kind of use turns an asset into a liability.

We have two means of bringing energy to use: by living things (plants, animals, our own bodies) and by tools (machines, energy-harnesses). For the use of these we have skills or techniques. All three together comprise our technology. Technology joins us to energy, to life. It is not, as many technologists would have us believe, a simple connection. Our technology is the practical aspect of our culture. By it we enact our religion, or our lack of it.

I began thinking about this by trying to make a clear distinction between the living organisms and skills of technology and its mechanisms, and to say that the living aspect was better than the mechanical. I found it impossible to make such a distinction. I thought of going back through history to a point at which such distinction would become possible, but found that the farther back I went the less possible it became. When people had no machines other than throwing stones and clubs, their technology was all of a piece. It stayed that way through their development of more sophisticated tools, their mastery of fire, their domestication of plants and animals. Lives, skills, and tools were culturally indivisible.

The question at issue, then, is not of distinction but of balance. The ideal seems to be that the living part of our technology should not be devalued or overpowered by the mechanical. Because the biological limits are probably narrower than the mechanical, this calls for restraint on the proliferation of machines.

At some point in history the balance between life and machinery was overthrown. I think this began to happen when people began to desire long-term stores or supplies of energy —that is, when they began to think of energy as volume as well as force—and when machines ceased to enhance or elaborate skill and began to replace it.

Though it seems impossible to distinguish between the living and the mechanical aspects of technology, it is possible to distinguish between two kinds of energy: that which is made available by living things and that which is made available by machines.

The energy that comes from living things is produced by combining the four elements of medieval science: earth, air, fire (sunlight), and water. This is current energy. Though it is possible to speak of a *reserve* of such energy, as Sir Albert Howard does, in the sense of a surplus of fertility, it is impossible to conceive of a *reservoir* of it. It is not available in long-term supplies in any form in which it can be preserved, as in humus, in the flesh of living animals, in cans or freezers or grain elevators, it still perishes fairly quickly in comparison, say, to coal or plutonium. It lasts over a long term only in the living cycle of birth, growth, maturity, death, and decay. The technology appropriate to the use of this energy, therefore, preserves its cycles. It is a technology that never escapes into its own logic but remains bound in analogy to natural law.

The energy that is made available, and consumed, by machines is typically energy that can be accumulated in stockpiles or reservoirs. Energy from wind and water obviously does not fit this category, but it suggests the possibility of bigger and better storage batteries, which one must assume will sooner or later be produced. And, of course, we already store water power behind hydroelectric dams. This mechanically derived energy is supposed to have set people free from work and other difficulties once considered native to the human condition. Whether or not it has done so in any meaningful sense is questionable—in my opinion, it is highly questionable. But there is no doubt that this sort of energy has freed machinery from the natural restraints that apply to the use of organic energy. We now have a purely mechanical technology that is very nearly a law unto itself.

And yet, in the long term, this liberation of the machine is illusory. Mechanical technology is based on quantities of materials and fuels that are finite. If the prophets of science foresee "limitless abundance" and "infinite resources," one must assume that they are speaking figuratively, meaning simply that they cannot comprehend how much there may be. In that sense, they are right: there are sources of energy that, given the necessary machinery, are inexhaustible *as far as we can see.*

The great difficulty, which these cheerful prophets do not acknowledge at all, is that we are trustworthy only so far as we can see. The length of our vision is our moral boundary. Even if these foreseen supplies *are* limitless, we can use them only within limits. We can bring the infinite to bear only within the finite bounds of our biological circumstance and our understanding. It is already certain that our planet alone—not to mention potential sources in space—can provide us with more energy and materials than we can use safely or well. By our abuse of our finite sources, our lives and all life are already in danger. What might we bring into danger by the abuse of "infinite" sources?

The difficulty with mechanically extractable energy is that so far we have been unable to make it available without serious geological and ecological damage, or to effectively restrain its use, or to use or even neutralize its wastes. From birth, right now, we are carrying the physical and the moral poisons produced by our crude and ignorant use of this sort of energy. And the more abundant the energy of this sort that we use, the more abounding must be the consequences.

It is typical of the mentality of our age that we cannot conceive of infinity except as an enormous quantity. We cannot conceive of it as orderly process, as pattern or cycle, as shapeliness. We conceive of it as inconceivable quantity—that is, as the immeasurable. Any quantity that we cannot measure we assume must be infinite. That is about as sophisticated as saying that the world is flat because it *looks* flat. The talk about "infinite" resources is thus a kind of scientific-sounding foolishness. And it involves some quaint paradoxes. If we think, for instance, of infinite energy as immeasurable fuel, we are committed in the same thought to its destruction, for fuel must be destroyed to be used. We thus arrive at the curious idea of a destructible infinity. Furthermore, we have become guilty not only of the demonstrably silly assumption that we know what to do with infinite energy, but also of the monstrous pride of thinking ourselves somehow entitled to undertake infinite destruction.

This mechanically rendered infinitude of energy is an ambition surrounded by terrific problems. Such energy cannot be used constructively without at the same time being used destructively. And which way the balance will finally fall is a question that baffles the best minds. Nobody knows what will be the ultimate consequences of our present use of fossil fuel, much less those of our future use of atomic fuel. The sun may prove an "infinite" source of energy—at least one that may last several billion years. But who will control the use of that energy? How and for what purposes will it be used?

How much can be used without overthrowing ecological or social or political balances? Nobody knows.

The energy that is made available to us by living things, on the other hand, is made available not as an inconceivable quantity, but as a conceivable pattern. And for the mastery of this pattern—that is, the ability to see its absolute importance and to preserve it in use—one does not need a Ph.D. or a laboratory or a computer. One can master it in this sense, in fact, without having any analytic or scientific understanding of it at all. It was mastered, better than our scientific experts have mastered it, by "primitive" peasants and tribesmen thousands of years before modern science. It is conceivable not so much to the analytic intelligence, to which it may always remain in part mysterious, as to the imagination, by which we perceive, value, and imitate order beyond our understanding.

We cannot create biological energy any more than we can create atomic or fossil fuel energy. But we *can* preserve it in *use*; we can probably even augment it in use, in the sense that, by proper care, we can "build" soil. We cannot do that with machine-derived energy. This is an extremely important difference, with respect both to the energy economy itself and to the moral order that is undoubtedly determined by, as much as it determines, the value we put on energy.

The moral order by which we use machine-derived energy is comparatively simple. Whatever uses this sort of energy works simply as a conduit that carries it beyond use: the energy goes in as "fuel" and comes out as "waste." This principle sustains a highly simplified economy having only two functions: production and consumption.

The moral order appropriate to the use of biological energy, on the other hand, requires the addition of a third term: production, consumption, *and return*. It is the principle of return that complicates matters, for it requires responsibility, care, of a different and higher order than that required by production and consumption alone, and it calls for methods and economies of a different kind. In an energy economy appropriate to the use of biological energy, all bodies, plant and animal and human, are joined in a kind of energy community. They are not divided from each other by greedy, "individualistic" efforts to produce and consume large quantities of energy, much less to store large quantities of it. They are indissolubly linked in complex patterns of energy exchange. They die into each other's life, live into each other's death. They do not consume in the sense of using up. They do not produce waste. What they take in they change, but they change it always into a form necessary for its use by a living body of another kind. And this exchange goes on and on, round and round, the Wheel of Life rising out of the soil, descending into it, through the bodies of creatures.

The soil is the great connector of lives, the source and destination of all. It is the healer and restorer and resurrector, by which disease passes into health, age into youth, death into life. Without proper care for it we can have no community, because without proper care for it we can have no life.

It is alive itself. It is a grave, too, of course. Or a healthy soil is. It is full of dead animals and plants, bodies that have passed through other bodies. For except for some humans—with their sealed coffins and vaults, their pathological fear of the earth—the only way into the soil is through other bodies. But no matter how finely the dead are broken down, or how many times they are eaten, they yet give into other life. If a healthy soil is full of death it is also full of life: worms, fungi, microorganisms of all kinds, for which, as for us humans, the dead bodies of the once living are a feast. Eventually this dead matter becomes soluble, available as food for plants, and life begins to rise up again, out of the soil into the light. Given only the health of the soil, nothing that dies is dead for very long. Within this powerful economy, it seems that death occurs only for the good of life. And having followed the cycle around, we see that we have not only a description of the fundamental biological process, but also a metaphor of great beauty and power. It is impossible to contemplate the life of the soil for very long without seeing it as analogous to the life of the spirit. No less than the faithful of religion is the good farmer mindful of the persistence of life through death, the passage of energy through changing forms.

And this living topsoil—living in both the biological sense and in the cultural sense, as metaphor—is the basic element in the technology of farming.

It is the nature of the soil to be highly complex and variable, to conform very inexactly to human conclusions and rules. It is itself a pattern of inexhaustible intricacy, and so it is easily damaged by the imposition of alien patterns. Out of the random grammar and lexicon of possibilities—geological, topographical, climatological, biological—the soil of any one place makes its own peculiar and inevitable sense. It makes an order, a pattern of forms, kinds, and processes, that includes any number of offsets and variables. By its permeability and absorbency, for example, the healthy soil corrects the irregularities of rainfall; by the diversity of its vegetation it protects against both disease and erosion. Most farms, even most fields, are made up of different kinds of soil patterns or soil sense. Good farmers have always known this and have used the land accordingly; they have been careful students of the natural vegetation, soil depth and structure, slope and drainage. They are not appliers of generalizations, theoretical or methodological or mechanical. Nor are they the active agents of their own economic will, working their way upon an inert and passive mass. They are responsive partners in an intimate and mutual relationship.

Because the soil is alive, various, intricate, and because its processes yield more readily to imitation than to analysis, more readily to care than to coercion, agriculture can never be an exact science. There is an inescapable kinship between farming and art, for farming depends as much on character, devotion, imagination, and the sense of structure, as on knowledge. It is a practical art.

But it is also a practical religion, a practice of religion, a rite. By farming we enact our fundamental connection with energy and matter, light and darkness. In the cycles of farming, which carry the elemental energy again and again through the seasons and the bodies of living things, we recognize the only infinitude within reach of the imagination. How long this cycling of energy will continue we do not know; it will have to end, at least here on this planet, sometime within the remaining life of the sun. But by aligning ourselves with it here, in our little time within the unimaginable time of the sun's burning, we touch infinity; we align ourselves with the universal law that brought the cycles into being and that will survive them.

The word *agriculture*, after all, does not mean "agriscience," much less "agribusiness." It means "cultivation of land." And *cultivation is* at the root of the sense both of *culture* and of *cult.* The ideas of tillage and worship are thus joined in *culture.* And these words all come from an Indo-European root meaning both "to revolve" and "to dwell." To live, to survive on the earth, to care for the soil, and to worship, all are bound at the root to the idea of a cycle. It is only by understanding the cultural complexity and largeness of the concept of agriculture that we can see the threatening diminishments implied by the term "agribusiness."

That agriculture is in so complex a sense a cultural endeavor —and that food is therefore a cultural product—would be regarded as heresy by most of the agencies, institutions, and publications of modern farming. The spokesmen of the official reckoning would doubtless respond that they are not cultural but scientific, that they are specialists of "agriscience." If agriculture is acknowledged to have anything to do with culture, then its study has to include people. But the agriculture experts ruled people out when they made their discipline a specialty —or, rather, when they sorted it into a collection of specialties —and moved it into its own "college" in the university. This specialty collection is interested in soils (in the limited sense of soil chemistry), in plants and animals, and in machines and chemicals. It is not interested in people.

But what respect is one to give to a science that parcels a unified discipline into discrete fragments, that has no interest in its effects if they are not immediately measurable in a laboratory, and that is founded upon the waste of topsoil, energy, and manpower, and upon the dissolution of communities? Not much. And it has been my experience that, with respect to this science, farmers are divided into two kinds: those who endanger their solvency, and often their sanity, by trusting it and those who hold it in contempt.

In the view of the experts, then, agriculture is not only not a concern of culture, but not even a concern of science, for they have abandoned interest in the health of the farming communities on the one hand and in the health of the land on the other. They appear to have concluded that agriculture is purely a commercial concern; its purpose is to provide as much food as quickly and cheaply and with as few man-hours as possible and to be a market for machines and chemicals. It is, after all, "agribusiness"—not the land or the farming people —that now benefits most from agricultural research and that can promote humble academicians to highly remunerative and powerful positions in corporations and in government. Former Secretary Earl Butz's career exemplifies the predominant direction of interest of the agriculture specialist. According to Lauren Soth, writing in the *Nation*, "Butz is the perfect example of the agribusiness, commercial-farming, agricultural-education establishment man. When dean of agriculture at Purdue University, he also sat on the boards of directors of the Ralston-Purina Co., the J. I. Case Co., International Minerals and Chemicals Corp., Stokely-Van Camp Co. and Standard Life Insurance Co. of Indiana." By such men and such careers the land-grant college system, originally meant to enhance the small-farm possibility, has been captured for the corporations.

The discipline of agriculture—the "great subject," as Sir Albert Howard called it, "of health in soil, plant, animal, and man"—has been reduced to fit first the views of a piecemeal "science" and then the purposes of corporate commerce. I can see no possibility of a doubt that this is true, though I cannot explain exactly how it happened. But it seems to me that the way was prepared when the specialized shapers or makers of agricultural thought simplified their understanding of energy and began to treat current, living, biological energy as if it were a *store* of energy extractable by machinery. At that point the living part of technology began to be overpowered by the mechanical. The machine was on its own, to follow its own logic of elaboration and growth apart from life, the standard that had previously defined its purposes and hence its limits. Let loose from any moral standard or limit, the machine was also let loose in another way: it replaced the Wheel of Life as the governing cultural metaphor. Life came to be seen as a road, to be traveled as fast as possible, never to return. Or, to put it another way, the Wheel of Life became an industrial metaphor; rather than turning in place, revolving in order to dwell, it began to roll on the "highway of progress" toward an ever-receding horizon. The idea, the responsibility, of return weakened and disappeared from agricultural discipline. Henceforth, any resource would be regarded as an ore.

If agriculture is founded upon life, upon the use of living energy to serve human life, and if its primary purpose must therefore be to preserve the integrity of the life cycle, then agricultural technology must be bound under the rule of life. It must conform to natural processes and limits rather than to mechanical or economic models. The culture that sustains agriculture and that it sustains must form its consciousness and its aspiration upon the correct metaphor of the Wheel of Life. The appropriate agricultural technology would therefore be diverse; it would aspire to diversity; it would enable the diversification of economies, methods, and species to conform to the diverse kinds of land. It would always use plants and animals together. It would be as attentive to decay as to growth, to maintenance as to production. It would return all wastes to the soil, control erosion, and conserve water. To enable care and devotion and to safeguard the local communities and cultures of agriculture, it would use the land in small holdings. It would aspire to make each farm so far as possible the source of its own operating energy, by the use of human energy, work animals, methane, wind or water or solar power.

The mechanical aspect of the technology would serve to harness or enhance the energy available on the farm. It would not be permitted to replace such energies with imported fuels, to replace people, or to replace or reduce human skills.

The damages of our present agriculture all come from the determination to use the life of the soil as if it were an extractable resource like coal, to use living things as if they were machines, to impose scientific (that is, laboratory) exactitude upon living complexities that are ultimately mysterious.

If animals are regarded as machines, they are confined in pens remote from the source of their food, where their excrement becomes, instead of a fertilizer, first a "waste" and then a pollutant. Furthermore, because confinement feeding depends so largely on grains, grass is removed from the rotation of crops and more land is exposed to erosion.

If plants are regarded as machines, we wind up with huge monocultures, productive of elaborate ecological mischiefs, which are in turn productive of agricultural mischief: mono-cultures are much more susceptible to pests and diseases than mixed cultures and are therefore more dependent on chemicals.

If the soil is regarded as a machine, then its life, its involve-ment in living systems and cycles, must perforce be ignored. It must be treated as a dead, inert chemical mass. If its life is ignored, then so must be the natural sources of its fertility —and not only ignored, but scorned. Alfalfa and the clovers, according to some of the most up-to-date practitioners, are "weeds"; the only legitimate source of nitrogen is the fertilizer manufacturer. And animal manures are "wastes"; "efficiency" cannot use them. Not long ago I found that the manure from a saddle-horse barn belonging to the University of Kentucky was simply being dumped. When I asked why it was not used somewhere on the farm, I was told that it would interfere with the College of Agriculture's experiments. The result is absurd: our agriculture, potentially capable of a large measure of in-dependence, is absolutely dependent on petroleum, on the oil companies, and on the vagaries of politics.

If people are regarded as machines, they must be regarded as replaceable by other machines. They are regarded, in other words, as dispensable. Their place on the farm is safe only as long as they are mechanically necessary.

In modern agriculture, then, the machine metaphor is al-lowed to usurp and wipe from consideration not merely *some* values, but the very *issue* of value. Once the expert's interest is focused on the question of "what will work" within the exclu-sive confines of his theoretical model, values are no longer of any concern whatever. The confines of his specialty enable him to impose a biological totalitarianism on—he thinks, since he is an agricultural expert—the farm. When he leaves his office or laboratory he will, he assumes, go "home" to value.

But then it must be asked if we can remove cultural value from one part of our lives without destroying it also in the other parts. Can we justify secrecy, lying, and burglary in our so-called intelligence organizations and yet preserve openness, honesty, and devotion to principle in the rest of our govern-ment? Can we subsidize mayhem in the military establishment and yet have peace, order, and respect for human life in the city streets? Can we degrade all forms of essential work and yet expect arts and graces to flourish on weekends? And can we ignore all questions of value on the farm and yet have them answered affirmatively in the grocery store and the household?

The answer is that, though such distinctions can be made theoretically, they cannot be preserved in practice. Values may be corrupted or abolished in only one discipline at the start, but the damage must sooner or later spread to all; it can no more be confined than air pollution. If we corrupt agriculture we corrupt culture, for in nature and within certain invariable social necessities we are one body, and what afflicts the hand will afflict the brain.

The effective knowledge of this unity must reside not so much in doctrine as in skill. Skill, in the best sense, is the enactment or the acknowledgment or the signature of responsibility to other lives; it is the practical understanding of value. Its opposite is not merely unskillfulness, but ignorance of sources, dependences, relationships.

Skill is the connection between life and tools, or life and machines. Once, skill was defined ultimately in qualitative terms: How *well* did a person work; how good, durable, and pleasing were his products? But as machines have grown larger and more complex, and as our awe of them and our desire for labor-saving have grown, we have tended more and more to define skill quantitatively: How speedily and cheaply can a person work? We have increasingly wanted a *measurable* skill. And the more quantifiable skills became, the easier they were to replace with machines. As machines replace skill, they disconnect themselves from life; they come between us and life. They begin to enact our ignorance of value—of essential sources, dependences, and relationships.

The catch is that we cannot live in machines. We can only live in the world, in life. To live, our contact with the sources of life must remain direct: we must eat, drink, breathe, move, mate, etc. When we let machines and machine skills obscure the values that represent these fundamental dependences, then we inevitably damage the world; we diminish life. We begin to "prosper" at the cost of a fundamental degradation.

The digging stick, for example, brought in a profound technological revolution: it made agriculture possible. Its use required skill. But its *effect* also required skill, and this kind of skill was higher and more complex than the first, for it involved restraint and responsibility. The digging stick made it possible to grow food; that was one thing. It also made it possible, and necessary, to disturb the earth; and that was another thing. The first skill required others that were its moral elaboration: the skill used in disturbing the earth called directly for other skills that would preserve the earth and restore its fertility.

Until fairly recently, as agricultural tools became more efficient or powerful or both, they required an increase of both kinds of skill. One could do more with stone implements than with sticks, and more with metal implements than with stone implements; the skilled use of these tools enabled one to disturb more ground and so called for further elaboration of the skills of responsibility.

This remained true after the beginning of the use of draft animals. The skills of use had to become much greater, for the human mind had to relate to the animal mind in a new way: not by the magic and cunning of the hunt, but in the practical intricacies of collaboration. And the skills of responsibility had to increase proportionately. More ground could now be disturbed, and so the technology of preservation had to become much larger. Also, the investment of life in work greatly increased; people had to take responsibility not only for their own appetites and excrements but for those of their animals as well.

It was only with the introduction of self-powering machines, and of machine-extracted energy, into the fields that something really new happened to agricultural skills: they began a radical diminishment.

In the first place, it requires more skill to use a team of horses or mules or oxen than to use a tractor. It is more difficult to learn to manage an animal than a machine; it takes longer. Two minds and two wills are involved. A relationship between a person and a work animal is analogous to a relationship between two people. Success depends upon the animal's willingness and upon its health; certain moral imperatives and restraints are therefore pragmatically essential. No such relationship is either necessary or possible with a machine. Within the range of the possible, a machine is directly responsive to human will; it neither starts nor stops because it wants to. A machine has no life, and for this reason it cannot of itself impose any restraint or any moral limit on behavior.

In the second place, the substitution of machines for work animals is justified mainly by their ability to increase the volume of work per man—that is, by their greater speed. But as speed increases, care declines. And so, necessarily, do the skills of responsibility. If this were not so, we would not restrict the speed of traffic in residential areas. We know that there is a limit to the capacity of attention, and that the faster we go the less we see. This law applies with equal force to work; the faster we work the less attention we can pay to its details, and the less skill we can apply to it.

This is true of *any* productive work, and it has great cultural importance; at present we are all suffering, in various ways, from dependence on goods that are poorly made. But its importance in agricultural production is probably more critical than elsewhere. In any biological system the first principle is restraint—that is, the natural or moral checks that maintain a balance between use and continuity. The life of one year must not be allowed to diminish the life of the next; nothing must live at the expense of the source. Thus, in nature, the food species is dependent on its predator, and pests and diseases are agents of health; so populations are controlled and balanced. In agriculture these natural checks are removed and therefore must be replaced by the skills of responsibility, which have to do with the prevention of erosion, the diversification and rotation of plant and animal species, the return of wastes to the soil, and all the other provisionings of the source. When productive power—that is, speed—in machines replaces the productive skills of people, there is a consequent narrowing of attention. The machines are expensive and they run on purchased fuels; they feed upon money. The work of production is immediately profitable, whereas the work of responsibility is not. Once the machine is in the field it creates an economic pressure that enforces haste; the machine concentrates all the energy of the farm and hurries it toward the marketplace. The demands of immediate use eclipse the demands of continuity. As the skills of production decline, the skills of responsibility perish.

To argue for a balance between people and their tools, between life and machinery, between biological and machine-produced energy, is to argue for restraint upon the use of machines. The arguments that rise out of the machine metaphor—arguments for cheapness, efficiency, labor-saving, economic growth, etc.—all point to infinite industrial growth and infinite energy consumption. The moral argument points to restraint; it is a conclusion that may be in some sense tragic, but there is no escaping it. Much as we long for infinities of power and duration, we have no evidence that these lie within our reach, much less within our responsibility. It is more likely that we will have either to live within our limits, within the human definition, or not live at all. And certainly the knowledge of these limits and of how to live within them is the most comely and graceful knowledge that we have, the most healing and the most whole.

The knowledge that purports to be leading us to transcendence of our limits has been with us a long time. It thrives by offering material means of fulfilling a spiritual, and therefore materially unappeasable, craving: we would all very much like to be immortal, infallible, free of doubt, at rest. It is because this need is so large, and so different in kind from all material means, that the knowledge of transcendence—our entire history of scientific "miracles"—is so tentative, fragmentary, and grotesque. Though there are undoubtedly mechanical limits, because there are human limits, there is no mechanical restraint. The only logic of the machine is to get bigger and more elaborate. In the absence of *moral* restraint—and we have never imposed adequate moral restraint upon our use of machines—the machine is out of control by definition. From the beginning of the history of machine-developed energy, we have been able to harness more power than we could use responsibly. From the beginning, these machines have created effects that society could absorb only at the cost of suffering and disorder.

And so the issue is not of supply but of use. The energy crisis is not a crisis of technology but of morality. We already have available more power than we have so far dared to use. If, like the strip-miners and the "agribusinessmen," we look on all the world as fuel or as extractable energy, we can do nothing but destroy it. The issue is restraint. The energy crisis reduces to a single question: Can we forbear to do anything that we are able to do? Or to put the question in the words of Ivan Illich: Can we, believing in "the effectiveness of power," see "the disproportionately greater effectiveness of abstaining from its use"?

The only people among us that I know of who have answered this question convincingly in the affirmative are the Amish. They alone, as a community, have carefully restricted their use of machine-developed energy, and so have become the only true masters of technology. They are mostly farmers, and they do most of their farm work by hand and by the use of horses and mules. They are pacifists, they operate their own local schools, and in other ways hold themselves aloof from the ambitions of a machine-based society. And by doing so they have maintained the integrity of their families, their community, their religion, and their way of life. They have escaped the mainstream American life of distraction, haste, aimlessness, violence, and disintegration. Their life is not idly wasteful, or destructive. The Amish no doubt have their problems; I do not wish to imply that they are perfect. But it cannot be denied that they have mastered one of the fundamental paradoxes of our condition: we can make ourselves whole only by accepting our partiality, by living within our limits, by being human—not by trying to be gods. By restraint they make themselves whole.

But just stop for a minute and think about what it means to live in a land where 95 percent of the people can be freed from the drudgery of preparing their own food.

JAMES E. BOSTIC, JR., FORMER DEPUTY ASSISTANT SECRETARY OF AGRICULTURE FOR RURAL DEVELOPMENT

Find the shortest, simplest way between the earth, the hands and the mouth.

LANZA DEL VASTO

CHAPTER SEVEN

The Body and the Earth

ON THE CLIFF

THE question of human limits, of the proper definition and place of human beings within the order of Creation, finally rests upon our attitude toward our biological existence, the life of the body in this world. What value and respect do we give to our bodies? What uses do we have for them? What relation do we see, if any, between body and mind, or body and soul? What connections or responsibilities do we maintain between our bodies and the earth? These are religious questions, obviously, for our bodies are part of the Creation, and they involve us in all the issues of mystery. But the questions are also agricultural, for no matter how urban our life, our bodies live by farming; we come from the earth and return to it, and so we live in agriculture as we live in flesh. While we live our bodies are moving particles of the earth, joined inextricably both to the soil and to the bodies of other living creatures. It is hardly surprising, then, that there should be some profound resemblances between our treatment of our bodies and our treatment of the earth.

That humans are small within the Creation is an ancient perception, represented often enough in art that it must be supposed to have an elemental importance. On one of the painted walls of the Lascaux cave (20,000–15,000 B.C.), surrounded by the exquisitely shaped, shaded, and colored bodies of animals, there is the childish stick figure of a man, a huntsman who, having cast his spear into the guts of a bison, is now weaponless and vulnerable, poignantly frail, exposed, and incomplete. The message seems essentially that of the voice out of the whirlwind in the Book of Job: the Creation is bounteous and mysterious, and humanity is only a part of it—not its equal, much less its master.

Old Chinese landscape paintings reveal, among towering mountains, the frail outline of a roof or a tiny human figure passing along a road on foot or horseback. These landscapes are almost always populated. There is no implication of a de-humanized interest in nature "for its own sake." What is represented is a world in which humans belong, but which does not belong to humans in any tidy economic sense; the Creation provides a place for humans, but it is greater than humanity and within it even great men are small. Such humility is the consequence of an accurate insight, ecological in its bearing, not a pious deference to "spiritual" value.

Closer to us is a passage from the fourth act of *King Lear*, describing the outlook from one of the Dover cliffs:

> The crows and choughs that wing the midway air
> Show scarce so gross as beetles. Halfway down
> Hangs one that gathers samphire, dreadful trade!
> Methinks he seems no bigger than his head.
> The fishermen that walk upon the beach
> Appear like mice, and yond tall anchoring bark
> Diminished to her cock—her cock, a buoy
> Almost too small for sight.

And this is no mere description of a scenic "view." It is part of a play-within-a-play, a sort of ritual of healing. In it Shakespeare is concerned with the curative power of the perception we are dealing with: by understanding accurately his proper place in Creation, a man may be made whole.

In the lines quoted, Edgar, disguised as a lunatic, a Bedlamite, is speaking to his father, the Earl of Gloucester. Gloucester, having been blinded by the treachery of his false son, Edmund, has despaired and has asked the supposed madman to lead him to the cliff's edge, where he intends to destroy himself. But Edgar's description is from memory; the two are not standing on any such dizzy verge. What we are witnessing is the working out of Edgar's strategy to save his father from false feeling—both the pride, the smug credulity, that led to his suffering and the despair that is its result. These emotions are perceived as madness; Gloucester's blindness is literally the result of the moral blindness of his pride, and it is symbolic of the spiritual blindness of his despair.

Thinking himself on the edge of a cliff, he renounces this world and throws himself down. Though he falls only to the level of his own feet, he is momentarily stunned. Edgar remains with him, but now represents himself as an innocent bystander at the foot of what Gloucester will continue to think is a tall cliff. As the old man recovers his senses, Edgar persuades him that the madman who led him to the cliff's edge was in reality a "fiend." And Gloucester repents his self-destructiveness, which he now recognizes as another kind of pride; a human has no right to destroy what he did not create:

> You ever-gentle gods, take my breath from me.
> Let not my worser spirit tempt me again
> To die before you please.

What Gloucester has passed through, then, is a rite of death and rebirth. In his new awakening he is finally able to recognize his true son. He escapes the unhuman conditions of godly pride and fiendish despair and dies "smilingly" in the truly human estate "'Twixt two extremes of passion, joy and grief . . ."

Until modern times, we focused a great deal of the best of our thought upon such rituals of return to the human condition. Seeking enlightenment or the Promised Land or the way home, a man would go or be forced to go into the wilderness, measure himself against the Creation, recognize finally his true place within it, and thus be saved both from pride and from despair. Seeing himself as a tiny member of a world he cannot comprehend or master or in any final sense possess, he cannot possibly think of himself as a god. And by the same token, since he shares in, depends upon, and is graced by all of which he is a part, neither can he become a fiend; he cannot descend into the final despair of destructiveness. Returning from the wilderness, he becomes a restorer of order, a preserver. He sees the truth, recognizes his true heir, honors his forebears and his heritage, and gives his blessing to his successors. He embodies the passing of human time, living and dying within the human limits of grief and joy.

ON THE TOWER

Apparently with the rise of industry, we began to romanticize the wilderness—which is to say we began to institutionalize it within the concept of the "scenic." Because of railroads and improved highways, the wilderness was no longer an arduous passage for the traveler, but something to be looked at as grand or beautiful from the high vantages of the roadside. We became viewers of "views." And because we no longer traveled in the wilderness as a matter of course, we forgot that wilderness still circumscribed civilization and persisted in domesticity. We forgot, indeed, that the civilized and the domestic continued to *depend* upon wilderness—that is, upon natural forces within the climate and within the soil that have never in any meaningful sense been controlled or conquered. Modern civilization has been built largely in this forgetfulness.

And as we transformed the wilderness into scenery, we began to feel in the presence of "nature" an awe that was increasingly statistical. We would not become appreciators of the Creation until we had taken its measure. Once we had climbed or driven to the mountain top, we were awed by the view, but it was an awe that we felt compelled to validate or prove by the knowledge of how high we stood and how far we saw. We are invited to "see seven states from atop Lookout Mountain," as if our political boundaries had been drawn in red on the third morning of Creation.

We became less and less capable of sensing ourselves as small within Creation, partly because we thought we could comprehend it statistically, but also because we were becoming creators, ourselves, of a mechanical creation by which we felt ourselves greatly magnified. We built bridges that stood imposingly in titanic settings, towers that stood around us like geologic presences, single machines that could do the work of hundreds of people. Why, after all, should one get excited about a mountain when one can see almost as far from the top of a building, much farther from an airplane, farther still from a space capsule? We have learned to be fascinated by the statistics of magnitude and power. There is apparently no limit in sight, no end, and so it is no wonder that our minds, dizzy with numbers, take refuge in a yearning for infinitudes of energy and materials.

And yet these works that so magnify us also dwarf us, reduce us to insignificance. They magnify us because we are capable of them. They diminish us because, say what we will, once we build beyond a human scale, once we conceive ourselves as

Titans or as gods, we are lost in magnitude; we cannot control or limit what we do. The statistics of magnitude call out like Sirens to the statistics of destruction. If we have built towering cities, we have raised even higher the cloud of megadeath. If people are as grass before God, they are as nothing before their machines.

If we are fascinated by the statistics of magnitude, we are no less fascinated by the statistics of our insignificance. We never tire of repeating the commonizing figures of population and population growth. We are entranced to think of ourselves as specks on the pages of our own overwhelming history. I remember that my high-school biology text dealt with the human body by listing its constituent elements, measuring their quantities, and giving their monetary worth—at that time a little less than a dollar. That was a bit of the typical fodder of the modern mind, at once sensational and belittling—no accidental product of the age of Dachau and Hiroshima.

In our time Shakespeare's cliff has become the tower of a bridge—not the scene of a wakening rite of symbolic death and rebirth, but of the real and final death of suicide. Hart Crane wrote its paradigm, as if against his will, in *The Bridge*:

> Out of some subway scuttle, cell or loft
> A bedlamite speeds to thy parapets,
> Tilting there momentarily, shrill shirt ballooning,
> A jest falls from the speechless caravan.

In Shakespeare, the real Bedlamite or madman is the desperate and suicidal Gloucester. The supposed Bedlamite is in reality his true son, and together they enact an eloquent ritual in which Edgar gives his father a vision of Creation. Gloucester abandons himself to this vision, literally casting himself into it, and is renewed; he finds his life by losing it. Gloucester is saved by a renewal of his sense of the world and of his proper place in it. And this is brought about by an enactment that is communal, both in the sense that he is accompanied in it by his son, who for the time being has assumed the disguise of a madman but the role of a priest, and in the sense that it is deeply traditional in its symbols and meanings. In Crane, on the other hand, the Bedlamite is alone, surrounded by speechlessness, cut off within the crowd from any saving or renewing vision.

The height, which in Shakespeare is the traditional place of vision, has become in Crane a place of blindness; the bridge, which Crane intended as a unifying symbol, has become the symbol of a final estrangement.

HEALTH

After I had begun to think about these things, I received a letter containing an account of a more recent suicide. The following sentences from that letter seem both to corroborate Crane's lines and to clarify them:

"My friend ＿＿ jumped off the Golden Gate Bridge two months ago. . . . She had been terribly depressed for years. There was no help for her. None that she could find that was sufficient. She was trying to get from one phase of her life to another, and couldn't make it. She had been terribly wounded as a child. . . . Her wound could not be healed. She destroyed herself."

The letter had already asked, "How does a human pass through youth to maturity without 'breaking down'?" And it had answered: "help from tradition, through ceremonies and rituals, rites of passage at the most difficult stages."

My correspondent went on to say: "Healing, it seems to me, is a necessary and useful word when we talk about agriculture." And a few paragraphs later he wrote: "The theme of suicide belongs in a book about agriculture . . ."

I agree. But I am also aware that many people will find it exceedingly strange that these themes should enter so forcibly into this book. It will be thought that I am off the subject. And so I want to take pains to show that I am on the subject —and on it, moreover, in the only way most people have of getting on it: by way of the issue of their own health. Indeed, it is when one approaches agriculture from any *other* issue than that of health that one may be said to be off the subject.

The difficulty probably lies in our narrowed understanding of the word *health*. That there is some connection between how we feel and what we eat, between our bodies and the earth, is acknowledged when we say that we must "eat right to keep fit" or that we should eat "a balanced diet." But by health we mean little more than how we feel. We are healthy, we think, if we do not feel any pain or too much pain, and if we are strong enough to do our work. If we become unhealthy, then we go to a doctor who we hope will "cure" us and restore us to health. By health, in other words, we mean merely the absence of disease. Our health professionals are interested almost exclusively in preventing disease (mainly by destroying germs) and in curing disease (mainly by surgery and by destroying germs).

But the concept of health is rooted in the concept of wholeness. To be healthy is to be whole. The word *health* belongs to a family of words, a listing of which will suggest how far the consideration of health must carry us: *heal, whole, wholesome, hale, hallow, holy.* And so it is possible to give a definition to health that is positive and far more elaborate than that given to it by most medical doctors and the officers of public health.

If the body is healthy, then it is whole. But how can it be whole and yet be dependent, as it obviously is, upon other bodies and upon the earth, upon all the rest of Creation, in fact? It becomes clear that the health or wholeness of the body is a vast subject, and that to preserve it calls for a vast enterprise. Blake said that "Man has no Body distinct from his Soul . . ." and thus acknowledged the convergence of health and holiness. In that, all the convergences and dependences of Creation are surely implied. Our bodies are also not distinct from the bodies of other people, on which they depend in a complexity of ways from biological to spiritual. They are not distinct from the bodies of plants and animals, with which we are involved in the cycles of feeding and in the intricate companionships of ecological systems and of the spirit. They are not distinct from the earth, the sun and moon, and the other heavenly bodies.

It is therefore absurd to approach the subject of health piecemeal with a departmentalized band of specialists. A medical doctor uninterested in nutrition, in agriculture, in the wholesomeness of mind and spirit is as absurd as a farmer who is uninterested in health. Our fragmentation of this subject cannot be our cure, because it is our disease. The body cannot be whole alone. Persons cannot be whole alone. It is wrong to think that bodily health is compatible with spiritual confusion or cultural disorder, or with polluted air and water or impoverished soil. Intellectually, we know that these patterns of interdependence exist; we understand them better now perhaps than we ever have before; yet modern social and cultural patterns contradict them and make it difficult or impossible to honor them in practice.

To try to heal the body alone is to collaborate in the destruction of the body. Healing is impossible in loneliness; it is the opposite of loneliness. Conviviality is healing. To be healed we must come with all the other creatures to the feast of Creation. Together, the above two descriptions of suicides suggest this very powerfully. The setting of both is urban, amid the gigantic works of modern humanity. The fatal sickness is despair, a wound that cannot be healed because it is encapsulated in loneliness, surrounded by speechlessness. Past the scale of the human, our works do not liberate us—they confine us. They cut off access to the wilderness of Creation where we must go to be reborn—to receive the awareness, at once humbling and exhilarating, grievous and joyful, that we are a part of Creation, one with all that we live from and all that, in turn, lives from us. They destroy the communal rites of passage that turn us toward the wilderness and bring us home again.

THE ISOLATION OF THE BODY

Perhaps the fundamental damage of the specialist system—the damage from which all other damages issue—has been the isolation of the body. At some point we began to assume that the life of the body would be the business of grocers and medical doctors, who need take no interest in the spirit, whereas the life of the spirit would be the business of churches, which would have at best only a negative interest in the body. In the same way we began to see nothing wrong with putting the body—most often somebody else's body, but frequently our own—to a task that insulted the mind and demeaned the spirit. And we began to find it easier than ever to prefer our own bodies to the bodies of other creatures and to abuse, exploit, and otherwise hold in contempt those other bodies for the greater good or comfort of our own.

The isolation of the body sets it into direct conflict with

everything else in Creation. It gives it a value that is destructive of every other value. That this has happened is paradoxical, for the body was set apart from the soul in order that the soul should triumph over the body. The aim is stated in Shakespeare's Sonnet 146 as plainly as anywhere:

> Poor soul, the center of my sinful earth,
> Lord of these rebel powers that thee array,
> Why dost thou pine within and suffer dearth,
> Painting thy outward walls so costly gay?
> Why so large cost, having so short a lease,
> Dost thou upon thy fading mansion spend?
> Shall worms, inheritors of this excess,
> Eat up thy charge? Is this thy body's end?
> Then, soul, live thou upon thy servant's loss,
> And let that pine to aggravate thy store;
> Buy terms divine in selling hours of dross;
> Within be fed, without be rich no more.
> So shalt thou feed on death, that feeds on men,
> And death once dead, there's no more dying then.

The soul is thus set against the body, to thrive at the body's expense. And so a spiritual economy is devised within which the only law is competition. If the soul is to live in this world only by denying the body, then its relation to worldly life becomes extremely simple and superficial. Too simple and superficial, in fact, to cope in any meaningful or useful way with the world. Spiritual value ceases to have any worldly purpose or force. To fail to employ the body in this world at once for its own good and the good of the soul is to issue an invitation to disorder of the most serious kind.

What was not foreseen in this simple-minded economics of religion was that it is not possible to devalue the body and value the soul. The body, cast loose from the soul, is on its own. Devalued and cast out of the temple, the body does not skulk off like a sick dog to die in the bushes. It sets up a counterpart economy of its own, based also on the law of competition, in which it devalues and exploits the spirit. These two economies maintain themselves at each other's expense, living upon each other's loss, collaborating without cease in mutual futility and absurdity.

You cannot devalue the body and value the soul—or value anything else. The prototypical act issuing from this division was to make a person a slave and then instruct him in religion —a "charity" more damaging to the master than to the slave. Contempt for the body is invariably manifested in contempt for other bodies—the bodies of slaves, laborers, women, animals, plants, the earth itself. Relationships with all other creatures become competitive and exploitive rather than collaborative and convivial. The world is seen and dealt with, not as an ecological community, but as a stock exchange, the ethics of which are based on the tragically misnamed "law of the jungle." This "jungle" law is a basic fallacy of modern culture. The body is degraded and saddened by being set in conflict against the Creation itself, of which all bodies are members, therefore members of each other. The body is thus sent to war against itself.

Divided, set against each other, body and soul drive each other to extremes of misapprehension and folly. Nothing could be more absurd than to despise the body and yet yearn for its resurrection. In reaction to this supposedly religious attitude, we get, not reverence or respect for the body, but another kind of contempt: the desire to comfort and indulge the body with equal disregard for its health. The "dialogue of body and soul" in our time is being carried on between those who despise the body for the sake of its resurrection and those, diseased by bodily extravagance and lack of exercise, who nevertheless desire longevity above all things. These think that they oppose each other, and yet they could not exist apart. They are locked in a conflict that is really their collaboration in the destruction of soul and body both.

What this conflict has done, among other things, is to make it extremely difficult to set a proper value on the life of the body in this world—to believe that it is good, howbeit short and imperfect. Until we are able to say this and know what we mean by it, we will not be able to live our lives in the human estate of grief and joy, but repeatedly will be cast outside in violent swings between pride and despair. Desires that cannot be fulfilled in health will keep us hopelessly restless and unsatisfied.

COMPETITION

By dividing body and soul, we divide both from all else. We thus condemn ourselves to a loneliness for which the only compensation is violence—against other creatures, against the earth, against ourselves. For no matter the distinctions we draw between body and soul, body and earth, ourselves and others—the connections, the dependences, the identities remain. And so we fail to contain or control our violence. It gets loose. Though there are categories of violence, or so we think, there are no categories of victims. Violence against one is ultimately violence against all. The willingness to abuse other bodies is the willingness to abuse one's own. To damage the earth is to damage your children. To despise the ground is to despise its fruit; to despise the fruit is to despise its eaters. The wholeness of health is broken by despite.

If competition is the correct relation of creatures to one another and to the earth, then we must ask why exploitation is not more successful than it is. Why, having lived so long at the expense of other creatures and the earth, are we not healthier and happier than we are? Why does modern society exist under constant threat of the same suffering, deprivation, spite, contempt, and obliteration that it has imposed on other people and other creatures? Why do the health of the body and the health of the earth decline together? And why, in consideration of this decline of our worldly flesh and household, our "sinful earth," are we not healthier in spirit?

It is not necessary to have recourse to statistics to see that the human estate is declining with the estate of nature, and that the corruption of the body is the corruption of the soul. I know that the country is full of "leaders" and experts of various sorts who are using statistics to prove the opposite: that we have more cars, more super-highways, more TV sets, motorboats, prepared foods, etc., than any people ever had before—and are therefore better off than any people ever were before. I can see the burgeoning of this "consumer economy" and can appreciate some of its attractions and comforts. But that economy has an inside and an outside; from the outside there are other things to be seen.

I am writing this in the north-central part of Kentucky on a morning near the end of June. We have had rain for two days, hard rain during the last several hours. From where I sit I can see the Kentucky River swiftening and rising, the water already yellow with mud. I know that inside this city-oriented consumer economy there are many people who will never see this muddy rise and many who will see it without knowing what it means. I know also that there are many who will see it, and know what it means, and not care. If it lasts until the weekend there will be people who will find it as good as clear water for motorboating and waterskiing.

In the past several days I have seen some of the worst-eroded corn fields that I have seen in this country in my life. This erosion is occurring on the cash-rented farms of farmers' widows and city farmers, absentee owners, the doctors and businessmen who buy a farm for the tax breaks or to have "a quiet place in the country" for the weekends. It is the direct result of economic and agricultural policy; it might be said to *be* an economic and agricultural policy. The signs of the "agridollar," big-business fantasy of the Butz mentality are all present: the absenteeism, the temporary and shallow interest of the land-renter, the row-cropping of slopes, the lack of rotation, the plowed-out waterways, the rows running up and down the hills. Looked at from the field's edge, this is ruin, criminal folly, moral idiocy. Looked at from Washington, D.C., from inside the "economy," it is called "free enterprise" and "full production."

And around me here, as everywhere else I have been in this country—in Nebraska, Iowa, Indiana, New York, New England, Tennessee—the farmland is in general decline: fields and whole farms abandoned, given up with their scars unmended, washing away under the weeds and bushes; fine land put to row crops year after year, without rest or rotation; buildings and fences going down; good houses standing empty, unpainted, their windows broken.

And it is clear to anyone who looks carefully at any crowd that we are wasting our bodies exactly as we are wasting our land. Our bodies are fat, weak, joyless, sickly, ugly, the virtual prey of the manufacturers of medicine and cosmetics. Our bodies have become marginal; they are growing useless like our "marginal" land because we have less and less use for them. After the games and idle flourishes of modern youth, we use them only as shipping cartons to transport our brains and our few employable muscles back and forth to work.

As for our spirits, they seem more and more to comfort themselves by buying things. No longer in need of the exalted drama of grief and joy, they feed now on little shocks of greed, scandal, and violence. For many of the churchly, the life of the spirit is reduced to a dull preoccupation with getting to Heaven. At best, the world is no more than an embarrassment and a trial to the spirit, which is otherwise radically separated from it. The true lover of God must not be burdened with any care or respect for His works. While the body goes about its business of destroying the earth, the soul is supposed to lie back and wait for Sunday, keeping itself free of earthly contaminants. While the body exploits other bodies, the soul stands aloof, free from sin, crying to the gawking bystanders: "I am not enjoying it!" As far as this sort of "religion" is concerned, the body is no more than the lusterless container of the soul, a mere "package," that will nevertheless light up in eternity, forever cool and shiny as a neon cross. This separation of the soul from the body and from the world is no disease of the fringe, no aberration, but a fracture that runs through the mentality of institutional religion like a geologic fault. And this rift in the mentality of religion continues to characterize the modern mind, no matter how secular or worldly it becomes.

But I have not stated my point exactly enough. This rift is not *like* a geologic fault; it *is* a geologic fault. It is a flaw in the mind that runs inevitably into the earth. Thought affects or afflicts substance neither by intention nor by accident, but because, occurring in the Creation that is unified and whole, it must; there is no help for it.

The soul, in its loneliness, hopes only for "salvation." And yet what is the burden of the Bible if not a sense of the mutuality of influence, rising out of an essential unity, among soul and body and community and world? These are all the works of God, and it is therefore the work of virtue to make or restore harmony among them. The world is certainly thought of as a place of spiritual trial, but it is also the confluence of soul and body, word and flesh, where thoughts must become deeds, where goodness is to be enacted. This is the great meeting place, the narrow passage where spirit and flesh, word and world, pass into each other. The Bible's aim, as I read it, is not the freeing of the spirit from the world. It is the handbook of their interaction. It says that they cannot be divided; that their mutuality, their unity, is inescapable; that they are not reconciled in division, but in harmony. What else can be meant by the resurrection of the body? The body should be "filled with light," perfected in understanding. And so everywhere there is the sense of consequence, fear and desire, grief and joy. What is desirable is repeatedly defined in the tensions of the sense of consequence. False prophets are to be known "by their fruits." We are to treat others as we would be treated; thought is thus barred from any easy escape into aspiration or ideal, is turned around and forced into action. The following verses from Proverbs are not very likely the original work of a philosopher-king; they are overheard from generations of agrarian grandparents whose experience taught them that spiritual qualities become earthly events:

> I went by the field of the slothful, and by the vineyard of the man void of understanding;
> And, lo, it was all grown over with thorns, and nettles had covered the face thereof, and the stone wall thereof was broken down.
> Then I saw, and considered it well. I looked upon it, and received instruction.
> Yet a little sleep, a little slumber, a little folding of the hands to sleep:
> So shall thy poverty come as one that traveleth; and thy want as an armed man.

CONNECTIONS

I do not want to speak of unity misleadingly or too simply. Obvious distinctions can be made between body and soul, one body and other bodies, body and world, etc. But these things that appear to be distinct are nevertheless caught in a network of mutual dependence and influence that is the substantiation of their unity. Body, soul (or mind or spirit), community, and world are all susceptible to each other's influence, and they are all conductors of each other's influence. The body is damaged by the bewilderment of the spirit, and it conducts the influence of that bewilderment into the earth, the earth conducts it into the community, and so on. If a farmer fails to understand what health is, his farm becomes unhealthy; it produces unhealthy food, which damages the health of the community. But this is a network, a spherical network, by which each part is connected to every other part. The farmer is a part of the community, and so it is as impossible to say exactly where the trouble began as to say where it will end. The influences go backward and forward, up and down, round and round, compounding and branching as they go. All that is certain is that an error introduced anywhere in the network ramifies beyond the scope of prediction; consequences occur all over the place, and each consequence breeds further consequences. But it seems unlikely that an error can ramify endlessly. It spreads by way of the connections in the network, but sooner or later it must also begin to break them. We are talking, obviously, about a circulatory system, and a disease of a circulatory system tends first to impair circulation and then to stop it altogether.

Healing, on the other hand, complicates the system by opening and restoring connections among the various parts—in this way restoring the ultimate simplicity of their union. When all the parts of the body are working together, are under each other's influence, we say that it is whole; it is healthy. The same is true of the world, of which our bodies are parts. The parts are healthy insofar as they are joined harmoniously to the whole.

What the specialization of our age suggests, in one example after another, is not only that fragmentation is a disease, but that the diseases of the disconnected parts are similar or analogous to one another. Thus they memorialize their lost unity, their relation persisting in their disconnection. Any severance produces two wounds that are, among other things, the record of how the severed parts once fitted together.

The so-called identity crisis, for instance, is a disease that seems to have become prevalent after the disconnection of body and soul and the other piecemealings of the modern period. One's "identity" is apparently the immaterial part of one's being—also known as psyche, soul, spirit, self, mind, etc. The

dividing of this principle from the body and from any particular worldly locality would seem reason enough for a crisis.
Treatment, it might be thought, would logically consist in the
restoration of these connections: the lost identity would find
itself by recognizing physical landmarks, by connecting itself
responsibly to practical circumstances; it would learn to stay
put in the body to which it belongs and in the place to which
preference or history or accident has brought it; it would, in
short, find itself in finding its work. But "finding yourself,"
the pseudo-ritual by which the identity crisis is supposed to be
resolved, makes use of no such immediate references. Leaving aside the obvious, and ancient, realities of doubt and self-
doubt, as well as the authentic madness that is often the result
of cultural disintegration, it seems likely that the identity crisis
is a conventional illusion, one of the genres of self-indulgence.
It can be an excuse for irresponsibility or a fashionable mode of
self-dramatization. It is the easiest form of self-flattery—a way
to construe procrastination as a virtue—based on the romantic
assumption that "who I really am" is better in some fundamental way than the available evidence would suggest.

The fashionable cure for this condition, if I understand the
lore of it correctly, has nothing to do with the assumption of
responsibilities or the renewal of connections. The cure is "autonomy," another illusory condition, suggesting that the self
can be self-determining and independent without regard for
any determining circumstance or any of the obvious dependences. This seems little more than a jargon term for indifference to the opinions and feelings of other people. There is, in
practice, no such thing as autonomy. Practically, there is only a
distinction between responsible and irresponsible dependence.
Inevitably failing this impossible standard of autonomy, the
modern self-seeker becomes a tourist of cures, submitting his
quest to the guidance of one guru after another. The "cure"
thus preserves the disease.

It is not surprising that this strange disease of the spirit—the
self's loss of self—should have its counterpart in an anguish
of the body. One of the commonplaces of modern experience
is dissatisfaction with the body—not as one has allowed it to
become, but as it naturally is. The hardship is perhaps greater
here because the body, unlike the self, is substantial and cannot

be supposed to be inherently better than it was born to be. It can only be thought inherently worse than it *ought* to be. For the appropriate standard for the body—that is, health—has been replaced, not even by another standard, but by very exclusive physical *models*. The concept of "model" here conforms very closely to the model of the scientists and planners: it is an exclusive, narrowly defined ideal which affects destructively whatever it does not include.

Thus our young people are offered the ideal of health only by what they know to be lip service. What they are made to feel forcibly, and to measure themselves by, is the exclusive desirability of a certain physical model. Girls are taught to want to be leggy, slender, large-breasted, curly-haired, unimposingly beautiful. Boys are instructed to be "athletic" in build, tall but not too tall, broad-shouldered, deep-chested, narrow-hipped, square-jawed, straight-nosed, not bald, unimposingly handsome. Both sexes should look what passes for "sexy" in a bathing suit. Neither, above all, should look old.

Though many people, in health, are beautiful, very few resemble these models. The result is widespread suffering that does immeasurable damage both to individual persons and to the society as a whole. The result is another absurd pseudoritual, "accepting one's body," which may take years or may be the distraction of a lifetime. Woe to the man who is short or skinny or bald. Woe to the man with a big nose. Woe, above all, to the woman with small breasts or a muscular body or strong features; Homer and Solomon might have thought her beautiful, but she will see her own beauty only by a difficult rebellion. And like the crisis of identity, this crisis of the body brings a helpless dependence on cures. One spends one's life dressing and "making up" to compensate for one's supposed deficiencies. Again, the cure preserves the disease. And the putative healer is the guru of style and beauty aid. The sufferer is by definition a customer.

SEXUAL DIVISION

To divide body and soul, or body and mind, is to inaugurate an expanding series of divisions—not, however, an *infinitely* expanding series, because it is apparently the nature of division sooner or later to destroy what is divided; the principle of durability is unity. The divisions issuing from the division of body and soul are first sexual and then ecological. Many other divisions branch out from those, but those are the most important because they have to do with the fundamental relationships —with each other and with the earth—that we all have in common.

To think of the body as separate from the soul or as soulless, either to subvert its appetites or to "free" them, is to make an object of it. As a thing, the body is denied any dimension or rightful presence or claim in the mind. The concerns of the body—all that is comprehended in the term *nurture*—are thus degraded, denied any respected place among the "higher things" and even among the more exigent practicalities.

The first sexual division comes about when nurture is made the exclusive concern of women. This cannot happen until a society becomes industrial; in hunting and gathering and in agricultural societies, men are of necessity also involved in nurture. In those societies there usually have been differences between the work of men and that of women. But the necessity here is to distinguish between sexual difference and sexual division.

In an industrial society, following the division of body and soul, we have at the "upper" or professional level a division between "culture" (in the specialized sense of religion, philosophy, art, the humanities, etc.) and "practicality," and both of these become increasingly abstract. Thinkers do not act. And the "practical" men do not work with their hands, but manipulate the abstract quantities and values that come from the work of "workers." Workers are simplified or specialized into machine parts to do the wage-work of the body, which they were initially permitted to think of as "manly" because for the most part women did not do it.

Women traditionally have performed the most confining— though not necessarily the least dignified—tasks of nurture: housekeeping, the care of young children, food preparation. In the urban-industrial situation the confinement of these traditional tasks divided women more and more from the "important" activities of the new economy. Furthermore, in this situation the traditional nurturing role of men—that of provisioning the household, which in an agricultural society had become as constant and as complex as the women's role —became completely abstract; the man's duty to the household came to be simply to provide money. The only remaining *task* of provisioning—purchasing food—was turned over to women. This determination that nurturing should become *exclusively* a concern of women served to signify to both sexes that neither nurture nor womanhood was very important.

But the assignment to women of a kind of work that was thought both onerous and trivial was only the beginning of their exploitation. As the persons exclusively in charge of the tasks of nurture, women often came into sole charge of the household budget; they became family purchasing agents. The time of the household barterer was past. Kitchens were now run on a cash economy. Women had become customers, a fact not long wasted on the salesmen, who saw that in these women they had customers of a new and most promising kind. The modern housewife was isolated from her husband, from her school-age children, and from other women. She was saddled with work from which much of the skill, hence much of the dignity, had been withdrawn, and which she herself was less and less able to consider important. She did not know what her husband did at work, or after work, and she knew that her life was passing in his regardlessness and in his absence. Such a woman was ripe for a sales talk: this was the great commercial insight of modern times. Such a woman must be told—or subtly made to understand—that she must not be a drudge, that she must not let her work affect her looks, that she must not become "unattractive," that she must always be fresh, cheerful, young, shapely, and pretty. All her sexual and mortal fears would thus be given voice, and she would be made to reach for money. What was implied was always the question that a certain bank finally asked outright in a billboard advertisement: "Is your husband losing interest?"

Motivated no longer by practical needs, but by loneliness and fear, women began to identify themselves by what they bought rather than by what they did. They bought labor-saving devices which worked, as most modern machines have tended to work, to devalue or replace the skills of those who used them. They bought manufactured foods, which did likewise.

They bought any product that offered to lighten the burdens of housework, to be "kind to hands," or to endear one to one's husband. And they furnished their houses, as they made up their faces and selected their clothes, neither by custom nor invention, but by the suggestion of articles and advertisements in "women's magazines." Thus housewifery, once a complex discipline acknowledged to be one of the bases of culture and economy, was reduced to the exercise of purchasing power.* The housewife's only remaining productive capacity was that of reproduction. But even as a mother she remained a consumer, subjecting herself to an all-presuming doctor and again to written instructions calculated to result in the purchase of merchandise. Breast-feeding of babies became unfashionable, one suspects, because it was the last form of home production; no way could be found to persuade a woman to purchase her own milk. All these "improvements" involved a radical simplification of mind that was bound to have complicated, and ironic, results. As housekeeping became simpler and easier, it also became more boring. A woman's work became less accomplished and less satisfying. It became easier for her to believe that what she did was not important. And this heightened her anxiety and made her even more avid and even less discriminating as a consumer. The cure not only preserved the disease, it compounded it.

There was, of course, a complementary development in the minds of men, but there is less to say about it. The man's mind was not simplified by a degenerative process, but by a kind of coup: as soon as he separated working and living and began to work away from home, the practical considerations of the household were excerpted from his mind all at once.

In modern marriage, then, what was once a difference of work became a division of work. And in this division the household was destroyed as a practical bond between husband and

---

*She did continue to do "housework," of course. But we must ask what this had come to mean. The industrial economy had changed the criterion of housekeeping from thrift to convenience. Thrift was a complex standard, requiring skill, intelligence, and moral character, and private thrift was rightly considered a public value. Once thrift was destroyed as a value, housekeeping became simply a corrupt function of a corrupt economy: its public "value" lay in the wearing out or using up of commodities.

wife. It was no longer a condition, but only a place. It was no longer a circumstance that required, dignified, and rewarded the enactment of mutual dependence, but the site of mutual estrangement. Home became a place for the husband to go when he was not working or amusing himself. It was the place where the wife was held in servitude.

A sexual difference is not a wound, or it need not be; a sexual division is. And it is important to recognize that this division—this destroyed household that now stands between the sexes—is a wound that is suffered inescapably by both men and women. Sometimes it is assumed that the estrangement of women in their circumscribed "women's world" can only be for the benefit of men. But that interpretation seems to be based on the law of competition that is modeled in the exploitive industrial economy. This law holds that for everything that is exploited or oppressed there must be something else that is proportionately improved; thus, men must be as happy as women are unhappy.

There is no doubt that women have been deformed by the degenerate housewifery that is now called their "role"—but not, I think, for any man's benefit. If women are deformed by their role, then, insofar as the roles are divided, men are deformed by theirs. Degenerate housewifery is indivisible from degenerate husbandry. There is no escape. This is the justice that we are learning from the ecologists: you cannot damage what you are dependent upon without damaging yourself. The suffering of women is noticed now, is noticeable now, because it is not given any considerable status or compensation. If we removed the status and compensation from the destructive exploits we classify as "manly," men would be found to be suffering as much as women. They would be found to be suffering for the same reason: they are in exile from the communion of men and women, which is their deepest connection with the communion of all creatures.

For example: a man who is in the traditional sense a *good* farmer is husbandman and husband, the begetter and conserver of the earth's bounty, but he is also midwife and motherer. He is a nurturer of life. His work is domestic; he is bound to the household. But let "progress" take such a man and transform him into a technologist of production (that is, sever his bonds to the household, make useless or pointless or "uneconomical" his impulse to conserve and to nurture), and it will have made of him a creature as deformed, and as pained, as it has notoriously made of his wife.

THE DISMEMBERMENT OF THE HOUSEHOLD

We are familiar with the concept of the disintegral life of our time as a dismembered cathedral, the various concerns of culture no longer existing in reference to each other or within the discipline of any understanding of their unity. It may also be conceived, and its strains more immediately felt, as a dismembered household. Without the household—not just as a unifying ideal, but as a practical circumstance of mutual dependence and obligation, requiring skill, moral discipline, and work—husband and wife find it less and less possible to imagine and enact their marriage. Without much in particular that they can *do* for each other, they have a scarcity of practical reasons to be together. They may "like each other's company," but that is a reason for friendship, not for marriage. Aside from affection for any children they may have and their abstract legal and economic obligations to each other, their union has to be empowered by sexual energy alone.

Perhaps the most dangerous, certainly the most immediately painful, consequence of the disintegration of the household is this isolation of sexuality. The division of sexual energy from the functions of household and community that it ought both to empower and to grace is analogous to that other modern division between hunger and the earth. When it is no longer allied by proximity and analogy to the nurturing disciplines that bound the household to the cycles of fertility and the seasons, life and death, then sexual love loses its symbolic or ritualistic force, its deepest solemnity and its highest joy. It loses its sense of consequence and responsibility. It becomes "autonomous," to be valued only for its own sake, therefore frivolous, therefore destructive—even of itself. Those who speak of sex as "recreation," thinking to claim for it "a new place," only acknowledge its displacement from Creation.

The isolation of sexuality makes it subject to two influences that dangerously oversimplify it: the lore of sexual romance and capitalist economics. By "sexual romance" I mean the sentimentalization of sexual love that for generations has been the work of popular songs and stories. By means of them, young people have been taught a series of extremely dangerous falsehoods:

1. That people in love ought to conform to the fashionable models of physical beauty, and that to be unbeautiful by these standards is to be unlovable.

2. That people in love are, or ought to be, young—even though love is said to last "forever."

3. That marriage is a solution—whereas the most misleading thing a love story can do is to end "happily" with a marriage, not because there is no such thing as a happy marriage, but because marriage cannot be happy except by being *made* happy.

4. That love, alone, regardless of circumstances, can make harmony and resolve serious differences.

5. That "love will find a way" and so finally triumph over any kind of practical difficulty.

6. That the "right" partners are "made for each other," or that "marriages are made in Heaven."

7. That lovers are "each other's all" or "all the world to each other."

8. That monogamous marriage is therefore logical and natural, and "forsaking all others" involves no difficulty.

Believing these things, a young couple could not be more cruelly exposed to the abrasions of experience—or better prepared to experience marriage as another of those grim and ironic modern competitions in which the victory of one is the defeat of both.

As experience frets away gullibility, the exclusiveness of the sentimental ideal gives way to the possessiveness of sexual capitalism. Failing, as they cannot help but fail, to be each other's all, the husband and wife become each other's only. The sacrament of sexual union, which in the time of the household was a communion of workmates, and afterward tried to be a lovers' paradise, has now become a kind of marketplace in which husband and wife represent each other as sexual property. Competitiveness and jealousy, imperfectly sweetened and disguised by the illusions of courtship, now become governing principles, and they work to isolate the couple inside their marriage. Marriage becomes a capsule of sexual fate. The man must look on other men, and the woman on other women, as threats. This seems to have become particularly damaging to women; because of the progressive degeneration and isolation of their "role," their worldly stock in trade has increasingly had to be "their" men. In the isolation of the resulting sexual "privacy," the disintegration of the community begins. The energy that is the most convivial and unifying loses its communal forms and becomes divisive. This dispersal was nowhere more poignantly exemplified than in the replacement of the old ring dances, in which all couples danced together, by the so-called ballroom dancing, in which each couple dances alone. A significant part of the etiquette of ballroom dancing is, or was, that the exchange of partners was accomplished by a "trade." It is no accident that this capitalization of love and marriage was followed by a divorce epidemic—and by fashions of dancing in which each one of the dancers moves alone.

The disintegration of marriage, which completes the disintegration of community, came about because the encapsulation of sexuality, meant to preserve marriage from competition, inevitably *enclosed* competition. The principle that fenced everyone else out fenced the couple in; it became a sexual cul-de-sac. The model of economic competition proved as false to marriage as to farming. As with other capsules, the narrowness of the selective principle proved destructive of what it excluded, and what it excluded was essential to the life of what it enclosed: the nature of sexuality itself. Sexual romance cannot bear to acknowledge the generality of instinct, whereas sexual capitalism cannot acknowledge its particularity. But sexuality appears to be *both* general and particular. One cannot love a particular woman, for instance, unless one loves womankind —if not all women, at least other women. The capsule of sexual romance leaves out this generality, this generosity of instinct; it excludes Aphrodite and Dionysus. And it fails for that reason. Though sexual love can endure between the same two people for a long time, it cannot do so on the basis of this pretense of the exclusiveness of affection. The sexual capitalist—that is, the disillusioned sexual romantic—in reaction to disillusion makes the opposite oversimplification; one acknowledges one's spouse as one of a general, necessarily troublesome kind or category.

Both these attitudes look on sexual love as ownership. The sexual romantic croons, "You be-long to me." The sexual capitalist believes the same thing but has stopped crooning. Each holds that a person's sexual property shall be sufficient unto him or unto her, and that the morality of that sufficiency is to be forever on guard against expropriation. Within the capsule of marriage, as in that of economics, one intends to exploit one's property and to protect it. Once the idea of property becomes abstract or economic, both these motives begin to rule over it. They are, of course, contradictory; all that one can really protect is one's "right" or intention to exploit. The proprieties and privacies used to encapsulate marriage may have come from the tacit recognition that exploitive sex, like exploitive economics, is a very dirty business. One makes a secret of the sexuality of one's marriage for the same reason that one posts "Keep Out/Private Property" on one's strip mine. The tragedy, more often felt than acknowledged, is that what is exploited becomes undesirable.

The protective capsule becomes a prison. It becomes a household of the living dead, each body a piece of incriminating evidence. Or a greenhouse excluding the neighbors and the weather for the sake of some alien and unnatural growth. The marriage shrinks to a dull vigil of duty and legality. Husband and wife become competitors necessarily, for their only freedom is to exploit each other or to escape.

It is possible to imagine a more generous enclosure—a household welcoming to neighbors and friends; a garden open to the weather, between the woods and the road. It is possible to imagine a marriage bond that would bind a woman and a man not only to each other, but to the community of marriage, the amorous communion at which all couples sit: the sexual feast and celebration that joins them to all living things and to the fertility of the earth, and the sexual responsibility that joins them to the human past and the human future. It is possible to imagine marriage as a grievous, joyous human bond, endlessly renewable and renewing, again and again rejoining memory and passion and hope.

FIDELITY

But it is extremely difficult, now, to imagine marriage in terms of such dignity and generosity, and this difficulty is explained by the failure of these possessive and competitive forms of sexual love that have been in use for so long. This failure raises unavoidably the issue of fidelity: What is it, and what does it mean—in marriage, and also, since marriage is a fundamental relationship and metaphor, in other relationships?

No one can be glad to have this issue so starkly raised, for any consideration of it now must necessarily involve one's own bewilderment. We are apparently near the end of a degenerative phase of an evolutionary process—a long way from any large-scale regeneration. For that reason it is necessary to be hesitant and cautious, respectful of the complexity and importance of the problem. Marriage is not going to change because somebody thinks about it and recommends an "answer"; it can change only as its necessities are felt and as its circumstances change.

The idea of fidelity is perverted beyond redemption by understanding it as a grim, literal duty enforced only by willpower. This is the "religious" insanity of making a victim of the body as a victory of the soul. Self-restraint that is so purely negative is self-hatred. And one cannot be good, anyhow, just by not being bad. To be faithful merely out of duty is to be blinded to the possibility of a better faithfulness for better reasons.

It is reasonable to suppose, if fidelity is a virtue, that it is a virtue with a purpose. A purposeless virtue is a contradiction in terms. Virtue, like harmony, cannot exist alone; a virtue must lead to harmony between one creature and another. To be good for nothing is just that. If a virtue has been thought a virtue long enough, it must be assumed to have practical justification—though the very longevity that proves its practicality may obscure it. That seems to be what happened with the idea of fidelity. We heard the words "forsaking all others" repeated over and over again for so long that we lost the sense of their practical justification. They assumed the force of superstition: people came to be faithful in marriage not out of any understanding of the meaning of faith or of marriage, but out of the same fear of obscure retribution that made one careful not to break a mirror or spill the salt. Like other superstitions, this one was weakened by the scientific, positivist intellectuality of modern times and by the popular "sophistication" that came with it. Our age could be characterized as a manifold experiment in faithlessness, and if it has as yet produced no effective understanding of the practicalities of faith, it has certainly produced massive evidence of the damage and disorder of its absence.

It is possible to open this issue of the practicality of fidelity by considering that the modern age was made possible by the freeing, and concurrently by the cheapening, of energy. It can be said, of course, that the modern age was made possible by technologies that *control* energy and thus make it usable at an unprecedented rate. But such control is at best extremely limited: the devices by which industrial and military energies are used control them only momentarily; their moment of usefulness sets them loose in the world as social, ecological, and geological *forces*. We can use these energies only as explosives; we can control the rate, intensity, and time of combustion, but our effective control ends with the use of the small amount of the released energy that we are able to harness. Past that, the effects are on their own, to compound themselves as they will. In modern times we have never been able to subject our use of energy to a sense of responsibility anywhere near complex enough to be equal to its effects.

It may be that the principle of sexual fidelity, once it is again fully understood, will provide us with as good an example as we can find of the responsible use of energy. Sexuality is, after all, a form of energy, one of the most powerful. If we see sexuality as energy, then it becomes impossible to see sexual fidelity as merely a "duty," a virtue for the sake of virtue, or a superstition. If we made a superstition of fidelity, and thereby weakened it, by thinking of it as purely a moral or spiritual virtue, then perhaps we can restore its strength by recovering an awareness of its practicality.

At the root of culture must be the realization that uncontrolled energy is disorderly—that in nature all energies move in forms; that, therefore, in a human order energies must be *given* forms. It must have been plain at the beginning, as cultural degeneracy has made it plain again and again, that one can be indiscriminately sexual but not indiscriminately responsible, and that irresponsible sexuality would undermine any possibility of culture since it implies a hierarchy based purely upon brute strength, cunning, regardlessness of value and of consequence. Fidelity can thus be seen as the necessary discipline of sexuality, the practical definition of sexual responsibility, or the definition of the moral limits within which such responsibility can be conceived and enacted. The forsaking of all others is a keeping of faith, not just with the chosen one, but with the ones forsaken. The marriage vow unites not just a woman and a man with each other; it unites each of them with the community in a vow of sexual responsibility toward all others. The whole community is married, realizes its essential unity, in each of its marriages.

Another use of fidelity is to preserve the possibility of devotion against the distractions of novelty. What marriage offers —and what fidelity is meant to protect—is the possibility of moments when what we have chosen and what we desire are the same. Such a convergence obviously cannot be continuous. No relationship can continue very long at its highest emotional pitch. But fidelity prepares us for the return of these moments, which give us the highest joy we can know: that of union, communion, atonement (in the root sense of at-one-ment). The principle is stated in these lines by William Butler Yeats (by "the world" he means the world after the Fall):

> Maybe the bride-bed brings despair,
> For each an imagined image brings
> And finds a real image there;
> Yet the world ends when these two things,
> Though several, are a single light . . .

To forsake all others does not mean—because it *cannot* mean —to ignore or neglect all others, to hide or be hidden from all others, or to desire or love no others. To live in marriage is a responsible way to live in sexuality, as to live in a household is a responsible way to live in the world. One cannot enact or fulfill one's love for womankind or mankind, or even for all the women or men to whom one is attracted. If one is to have the power and delight of one's sexuality, then the generality of instinct must be resolved in a responsible relationship to a particular person. Similarly, one cannot live in the world; that is, one cannot become, in the easy, generalizing sense with which the phrase is commonly used, a "world citizen." There can be no such thing as a "global village." No matter how much one may love the world as a whole, one can live fully in it only by living responsibly in some small part of it. Where we live and who we live there with define the terms of our relationship to the world and to humanity. We thus come again to the paradox that one can become whole only by the responsible acceptance of one's partiality.

But to encapsulate these partial relationships is to entrap and condemn them in their partiality; it is to endanger them and to make them dangerous. They are enlivened and given the possibility of renewal by the double sense of particularity and generality: one lives in marriage *and* in sexuality, at home *and* in the world. It is impossible, for instance, to conceive that a man could despise women and yet love his wife, or love his own place in the world and yet deal destructively with other places.

HOME LAND AND HOUSE HOLD

What I have been trying to do is to define a pattern of disintegration that is at once cultural and agricultural. I have been groping for connections—that I think are indissoluble, though obscured by modern ambitions—between the spirit and the body, the body and other bodies, the body and the earth. If these connections do necessarily exist, as I believe they do, then it is impossible for material order to exist side by side with spiritual disorder, or vice versa, and impossible for one to thrive long at the expense of the other; it is impossible, ultimately, to preserve ourselves apart from our willingness to preserve other creatures, or to respect and care for ourselves except as we respect and care for other creatures; and, most to the point of this book, it is impossible to care for each other more or differently than we care for the earth.

This last statement becomes obvious enough when it is considered that the earth is what we all have in common, that it is

what we are made of and what we live from, and that we therefore cannot damage it without damaging those with whom
we share it. But I believe it goes farther and deeper than that.
There is an uncanny *resemblance* between our behavior toward
each other and our behavior toward the earth. Between our
relation to our own sexuality and our relation to the reproduc-
tivity of the earth, for instance, the resemblance is plain and
strong and apparently inescapable. By some connection that
we do not recognize, the willingness to exploit one becomes
the willingness to exploit the other. The conditions and the
means of exploitation are likewise similar.

The modern failure of marriage that has so estranged the
sexes from each other seems analogous to the "social mobility"
that has estranged us from our land, and the two are historically
parallel. It may even be argued that these two estrangements
are very close to being one, both of them having been caused
by the disintegration of the household, which was the formal
bond between marriage and the earth, between human sexuality and its sources in the sexuality of Creation. The importance
of this practical bond has not been often or very openly recognized in our tradition; in modern times it has almost disappeared under the burden of adverse fashion and economics. It
is necessary to go far back to find it clearly exemplified.

To my mind, one of the best examples that we have is in Homer's *Odyssey*. Nowhere else that I know are the connections
between marriage and household and the earth so fully and so
carefully understood.

At the opening of the story Odysseus, after a twenty-year
absence, is about to begin the last leg of his homeward journey.
The sole survivor of all his company of warriors, having lived
through terrible trials and losses, Odysseus is now a castaway
on the island of the goddess Kalypso. He is Kalypso's lover
but also virtually her prisoner. At night he sleeps with Kalypso
in her cave; by day he looks across the sea toward Ithaka, his
home, and weeps. Homer does not stint either feeling—the
delights of Kalypso's cave, where the lovers "revel and rest
softly, side by side," or the grief and longing of exile.

But now Zeus commands Kalypso to allow Odysseus to depart; she comes to tell him that he is free to go. And yet it is
a tragic choice that she offers him: he must choose between

her and Penélopê, his wife. If he chooses Kalypso, he will be immortal, but remain in exile; if he chooses Penélopê, he will return home at last, but will die in his time like other men:

> If you could see it all, before you go—
> all the adversity you face at sea—
> you would stay here, and guard this house, and be
> immortal—though you wanted her forever,
> that bride for whom you pine each day.
> Can I be less desirable than she is?
> Less interesting? Less beautiful? Can mortals
> compare with goddesses in grace and form?

And Odysseus answers:

> My quiet Penélopê—how well I know—
> would seem a shade before your majesty,
> death and old age being unknown to you,
> while she must die. Yet, it is true, each day
> I long for home . . .

This is, in effect, a wedding ritual much like our own, in which Odysseus forsakes all others, in renouncing the immortal womanhood of the goddess, and renews his pledge to the mortal terms of his marriage. But unlike our ritual, this one involves an explicit loyalty to a home. Odysseus' far-wandering through the wilderness of the sea is not merely the return of a husband; it is a journey home. And a great deal of the power as well as the moral complexity of *The Odyssey* rises out of the richness of its sense of home.

By the end of Book XXIII, it is clear that the action of the narrative, Odysseus' journey from the cave of Kalypso to the bed of Penélopê, has revealed a structure that is at once geographical and moral. This structure may be graphed as a series of diminishing circles centered on one of the posts of the marriage bed. Odysseus makes his way from the periphery toward that center.

All around, this structure verges on the sea, which is the wilderness, ruled by the forces of nature and by the gods. In spite of the excellence of his ship and crew and his skill in navigation, a man is alien there. Only when he steps ashore does he enter a human order. From the shoreline of his island of

Ithaka, Odysseus makes his way across a succession of boundaries, enclosed and enclosing, with the concentricity of a blos-
som around its pistil, a human pattern resembling a pattern of
nature. He comes to his island, to his own lands, to his town,
to his household and house, to his bedroom, to his bed.

As he moves toward this center he moves also through a series of recognitions, tests of identity and devotion. By these, his
homecoming becomes at the same time a restoration of order.
At first, having been for a while uncertain of his whereabouts,
he recognizes his homeland by the conformation of the countryside and by a certain olive tree. He then becomes the guest
of his swineherd, Eumaios, and tests his loyalty, though Eumaios will not be permitted to recognize his master until the
story approaches its crisis. In the house of Eumaios, Odysseus
meets and makes himself known to his son, Telémakhos. As he
comes, disguised as a beggar, into his own house, he is recognized by Argus, his old hunting dog. That night, as the guest of
Penélopê, who does not yet know who he is, he is recognized
by his aged nurse, Eurýkleia, who sees a well-remembered scar
on his thigh as she is bathing his feet.

He is scorned and abused as a vagabond by the band of
suitors who, believing him dead, have been courting his wife,
consuming his meat and wine, desecrating his household, and
plotting the murder of his son. Penelope proposes a trial by
which the suitors will compete for her: she will become the
bride of whichever one can string the bow of her supposedly
dead husband and shoot an arrow through the aligned helve-
sockets of twelve axe heads. The suitors fail. Odysseus per-
forms the feat easily and is thereby recognized as "the great
husband" himself. And then, with the help of the swineherd,
the cowherd, and Telémakhos, he proceeds to trap the suitors and slaughter them all without mercy. To so distinguished
a commentator as Richmond Lattimore, their punishment
"seems excessive." But granting the acceptability of violent
means to a warrior such as Odysseus, this outcome seems to
me appropriate to the moral terms of the poem. It is made
clear that the punishment is not merely the caprice of a human
passion: Odysseus enacts the will of the gods; he is the agent of
a divine judgment. The suitors' sin is their utter contempt for

the domestic order that the poem affirms. They do not respect or honor the meaning of the household, and in *The Odyssey* this meaning is paramount.

It is therefore the recognition of Odysseus by Penélopê that is the most interesting and the most crucial. By the time Odysseus' vengeance and his purification of the house are complete, Penélopê is the only one in the household who has not acknowledged him. It is only reasonable that she should delay this until she is absolutely certain. After all, she has waited twenty years; it is not to be expected that she would be less than cautious now. Her faith has been equal and more than equal to his, and now she proves his equal also in cunning. She tells Euríkleia to move their bed outside their bedroom and to make it up for Odysseus there. Odysseus' rage at hearing that identifies him beyond doubt, for she knew that only Odysseus would know—it is their "pact and pledge" and "secret sign" —that the bed could not be moved without destroying it. He built their bedroom with his own hands, and an old olive tree, as he says,

> grew like a pillar on the building plot,
> and I laid out our bedroom round that tree . . .
> . . . I lopped off the silvery leaves and branches,
> hewed and shaped that stump from the roots up
> into a bedpost . . .

She acknowledges him then, and only then does she give herself to his embrace.

> Now from his breast into his eyes the ache
> of longing mounted, and he wept at last,
> his dear wife, clear and faithful in his arms,
> longed for
> as the sunwarmed earth is longed for by a swimmer
> spent in rough water where his ship went down . . .

And so in the renewal of his marriage, the return of Odysseus and the restoration of order are complete. The order of the kingdom is centered on the marriage bed of the king and queen, and that bed is rooted in the earth. The figure last quoted makes explicit at last the long-hinted analogy between

Odysseus' fidelity to his wife and his fidelity to his homeland. In Penélopê's welcoming embrace his two fidelities become one.

For Odysseus, then, marriage was not merely a legal bond, nor even merely a sacred bond, between himself and Penélopê. It was part of a complex practical circumstance involving, in addition to husband and wife, their family of both descendants and forebears, their household, their community, and the sources of all these lives in memory and tradition, in the countryside, and in the earth. These things, wedded together in his marriage, he thought of as his home, and it held his love and faith so strongly that sleeping with a goddess could not divert or console him in his exile.

In Odysseus' return, then, we see a complete marriage and a complete fidelity. To reduce marriage, as we have done, to a mere contract of sexual exclusiveness is at once to degrade it and to make it impossible. That is to take away its dignity and its potency of joy, and to make it only a pitiful little duty—not a union, but a division and a solitude.

The Odyssey's understanding of marriage as the vital link which joins the human community and the earth is obviously full of political implication. In this it will remind us of the Confucian principle that "The government of the state is rooted in family order." But *The Odyssey* goes further than the Confucian texts, it seems to me, in its understanding of agricultural value as the foundation of domestic order and peace.

I have considered the poem so far as describing a journey from the non-human order of the sea wilderness to the human order of the cleansed and reunited household. But it is also a journey between two kinds of human value; it moves from the battlefield of Troy to the terraced fields of Ithaka, which, through all the years and great deeds of Odysseus' absence, the peasants have not ceased to farm.

The Odyssey begins in the world of *The Iliad*, a world which, like our own, is war-obsessed, preoccupied with "manly" deeds of exploitation, anger, aggression, pillage, and the disorder, uprootedness, and vagabondage that are their result. At the end of the poem, Odysseus moves away from the values of that world toward the values of domesticity and peace. He restores order to his household by an awesome violence, it is true. But that finished and the house purified, he re-enters his marriage, the bedchamber and the marriage bed rooted in the earth. From there he goes into the fields.

The final recognition scene occurs between Odysseus and his old father, Laërtês:

> Odysseus found his father in solitude
> spading the earth around a young fruit tree.
>
> He wore a tunic, patched and soiled, and leggings—
> oxhide patches, bound below his knees
> against the brambles . . .

The point is not stated—the story is moving so evenly now toward its conclusion that it will not trouble to remind us that the man thus dressed is a *king*—but it is clear that Laërtês has survived his son's absence and the consequent grief and disorder *as a peasant*. Although Odysseus jokes about his father's appearance, the appropriateness of what he is doing is never questioned. In a time of disorder he has returned to the care of the earth, the foundation of life and hope. And Odysseus finds him in an act emblematic of the best and most responsible kind of agriculture: an old man caring for a young tree.

But the homecoming of Odysseus is still not complete. During his wanderings, he was instructed by the ghost of the seer Teirêsias to perform what is apparently to be a ritual of atonement. As the poem ends he still has this before him. Carrying an oar on his shoulder, he must walk inland until he comes to a place where men have no knowledge of the sea or ships, where a passerby will mistake his oar for a winnowing fan. There he must "plant" his oar in the ground and make a sacrifice to the sea god, Poseidon. Home again, he must sacrifice to all the gods. Like those people of the Biblical prophecy who will "beat their swords into plowshares, and their spears into pruning hooks" and not "learn war any more," Odysseus will not know rest until he has carried the instrument of his sea wanderings inland and planted it like a tree, until he has seen the symbol of his warrior life as a farming tool. But after his atonement has been made, a gentle death will come to him when he is weary with age, his countrymen around him "in blessed peace."

The Odyssey, then, is in a sense an anti-*Iliad*, posing against the warrior values of the other epic—the glories of battle and foreign adventuring—an affirmation of the values of domesticity and farming. But at the same time *The Odyssey* is too bountiful and wise to set these two kinds of value against each other in any purity or exclusiveness of opposition. Even less does it set into such opposition the two kinds of experience. The point seems to be that these apparently opposed experiences are linked together. The higher value may be given to domesticity, but this cannot be valued or understood alone. Odysseus' fidelity and his homecoming are as moving and instructive as they are precisely because they are the result of *choice*. We know—as Odysseus undoubtedly does also—the extent of his love for Penélopê because he can return to her only by choosing her, at the price of death, over Kalypso. We feel and understand, with Odysseus, the value of Ithaka as a homeland, because bound inextricably to the experience of his return is the memory of his absence, of his long wandering at sea, and even of the excitement of his adventures. The prophecy of the peaceful death that is to come to him is so deeply touching because the poem has so fully realized the experiences of discord and violent death. The farm life of the island seems so sweet and orderly because we know the dark wilderness of natural force and mystery within which the fields are cleared and lighted.

THE NECESSITY OF WILDNESS

Domestic order is obviously threatened by the margin of wilderness that surrounds it. Marriage may be destroyed by instinctive sexuality; the husband may choose to remain with Kalypso or the wife may run away with godlike Paris. And the forest is always waiting to overrun the fields. These are real possibilities. They must be considered, respected, even feared.

And yet I think that no culture that hopes to endure can afford to destroy them or to set up absolute safeguards against them. Invariably the failure of organized religions, by which they cut themselves off from mystery and therefore from sanctity, lies in the attempt to impose an absolute division between faith and doubt, to make belief perform as knowledge; when they forbid their prophets to go into the wilderness, they lose the possibility of renewal. And the most dangerous tendency in modern society, now rapidly emerging as a scientific-industrial ambition, is the tendency toward encapsulation of human order—the severance, once and for all, of the umbilical cord fastening us to the wilderness or the Creation. The threat is not only in the totalitarian desire for absolute control. It lies in the willingness to ignore an essential paradox: the natural forces that so threaten us are the same forces that preserve and renew us.

An enduring agriculture must never cease to consider and respect and preserve wildness. The farm can exist only within the wilderness of mystery and natural force. And if the farm is to last and remain in health, the wilderness must survive within the farm. That is what agricultural fertility *is*: the survival of natural process in the human order. To learn to preserve the fertility of the farm, Sir Albert Howard wrote, we must study the forest.

Similarly, the instinctive sexuality within which marriage exists must somehow be made to thrive within marriage. To divide one from the other is to degrade both and ultimately to destroy marriage.

Fidelity to human order, then, if it is fully responsible, implies fidelity also to natural order. Fidelity to human order makes devotion possible. Fidelity to natural order preserves the possibility of choice, the possibility of the renewal of devotion. Where there is no possibility of choice, there is no possibility of faith. One who returns home—to one's marriage and household and place in the world—desiring anew what was previously chosen, is neither the world's stranger nor its prisoner, but is at once in place and free.

The relation between these two fidelities, inasmuch as they sometimes appear to contradict one another, cannot help but be complex and tricky. In our present stage of cultural evolution, it cannot help but be baffling as well. And yet it is only the double faith that is adequate to our need. If we are to have a culture as resilient and competent in the face of necessity as it needs to be, then it must somehow involve within itself a ceremonious generosity toward the wilderness of natural force and instinct. The farm must yield a place to the forest, not as a wood lot, or even as a necessary agricultural principle, but as a sacred grove—a place where the Creation is let alone, to serve as instruction, example, refuge; a place for people to go, free of work and presumption, to let themselves alone. And marriage must recognize that it survives because of, as well as in spite of, Kalypso and Paris and the generosity of instinct that they represent. It must give some ceremonially acknowledged place to the sexual energies that now thrive outside all established forms, in the destructive freedom of moral ignorance or disregard. Without these accommodations we will remain divided: some of us will continue to destroy the world for purely human ends, while others, for the sake of nature, will abandon the task of human order.

What forms or revisions of forms may be adequate to this double faith, I do not know. Cultural solutions are organisms, not machines, and they cannot be invented deliberately or imposed by prescription. Perhaps all that one can do is to clarify as well as possible the needs and pressures that bear upon the process of cultural evolution. I am certain, however, that no satisfactory solution can come from considering marriage alone or agriculture alone. These are our basic connections to each other and to the earth, and they tend to relate analogically and to be reciprocally defining: our demands upon the earth are determined by our ways of living with one another; our regard for one another is brought to light in our ways of using the earth. And I am certain that neither can be changed for the better in the experimental, prescriptive ways we have been using. Ways of life change only in living. To live by expert advice is to abandon one's life.

"FREEDOM" FROM FERTILITY

The household is the bond of marriage that is most native to it, that grows with it and gives it substantial being in the world. It is the practical condition within which husband and wife can enact devotion and loyalty to each other. The motive power of sexual love is thus joined directly to constructive work and is given communal and ecological value. Without the particular demands and satisfactions of the making and keeping of a household, the sanctity and legality of marriage remain abstract, in effect theoretical, and its sexuality becomes a danger. Work is the health of love. To last, love must enflesh itself in the materiality of the world—produce food, shelter, warmth or shade, surround itself with careful acts, well-made things. This, I think, is what Millen Brand means in *Local Lives* when he speaks of the "threat" of love—"so that perhaps acres of earth and its stones are needed and drawn-out work and monotony/ to balance that danger. . . ."

Marriage and the care of the earth are each other's disciplines. Each makes possible the enactment of fidelity toward the other. As the household has become increasingly generalized as a function of the economy and, as a consequence, has become increasingly "mobile" and temporary, these vital connections have been weakened and finally broken. And whatever has been thus disconnected has become a ground of exploitation for some breed of salesman, specialist, or expert.

A direct result of the disintegration of the household is the division of sexuality from fertility and their virtual takeover by specialists. The specialists of human sexuality are the sexual clinicians and the pornographers, both of whom subsist on the increasing possibility of sex between people who neither know nor care about each other. The specialists of human fertility are the evangelists, technicians, and salesmen of birth control, who subsist upon our failure to see any purpose or virtue in sexual discipline. In this, as in our use of every other kind of energy, our inability to contemplate any measure of restraint or forbearance has been ruinous. Here the impulse is characteristically that of the laboratory scientist: to encapsulate sexuality by separating it absolutely from the problems of fertility.

This division occurs, it seems to me, in a profound cultural failure: the loss of any sense that sexuality and fertility might exist together compatibly in this world. We have lost this possibility because we do not understand, because we cannot bear to consider the meaning of restraint.* The sort of restraint I am talking about is illustrated in a recent *National Geographic*

---

*At the root of this failure is probably another sexual division: the assignment to women of virtually all responsibility for sexual discipline.

article about the people of Hunza in northern Pakistan. The author is a woman, Sabrina Michaud, and she is talking with a Hunza woman in her kitchen:

"'What have you done to have only one child?' she asks me. Her own children range from 12 to 30 years of age, and seem evenly spaced, four to five years apart. 'We leave our husband's bed until each child is weaned,' she explains simply. But this natural means of birth control has declined, and population has soared."

The woman's remark is thus passed over and not returned to; but if I understand the significance of this paragraph, it is of great importance. The decline of "this natural means of birth control" seems to have been contemporaneous with the coming of roads and "progress" and the opening up of a previously isolated country. What is of interest is that in their isolation in arid, narrow valleys surrounded by the stone and ice of the Karakoram Mountains, these people had practiced sexual restraint as a form of birth control. They had neither our statistical expertise nor our doom-prophets of population growth; it just happened that, placed geographically as they were, they lived always in sight of their agricultural or ecological limits, and they made a competent response.

We have been unable to see the difference between this kind of restraint—a cultural response to an understood practical limit—and the obscure, self-hating, self-congratulating Victorian self-restraint, of which our attitudes and technologies of sexual "freedom" are merely the equally obscure other side. This so-called freedom fragments us and turns us more vehemently and violently than before against our own bodies and against the bodies of other people.

For the care or control of fertility, both that of the earth and that of our bodies, we have allowed a technology of chemicals and devices to replace entirely the cultural means of ceremonial forms, disciplines, and restraints. We have gathered up the immense questions that surround the coming of life into the world and reduced them to simple problems for which we have manufactured and marketed simple solutions. An infertile woman and an infertile field both receive a dose of chemicals, at the calculated risk of undesirable consequences, and are thus equally reduced to the status of productive machines.

And for unwanted life—sperm, ova, embryos, weeds, insects, etc.—we have the same sort of ready remedies, for sale, of course, and characteristically popularized by advertisements that speak much of advantages but little of problems.

The result is that we are bringing up a generation of young people who feel that they are "free from worry" about fertility. The pharmacist or the doctor will look after the fertility of the body, and the farming experts and agribusinessmen will look after the fertility of the earth. This is to short-circuit human culture at its source. It is, in effect, to remove from consciousness the two fundamental issues of human life. It permits two great powers to be regarded and used as if they were unimportant.

More serious is the resort to "authorized" modes of direct violence. In land use, this is the permanent diminishment or destruction of fertility as an allowable cost of production, as in strip-mining or in the sort of agriculture that good farmers have long referred to also as "mining." This use of technological means cuts across all issues of health and culture for the sake of an annual quota of production.

The human analogue is in the "harmless" and "simple" surgeries of permanent sterilization, which are now being promoted by a propaganda of extreme oversimplification. The publicity on this subject is typically evangelical in tone and simplistically moral; the operations are recommended like commercial products by advertisings complete with exuberant testimonials of satisfied customers and appeals to the prospective customer's maturity, sexual pride, and desire for "freedom"; and the possible physical and psychological complications are played down, misrepresented, not mentioned at all, or simply not known. It is altogether possible that the operations will be performed by doctors as perfunctory, simplistic, presumptuous, and uninforming as the public literature.

I am fully aware of the problem of overpopulation, and I do not mean to say that birth control is unnecessary. What I do mean to say is that any means of birth control is a serious matter, both culturally and biologically, and that sterilization is the most serious of all: to give up fertility is a major change, as important as birth, puberty, marriage, or death.

The great changes having to do with a woman's fertility—puberty, childbirth, and menopause—have, like sexual desire, the unarguable sanction of biological determinism. They belong to a kind of natural tradition. As a result, they are not only endurable, but they belong to a process—the life process or the Wheel of Life—that we have learned to affirm with some understanding. We know, among other things, that this process includes tragedy and survives it, even triumphs over it. The same applies to the occasions of a man's fertility, although not so formidably, a man being less involved, physically, in the *predicament* of fertility and consciously involved in it only if he wants to be. Nevertheless, he comes to fertility and, if he is a moral person, to the same issues of responsibility that it poses for women.

One of the fundamental interests of human culture is to impose this responsibility, to subject fertility to moral will. Culture articulates needs and forms for sexual restraint and involves issues of value in the process of mating. It is possible to imagine that the resulting tension creates a distinctly human form of energy, highly productive of works of the hands and the mind. But until recently there was no division between sexuality and fertility, because none was possible.

This division was made possible by modern technology, which subjected human fertility, like the fertility of the earth, to a new kind of will: the technological will, which may not *necessarily* oppose the moral will, but which has not only tended to do so, but has tended to replace it. Simply because it became possible—and simultaneously profitable—we have cut the cultural ties between sexuality and fertility, just as we have cut those between eating and farming. By "freeing" food and sex from worry, we have also set them apart from thought, responsibility, and the issue of quality. The introduction of "chemical additives" has tended to do away with the issue of taste or preference; the specialist of sex, like the specialist of food, is dealing with a commodity, which he can measure but cannot value.

What is horrifying is not only that we are relying so exclusively on a technology of birth control that is still experimental, but that we are using it *casually*, in utter cultural nakedness, unceremoniously, without sufficient understanding, and as a substitute for cultural solutions—exactly as we now employ the technology of land use. And to promote these means without cultural and ecological insight, as merely a way to divorce sexuality from fertility, pleasure from responsibility —or to *sell* them that way for ulterior "moral" motives—is to try to cure a disease by another disease. That is only a new battle in the old war between body and soul—as if we were living in front of a chorus of the most literal fanatics chanting: "If thy right eye offend thee, pluck it out! If thy right hand offend thee, cut it off!"

The technologists of fertility exercise the powers of gods and the social function of priests without community ties or cultural responsibilities. The clinicians of sex change the lives of people —as the clinicians of agriculture change the lives of places and communities—to whom they are strangers and whom they do not know. These specialists thrive in a profound cultural rift, and they are always accompanied by the exploiters who mine that rift for gold. The pornographer exploits sexual division. And working the similar division between us and our land we have the "agribusinessmen," the pornographers of agriculture.

FERTILITY AS WASTE

But there is yet another and more direct way in which the isolation of the body has serious agricultural effects. That is in our society's extreme oversimplification of the relation between the body and its food. By regarding it as merely a consumer of food, we reduce the function of the body to that of a conduit which channels the nutrients of the earth from the supermarket to the sewer. Or we make it a little factory which transforms fertility into pollution—to the enormous profit of "agribusiness" and to the impoverishment of the earth. This is another technological and economic interruption of the cycle of fertility.

Much has already been said here about the division between the body and its food in the productive phase of the cycle. It is the alleged wonder of the Modern World that so many people take energy from food in which they have invested no energy, or very little. Ninety-five percent of our people, boasted the former deputy assistant secretary of agriculture, are now free of the "drudgery" of food production. The meanings of that division, as I have been trying to show, are intricate and degenerative. But that is only half of it. Ninety-five percent (at least) of our people are also free of any involvement or interest in the maintenance phase of the cycle. As their bodies take in and use the nutrients of the soil, those nutrients are transformed into what we are pleased to regard as "wastes"—and are duly wasted.

This waste also has its cause in the old "religious" division between body and soul, by which the body and its products are judged offensive. Once, living with this offensiveness was considered a condemnation, and that was bad enough. But modern technology "saved" us with the flush toilet and the waterborne sewage system. These devices deal with the "wastes" of our bodies by simply removing them from consideration. The irony is that this technological purification of the body requires the pollution of the rivers and the starvation of the fields. It makes the alleged offensiveness of the body truly and inescapably offensive and blinds an entire society to the knowledge that these "offensive wastes" are readily purified in the topsoil—that, indeed, from an ecological point of view, these are not wastes and are not offensive, but are valuable agricultural products essential both to the health of the land and to that of the "consumers."

Our system of agriculture, by modeling itself on economics rather than biology, thus removes food from the *cycle* of its production and puts it into a finite, linear process that in effect destroys it by trans-forming it into waste. That is, it transforms food into fuel, a form of energy that is usable only once, and in doing so it transforms the body into a consumptive machine.

It is strange, but only apparently so, that this system of agriculture is institutionalized, not in any form of rural life or culture, but in what we call our "urban civilization." The cities subsist in competition with the country; they live upon a one-way movement of energies out of the countryside—food and fuel, manufacturing materials, human labor, intelligence, and talent. Very little of this energy is ever returned. Instead of gathering these energies up into coherence, a cultural consummation that would not only return to the countryside what belongs to it, but also give back generosities of learning and art, conviviality and order, the modern city dissipates and wastes them. Along with its glittering "consumer goods," the modern city produces an equally characteristic outpouring of garbage and pollution—just as it produces and/or collects unemployed, unemployable, and otherwise wasted people.

Once again it must be asked, if competition is the appropriate relationship, then why, after generations of this inpouring of rural wealth, materials, and humanity into the cities, are the cities and the countryside in equal states of disintegration and disrepair? Why have the rural and urban communities *both* fallen to pieces?

HEALTH AND WORK

The modern urban-industrial society is based on a series of radical disconnections between body and soul, husband and wife, marriage and community, community and the earth. At each of these points of disconnection the collaboration of corporation, government, and expert sets up a profit-making enterprise that results in the further dismemberment and impoverishment of the Creation.

Together, these disconnections add up to a condition of critical ill health, which we suffer in common—not just with each other, but with all other creatures. Our economy is based upon this disease. Its aim is to separate us as far as possible from the sources of life (material, social, and spiritual), to put these sources under the control of corporations and specialized professionals, and to sell them to us at the highest profit. It fragments the Creation and sets the fragments into conflict with one another. For the relief of the suffering that comes of this fragmentation and conflict, our economy proposes, not health, but vast "cures" that further centralize power and increase profits: wars, wars on crime, wars on poverty, national schemes of medical aid, insurance, immunization, further industrial and economic "growth," etc.; and these, of course, are followed by more regulatory laws and agencies to see that our health is protected, our freedom preserved, and our money well spent. Although there may be some "good intention" in this, there is little honesty and no hope.

Only by restoring the broken connections can we be healed. Connection *is* health. And what our society does its best to disguise from us is how ordinary, how commonly attainable, health is. We lose our health—and create profitable diseases and dependences—by failing to see the direct connections between living and eating, eating and working, working and loving. In gardening, for instance, one works with the body to feed the body. The work, if it is knowledgeable, makes for excellent food. And it makes one hungry. The work thus makes eating both nourishing and joyful, not consumptive, and keeps the eater from getting fat and weak. This is health, wholeness, a source of delight. And such a solution, unlike the typical industrial solution, does not cause new problems.

The "drudgery" of growing one's own food, then, is not drudgery at all. (If we make the growing of food a drudgery, which is what "agribusiness" does make of it, then we also make a drudgery of eating and of living.) It is—in addition to being the appropriate fulfillment of a practical need—a sacrament, as eating is also, by which we enact and understand our oneness with the Creation, the conviviality of one body with all bodies. This is what we learn from the hunting and farming rituals of tribal cultures.

As the connections have been broken by the fragmentation and isolation of work, they can be restored by restoring the wholeness of work. There is work that is isolating, harsh, destructive, specialized or trivialized into meaninglessness. And there is work that is restorative, convivial, dignified and dignifying, and pleasing. Good work is not just the maintenance of connections—as one is now said to work "for a living" or "to support a family"—but the *enactment* of connections. It *is* living, and a way of living; it is not support for a family in the sense of an exterior brace or prop, but is one of the forms and acts of love.

To boast that now "95 percent of the people can be freed from the drudgery of preparing their own food" is possible only to one who cannot distinguish between these kinds of work. The former deputy assistant secretary cannot see work as a vital connection; he can see it only as a trade of time for money, and so of course he believes in doing as little of it as possible, especially if it involves the use of the body. His ideal is apparently the same as that of a real-estate agency which promotes a rural subdivision by advertising "A homelife of endless vacation." But the society that is so glad to be free of the drudgery of growing and preparing food also boasts a thriving medical industry to which it is paying $500 per person per year. And that is only the down payment.

We embrace this curious freedom and pay its exorbitant cost because of our hatred of bodily labor. We do not want to work "like a dog" or "like an ox" or "like a horse"—that is, we do not want to use ourselves as beasts. This as much as anything is the cause of our disrespect for farming and our abandonment of it to businessmen and experts. We remember, as we should, that there have been agricultural economies that used people as beasts. But that cannot be remedied, as we have attempted to do, by using people as machines, or by not using them at all.

Perhaps the trouble began when we started using animals disrespectfully: as "beasts"—that is, as if they had no more feeling than a machine. Perhaps the destructiveness of our use of machines was prepared in our willingness to abuse animals. That it was never necessary to abuse animals in order to use them is suggested by a passage in *The Horse in the Furrow*, by George Ewart Evans. He is speaking of how the medieval ox teams were worked at the plow: ". . . the ploughman at the handles, the team of oxen—yoked in pairs or four abreast—and the driver who walked alongside with his goad." And then he says: "It is also worth noting that in the Welsh organization . . . the counterpart of the driver was termed *y geilwad* or the *caller*. He walked *backwards* in front of the oxen singing to them as they worked. Songs were specially composed to suit the rhythm of the oxen's work . . ."

That seems to me to differ radically from our present customary use of any living thing. The oxen were not used as beasts or machines, but as fellow creatures. It may be presumed that this work used people the same way. It is possible, then, to believe that there is a kind of work that does not require abuse or misuse, that does not use anything as a substitute for anything else. We are working well when we use ourselves as the fellow creatures of the plants, animals, materials, and other people we are working with. Such work is unifying, healing. It brings us home from pride and from despair, and places us responsibly within the human estate. It defines us as we are: not too good to work with our bodies, but too good to work poorly or joylessly or selfishly or alone.

Instead of sending the experimenter into the fields and meadows to question the farmer and the land worker so as to understand how important quality is, and above all to take up a piece of land himself, the new authoritarian doctrine demands that he shut himself up in a study . . .

SIR ALBERT HOWARD, *The Soil and Health*

. . . his education had had the curious effect of making things that he read and wrote more real to him than things he saw. Statistics about agricultural labourers were the substance; any real ditcher, ploughman, or farmer's boy, was the shadow. Though he had never noticed it himself, he had a great reluctance, in his work, ever to use such words as "man" or "woman." He preferred to write about "vocational groups," "elements," "classes" and "populations": for, in his own way, he believed as firmly as any mystic in the superior reality of the things that are not seen.

C. S. LEWIS, *That Hideous Strength*

CHAPTER EIGHT

Jefferson, Morrill, and the Upper Crust

THE CONVICTION OF THOMAS JEFFERSON

IN the mind of Thomas Jefferson, farming, education, and democratic liberty were indissolubly linked. The great conviction of his life, which he staked his life upon and celebrated in a final letter two weeks before his death, was "that the mass of mankind has not been born with saddles on their backs, nor a favored few booted and spurred, ready to ride them legitimately, by the grace of God." But if liberty was in that sense a right, it was nevertheless also a privilege to be earned, deserved, and strenuously kept; to keep themselves free, he thought, a people must be stable, economically independent, and virtuous. He believed—on the basis, it should be remembered, of extensive experience both in this country and abroad—that these qualities were most dependably found in the farming people: "Cultivators of the earth are the most valuable citizens. They are the most vigorous, the most independent, the most virtuous, and they are tied to their country, and wedded to its liberty and interests by the most lasting bonds." These bonds were not merely those of economics and property, but those, at once more feeling and more practical, that come from the investment in a place and a community of work, devotion, knowledge, memory, and association.

By contrast, Jefferson wrote: "I consider the class of artificers as the panders of vice, and the instruments by which the liberties of a country are generally overturned." By "artificers" he meant manufacturers, and he made no distinction between "management" and "labor." The last-quoted sentence is followed by no explanation, but its juxtaposition with the one first quoted suggests that he held manufacturers in suspicion because their values were already becoming abstract, enabling them to be "socially mobile" and therefore subject preeminently to the motives of self-interest.

To foster the strengths and virtues necessary to citizenship in a democracy, public education was obviously necessary, and

Jefferson never ceased to be thoughtful of that necessity: ". . . I do most anxiously wish to see the highest degrees of education given to the higher degrees of genius, and to all degrees of it, so much as may enable them to read and understand what is going on in the world, and to keep their part of it going on right for nothing can keep it right but their own vigilant and distrustful superintendence."

And all these statements must be read in the light of Jefferson's apprehension of the disarray of agriculture and of agricultural communities in his time: ". . . the long succession of years of stunted crops, of reduced prices, the general prostration of the farming business, under levies for the support of manufacturers, etc., with the calamitous fluctuations of value in our paper medium, have kept agriculture in a state of abject depression, which has peopled the Western States by silently breaking up those on the Atlantic . . ."

JUSTIN MORRILL AND THE LAND-GRANT COLLEGE ACTS

On July 2, 1862, two days less than thirty-six years after the death of Jefferson, the first of the land-grant college acts became law. This was the Morrill Act, which granted "an amount of public land, to be apportioned to each State a quantity equal to thirty thousand acres for each Senator and Representative in Congress . . ." The interest on the money from the sale of these lands was to be applied by each state "to the endowment, support, and maintenance of at least one college where the leading object shall be . . . to teach such branches of learning as are related to agriculture and the mechanic arts . . . in order to promote the liberal and practical education of the industrial classes in the several pursuits and professions in life."

In 1887 Congress passed the Hatch Act, which created the state agricultural experiment stations, with the purpose, among others, of promoting "a sound and prosperous agriculture and rural life as indispensable to the maintenance of maximum employment and national prosperity and security." This act states that "It is also the intent of Congress to assure agriculture a position in research *equal to that of industry*, which will aid in maintaining an equitable balance between agriculture and other segments of the economy." (Emphasis mine—to call attention to the distinction made between agriculture and industry.) The act declares, further, that "It shall be the object and duty of the State agricultural experiment stations . . . to conduct . . . researches, investigations, and experiments bearing directly on and contributing to the establishment and maintenance of a permanent and effective agricultural industry . . . including . . . such investigations as have for their purpose the development and improvement of the rural home and rural life . . ."

And in 1914 the Smith-Lever Act created the cooperative extension service "In order to aid in diffusing among the people . . . useful and practical information on subjects relating to agriculture and home economics, and to encourage the application of the same . . ."

Together, these acts provide for what is known as the land-grant college complex. They fulfill the intention of Justin Smith Morrill, representative and later senator from Vermont. In clarification of the historical pertinence and the aims of the language of the several bills, it is useful to have Morrill's statement of his intentions in a memoir written "apparently in 1874."

Morrill was aware, as Jefferson had been, of an agricultural disorder manifested both by soil depletion and by the unsettlement of population: ". . . the very cheapness of our public lands, and the facility of purchase and transfer, tended to a system of bad-farming or strip and waste of the soil, by encouraging short occupancy and a speedy search for new homes, entailing upon the first and older settlements a rapid deterioration of the soil, which would not be likely to be arrested except by more thorough and scientific knowledge of agriculture and by a higher education of those who were devoted to its pursuit."

But Morrill, unlike Jefferson, had personal reason to be generously concerned for "the class of artificers": ". . . being myself the son of a hard-handed blacksmith . . . who felt his own deprivation of schools . . . I could not overlook mechanics in any measure intended to aid the industrial classes in the procurement of an education that might exalt their usefulness."

And he wished to break what seemed to him "a monopoly of education": ". . . most of the existing collegiate institutions and their feeders were based upon the classic plan of teaching those only destined to pursue the so-called learned professions, leaving farmers and mechanics and all those who must win their bread by labor, to the haphazard of being self-taught or not scientifically taught at all, and restricting the number of those who might be supposed to be qualified to fill places of higher consideration in private or public employments to the limited number of the graduates of the literary institutions."

THE LAND-GRANT COLLEGES

To understand what eventually became of the land-grant college complex, it will be worthwhile to consider certain significant differences between the thinking of Jefferson and that of Morrill. The most important of these is the apparent absence from Morrill's mind of Jefferson's complex sense of the dependence of democratic citizenship upon education. For Jefferson, the ideals and aims of education appear to have been defined directly by the requirements of political liberty. He envisioned a local system of education with a double purpose: to foster in the general population the critical alertness necessary to good citizenship and to seek out and prepare a "natural aristocracy" of "virtue and talents" for the duties and trusts of leadership. His plan of education for Virginia did not include any form of specialized or vocational training. He apparently assumed that if communities could be stabilized and preserved by the virtues of citizenship and leadership, then the "practical arts" would be improved as a matter of course by local example, reading, etc. Morrill, on the other hand, looked at education from a strictly practical or utilitarian viewpoint. He believed that the primary aims of education were to correct the work of farmers and mechanics and "exalt their usefulness." His wish to break the educational monopoly of the professional class was Jeffersonian only in a very limited sense: he wished to open the professional class to the children of laborers. In distinguishing among the levels of education, he did not distinguish, as Jefferson did, among "degrees of genius."

Again, whereas Jefferson regarded farmers as "the most valuable citizens," Morrill looked upon the professions as "places of higher consideration." We are thus faced with a difficulty in understanding Morrill's wish to "exalt the usefulness" of "those who must win their bread by labor." Would education exalt their usefulness by raising the quality of their work or by making them eligible for promotion to "places of higher consideration"?

Those differences and difficulties notwithstanding, the apparent intention in regard to agriculture remains the same from Jefferson to Morrill to the land-grant college acts. That intention was to promote the stabilization of farming populations and communities and to establish in that way a "permanent" agriculture, enabled by better education to preserve both the land and the people.

The failure of this intention, and the promotion by the land-grant colleges of an *impermanent* agriculture destructive of land and people, was caused in part by the lowering of the educational standard from Jefferson's ideal of public or community responsibility to the utilitarianism of Morrill, insofar as this difference in the aims of the two men represented a shift of public value. The land-grant colleges have, in fact, been very little—and have been less and less—concerned "to promote the liberal and practical education of the industrial classes" or of any other classes. Their history has been largely that of the whittling down of this aim—from education in the broad, "liberal" sense to "practical" preparation for earning a living to various "programs" for certification. They first reduced "liberal and practical" to "practical," and then for "practical" they substituted "specialized." And the standard of their purpose has shifted from usefulness to careerism. And if this has not been caused by, it has certainly accompanied a degeneration of faculty standards, by which professors and teachers of disciplines become first upholders of "professional standards" and then careerists in pursuit of power, money, and prestige.

The land-grant college legislation obviously calls for a system of local institutions responding to local needs and local problems. What we have instead is a system of institutions which more and more resemble one another, like airports and motels, made increasingly uniform by the transience or rootlessness of their career-oriented faculties and the consequent inability to respond to local conditions. The professor lives in his career, in a ghetto of career-oriented fellow professors. Where he may be geographically is of little interest to him. One's career is a vehicle, not a dwelling; one is concerned less for where it is than for where it will go.

The careerist professor is by definition a specialist professor. Utterly dependent upon his institution, he blunts his critical intelligence and blurs his language so as to exist "harmoniously" within it—and so serves his school with an emasculated and fragmentary intelligence, deferring "realistically" to the redundant procedures and meaningless demands of an inflated administrative bureaucracy whose educational purpose is written on its paychecks.

But just as he is dependent on his institution, the specialist professor is also dependent on his students. In order to earn a living, he must teach; in order to teach, he must have students. And so the tendency is to make a commodity of education: to package it attractively, reduce requirements, reduce homework, inflate grades, lower standards, and deal expensively in "public relations."

As self-interest, laziness, and lack of conviction augment the general confusion about what an education is or ought to be, and as standards of excellence are replaced by sliding scales of adequacy, these schools begin to depend upon, and so to institutionalize, the local problems that they were founded to solve. They begin to need, and so to promote, the mobility, careerism, and moral confusion that are victimizing the local population and destroying the local communities. The stock in trade of the "man of learning" comes to be ignorance.

The colleges of agriculture are focused somewhat more upon their whereabouts than, say, the colleges of arts and sciences because of the local exigencies of climate, soils, and crop varieties; but like the rest they tend to orient themselves within the university rather than within the communities they were intended to serve. The impression is unavoidable that the academic specialists of agriculture tend to validate their work experimentally rather than practically, that they would rather be professionally reputable than locally effective, and that they pay little attention, if any, to the social, cultural, and political consequences of their work. Indeed, it sometimes appears that they pay very little attention to its economic consequences.

There is nothing more characteristic of modern agricultural research than its divorcement from the sense of consequence and from all issues of value.

This is facilitated on the one hand by the academic ideal of "objectivity" and on the other by a strange doctrine of the "inevitability" of undisciplined technological growth and change. "Objectivity" has come to be simply the academic uniform of moral cowardice: one who is "objective" never takes a stand. And in the fashionable "realism" of technological determinism, one is shed of the embarrassment of moral and intellectual standards and of any need to define what is excellent or desirable. Education is relieved of its concern for truth in order to prepare students to live in "a changing world." As soon as educational standards begin to be dictated by "a changing world" (changing, of course, to a tune called by the governmental-military-academic-industrial complex), then one is justified in teaching virtually anything in any way—for, after all, one never knows for sure what "a changing world" is going to become. The way is thus opened to run a university as a business, the main purpose of which is to sell diplomas—after a complicated but undemanding four-year ritual—and thereby give employment to professors.

COLLEGES OF "AGRIBUSINESS" AND UNSETTLEMENT

That the land-grant college complex has fulfilled its obligation "to assure agriculture a position in research equal to that of industry" simply by failing to distinguish between the two is acknowledged in the term "agribusiness." The word does not denote any real identity either of function or interest, but only an expedient confusion by which the interests of industry have subjugated those of agriculture. This confusion of agriculture with industry has utterly perverted the intent of the land-grant college acts. The case has been persuasively documented by a task force of the Agribusiness Accountability Project. In the following paragraphs, Jim Hightower and Susan DeMarco give the task force's central argument:

"Who is helped and who is hurt by this research?

"It is the largest-scale growers, the farm machinery and chemicals input companies and the processors who are the primary beneficiaries. Machinery companies such as John Deere, International Harvester, Massey-Ferguson, Allis-Chalmers and J. I. Case almost continually engage in cooperative research efforts at land grant colleges. These corporations contribute money and some of their own research personnel to help land grant scientists develop machinery. In return, they are able to incorporate technological advances in their own products. In some cases they actually receive exclusive licences to manufacture and sell the products of tax-paid research.

"If mechanization has been a boon to agribusiness, it has been a bane to millions of rural Americans. Farmworkers have been the earliest victims. There were 4.3 million hired farmworkers in 1950. Twenty years later that number had fallen to 3.5 million . . .

"Farmworkers have not been compensated for jobs lost to mechanized research. They were not consulted when that work was designed, and their needs were not a part of the research that resulted. They simply were left to fend on their own—no re-training, no unemployment compensation, no research to help them adjust to the changes that came out of the land grant colleges.

"Independent family farmers also have been largely ignored by the land grant colleges. Mechanization research by land grant colleges is either irrelevant or only incidentally adaptable to the needs of 87 to 99 percent of America's farmers. The public subsidy for mechanization actually has weakened the competitive position of the family farmer. Taxpayers, through the land grant college complex, have given corporate producers a technological arsenal specifically suited to their scale of operation and designed to increase their efficiency and profits. The independent family farmer is left to strain his private resources to the breaking point in a desperate effort to clamber aboard the technological treadmill."

The task force also raised the issue of academic featherbedding —irrelevant or frivolous research or instruction carried on by colleges of agriculture, experiment stations, and extension services. Evidently, people in many states may expect to be "served" by such studies as one at Cornell that discovered that "employed home-makers have less time for housekeeping tasks than non-employed homemakers." An article in the *Louisville*

Courier-Journal lately revealed, for example, that "a 20-year-old waitress . . . recently attended a class where she learned 'how to set a real good table.'

"She got some tips on how to save steps and give faster service by 'carrying quite a few things' on the same tray. And she learned most of the highway numbers in the area, so she could give better directions to confused tourists.

"She learned all of that from the University of Kentucky College of Agriculture. Specialists in restaurant management left the Lexington campus to give the training to waitresses . . .

"The UK College of Agriculture promotes tourism.

"The college also helps to plan highways, housing projects, sewer systems and industrial developments throughout the state.

"It offers training in babysitting, 'family living' . . ."

This sort of "agricultural" service is justified under the Smith-Lever Act, Section 347a, inserted by amendment in 1955, and by Representative Lever's "charge" to the Extension Service in 1913. Both contain language that requires some looking at.

Section 347a is based mainly upon the following congressional insight: that "in certain agricultural areas," "there is concentration of farm families on farms either too small or too unproductive or both . . ." For these "disadvantaged farms" the following remedies were provided: "(1) Intensive on-the-farm educational assistance to the farm family in appraising and resolving its problems; (2) assistance and counseling to local groups in appraising resources for capability of improvement in agriculture or introduction of industry designed to supplement farm income; (3) cooperation with other agencies and groups in furnishing all possible information as to existing employment opportunities, particularly to farm families having underemployed workers; and (4) in cases where the farm family, after analysis of its opportunities and existing resources, finds it advisable to seek a new farming venture, the providing of information, advice, and counsel in connection with making such change."

The pertinent language of Representative Lever's "charge," which is apparently regarded as having the force of law, at least by the University of Kentucky Cooperative Extension Service, places upon extension agents the responsibility "to assume leadership in every movement, whatever it may be, the aim of which is better farming, better living, more happiness, more education and better citizenship."

If Section 347a is an example—as it certainly is—of special-interest legislation, its special interest is only ostensibly and vaguely in the welfare of small ("disadvantaged") farmers. To begin with, it introduces into law and into land-grant philosophy the startling concept that a farm can be "too small" or "too unproductive." The only standard for this judgment is implied in the clauses that follow it: the farmers of such farms "are unable to make adjustments and investments required to establish profitable operations"; such a farm "does not permit profitable employment of available labor"; and—most revealing—"many of these farm families are not able to make full use of current extension programs . . ."

The first two of these definitions of a "too small" or "too unproductive" farm are not agricultural but economic: the farm must provide, not a living, but a profit. And it must be profitable, moreover, in an economy that—in 1955, as now—favors "agribusiness." (Section 347a is a product of the era in which then Assistant Secretaries of Agriculture John Davis and Earl Butz were advocating "corporate control to 'rationalize' agriculture production"; in which Mr. Davis himself invented the term "agribusiness"; in which then Secretary of Agriculture Ezra Taft Benson told farmers to "Get big or get out.") Profitability may be a standard of a sort, but a most relative sort and by no means sufficient. It leaves out of consideration, for instance, the possibility that a family might farm a small acreage, take excellent care of it, make a decent, honorable, and independent living from it, and yet fail to make what the authors of Section 347a would consider a profit.

But the third definition is, if possible, even more insidious: a farm is "too small" or "too unproductive" if it cannot "make full use of current extension programs." The farm is not to be the measure of the service; the service is to be the measure of the farm.

It will be argued that Section 347a was passed in response to real conditions of economic hardship on the farm and that the aim of the law was to permit the development of *new* extension programs as remedies. But that is at best only half true. There certainly were economic hardships on the farm in 1955; we have proof of that in the drastic decline in the number of farms and farmers since then. But there was plenty of land-grant legislation at that time to permit the extension service to devise any program necessary to deal with agricultural problems *as such*. What is remarkable about Section 347a is that it permitted the land-grant colleges to abandon these problems as such, to accept the "agribusiness" revolution as inevitable, and to undertake non-agricultural solutions to agricultural problems. And the assistances provided for in Section 347a are so general and vague as to allow the colleges to be most inventive. After 1955, the agricultural academicians would have a vested interest, not in the welfare of farmers, but in virtually anything at all that might happen to ex-farmers, their families, and their descendants forevermore. They have, in other words, a vested interest in their own failure—foolproof job security.

But it is hard to see how the language of Section 347a, loose as it is, justifies the teaching of highway numbers to waitresses, the promotion of tourism, and the planning of industrial developments, sewer systems, and housing projects. For justification of these programs we apparently must look to the language of Representative Lever's "charge," which in effect tells the extension agents to do anything they can think of.

These new "services" seem little more than desperate maneuvers on the part of the land-grant colleges to deal with the drastic reduction in the last thirty years of their lawful clientele —a reduction for which the colleges themselves are in large part responsible because of their eager collaboration with "agribusiness." As the conversion of farming into agribusiness has depopulated the farmland, it has become necessary for the agriculture specialists to develop "programs" with which to follow their erstwhile beneficiaries into the cities—either that or lose their meal ticket in the colleges. If the colleges of agriculture have so assiduously promoted the industrialization of farming and the urbanization of farmers that now "96 percent of America's manpower is freed from food production," then the necessary trick of survival is to become colleges of industrialization and urbanization—that is, colleges of "agribusiness"

—which, in fact, is what they have been for a long time. Their success has been stupendous: as the number of farmers has decreased, the colleges of agriculture have grown larger.

The bad faith of the program-mongering under Section 347a may be suggested by several questions:

Why did land-grant colleges not address themselves to the *agricultural* problems of small or "disadvantaged" farmers?

Why did they not undertake the development of small-scale technologies and methods appropriate to the small farm?

Why have they assumed that the turn to "agribusiness" and big technology was "inevitable"?

Why, if they can promote tourism and plan sewer systems, have they not promoted cooperatives to give small farmers some measure of protection against corporate suppliers and purchasers?

Why have they watched in silence the destruction of the markets of the small producers of poultry, eggs, butter, cream, and milk—once the mainstays of the small-farm economy?

Why have they never studied or questioned the necessity or the justice of the sanitation laws that have been used to destroy such markets?

Why have they not tried to calculate the real (urban and rural) costs of the migration from farm to city?

Why have they raised no questions of social, political, or cultural value?

That the colleges of agriculture should have become colleges of "agribusiness"—working, in effect, *against* the interests of the small farmers, the farm communities, and the farmland—can only be explained by the isolation of specialization.

First we have the division of the study of agriculture into specialties. And then, within the structure of the university, we have the separation of these specialties from specialties of other kinds. This problem is outlined with forceful insight by Andre Mayer and Jean Mayer in an article entitled "Agriculture, the Island Empire," published in the summer 1974 issue of *Daedalus*. Like other academic professions, agriculture has gone its separate way and aggrandized itself in its own fashion: "As it developed into an intellectual discipline in the nineteenth century, it did so in academic divisions which were isolated from the liberal arts center of the university . . ." It "produced ancillary disciplines parallel to those in the arts and sciences . . ." And it "developed its own scientific organizations; its own professional, trade, and social organizations; its own technical and popular magazines; and its own public. It even has a separate political system . . ."

The founding fathers, these authors point out, "placed agriculture at the center of an Enlightenment concept of science broad enough to include society, politics, and sometimes even theology." But the modern academic structure has alienated agriculture from such concerns. The result is an absurd "independence" which has produced genetic research "without attention to nutritional values," which has undertaken the so-called Green Revolution without concern for its genetic oversimplification or its social, political, and cultural dangers, and which keeps agriculture in a separate "field" from ecology.

A BETRAYAL OF TRUST

The educational *ideal* that concerns us here was held clearly in the mind of Thomas Jefferson, was somewhat diminished or obscured in the mind of Justin Morrill, but survived indisputably in the original language of the land-grant college acts. We see it in the intention that education should be "liberal" as well as "practical," in the wish to foster "a sound and prosperous agriculture and rural life," in the distinction between agriculture and industry, in the purpose of establishing and maintaining a "permanent" agriculture, in the implied perception that this permanence would depend on the stability of "the rural home and rural life." This ideal is simply that farmers should be educated, liberally and practically, *as farmers*; education should be given and acquired with the understanding that those so educated would return to their home communities, not merely to be farmers, corrected and improved by their learning, but also to assume the trusts and obligations of community leadership, the highest form of that "vigilant and distrustful superintendence" without which the communities could not preserve themselves. This leadership, moreover, would tend to safeguard agriculture's distinction from and competitiveness with industry. Conceivably, had it existed, this leadership might have resulted in community-imposed restraints upon technology, such as those practiced by the Amish.

Having stated the ideal, it becomes possible not merely to perceive the degeneracy and incoherence of the land-grant colleges within themselves, but to understand their degenerative influence on the farming communities. It becomes possible to see that their failure goes beyond the disintegration of intellectual and educational standards; it is the betrayal of a trust.

The land-grant acts gave to the colleges not just government funds and a commission to teach and to do research, but also a purpose which may be generally stated as the preservation of agriculture and rural life. That this purpose is a practical one is obvious from the language of the acts; no one, I dare say, would deny that this is so. It is equally clear, though far less acknowledged, that the purpose is also moral, insofar as it raises issues of value and of feeling. It may be that pure practicality can deal with agriculture so long as agriculture is defined as a set of problems that are purely technological (though such a definition is in itself a gross falsification), but it inevitably falters at the meanings of "liberal," "sound and prosperous," "permanent and effective," "development and improvement"; and it fails altogether to address the concepts of "the rural home and rural life." When the Hatch Act, for instance, imposed upon the colleges the goals of "a permanent and effective agricultural industry" and "the development and improvement of the rural home and rural life," it implicitly required of them an allegiance to the agrarian values that have constituted one of the dominant themes of American history and thought.

The tragedy of the land-grant acts is that their moral imperative came finally to have nowhere to rest except on the careers of specialists whose standards and operating procedures were amoral: the "objective" practitioners of the "science" of agriculture, whose minds have no direction other than that laid out by career necessity and the logic of experimentation. They have no apparent moral allegiances or bearings or limits. Their work thus inevitably serves whatever power is greatest. That power at present is the industrial economy, of which "agribusiness" is a part. Lacking any moral force or vision of its own, the "objective" expertise of the agriculture specialist points like a compass needle toward the greater good of the "agribusiness" corporations. The objectivity of the laboratory functions in the world as indifference; knowledge without responsibility is merchandise, and greed provides its applications. Far from developing and improving the rural home and rural life, the land-grant colleges have blindly followed the drift of virtually the whole population away from home, blindly documenting or "serving" the consequent disorder and blindly rationalizing this disorder as "progress" or "miraculous development."

At this point one can begin to understand the violence that has been done to the Morrill Act's provision for a "liberal and practical education." One imagines that Jefferson might have objected to the inclusion of the phrase "and practical," and indeed in retrospect the danger in it is clearly visible. Nevertheless, the law evidently sees "liberal and practical" as a description of *one* education, not two. And as long as the two terms are thus associated, the combination remains thinkable: the "liberal" side, for instance, might offer necessary restraints of value to the "practical"; the "practical" interest might direct the "liberal" to crucial issues of use and effect.

In practice, however, the Morrill Act's formula has been neatly bisected and carried out as if it read "a liberal or a practical education." But though these two kinds of education may theoretically be divided and given equal importance, in fact they are no sooner divided than they are opposed. They enter into competition with one another, and by a kind of educational Gresham's Law the practical curriculum drives out the liberal.

This happens because the *standards* of the two kinds of education are fundamentally different and fundamentally opposed. The standard of liberal education is based upon definitions of excellence in the various disciplines. These definitions are in turn based upon example. One learns to order one's thoughts and to speak and write coherently by studying exemplary thinkers, speakers, and writers of the past.

One studies *The Divine Comedy* and the Pythagorean theorem not to acquire something to be exchanged for something else, but to understand the orders and the kinds of thought and to furnish the mind with subjects and examples. Because the standards are rooted in examples, they do not change.

The standard of practical education, on the other hand, is based upon the question of what will work, and because the practical is by definition of the curriculum set aside from issues of value, the question tends to be resolved in the most shallow and immediate fashion: what is practical is what makes money; what is most practical is what makes the most money. Practical education is an "investment," something acquired to be exchanged for something else—a "good" job, money, prestige. It is oriented entirely toward the future, toward what *will* work in the "changing world" in which the student is supposedly being prepared to "compete." The standard of practicality, as used, is inherently a degenerative standard. There is nothing to correct it except suppositions about what the world will be like and what the student will therefore need to know. Because the future is by definition unknown, one person's supposition about the future tends to be as good, or as forceful, as another's. And so the standard of practicality tends to revise itself downward to meet, not the needs, but the desires of the student who, for instance, does not want to learn a science because he *intends* to pursue a career in which he does not *think* a knowledge of science will be necessary.

It could be said that a liberal education has the nature of a bequest, in that it looks upon the student as the potential heir of a cultural birthright, whereas a practical education has the nature of a commodity to be exchanged for position, status, wealth, etc., *in the future*. A liberal education rests on the assumption that nature and human nature do not change very much or very fast and that one therefore needs to understand the past. The practical educators assume that human society itself is the only significant context, that change is therefore fundamental, constant, and necessary, that the future will be wholly unlike the past, that the past is outmoded, irrelevant, and an encumbrance upon the future—the present being only a time for dividing past from future, for getting ready.

But these definitions, based on division and opposition, are too simple. It is easy, accepting the viewpoint of either side, to find fault with the other. But the wrong is on neither side; it is in their division. One of the purposes of this book is to show how the practical, divorced from the discipline of value, tends to be defined by the immediate interests of the practitioner, and so becomes destructive of value, practical and otherwise. But it must not be forgotten that, divorced from the practical, the liberal disciplines lose their sense of use and influence and become attenuated and aimless. The purity of "pure" science is then ritualized as a highly competitive intellectual game without awareness of use, responsibility, or consequence, such as that described in *The Double Helix*, James D. Watson's book about the discovery of the structure of DNA. And the so-called humanities become a world of their own, a collection of "professional" sub-languages, complicated circuitries of abstruse interpretation, feckless exercises of sensibility. Without the balance of historic value, practical education gives us that most absurd of standards: "relevance," based upon the suppositional needs of a theoretical future. But liberal education, divorced from practicality, gives something no less absurd: the specialist professor of one or another of the liberal arts, the custodian of an inheritance he has learned much about, but nothing from.

And in the face of competition from the practical curriculum, the liberal has found it impossible to maintain its own standards and so has become practical—that is, career-oriented —also. It is now widely assumed that the only good reason to study literature or philosophy is to become a teacher of literature or philosophy—in order, that is, to get an income from it. I recently received in the mail a textbook of rhetoric in which the author stated that "there is no need for anyone except a professional linguist to be able to explain language operations specifically and accurately." Maybe so, but how does one escape the implicit absurdity that linguists should study the language only to teach aspiring linguists?

The education of the student of agriculture is almost as absurd, and it is more dangerous: he is taught a course of practical knowledge and procedures for which uses do indeed exist, but these uses lie outside the purview and interest of the school. The colleges of agriculture produce agriculture specialists and "agribusinessmen" as readily as farmers, and they are producing far more of them. Public funds originally voted to provide for "the liberal and practical education" of farmers thus become, by moral default, an educational subsidy given to the farmers' competitors.

THE VAGRANT ARISTOCRACY

But in order to complete an understanding of the modern disconnection between work and value, it is necessary to see how certain "aristocratic" ideas of status and leisure have been institutionalized in this system of education. This is one of the liabilities of the social and political origins not only of our own nation, but of most of the "advanced" nations of the world. Democracy has involved more than the enfranchisement of the lower classes; it has meant also the popularization of the more superficial upper-class values: leisure, etiquette (as opposed to good manners), fashion, everyday dressing up, and a kind of dietary persnicketiness. We have given a highly inflated value to "days off" and to the wearing of a necktie; we pay an exorbitant price for the *looks* of our automobiles; we pay dearly, in both money and health, for our predilection for white bread. We attach much the same values to kinds of profession and levels of income that were once attached to hereditary classes.

It is extremely difficult to exalt the usefulness of any productive discipline *as such* in a society that is at once highly stratified and highly mobile. Both the stratification and the mobility are based upon notions of prestige, which are in turn based upon these reliquary social fashions. Thus doctors are given higher status than farmers, not because they are more necessary, more useful, more able, more talented, or more virtuous, but because they are *thought* to be "better"—one assumes because they talk a learned jargon, wear good clothes all the time, and make a lot of money. And this is true generally of "office people" as opposed to those who work with their hands. Thus an industrial worker does not aspire to become a master craftsman, but rather a foreman or manager. Thus a farmer's son does not usually think to "better" himself by becoming a better farmer than his father, but by becoming, professionally, a better *kind* of man than his father.

It is characteristic of our present society that one does not think to improve oneself by becoming better at what one is doing or by assuming some measure of public responsibility in order to improve local conditions; one thinks to improve oneself by becoming different, by "moving up" to a "place of higher consideration." Thinkable changes, in other words, tend to be quantitative rather than qualitative, and they tend to involve movement that is both social and geographic. The unsettlement at once of population and of values is virtually required by the only generally acceptable forms of aspiration. The typical American "success story" moves from a modest rural beginning to urban affluence, from manual labor to office work.

We must ask, then, what must be the educational effect, the influence, of a farmer's son who believes, with the absolute authorization of his society, that he has mightily improved himself by becoming a professor of agriculture. Has he not improved himself by an "upward" motivation which by its nature avoids the issue of quality—which assumes simply that an agriculture specialist is better than a farmer? And does he not exemplify to his students the proposition that "the way up" leads away from home? How could he, who has "succeeded" by earning a Ph.D. and a nice place in town, advise his best students to go home and farm, or even assume that they might find good reasons for doing so?

I am suggesting that our university-based structures of success, as they have come to be formed upon quantitative measures, virtually require the degeneration of qualitative measures and the disintegration of culture. The university accumulates information at a rate that is literally inconceivable, yet its structure and its self-esteem institutionalize the likelihood that not much of this information will ever be taken *home*. We do not work where we live, and if we are to hold up our heads in the presence of our teachers and classmates, we must not live where we come from.

THE STATUS QUO

So far, in tracing the changes of an American educational ambition, this chapter has necessarily been to some extent conjectural. As elsewhere in this book, I have been writing what my experience has made it possible for me to say—with the understanding that it must then await confirmation, amplification, or contradiction from the experience of other people. I have intentionally placed experience ahead of "proof," feeling that the ordinary visibility of the deterioration of rural life ought to take precedence over statistics and expert testimony.

Nevertheless, the testimony of experts must be taken into account. It seems appropriate that I should conclude this chapter by examining in some detail a prominent expert's justification of the agricultural status quo. The article, "The Agriculture of the U.S.," comes from the September 1976 issue of *Scientific American*. Its author is Earl O. Heady, Curtiss Distinguished Professor at Iowa State University and director of that university's Center for Agriculture and Rural Development. Professor Heady "was born and raised on a farm in Nebraska" and received his degrees from the University of Nebraska and from Iowa State. He is author or co-author of "17 books and more than 725 journal articles, research bulletins and monographs." He is vice-president of the American Association of Agricultural Economists, vice-president of the Canadian Agricultural Economics Association, and permanent chairman of the East-West Seminars for Agricultural Economists. His biographical note quotes him as follows: "I do a lot of work in developing countries, consulting with planners, evaluating policies for economic and agricultural development and analyzing development in general."

Professor Heady begins his account with this statement: "Over the past 200 years the U.S. has had the best, the most logical and the most successful program of agricultural development anywhere in the world. Other countries would do well to copy it." The occurrence of such an absolute assertion at the *beginning* of a scientific article by an objective scientist can only strike one as remarkable. And a little consideration makes it even more so. Has he forgotten, or did he ever know, for instance, that in 1907, F. H. King, also an American professor of agriculture and chief of Division of Soil Management, United States Department of Agriculture, was traveling in China, Korea, and Japan, studying the ancient agricultural practices of those countries and finding them exemplary? Does Professor Heady know, for that matter, of the work of any critic of his assumptions? And who is he trying to convince? Surely not the readers of *Scientific American*, most of whom will at least wish to see his evidence. But, in fact, for the supremacy of American agriculture over that of all other countries, Professor Heady's article offers no evidence whatsoever. And the evidence he does supply leaves the logic and success of the American program very much in doubt.

"At the beginning of the nation's agricultural development," Professor Heady writes, "land was abundant and labor was cheap. Capital inputs such as farm machinery, fertilizer and food for the farmer's family were relatively modest, and most of them were produced on the farm. Farmers created their own power in the form of the physical work of family members and of animals raised on the farm. They also harnessed energy from the sun for that work in the form of crops grown on the farm and eaten by the people or the animals. The farmers generated their own fertilizer by rotating crops and by utilizing the wastes from the animals. The rotation of crops also controlled insects to some extent."

That description is not critical enough. In its general outline it describes the agriculture of many parts of this country as late as World War II. The greatest weakness of that agriculture was undoubtedly its wastefulness of the soil itself, but there were other weaknesses also. It was the knowledge of these weaknesses that sent F. H. King to the Orient, and his discoveries there, had they taken root here, might have made our farmers more solvent and productive, and much kinder in their use of the soil. But Professor Heady's description may be allowed to stand; it does represent accurately enough the possibility of a thrifty, independent, diversified, farm-based agriculture that remained easily within our reach until a generation ago.

That possibility and a virgin continent were the endowment that we started with. In the rest of his article Professor Heady tells what we have done with, and to, that endowment.

In the nineteenth century, he tells us, after the United States had expanded to its westward limits and the public land grants had all been taken up, the government's agricultural policy shifted its emphasis from expansion to productivity. The land-grant college system was created "to encourage research and to extend new technical knowledge to farmers." Science and technology became "an effective substitute for land." As a result, production "approximately doubled" in the period from 1910 · to 1970, and "by 1970 the nation was producing its food on considerably fewer acres than it had been in 1910." Rapidly put into use, the new technology "became an effective substitute not only for land but also for labor. The result was that between 1950 and 1955 more than a million workers migrated out of the agricultural sector into other sectors of the economy."

We are asked to accept that our agricultural policy-makers displayed profound wisdom in shifting their emphasis from expansion to productivity—as if, after the possibility of expansion had ended, the choice was difficult. And we are asked to accept productivity as a sufficient criterion; nothing is said, here or elsewhere in Professor Heady's article, about the issues of restoration and maintenance. The displacement of a million workers in five years is cited merely as evidence of the efficacy of technology. One wonders what may have been the social and economic costs of that "migration." Into what "sectors of the economy" did those workers move? And it may not be impertinent in a democracy to ask, Did they want to go?

Next Professor Heady focuses on the period from 1950 to 1970: "Farms became larger and more specialized, handling either crops or livestock instead of both. Farms growing crops greatly increased their utilization of fertilizers, pesticides, farm machinery and other capital items . . . the use of fertilizer increased by 276 percent. . . . The use of powered machines increased by only 30 percent, but in 1972 there were substantially fewer farms than there were in 1950. The result was that farm labor declined by 54 percent over that period as labor productivity quadrupled and total farm output increased by 55 percent."

Again, highly problematic changes are cited solely as evidence of the advance of technology, which we are evidently expected to regard as simply good. And again a massive displacement of "labor" is treated as if people were merely underpowered, slow machines, now happily replaced by machines of a better make.

In 1974 and 1975, Professor Heady tells us, American farmers produced "record" yields, which brought them a "record" income. Records, as we know, are made by champions and are good beyond question. But Professor Heady goes on: "The rapid upward movement in income has put farmers in a highly favorable position with regard to capital assets. Although some farmers took advantage of the opportunity to repay their mortgage before it came due, the majority put their higher earnings into acquiring new farm equipment, upgrading their living facilities and enlarging their farms by buying more land. As a result farm real estate values more than doubled between 1970 and 1973."

This is the second time Professor Heady's article has spoken of the recent increases in the value of farmland to "record levels," as if this is some kind of grand agricultural achievement. But is this increase entirely due to competition among farmers for the land, or do inflation, urban development, and speculation have something to do with it? And are there dangers in these high prices? Although the fact of inflation is rather casually mentioned later in the article, the first question is really neither answered nor asked. The second question is answered later on, but the dangers are not admitted.

Meanwhile, Professor Heady acknowledges the existence of certain other problems: "The change in the very nature of farming, with its higher productivity and greater degree of mechanization, has severely affected rural communities. . . . With the decline in the farm population the demand for the goods and services of businesses in the country has been eroded. Employment and income opportunities in typical rural communities have therefore declined markedly. As people migrated out of the rural communities, there were fewer people left to participate in the services of schools, medical facilities and other institutions. With the lessened demand such services retreated in quantity and quality and advanced in cost.

"Nonfarm groups in the rural communities took large capital losses. . . ."

Professor Heady further acknowledges that "Rapid agricultural development . . . has also had a heavy impact on the environment." The larger and more specialized farms are "depleting the soil of certain specific nutrients and thus requiring larger amounts of fertilizer." This increase in the use of fertilizer has been accompanied by an increased use of pesticides and more intensive (that is, more continuous) cultivation. "The burden placed on streams and lakes by the runoff of silt and farm chemicals has therefore increased."

"On the other hand," he says, "the development of American

agriculture has fostered the growth of an entire agricultural industry—'agribusiness'—of which farming is only a small part."

Anyone who cares at all for the welfare of the rural home and rural life and for the good health of the farmland will see the arrogance of that phrase "on the other hand." It is the balancing point of a monstrous equation. Professor Heady has just described a serious impairment of rural life that is social, economic, and ecological, and he has said that it is justified and compensated by the growth of "agribusiness." The sacrifice of many and of much for the enrichment of a few is thus justified as if the Declaration of Independence had never been written.

The "industry" of modern agriculture, according to Professor Heady, has "three major components": "the input-processing industry," "the farm itself," and "the food-processing industry." I will quote, nearly complete, Professor Heady's description of the first and last of these, asking the reader to bear in mind the professor's earlier description of the kind of farming we had at the beginning of our "agricultural development."

"The input-processing industry now supplies many things that were once produced on the farm. Today tractors substitute for draft animals, fossil fuels for animal feeds, chemical fertilizers for manure and nitrogen-fixing crops. Such developments not only have shifted a greater proportion of the agricultural work force from the farms into the input-processing sector but also have increased the cash cost of farming. . . . The greater proportion of cash cost has made farm profits much more vulnerable to price fluctuations than they used to be.

"The food-processing sector has in recent years come to represent a larger proportion of the total agricultural industry than farming itself. In 1975, 42 cents of each consumer dollar spent for food at retail prices went to the farmer and 58 cents went to the food processor. Even the typical commercial farm family now buys frozen, packaged and ready-to-serve foods from the supermarket rather than consuming products raised and prepared on the farm."

So much for the ideal—and the practical values—of independence. If the farmer sells his foodstuff to "agribusiness" at a narrow profit, if any, and buys it back ready-to-serve from "agribusiness" to its great profit, then the cash flow has at that point deftly inserted its tail into its mouth, a wonder of sorts has been accomplished, and a reverent "Golly!" is heard from certain agricultural economists.

And now, sufficiently far from the question, Professor Heady gives us an answer as to the dangers of high land prices: "The change in the nature of agriculture has greatly enhanced the financial position of established farmers with large holdings. . . . The situation is not as favorable for farmers who are starting from scratch. . . . One can therefore expect to see an increasing trend toward more large commercial farms and fewer small ones."

Professor Heady's "therefore" is nearly as irresponsible as his "on the other hand." By various inequities, abuses, and misconceptions, a condition has come to exist in which big farmers thrive by the ruin of smaller ones. And Professor Heady enjoins this condition upon the future by a simple "therefore."

Aside from the urgent social and political questions that are obviously raised by Professor Heady's observation, it raises, in fact, some agricultural and economic questions that are also extremely serious. I shall mention two.

First, if hunger and malnutrition are now in prospect for many of the world's people, as hardly anyone (including Professor Heady) denies, and if productivity is therefore the major issue, can we afford this trend toward bigger and bigger farms? The question rises from the awareness, now shared by many experts, that large farms do not produce as abundantly or efficiently as small ones. Sterling Wortman, for instance, writing in the same issue of *Scientific American*, says that "mechanized agriculture is very productive in terms of output per man-year, but it is not as productive per unit of land as the highly intensive systems are." Why, then, does it not make sense to advocate a return to smaller, family-type farms, on which human and animal labor can be effectively substituted for machines?

Second, if the size of farms continues to increase, and the farm population proportionately decreases, will not that population become at the same time more vulnerable, less surely able to reproduce itself? According to Professor Heady, it is one of the grand achievements of American agriculture that it now employs "only 4.4 percent of the nation's population." But at what level does a population—especially one in precipitous decline—become threatened with extinction? I assume, as perhaps Professor Heady does not, that in order to run our farms productively we will have to have farmers, that a knowledge of farming and of land stewardship are of direct value to those who farm, and that the most obvious and economical way to get farmers with this knowledge is to raise them. By this accounting, the knowledge and interest of the many young farmers who are now being priced off the land amounts to a sizable loss.

According to Professor Heady, American agriculture still has plenty of room to expand: if necessary, in order to increase production, we can plow and plant some hundreds of millions of acres of fallows, pastures, forests, range lands, and wetlands. Land, then, so far as he is concerned, is not an agricultural problem. And he evidently has no doubt that we will continue to have plenty of farmers. His worries come from another direction: "The future of American agriculture will depend on a number of factors in addition to its productive capacity. The two most important factors will be the extent to which recent international conditions continue to prevail and the presence or absence of Government policies affecting output either through future supply-control programs or environmental limits on fertilizers, pesticides and soil erosion." In other words, American agriculture will continue to prosper so long as hunger remains an international threat, so long as "agribusiness" is not restrained, and so long as "established farmers with large holdings" are left free to continue the pollution and soil erosion that are the inevitable by-products of industrial agriculture.

By this "most logical" of developments, then, we have passed from a farm-based, family-based, independent agriculture to an agriculture abjectly dependent upon many kinds of industrial "inputs" and firmly based upon several kinds of disaster. We are producing, at an incalculable waste of topsoil and of human life and energy, and at the cost of destroying communities and poisoning the land and the streams, food to be used against the hungry as a weapon.*

---

*At this point one thinks with some solicitude of the "developing countries" in which Professor Heady does "a lot of work." They are apparently in danger of taking the advice of an agricultural consultant the success of whose policy requires them to get hungry.

EXPERIENCE AND EXPERIMENT

Having for some years attentively read and listened to the statements of agriculture experts, I cannot have the comfort of looking upon Professor Heady as an anomaly. I am constrained to regard him as representative of that academic upper crust that has provided a species of agricultural vandalism with the prestige of its professorships and the justifications of a bogus intellectuality, incomprehensible to any order of thought, but decked out in statistics, charts, and graphs to silence unspecialized skepticism and astonish gullibility.

In spite of his eagerness to defend what he calls a "logical" program, there is no logic in Professor Heady's defense. His defense is deduction *without* logic, a kind of disordered scholasticism that proceeds merely by flinging statistics at a premise. That his premise is called into serious question—if not disproved—by his "proof" does not cause Professor Heady to hesitate.

If Professor Heady and his kind had not so much power, they would deserve far less attention. But because they do have power, because they belong to that association of industrial conquistadors who would claim our future as their colony, it is important to understand how, and how poorly, they think.

Like most of that association, Professor Heady is a specialist. Within the enclosure of his specialty he is no doubt capable of order and sense of a very formidable kind. But when he tries to justify these in terms of value and to say that they and the assumptions on which they rest are "good," then he produces disorder and nonsense, because the order of his specialty does not comprehend a ground large enough to permit such a justification. The calculations that prove the efficacy of technology as a "substitute not only for land but also for labor" can do so convincingly only by ignoring the human and ecological contexts of the substitution. It would be possible to calculate the probable monetary cost of the unemployment, community and family breakdown, crime, vandalism, pollution, and soil loss that are the results of overwhelming "inputs" of technology—but apparently an agricultural economist is not expected to look either so widely around or so far ahead. Nor is any other agriculture expert. They are free to argue with the blind determination of fanatics from the premises that they prefer to the conclusions that they desire. It is an irony that would be amusing, were it not so frightening, that the prestigious "positions" that have relieved them of the necessity to use their hands have cost them the use of their heads.

No wonder they look forward so eagerly to the future. *We*, with our awkwardly divergent and valued lives, our bothersome rights and meanings, are not yet there. Only posterity is native there, and they have as yet produced nothing; they have no claim recognizable by an expert. The future is already surveyed and ribboned according to the claims of these people and their clients, the corporate industrialists and big businessmen. It is their New World, and they are its self-elected ruling class.

The expert knowledge of agriculture developed in the universities, like other such knowledges, is typical of the alien order imposed on a conquered land. We can never produce a native economy, much less a native culture, with this knowledge. It can only make us the imperialist invaders of our own country.

The reason is that this knowledge has no cultural depth or complexity whatever. It is concerned only with the most immediate practical (that is, economic and *sometimes* political) results. It has, for instance, never mastered the crucial distinction between experiment and experience. Experience, which is the basis of culture, tends always toward wholeness because it is interested in the *meaning* of what has happened; it is necessarily as interested in what does not work as in what does. It cannot hope or desire without remembering. Its approach to possibility is always conditioned by its remembrance of failure. It is therefore not "objective," but is at once personal and communal. The experimental intelligence, on the other hand, is only interested in what works; what doesn't work is ruled out of consideration. This sort of intelligence tends to be shallow in that it tends to impose upon experience the metaphor of experiment. It invariably sees innovation, not as adding to, but as replacing what existed or was used before. Thus machine technology is seen as a *substitute* for human or animal labor, requiring the "old way" to be looked upon henceforth with contempt. In technology, as in genetics, the experimental intelligence tends toward radical oversimplification, reducing the number of possibilities. Whereas the voice of experience, of culture, counsels, "Don't put all your eggs in one basket," the experimental intelligence, which behaves strangely like the intelligence of imperialists and religious fanatics, says, "This is the *only* truc way."

And this intelligence protects itself from the disruptive memories and questions of experience by building around itself the compartmental structure of the modern university, in which effects and causes need never meet. The experimental intelligence is a tyrant that is saved from the necessity of killing bearers of bad news because it lives at the center of a maze in which the bearers of bad news are lost before they can arrive.

But it is imperative to understand that this sort of intelligence *is* tyrannical. It is at least potentially totalitarian. To think or act without cultural value, and the restraints invariably implicit in cultural value, is simply to wait upon force. This sort of behavior is founded in the cultural disintegration and despair which are also the foundation of political totalitarianism. Whether recognized or not, there is in the workings of agricultural specialization an implicit waiting for the total state power that will permit experimentally derived, technologically pure solutions to be imposed by force.

Woe to those who add house to house
and join field to field
until everywhere belongs to them
and they are the sole inhabitants of the land.

ISAIAH 5:8

. . . it is not too soon to provide by every possible means
that as few as possible shall be without a little portion of
land. The small landholders are the most precious part of
a state. . . .

THOMAS JEFFERSON,
LETTER TO REVEREND JAMES MADISON,
OCTOBER 28, 1785

CHAPTER NINE

Margins

"AGRIBUSINESS" AS ORTHODOXY

Not all agricultural economists are blind to the human and ecological consequences of "agribusiness" economics. On March 1, 1972, Professor Philip M. Raup, of the University of Minnesota at St. Paul, testified as follows before the Subcommittee on Monopoly of the United States Senate Small Business Committee:

"Only in the past decade has serious attention been given to the fact that the large agricultural firm is . . . able to achieve benefits by externalizing certain costs. The disadvantages of large scale operation fall largely outside the decision-making framework of the large farm firm. Problems of waste disposal, pollution control, added burdens on public services, deterioration of rural social structures, impairment of the tax base, and the political consequences of a concentration of economic power have typically not been considered as costs of large scale, by the firm. They are unquestionably costs to the larger community.

"In theory, large-scale operation should enable the firm to bring a wide range of both benefits and costs within its internal decision-making framework. In practice, the economic and political power that accompanies large size provides a constant temptation to the large firm to take the benefits and pass on the costs.

"The rural community receives the immediate impact of this ability of large farm firms to practice selective internalization of benefits and externalization of costs. One of the most pervasive consequences is that the occupational composition of the population changes. Instead of a large number of small entrepreneurs, combining the functions of manager and laborer, the occupational structure includes a small number of managers and a large number of workers. In rural communities dominated by very large firms, the settlement and housing patterns reflect the increasingly transient nature of the labor force.

The symbol of the large corporate farm becomes the trailer house. Community institutions suffer from lack of leadership, and from the lack of a sense of commitment on the part of the labor force to long-run community welfare. Those institutions that survive take on a dependent character, reflecting the paternalistic role of the dominant firms. Income levels may stabilize, but at the expense of a decline in local capacity for risk-taking, decision-making, and investment of family labor in farms and local businesses."

And later in his testimony, Professor Raup spoke of the most ironic of these "externalized" costs: "Farmers who have succeeded in increasing their farm size to a scale that will enable them to achieve almost all of the economics of size in production now find that their capital structure is so large that their sons cannot finance a takeover of the family farm."

Professor Raup's distinction between internal and external accounting is of great usefulness in understanding our problem. It is by internal accounting that the modern American agricultural program may be thought "the best, the most logical and the most successful." External accounting brings us to a very different conclusion. External accounting pushes us back into our moral tradition, which asks us to consider that we are members of the human community and are therefore bound to help or harm it by our behavior. This sort of accounting involves much more than economics. It is broad and difficult, and it eludes quantification.

Modern American agriculture has made itself a "science" and has preserved itself within its grandiose and destructive assumptions by cutting itself off from the moral tradition (as it has done also from the agricultural tradition) and confining its vision and its thought within the bounds of internal accounting. Agriculture experts and "agribusinessmen" are free to believe that their system works because they have accepted a convention which makes "external," and therefore irrelevant, all evidence that it does not work. "External" questions are not asked or heard, much less answered.

But these people are human beings who inherit a community awareness, to whatever extent it may be suppressed, distorted, or ignored. Many, if not most, of them come from family farms, for which they feel some nostalgia, if no loyalty. And so it must be assumed that the claims of external accounting are still obscurely felt in the backs or the depths of their minds. Some of them may occasionally overhear their critics with a tremor of recognition; from time to time some of them may even come face to face with bad external results of internal purposes and recognize them as such. Internal accounting, then, must cohere under some pressure from the external. This obviously defines the necessary condition for a fierce and self-protective orthodoxy—a science-as-superstition, by which one clings to the assumption of the goodness of one kind of knowledge out of fear of knowledge of another kind. This fear makes the specialist scientist not merely willing to define a possibility, but *desperate* to define the *only* possibility. Only this desperation can explain the venomous contempt with which agricultural establishmentarians dismiss suggestions of other possibilities, old or new. These "objective" scientists exhibit an intense craving to be *right*—a craving hardly diminished by the profitability of their faith.

ORTHODOXY, MARGINS, AND CHANGE

Our history forbids us to be surprised that an orthodoxy of thought should become narrow, rigid, mercenary, morally corrupt, and vengeful against dissenters. This has happened over and over again. It might be thought the maturity of orthodoxy; it is what finally happens to a mind once it has consented to be orthodox. But one may be permitted a little amusement, if not surprise, that this should have befallen a modern science, which was set up, as it never tires of advertising, to *pursue* truth, not *protect* it.

But since what we now have in agriculture—as in several other "objective" disciplines—is a modern scientific orthodoxy as purblind, self-righteous, cocksure, and ill-humored as Cotton Mather's, our history also forbids us to expect it to change from within itself. Like many another orthodoxy, it would rather die than change, and may change only by dying.* This determination is enforced both from within and

---

*Orthodox agriculture is part of the larger orthodoxy of industrial progress and economic growth, which argues the necessity of pollution, unemployment, war, land spoliation, the exploitation of space, etc. And so the question is: Must we all die with it in order for it to change?

from without. It is enforced from within simply by prosperity: the professors, experts, and executives of the agrifaith do not want agricultural policy to change because they are eating very well off of it as it is. From without, it is enforced by the mistaken conviction of millions of believers that it is the only true way, that they have no choice but to accept the agribusiness philosophy or starve. But it is also enforced by the very nature of orthodoxy: one who presumes to *know* the truth does not *look* for it.

If change is to come, then, it will have to come from the outside. It will have to come from the margins. As an orthodoxy loses its standards, becomes unable to measure itself by what it ought to be, it comes to be measured by what it is not. The margins begin to close in on it, to break down the confidence that supports it, to set up standards clarified by a broadened sense of purpose and necessity, and to demonstrate better possibilities. Though it does not necessarily or always work for the better—though indeed this swing from the center to the margins and back again may be in itself a condemnation—this sort of change is a dominant theme of our tradition, whose "central" figures have often worked their way inward from the margins. It was the desert, not the temple, that gave us the prophets; the colonies, not the motherland, that gave us Adams and Jefferson.

The pattern of orthodoxy in religion, because it is well known, gives us a useful paradigm. The encrusted religious structure is not changed by its institutional dependents—they are part of the crust. It is changed by one who goes alone to the wilderness, where he fasts and prays, and returns with cleansed vision. In going alone, he goes independent of institutions, forswearing orthodoxy ("right opinion"). In going to the wilderness he goes to the margin, where he is surrounded by the possibilities—by no means all good—that orthodoxy has excluded. By fasting he disengages his thoughts from the immediate issues of livelihood; his willing hunger takes his mind off the payroll, so to speak. And by praying he acknowledges ignorance; the orthodox presume to *know*, whereas the marginal person is trying to find out. He returns to the community, not necessarily with new truth, but with a new vision of the truth; he sees it more whole than before.

In applying this pattern to agriculture, one is startled to realize that this is the first time it has been necessary, or possible, to do so. Not until recently have we had a widespread orthodoxy of agriculture in the same sense that we have had widespread orthodoxies of religion—an agriculture, that is to say, which is nearly uniform in technology and in its general assumptions and ambitions over a whole continent, and which, like many religions, aspires to become "universal" by means of a sort of evangelism, proclaiming that "Other countries would do well to copy it."

In agriculture there have always been prevalent patterns of technology, practice, and attitude that may have had the customary force of orthodoxy. But these patterns were local; they varied in response to local conditions. And, unlike orthodoxies, they were not imposed by external authority, but grew as part of a complex relationship between the human community and natural conditions. Until the triumph of the industrial values of the "agribusiness" vision, agriculture was very much a regional affair, a response at once to human need and to regional possibilities and limits, and it was successful and long-lasting in proportion to the sensitivity of that response.

A PRE-INDUSTRIAL EXAMPLE

By looking at an example of a sound pre-industrial agriculture we can get a sense of its ecological and cultural coherence and its geographical responsiveness and also a sense of its careful relationship to its margins. Perhaps no more vivid example exists in our time than that of the native agriculture of the Peruvian Andes. I take the following summary from an unpublished paper by Professor Stephen B. Brush, of the Department of Anthropology at the College of William and Mary.

Professor Brush's study focuses on the village of Uchucmarca in a valley in northern Peru. This valley has "one of the steepest environmental gradients in the world." Like other Andean farmers, the people of Uchucmarca farm in four different climatic zones, requiring four different kinds of agriculture:

"1. a tropical zone . . . which produces fruit (such as oranges and bananas), tropical root plants (such as manioc), chile peppers, and perhaps most importantly *coca* . . .

"2. a middle level mountain zone . . . where maize and wheat are grown . . .

"3. a relatively high mountain zone . . . where the staple of the Andean diet, potatoes, and other Andean tubers are grown . . .

"4. a high mountain zone . . . where llamas, alpacas, sheep, horses, and cattle are grazed on natural pasture . . ."

Within a distance of forty miles the valley rises from "roughly 3,200 feet" to "over 14,700 feet"—thus including a diversity of climates as great as that from west Texas to Alaska. The natives of the valley "recognize and name seven different production zones . . . which are variations on the four major zones."

The agriculture of the valley is based upon a highly evolved awareness of the nature of each of these climatic zones and of the differences among them. It involves a careful balance between the use and the maintenance of productivity. Professor Brush says that "The village economy may be understood as a set of subsistence strategies designed to provision each household with adequate food. . . . One of the most important features of the local economy is that it is able to function as a largely nonmonetized economy. The average family in Uchucmarca needs less than $100 yearly. . . ." It is significant that the verb in that last sentence is "needs," where "agribusiness" assumptions would require "has only." The governing concept of the agriculture of these Andean peasants, then, is *enough*, a long-term sufficiency, whereas the governing concept of ours is *profit* or *affluence*, without regard for long-term needs.*

Like most farmers, those of Uchucmarca must cope with the hazards of erosion, frost, too much or too little rain, pests, and diseases. They do this very effectively and without recourse to the industrial technology of machines and chemicals.

"The danger of erosion is avoided in three ways. First, fields are kept small—usually less than one acre. . . . A typical family

---

*In a letter to me, dated February 15, 1977, Professor Brush wrote as follows: ". . . I calculate that with their 'primitive' agriculture, the farmers of Uchucmarca produce 2700 calories and 80 grams of protein (vegetable) per capita per day. A very good diet and a well fed population. The worst malnutrition occurs in cities where people must depend on 'modern' agriculture."

of four to five persons cultivates between three and four acres, spread among as many fields. The small size of individual plots retards run-off and erosion. Second, each field is surrounded by a hedgerow constructed of rocks, brush, and living plants. Ostensibly built to keep out destructive livestock, these hedgerows effectively limit erosion. Their roots hold the soil, and horizontal plowing behind them tends to build up soil at the lower side of the field. This creates a quasi-terrace or lynchet. . . . Third, field rotation is practiced in the highest and steepest part of the valley where rainfall is heaviest and erosion most likely. Potatoes are cultivated under a regime of shifting cultivation in which fields are only planted for two or three years before being returned to a long fallow of five years or more. By using this method, the amount of soil washed off of fields is limited, and organic material is allowed to reaccumulate."

The people of Uchucmarca cope with climatic variations by "the exploitation of multiple zones and crops," so that if one crop fails they may rely on one of a different kind. "Another way is to plant several different fields of the same crop, hoping that if one field is destroyed, the other will survive. These means are reinforced by systems of economic reciprocity and mutual dependence which rely primarily on the kinship system." Within families, "individual households protect themselves from privation by exchanging land, labor, and goods."

Against insects and diseases, the main weapon of the Andean peasants is genetic diversity: "Botanists estimate that there are well over 2,000 potato varieties in Peru alone. In single villages like Uchucmarca people identify some fifty varieties . . ." And here we arrive at the greatest complexity, versatility, and responsiveness of this agriculture, as well as its most intense sensitivity to place. For these varieties are not used at random, but are delicately fitted into their appropriate ecological niches. "In Uchucmarca, a common practice is to plant fast growing varieties during the drier part of the year so as to avoid late blight which increases during the months of heavy rain. Another practice is to cultivate certain varieties believed to be somewhat frost resistant in flat, bottom areas of the high valley where frost but not late blight is common. Other varieties are chosen for hillside cultivation where late blight but not frost is common." Varieties are also chosen according to how well they do at certain altitudes or according to whether or not they need a soil that drains well. Of course, this description gives only a rough idea of the intricacy of possible adjustments among so many varieties and so many kinds of ground.

Professor Brush's work makes it plain that nearly all the methods of the Andean farmers are based upon the one principle of diversity. In their understanding and use of this principle, they have developed an agriculture much more sophisticated, efficient, and conservative of the soil than our own—and one that is also much more likely to survive a crisis. How finely this agriculture is attuned to the needs and circumstances of the community becomes apparent when Professor Brush describes recent attempts to change it by the introduction of industrial technology and "improved" potato varieties. Such change involves a gross simplification of the agriculture itself as well as a drastic complication of the economy. It requires a cash economy and credit, favors the larger producers, and threatens to destroy both the human community and the ecological viability of a farming system that is "the result of thousands of years of natural and human selection."

But the sophistication and durability of Andean agriculture cannot be fully appreciated until one has understood the way it utilizes—indeed, depends upon—its margins. The fifty potato varieties used in Uchucmarca are not a stable quantity, but rather a sort of genetic vocabulary in a state of continuous revision. Professor Brush says that "new varieties are constantly being created through cross-pollination between cultivated, wild and semidomesticated (weedy) species. . . . These wild and semidomesticated species thrive in the hedgerows around fields, and birds and insects living there assist cross-pollination." Thus, if an Andean farmer loses a crop because of an extremity of the weather or an infestation of insects or disease, he may find a plant of a new variety that has survived the calamity and produced in spite of it. If he finds such a plant, he may add it to his collection of domesticated varieties or substitute it for the one that has failed.

This Andean agriculture, then, does not push its margins back to land unsuitable for farming, as ours does, but incorporates them into the very structure of its farms. The hedgerows are marginal areas, little thoroughfares of wilderness closely crisscrossing the farmland, and in them agriculture is constantly renewing itself in direct response to what threatens it. This network of wilderness threading through the fields serves the Andean farmer as a college of agriculture and experiment station. And in at least one respect it serves him better: whatever is discovered there has already been tested in the circumstances of the farm itself, and its worth or worthlessness proven. The farmer, in whose mind culture and agriculture are wedded, acts as both teacher-researcher and student, both extension agent and client. Set thus in the light of a truly healthy agriculture, our land-grant college complex may be seen less as a symbol of our agricultural success than as a symptom of our failure.

And this integration of Andean farming with its margins may serve us in another way. It offers an example of a sort of reconciliation by which we might escape the endless swinging between center and margins, rigidity and revolt, that has dominated our culture for so long. The remedy is to accommodate the margin within the form, to allow the wilderness or nature to thrive in domesticity, to accommodate diversity within unity. It is surely by this means—this graceful, practical generosity toward the possible and the unexpected, toward time and history—that Andean agriculture has survived for so long, cohering even through the severe disturbances of the Spanish Conquest. By responding competently to whatever has threatened it, and by doing so in the most local and immediate fashion, it has kept its hold on the world, much as life itself has kept its hold. Having understood this reconciliation or integration of the human community with its natural margins, we may see how crude and dangerous are our absolute divisions between city and farmland, farmland and wilderness, by which we seek to exclude from our domestic enclosures everything for which we have foreseen no use or market.

This principle of accommodating the margins, of diversity within unity, underlies our Constitution and Bill of Rights. But we live by this principle only negatively and grudgingly: we *permit* or *tolerate* dissent and divergence because the law requires us to. And the law can do no more than that. To put dissent and divergence to use, to turn a curious eye to the margins, eager to see what may have been tried and proven there, we will need a sounder, saner culture than we have.

MARGINS AND HEALTH

By narrowing itself so fanatically, orthodox agriculture has, in one sense, left its margins extremely wide. For motive power, it has made itself almost exclusively dependent on the internal combustion engine, and its ambition is to become completely so—leaving out of use or consideration the large variety of tools and techniques for the employment of human and animal power. Its earlier dependence on wind and water power—for pumping, milling, etc.—has now been shifted to electricity. It has little interest in on-the-farm collection and use of methane gas or solar energy.

It has greatly reduced regional differences in technology, methods of tillage, soil husbandry, etc. At the same time, it has reduced the variety of production within regions. This is, as Maurice Telleen says, "the regional specialization, that inevitably flows from individual specialization." And, just as dangerously, it has reduced the genetic diversity of both field crops and animals.

It has drawn an ever straighter, stricter line between the domestic and the wild, crowding nature itself into the margins. For the complex biological wilderness of a healthy topsoil it has substituted a simple chemistry. It has plowed up fence rows and roadsides and waterways, bulldozed woodlands, drained and plowed marshes. It has made itself not only inhospitable but dangerous to wild animals, birds, and harmless or beneficial insects.

It has made a margin even of the agricultural past, which is no longer regarded as a resource, a fund of experience, or a lexicon of proven possibilities and understood mistakes, but only as an amusement for the idly curious or, in advertisements, a measure of "how far we have come." Farm-equipment corporations are fond of printing old photographs to show the "drawbacks" of the agricultural past in comparison to the shiny "sophistication" of modern times. But as a working principle, whatever has been displaced or outmoded is simply ignored. About anything "old-fashioned," whatever its worth, the invariable comment is that "You can't go back."

For the principle of diversity, in nature and in earlier agriculture, and for the principle of unity that includes and depends upon diversity, orthodox agriculture has substituted a dull, tight uniformity, not only ignorant of other possibilities, but scared of them, and vengeful in its ignorance.

People who remove their minds from this shadowy twilight of agribigotry find that they are surrounded by an abundance of divergent possibilities—from our own past, from the history and present practice of other peoples, from new technology, from new understandings of biology and ecology. But they soon become aware, especially if their interest in agriculture is personal and practical, that this wide margin is only a margin in the mind, seriously beset by speculations, questions, and doubts. The possibilities obviously do exist as possibilities, but where do they exist in proof? Where are they being enacted by a living farmer on a living farm? Having arrived at these questions, one realizes that as the margin of divergent possibility has widened around orthodox agriculture, the margins of geography and practice have been drastically narrowed. Who are the people who know how to farm in these other—and, one believes, better—ways? And where are they? They are few, as the saying goes, and they are far between.

In the last few years, I have made an effort to do a little traveling along the agricultural margins, to visit farms where unorthodox ways are working, to see for myself what these dissident farmers are doing, and to listen to what they have to say. In telling about them, I wish to respect their privacy, and so I will not give their names or say very specifically where they live.

Nor, except for some merely descriptive figures, will I deal very much in statistics. I have chosen instead to rely on the evidence that I have seen, and that other people can see, too, if they will look carefully. One need not be a specialist to understand the difference between good and bad farming. There is nothing mysterious or abstruse about it. It only requires enough acquaintance with land and people to have some sense of what a prospering farm and a prospering farm community ought to look like and the same acquaintance with the signs of greed, hopelessness, neglect, and abandonment.

The health of a farm is as apparent to the eye as the health of a person. To look at a farm in full health gives the same complex pleasure as looking at a fully healthy person or animal. It will give the same impression of abounding life. What grows on it will be thriving. It will seem to belong where it is; the form of it will be a considerate response to the nature of its place; it will not have the look of an abstract idea of a farm imposed upon an area somewhere or other. It will look cared for—groomed, so to speak—like a healthy person or animal; it will look lived in by people who care where they live. It will show no gullies or galls or other signs of erosion. The waterways and field edges and areas around buildings will be grassed, something that becomes more necessary the steeper the ground is.

The place will look well maintained. Buildings, fences, equipment, etc., will have been kept in good repair, carefully used, protected from the weather. One of the commonest sights associated with orthodox farming is a lot of huge, expensive farm machinery left sitting out in the weather, having, like the economy that produced it, outgrown the possibility of care. Like the land itself, the equipment is used but not protected. This is one of the first and most ironic results of the high costs of industrial agriculture. The farmer is forced to protect his investment at the expense of what he has invested in. He plows out his waterways, abusing the land to get the maximum use of it and his machinery, and then allows the machinery to rust to save the cost of the necessary buildings. Such an economy will make a difference very quickly in the looks of a farm, as it will make a difference in the looks of a person or a nation.

A healthy farm will have trees on it—woodlands, where forest trees are native, but also fruit and nut trees, trees for shade and for windbreaks. Trees will be there for their usefulness: for food, lumber, fence posts, firewood, shade, and shelter. But they will also be there for comfort and pleasure, for the wildlife that they will harbor, and for their beauty. The woodlands bespeak the willingness to let live that keeps wildness flourishing in the settled place. A part of the health of a farm is the farmer's wish to remain there. His long-term good intention toward the place is signified by the presence of trees. A family is married to a farm more by their planting and protecting of trees than by their memories or their knowledge, for the trees stand for their fidelity and kindness to what they do not know. The most revealing sign of the ill health of industrial agriculture —its greed, its short-term ambitions—is its inclination to see trees as obstructions and to strip the land bare of them.

Woodlands, orchards, and shade trees are part of the diversity of life that is another of the prime characteristics of a healthy farm. And this principle will extend to cropland and pasture. The aim of a healthy farm will be to produce as many kinds of plants and animals as it sensibly can. This will be an *ordered* diversity, the various species moving in rotation over the fields. The land will be fenced for livestock, and its aspect will change from field to field.

Related to the principle of diversity is that of carrying capacity: the various crops and animals will be sensibly proportionate to one another; the farm will strive as far as possible toward the balance, the symmetry, of an ecological system; there will not be too much of anything. The fields will not be overcropped; the pastures will not be overgrazed. It will be understood that the plants growing on a farm are not just its produce, but also its protection, and so a row crop will be followed by a cover crop, the cover crop by a sod of grass and clover.

And a healthy farm not only will have the right proportion of plants and animals; it will have the right proportion of people. There will not be so many as to impoverish themselves and the farm, but there will be enough to care for it fully and properly without overwork. On a healthy farm there will be the right proportion between work and rest. Outside the Amish communities I do not know where in American agriculture one can find people and land in healthy balance. As far as I know, the Amish are the only American community to have formed deliberate strategies to keep enough people on the farms. All the non-Amish, full-time working farms that I have seen in this country have showed the need of more human hands.

Finally, a healthy farm will be so far as possible independent and self-sustaining. It is necessary to say "so far as possible," for we are by no means talking here about a "closed system." Simply by selling produce, a farm involves itself with other places both economically and biologically. And unless it encapsulates itself under a glass roof—which is really to become less independent—a farm cannot produce its own weather. Many farms cannot provide their own water. The wild plants, animals, birds, and insects upon which a farm's health depends will not respect its boundaries any more than the rain. And, of course, the people of a farm will belong complexly to a larger human community. Nevertheless, a certain kind and a certain measure of independence is a practicable ambition for a farm, and it is a necessity of agricultural health and longevity.

For one thing, fertility, the major capital of any farm, can be largely renewed and maintained from sources on the farm itself—assuming that all else is in balance. By proper tillage, rotation, the use of legumes, and the return of manure and other organic wastes to the soil, the fields can be kept productive with minimal recourse to fertilizers from outside sources. If the organic or decayable wastes of the cities, which have their source on the farm, could be returned to the farm, that would greatly increase both the health of the land and the independence, if not of the individual farm, at least of agriculture.

Equally important, by the good use of human power, animal power, solar, wind, and water power, methane gas, firewood from its own woodlands, etc., a farm can produce by far the major part of its own energy. This, of course, calls for a revitalization of local skills. But given the skills, these sources of power are possible. They come from the past and/or from new technology.

As a farm measures up in these various ways to the standard of health, its troubles from pests and diseases will radically diminish, and so consequently will its dependence on chemicals. A healthy farm will have no more need for these expensive remedies than a healthy person has for medicine.

Health, then, does not "come from" independence or "lead to" it. Health *is* independence. The healthy farm sustains itself the same way that a healthy tree does: by belonging where it is, by maintaining a proper relationship to the ground. It is by this standard of health or independence that one recognizes the absurdity of a farm absolutely dependent upon a complex of industrial corporations, which are in turn dependent upon the actions of foreign governments and politicians whom the farmer did not vote for or against and cannot influence.

The ultimate good health of a farm is in its ability to produce independently of the ups and downs of the Dow Jones industrial averages or the vagaries of politics. (When I visit a farm I always look to see how many trademarks and brand names are in sight. The orthodox industrial farm is, among other things, an advertising space for any number of corporations.) Those who pride themselves on the "science" that has made agriculture an industry have found this sort of independence simply beneath their notice. But I have watched, in Tuscany, a plowman driving a team of white cattle to a wooden plow, and realized that I was seeing the continuance of a motion and a way and a preoccupation begun before the rise of Rome. It is not nostalgia or sentimentality or wishful thinking to say that that man and his plow and team on the hand-built terrace under the olive trees represented a value, perhaps an immeasurable value, that modern agriculture has superseded but has by no means replaced.

A MARGINAL PLACE

But one's travels should begin at home. Before speaking of my travels on the margins in other places, I would like to say something about the margins I live among. Perhaps that will give an idea of what I have had in my mind as a sort of basis, and of the meanings and possibilities I have been looking for.

Not far from my house there is a hillside whose soil, declivity, and history are fairly representative of much of the hillside land in my part of the country. At one time this hillside was covered with a fine hardwood forest, which was no doubt cut soon after the establishment of the white people's tenure. The logs may have been sawed into lumber, but more likely they were burned simply to rid the land of them. The land was used agriculturally, for both row crops and pasture, with results that will remain visible for many more generations than the land was in use. Around the time of the Second World War, when machines began to replace the horse and mule teams as well as the people, the hillside began to "go back to the bushes." The thicket growth that follows agriculture began to take it over.

It is still "in bushes." In some places the better forest hardwoods have begun to establish themselves again among the weed trees. In other places there are still tangles of briars, cedars, thorns, sumac, box elder, elm. Under the trees are the slowly healing scoops and gullies of old erosion—part of the "investment" in a way of farming unsuited to the place, which no generation's income will ever redeem.

Walking along the contour of the slope, one crosses at intervals a series of natural waterways cut to the rock and running straight down the hill. The plows stopped short of these places by somewhat more than the length of a horse. The land here is whole; one supposes that it is virgin. The trees here are larger, and species grow here that do not grow in the abandoned fields. Beside one of these hollows, high up the slope, there is a big tulip poplar, a loam-loving tree rarely found in the uplands. It is two feet thick at the butt, and its trunk rises thirty or forty feet to the first branch. It is comparatively young, not by many years a survivor of the original forest, but in its proportions and its great health it is a reminder of that forest. It stands there on the edge of the hollow, not just because it has been spared, but because it is growing in excellent soil.

And so on the one hillside you are aware of crossing agricultural margins of two radically different kinds: one that farming damaged and has virtually abandoned and one that farming never came to. The second is not only the indispensable measure of the first, telling us by how much our history here has failed, but it shows us just as exactly what we must aspire to. It is an indispensable example, a little border of health along the edge of bewilderment and defeat.

But what of the abandoned fields, hidden with their scars under the bushes? What are we to think of them? Many people would say that we should not think of them at all, that they are fit only for growing bushes, as they are doing. But I disagree. If we are to have a respectable agriculture we will have to think competently and kindly of lands of all sorts, even the apparently useless. But, in fact, this hillside is not even apparently useless. Its soil, even where badly eroded, is fertile and readily responsive to good treatment. Such hillsides can be made to produce excellent pasture. I know that this is so because I have seen it done and I have done it myself.

And pasture is not all that such slopes are good for. They might, with care, be made to support a kind of mixed or "two-story" agriculture of both pasture (with selectively located hay crops) and trees. Natural stands of walnut trees are already established and thriving on many of these overgrown hillsides. These stands can be managed for their yield of nuts, for timber, or for both. And they could be augmented by planting grafted varieties of walnuts and perhaps of other native nut and fruit trees. If these plantings were done on the contour, perhaps along the backslopes of terraces, they would be perfectly compatible with the use of the land for pasture and hay.

The fertility of these slopes is by no means unknown to local farmers. But at present the use of them is problematic. The almost invariable method of clearing them nowadays is to bulldoze all the forest or thicket growth off the entire hillside at once, occasionally leaving an exceptional walnut or shade tree, and then either pile the brush and burn it or shove it off into the hollows, where much of the topsoil that comes in with it may be washed away and wasted. The cleared land is usually sowed in fescue or a mixture of fescue and clover. Rarely can a farmer afford the time and expense of sowing rye or another quick-growing crop along with the grass. Until a sod is established, the slope is seriously vulnerable to erosion, and soil loss is frequently added to the other expenses of the job. Some farmers mow these cleared fields once a year with tractors and rotary mowers, and so keep the bushes from returning. You can find some hill pastures kept in good shape in this way. But mowing them with a tractor is both dangerous and expensive, and far more time-consuming than the same work on leveler land. And the use of a tractor tends to work against any hospitality the farmer may feel toward trees; a tractor driver on a steep slope will look at a tree as at best an obstacle and at worst a hazard.

Another common practice, used to save time and expense, is simply to bulldoze the trees and bushes off the land, sow it to pasture, and then stock it heavily with cattle. As the pasture becomes stale and overgrazed, the cattle turn to browsing on the sprouts that come up from the old tree roots, and so for a while the bushes are controlled. But the cost of this practice is high, for the hillside suffers serious erosion from the combination of overgrazing and heavy trampling in wet weather and is finally grown over by the thorns and other rough trees that the cattle refuse to eat.

The good use of such land (use that is at once full, efficient, and careful) requires something altogether different and is probably unthinkable in terms of our present agricultural economy and cultural values. Good use calls, first, for great care in clearing, minimal groundbreaking, minimal bulldozing. Clearing probably should be done in narrow strips on the contour, working from the top of the slope downward in successive years. Terracing should be considered, wherever feasible; it seems to me that slowing and retaining the run-off behind terraces might make excellent sense in combination with the planting of tree crops. In some situations, when there is time and when earth does not have to be moved to repair washes, the overgrowth may be taken off by sawing; the best thrift would salvage a great quantity of fence posts and firewood from this sort of clearing. The steeper slopes, of course, should not be cleared at all, but should be left in trees to be selectively logged for posts, firewood, or lumber. The rule would be to clear only what can safely be kept clear.

Second, after clearing, the land should be sowed as quickly as possible. This sowing should include as great a diversity of clovers and grasses as makes sense for the location. (I have lately been using both bluegrass and fescue, as well as a clover mixture consisting of red, ladino, and sweet clovers, and Korean lespedeza.) A quick-growing "shelter crop" should be sowed with the pasture mixture to hold the ground until the grass and clover can get established. The seed can be sowed right onto the disturbed ground, which then ought to be passed over with a light harrow to cover the seed a little and to smooth the surface.

Third, such land needs to be managed intensively and in small fields. Steep land requires close attention, thorough understanding, and selfless care. It must be mowed at least once a year to control weeds and bushes, to stimulate new growth, and to encourage uniform grazing. Stock should be rotated from field to field, both to keep enough growth on the ground to protect it and to prevent the wearing of paths. Grazing such land too closely endangers it, and paths can be disastrous, especially if they run up and down the hill. For these reasons, large numbers of animals are incompatible with the good use of hill land; a big herd can do severe damage to a slope when the animals all must converge daily on the same watering or feeding places, gates, or milking barns. A good hill farm, if it is located where climate and soil permit intensive use, is almost by definition a small farm; and, insofar as it benefits from long-standing knowledge and devoted care, it is almost by definition a family farm. Nothing could be more alien to healthy agriculture than a large, production- or profit-oriented hill farm whose owner or owners do not live on it. In such a situation the balance between use and care is overthrown, and waste is the result. The small differences may be the most important. A family farmer, for instance, will walk his fields out of interest, the industrial farmer or manager only out of necessity.

And, finally, the good use of hill land requires a technology appropriate to it in scale and cost. Here we approach what most of the agriculture specialists and all of the "agribusinessmen" would be quick to describe as nostalgia or fantasy or craziness. They would do this to protect themselves and their assumptions and to disguise their most serious error. For the true measure of agriculture is not the sophistication of its equipment, the size of its income, or even the statistics of its productivity, but the good health of the land. And we are talking here about seriously damaged but potentially useful land, where American agriculture has so far failed. One must assume that if these hills *could* be farmed well with big, expensive, "modern" technology, they *would* be. That they are not suggests both that the technology is ill-suited to the terrain and that the cost cannot be afforded.

What sort of technology might make sense on such land? It will be at least strongly suggestive at this point to quote Thomas P. Cooper, one-time dean and director of the Extension Division of the University of Kentucky College of Agriculture:

"In Kentucky and many other states there are farming areas where work animals are indispensable. Small farms, hillside fields, rolling land and poorly drained areas can be successfully and economically farmed only by the use of horses and mules. The economic advantage of the use of workstock instead of power machinery for farm use is that horses and mules can be raised on farm-produced grasses and grains and maintained, while at work, in the same way. Farmers who own rolling or infertile farms, or who are farming on a subsistence basis are unable to purchase tractors or other power machinery. Such farmers are able to raise horses or mules and to use them to do farm work with the outlay of but little money. . . .

"There is much work on large as well as on small farms that can be done successfully and economically by horses or mules."

That circular is dated November 1937. I would not argue that we ought to "go back" to 1937. That, I am sure, *would* be nostalgic, fantastical, and crazy. But I am not so sure that what was considered "indispensable" in 1937 can be simply dismissed as "out-of-date" in 1977. My doubt is strengthened by the fact that in the intervening forty years, on thousands of acres of such land as I have just described, Dean Cooper's successors have produced, not a better agriculture or even a different one, but virtually none at all. That is, they have removed from consideration a way of farming suited to certain kinds of land and have replaced it merely with neglect and waste. It is notable that Dean Cooper's approach is to look at both the land and the farmer and then to suggest a suitable technology, whereas the approach of his successors has been to focus on the most "up-to-date" technology and expect the land and the farmer to conform to it. They seem to have answered Dean Cooper's argument that horses and mules were indispensable for certain farmers on certain lands by declaring that those farmers and those lands were themselves dispensable. I suggest that in light of the staggering losses of both farmers and land since 1937, and in light of the social problems and food needs of 1977, this assumption may be seen to be what it has always been: an extremely serious error of judgment.

A MARGINAL PERSON

The hillside that I have described, then, represents both a marginal place and a marginal possibility. As such, it is a measure both of local agricultural history and of the capacities and limits of prevailing agricultural technology and practice. But the full force of the necessary judgment will not be felt until I have also described a marginal person.

Some years ago I frequently used to drive past a farm in a creek valley of narrow, scarce bottomlands and hillsides rougher and less fertile than the one I have been talking about. The farm was small, mostly hillside, with a few narrow ridges and a creek bottom that could not have been larger than an acre and a half. In an area of semi-abandoned land, this farm was outstanding, not because of its "improvements," which were old and few, but because it was clearly both well used and well cared for. It was farmed by an old man and woman and a team of Percheron horses.

Everything about the place was neatly kept. House and yard and barn always showed a resident pride. There was an orderly, abundant vegetable garden beside the house. The pastures were mowed every summer. The tiny bottomland where the old man grew his tobacco crop was cut into three or four pieces by waterways which were grassed and bridged. More than anything else, those little timber bridges bespoke the old man's care; the usual thing would have been to drive regardlessly across such shallow drains and so wear the banks away. In addition to the team of horses, the pastures were stocked with a little herd of excellent beef cows.

This place interested me because it was a *good* marginal farm and because it was obviously a relic, the lone survivor within hundreds of square miles of a kind of farm that had been commonplace only thirty or thirty-five years ago. And finally it, too, went the way of the rest of them.

As I watched the old man's farm, driving by it at intervals, I saw it suddenly begin to change. The yard began to look unkept. Disorder began to spread around the house. The team of horses disappeared. I learned a little of the story. The old man had died. His wife had moved to town to live with her children. The house had been rented to people who, though they had technically become its residents, clearly did not *live* there. The farm also had begun to be used by someone who did not belong to it.

I had stopped once and talked a while with the old man. He was busy fixing a fence at the time, and though he received me courteously enough, he did not permit himself to be much interrupted. I told him that I admired his farm. He thanked me, but without enthusiasm, obviously having spent little time yearning to be complimented by strangers. I said his team of horses looked like a good one. He said that they did very well.

One morning after I had learned of his death, I stopped at the farm again—in his honor, maybe, or in honor of my own sense of loss. It was a gray, wintery day. The place looked and felt forgotten. It had gone out of mind. Absence was in it like a force. The barn was closed, empty, the doors tied shut by someone who did not intend to come back very soon. Peering in through a crack, I found that I was looking into a milking room with homemade wooden stanchions, unused for years. I knew why: it had become impossible to be a *small* dairyman. I spent some time looking at the old man's horse-drawn equipment. Some antique collector had taken the metal seats off several of the machines; these had become bar stools, perhaps, in somebody's suburban ranch house. For the rest apparently nobody now had a use. Examining the pieces of equipment, I saw that they were nearly completely worn out, patched and wired together like the fences and buildings, made to do—the forlorn tools of a man who had heirs, but no successors.

By the standards of orthodox agriculture, as well as by those of the present economy and culture, this old man and his farm were merely anachronisms, leftovers. The possibility of their existence would seem contemptible, not just to the majority of agriculture experts, but to the majority of influential people of other kinds. And yet we must ask *why*. And we must be careful not to accept too hasty or easy an answer. For no matter what may be said by the current standards of economics or technology or cultural fashion about this old man's life, there is still no legitimate way of withholding respect from him. In a time when millions of people, including very able and expensively educated young people, are finding it easy to accept a dependence on welfare or unemployment, and when millions more are dependent on social security and other public means of support, here was a man who worked until he died, taking care of himself and of his part of the earth.

The curious thing is that many agriculture specialists and "agribusinessmen" see themselves as conservatives. They look with contempt upon governmental "indulgence" of those who have no more "moral fiber" than to accept "handouts" from the public treasury—but they look with equal contempt upon the most traditional and appropriate means of independence. What do such conservatives wish to conserve? Evidently nothing less than the great corporate blocks of wealth and power, in whose every interest is implied the moral degeneracy and economic dependence of the people. They do not esteem the possibility of a prospering, independent class of small owners because they are, in fact, not conservatives at all, but the most doctrinaire and disruptive of revolutionaries.

Nevertheless, the old man and his farm together made a sort of cultural unit, recognized and valued in this country from colonial times. And it is still a perfectly respectable human possibility. All it requires is the proper humanity.

TRADITION AND EXPERIENCE

It is of great importance to understand that the marginal possibility, the marginal place, and the marginal humanity that I have been describing are reinforced by a marginal way of thinking—until now a sort of counter-theme in our history, so far always subordinate to the theme of exploitation, but unbroken and still alive. This is the theme of settlement, of kindness to the ground, of nurture.

To exhibit this theme, in both its articulateness and its commonness, I offer the following quotations from the *Farmers Home Journal*, a regional farm magazine once published in Louisville, Kentucky. The quotations are taken from the issue of January 2, 1892.

One correspondent writes "to urge every man in Kentucky to set out nut-bearing trees." And this purpose is urged upon the writer by his sense of the necessity of *settling* on the land: "The first thing a young man should do is to get him a home; the next thing get him a wife, and next set him an orchard, but do not think an orchard complete till you have set a few nut-bearing trees."

Another writes that "No man . . . should spend his labor and time over so large an acreage as to fail in making a first class garden." (The reader should be reminded here that the agricultural orthodoxy boasts that farm families have become patrons of the supermarkets.)

Even as early as 1892, we meet industrial arrogance, already fully inflated: "That farmers do not apply more commercial manures to their gardens is mainly because they do not think."

But we also have an example of such not-thinking in a letter from "W. C." of Rural Neck, Kentucky, a place no longer on the map. W. C.'s letter is an exuberant essay on the economy of the soil, and he makes a direct connection between that economy and the economy of money. He recommends the use, as fertilizer, of manure fresh from the barn, and also of scrapings from the barn lot, rotten straw from last year's threshing, old piles of chips and ashes, anything that will rot. "Yes, rot is the word. Rot means death, and without death and rot there can be no new life." He says that one can even use bone dust or superphosphate. "But it won't do for a farmer to go in debt for special or commercial fertilizers, as a rule. You can more safely go in debt for a good stable manure. . . . Nature never loses anything: she preserves and protects herself. It is only a fool man who squanders his substance and makes himself poor, and everybody around him, and the land that he lives on too." He follows this with an attack on soil erosion and praise for manure, industry, and brains. And he concludes: "When people learn to preserve the richness of the land that God has given them, and the rights to enjoy the fruits of their own labors, then will be the time when all shall have meat in the smokehouse, corn in the crib and time to go to the election."

It is a remarkable letter. W. C.'s argument is the one we get —howbeit with greatly increased scientific authority—from Sir Albert Howard, but W. C. is stating it plainly enough fifty years before Howard's books were published. "Rot means death, and without death and rot there can be no new life." This is a principle as new and common as biology, as old and exalted as the Bible: "Except a corn of wheat fall into the ground and die, it abideth alone: but if it die, it bringeth forth much fruit." And W. C.'s voice is seamlessly joined to those of his fellow correspondents who were insisting on the importance of home, household and family, orchard and garden.

What are we to make of these undistinguished men from out-of-the-way places, who pled their cause with the eloquence of good sense and the exuberance of conviction? Jefferson spoke for them in politics. Albert Howard would speak for them, later on, in science. But they speak out of a much more particular engagement with the life of farming than Jefferson ever did. And Howard was still half a century ahead of them. We have to conclude, I think, that they were speaking out of tradition (the yeoman's or the agrarian tradition, which grew

out of a peasant tradition still older) and out of experience
—out of tradition proved and upheld by experience. This association of tradition and experience in the intelligence of a
living person is humanly broad and deep. It is biologically, agriculturally, economically, politically, and culturally sound. It
is deeply founded, solid enough to build a civilization upon,
whereas the orthodox agriculture can support nothing but the
shallow expansion of a bookkeeper's economy.

ORGANIC FARMS

The attitudes and values of traditional agriculture still survive
in our time and are supported by the experience of our time.
Their survival is marginal and is mostly ignored both by the
colleges of agriculture and by the agricultural press, which, if
they acknowledge it at all, do so in order to treat it with contempt. But survivors do exist. They are connected by a sort of
network that one travels by hearsay and friendship. By now I
have encountered a good many of them, and have been impressed as often by the excellence of their characters as by the
excellence of their farms. They are people of principle, both
stubborn and adventurous, independent enough to trust their
own experience and strong enough to hold in considerable isolation to truths not officially or popularly favored. Their farms
stand for their principles and prove them; one has only to notice their example, or their examples, to understand that the
orthodox agriculture has founded its "scientific proofs" upon
shallow assumptions.

In spite of some public notice in the last year or two, it probably still is not generally known that there are a number of
large-scale, highly mechanized farms that do not use chemical
fertilizers or pesticides. When I first began my search for examples of healthy agriculture, I did not realize how compatible
organic soil management could be with a large scale of operation. And then in the spring of 1974, I visited a 900-acre
organic farm in Iowa. This farm made extensive use of a commercial organic fertilizer. But that seemed to me probably the
least important element of the farming there. More important,
I thought, were a careful plan of crop rotation (corn for a year
or two, oats, soybeans, and then two years of pasture), the use

of animal manure on corn ground every year, and the use of a
chisel plow rather than the conventional turning plow for the
preparation of crop ground.

This system was said to have the following advantages: within
the first year or two of its use, earthworms and other forms of
life had again become abundant in the soil; the ground had
become darker, looser, easier to work each year; crops could be
planted earlier because the increased humus in the soil permitted it to dry more quickly; stock feed went farther every year,
because as it became more palatable and nutritious it took less
to satisfy the cattle; the farm had no insecticide program at all,
either for crops or stock.

In its machinery, buildings, etc., this farm was as "modern"
as any other of comparable size. Even though it was far more
diversified than most large farms of these times, and did not
use chemical shortcuts, it required only four full-time workers.
Late in 1975 I visited another highly mechanized organic farm
—this one a 700-acre farm in Nebraska—another extremely
impressive example of organic farming on a large scale. The
existence of such farms as these, on which crops, animals, and
the farmers themselves are obviously thriving, invalidates out-
of-hand the contempt of orthodox agriculturists and suggests
strongly that their contempt must rest on ignorance or fright.

If all the farms in the country were managed organically,
both our people and our land would undoubtedly be healthier
and there would be a considerable ramification of the benefits. And yet the 700- or 800-acre organic farm equipped
with up-to-the-minute machine technology cannot be considered the solution to all of our agricultural problems, or to
the problems that grow out of our agricultural problems. If
we accept this as a solution, we forswear, for one thing, any
further discussion of the cultural and political importance of
the small landowner.

Much more suggestive, in this light, was another Iowa farm
that I visited, this one a family-size holding of 175 acres. Of this,
50 acres were in permanent hillside pasture for twenty-eight
Charolais cows. On the remaining 125 acres, the farmer grew
corn, oats, wheat, soybeans, and hay. In addition to his cow
herd, he kept twelve brood sows and a laying flock of 200 hens.

This farm had been under a completely organic system of soil management for eleven years at the time I saw it in 1974. Here again some commercial organic fertilizer was used to supplement a careful plan of soil husbandry. The cycle of crop rotation was as follows: oats and/or wheat, legumes for hay, soybeans, corn. The application of animal manure was estimated (conservatively) at two tons per acre, and this was put on the bean ground before planting it in corn. The expenditure for commercial fertilizer, which was used only on the corn ground, came to twelve dollars per acre. Every three years or so the pastures were dressed with 300 pounds per acre of a natural mineral fertilizer.

The farmer here was a man of impressive intelligence and judgment and impressively independent in both. Prescribed measures had been altered as he felt necessary to fit his place and his needs. The secretary of agriculture had called for all-out production that year, and on many farms the plowlands had begun to edge out dangerously into waterways and hillside pastures. I asked this farmer if the secretary's recommendation had affected his program. He answered that it had not done so in the least.

On this farm I first had a chance to watch a chisel plow at work and to see the ground it had prepared. This is a favorite tool of many mechanized organic farmers, who give it enthusiastic praise. I could see why. To begin with, it does all the work of seed-bed preparation, replacing both turning plow and harrow. But its great advantage is that it leaves the top layer of soil on top, which is where it belongs. Loosely stirred into this top layer, animal manures and plant residues decompose aerobically. The resulting high content of organic matter causes the surface of the field to act as a sponge, readily absorbing and retaining water and also allowing it to percolate downward into the lower layers. Another advantage of this plow is that it does not cause a hardpan; it does not interfere with the downward course of water through the pores of the soil, worm holes, and old root channels deep into the subsoil. The result is that the soil becomes at once less drouthy and less subject to erosion. It also becomes looser, easier, and cheaper to work, and so operating money goes farther and machinery lasts longer.

On this farm, sod ground to be broken is plowed once with the straight chisels in the fall and is then plowed twice again with sixteen-inch sweeps in the spring before planting. These sweeps are very good for destroying deep-rooted weeds.

This farmer used no herbicides. The reason he gave was that he did not want to contaminate the streams. But he also appeared to have no great need for such chemicals. He tried to plant in the latter part of the planting season so as to allow more weeds to germinate and be killed in the preparation of the ground. He cultivated his row crops to remove large-stemmed weeds, and he found that taking three cuttings a year from his hay fields helped considerably to control weeds in the row crops that followed.

As for crop yields on this farm, I quote the following from a letter that the farmer wrote to me several months after my visit: "I would say our soy beans average 40 or more bu. per acre in an average year. The state average is 30 to 33 bu. Our corn has been yielding 90 to 100 bu. the past 5 years. Neighbors' yields are about the same for the same soil type and lay of land. Our wheat yielded over 25 bu. per acre which we feel is very good for this area. . . . Our oats have been 60 bu. on the average."

The cow herd on this farm was given a balanced mineral mixture as a supplement but was wintered on hay alone, without grain. No insecticides were used on the cattle. The farmer wrote that although his cattle have flies on them in the summertime, they do not have pinkeye or other eye problems usually associated with flies. He attributed this to their extraordinary good health. In December of 1974, he wrote me that the twenty-three March and April calves off this herd weighed variously from 400 to 700 pounds per head. These calves were in robust health, without pinkeye or any other disease.

Another mechanized organic farm is the new experimental farm belonging to Rodale Press, publisher of *Organic Gardening and Farming*. In 1972, 290 acres of this farm were rented to an excellent Mennonite farmer, who agreed to operate it according to strict organic principles. A five-year rotation cycle was set up (corn to rye to barley to wheat to timothy and clover), with twelve tons of manure per acre to be applied to the corn ground. The crops were planted in strips on the contour. Before 1972 this farm was cropped in the orthodox fashion, using heavy applications of chemicals, and the following corn-production figures are especially interesting for that reason. (Figures are available only for corn.) In the first year the yield was 40 bushels per acre; in the second year, 60; in the third year, 80; in the fourth year, 140. In that fourth year the top yield in the same county was 157 bushels per acre—obtained with an application of 190 pounds of nitrogen, 230 pounds of phosphorus, and 673 pounds of potassium.

DR. COMMONER'S ARGUMENT

There is, then, no way to deny that crops and animals can be produced in respectable yields by the methods generally designated as "organic." These methods work on large farms and on small ones. Available evidence indicates that they work at least as well as orthodox methods within the economy of the individual farm, and they will undoubtedly work better as the costs of chemical fertilizers, pesticides, and herbicides rise with the cost of petroleum. But perhaps the greatest benefits from the widespread adoption of organic methods of soil management would go to the general public—in greatly reduced soil and water pollution, in reduced public expenditures for pollution control, in better health, and at least eventually in cheaper food.

The abounding good health of the farms I have described is dramatically evident to an experienced observer. I believe that it would be just as evident to an inexperienced observer who would spend a few hours looking closely and comparing. But in support of the visual impression we now have some evidence from the Center for the Biology of Natural Systems at Washington University—a report published in 1975 and entitled *A Comparison of the Production, Economic Returns, and Energy Intensiveness of Corn Belt Farms That Do and Do Not Use Inorganic Fertilizers and Pesticides*. In *The Poverty of Power*, Barry Commoner makes this study the fulcrum of a powerful argument for organic soil management. I am going to make extensive reference to Dr. Commoner's argument both because it supports and completes my own and because I want to take issue, a little further on, with one of his assumptions.

Dr. Commoner begins by going over some ground often

traversed by the specialists and apologists of orthodox agriculture, but he goes further and sees much more clearly. From
1950 to 1970, he acknowledges, American agriculture made
some impressive increases in productivity: corn production per
acre tripled; "a broiler chicken gained nearly 50 percent more
weight from its feed"; egg production increased by twenty-five
percent; overall farm production "increased by 40 percent."
But during that period the real farm income "*decreased* from
about $18 billion in 1950 to $13 billion in 1971. . . . Because
the number of farms also decreased by 50 percent, the income
per farm rose by 46 percent. . . . However . . . the average
increase in the family income of *all* U. S. families in that period
[was] 76 percent. Meanwhile, the total mortgage debt of U. S.
farms rose from about $8 billion in 1950 to $24 billion in 1971."
During this time there was also a massive shift from diversified
farming to monoculture, which reduced the time that the farmland was covered with plant growth, which in turn reduced the
amount of solar energy put to use on the farm. The removal
of animals from farms growing crops in monocultures reduced
the amount of organic waste returned to the fields. And there
was a shift from the use of nitrogen-fixing legumes to the use
of commercial nitrogen fertilizers. From 1959 to 1973 there was
a sixty-percent decrease in the production of legume seed. By
these and other changes, "The farm's link to the sun has been
weakened, replaced by a new and . . . dangerous liaison with
industry." And this dependence on sources of energy off the
farm explains the decline of farm income. The net farm income
decreased from 1950 to 1970, not in spite of, but *because of* the
new technology of machines and chemicals. Dr. Commoner
concludes his analysis of the effects of this technology with the
following indictment:

"One can almost admire the enterprise and clever salesmanship of the petrochemical industry. Somehow it has managed
to convince the farmer that he should give up the free solar
energy that drives the natural cycles and, instead, buy the
needed energy—in the form of fertilizer and fuel—from the
petrochemical industry. Not content with that commercial
coup, these industrial giants have completed their conquest
of the farmer by going into competition with what the farm
produces. They have introduced into the market a series of

competing synthetics: synthetic fiber, which competes with cotton and wool; detergents, which compete with soap made of natural oils and fat; plastics, which compete with wood; and pesticides that compete with birds and ladybugs, which used to be free.

"The giant corporations have made a colony out of rural America."

Dr. Commoner then turns to the organic farmers studied by the Washington University research group, of which he was a member. He sees in the methods of these farmers the obvious solution to the problem:

"The group analyzed the production of these farms for the 1974 season. The market value of the crops produced by the conventional farms was an average of $179 per acre, while the average value for the organic farm was $165 per acre. However, the operating costs of the conventional farms averaged $47 per acre, and those of the organic farms $31 per acre (the difference is largely due to the cost of the nitrogen fertilizer and pesticides used by the conventional farmers). As a result, the net income per acre of crop for the two types of farms is essentially the same . . . The yields of different crops obtained by the two groups of farms are about equal, except for a small excess (12 percent) of corn yields on conventional farms as compared with organic farms.

"The organic farms used only 6,800 BTU of energy to produce a dollar of output, while the conventional farms used 18,400 BTU. Thus, organic farms appear to yield about the same economic returns as the conventional ones, but do so by using about one-third as much energy."

THE USE OF DRAFT ANIMALS

Because "U.S. agriculture now consumes only about 4 percent of the total national energy budget," Dr. Commoner correctly perceives that the overriding issue here is not that of energy conservation, or even that of pollution resulting from farm use of fossil fuel energy. The overriding issue is economic: the colonization of the farmland by the petrochemical industry. But it seems to me that this perception is not carried far enough. Speaking of the adverse energy economy of the conventional farm, Dr. Commoner says that "when a farmer uses commercial nitrogen fertilizer, the amount of thermodynamic work expended to produce it is seven times greater than the minimum amount of work that is needed to accomplish the same result by planting vetch. But the external energy required to grow vetch *could* after all be reduced to essentially zero (for example, by using a horse fed on farm grown corn). On this albeit impractical standard, the fertilizer's thermodynamic efficiency is zero."

It is the qualifier in that last sentence that concerns me. Dr. Commoner is saying that he is willing to advocate only half the remedy that is called for by his argument. That is, he wishes to do away with agriculture's dependence on petroleum-derived fertilizers, pesticides, and herbicides, but he will not contemplate the reduction of its dependence on petroleum fuels. In the midst of an argument everywhere else incisively intelligent, he suddenly makes this perfunctory bow before the golden calf of "agribusiness"—this spurious standard of "practicality" by which any unorthodox technology may be loftily waved away. To suggest that anything besides a tractor could be used for motive power on the farm is like setting fire to the church—the righteous not only do not *do* it, they do not *think* about it.

But Dr. Commoner's routine refusal to defile the sanctuary is mild indeed in comparison to the official reaction to the same idea. In August of 1975, the *Farm Index*, a publication of the United States Department of Agriculture, carried an article entitled "Wanted[:] 61,000,000 Horses & Mules[,] 31,000,000 Farm Workers." This by now widely circulated article is "based on" a speech delivered by Earle E. Gavett of the National Resource Economics Division.

Mr. Gavett's purpose is to confound "some critics of today's farming practices" who, the article says, have advocated "an anti-technological revolution" involving an immediate and complete return to the use of horse and mule teams on American farms. This alleged proposal, the article is relieved to note, has "some serious—if not insurmountable—drawbacks" in that it requires sixty-one million horses and mules, of which there were only three million in the United States in 1975, and thirty-one million farm workers, of whom only four million were available. These figures were derived in the following way:

"The 1967 index was the yardstick. The 1918 crop had an index of 48—that is, 48 percent as large as the 1967 crop—compared with 109 in 1974. Thus 1974 production was about 2¼ times greater.

"As a peak year of nonmechanized farming, 1918 is an ideal choice in the comparison.

"A straight projection of 1918 resources to meet 1974 production can be made by simply multiplying the 26.7 million mules and horses and the 13½ million farm workers carrying on farming in 1918 by the 2.27 times larger output in 1974."

The article concedes that this is "only a guideline projection. Obviously, nonmechanical and nonchemical technology improvements . . . since 1918, such as hybrid seeds, would lessen the manpower and horsepower requirements by allowing greater yields for less work.

"But agricultural economists quickly emphasize that the point of the projection is valid: a complete abandonment of mechanized technology is a biologically impossible and sociologically impractical idea."

The necessary animals could not be produced, the article continues, before 1992 or 1993. To grow feed for these animals would require "180 million acres of prime farmland." And there would be "questions over feeding so many horses in this country while people abroad are starving."* Moreover, the necessary people are also in short supply, and "A movement of 26 million workers from city to farm would provide mind-boggling problems."

That gives the main line of the argument, which gets considerably more elaborate without ever becoming more intelligent. Like many another, this document would merit no more attention than it merits respect if it were not for its influence. It happens, however, that this argument was given the status of official policy of the United States Department of Agriculture in a speech by no other than former Secretary Butz himself.

---

*It is fascinating to observe the agriculture specialists' flexible mindfulness of the hungry, who are, according to the argument at hand, either to be compassionately fed or starved into compliance. Either way, their fate is directly bound to the ambitions of the "agribusiness" corporations, who have thus added an enlightened versatility to the originally narrow and primitive Christian concept of charity.

And so we have before us one of the characteristic political necessities of our time: to take seriously what we cannot respect.

The chief objection to this argument is that there was never a reason or an occasion for it. There are simply no serious critics of conventional agriculture who have advocated "a complete abandonment of mechanized technology." As the *Draft Horse Journal* noted in an editorial, "Most of the critics of today's agriculture . . . don't talk about any such complete anything, but rather a picking and choosing of techniques and tools to get the job done in the most energy conserving way possible." The key phrase here is "picking and choosing." There are indeed critics who believe that a much larger range of technological choices and alternatives ought to be available, that we will have neither a healthy agriculture nor a dependable food supply until such choices and alternatives are available—that, in short, the strength of agriculture is in diversity, of technology as of other things, and that the present agricultural orthodoxy ignores the principle of diversity altogether. A few of these critics have published articles in such magazines as *Organic Gardening and Farming*, *Mother Earth News*, and the *Draft Horse Journal*, in which they have pointed out that there are presently places in agriculture and forestry that can be competently and economically filled by horses or mules. Some have said that, given our difficult economies of both energy and money, much wider use might reasonably be made of draft animals in the future. No one, as far as I know, has *ever* proposed that such a change could, or should, be either complete or rapid.

That this small advocacy of a small diversity should have drawn a full-scale attack from the Department of Agriculture bespeaks both the totalitarianism and the paranoia of the "agribusiness" mentality. What can be the excuse for all this carrying on? If these critics are right, then as scientists, the agriculture experts might be expected simply to agree. If the critics are wrong, then it appears that they might safely be ignored, for orthodox farming is far too widely accepted to be seriously threatened by a bad idea. The truth is that these critics have offended, not by being either right or wrong, but by being *different*.

Even if we could grant that we are indeed threatened with "an anti-technological revolution," the competence of Mr.

Gavett's argument is still in question. As the *Draft Horse Journal* pointed out, the arithmetic of his "projection" is far too simple, assuming, as it does, "that it takes three times as many horses and mules to cultivate corn yielding 120 to 150 bushels to the acre as corn yielding 40 to 60." And of his assertion that it would require three acres of "prime farmland" to feed one horse, it can only be said that he does not know what he is talking about. By my figures, using the rations recommended in the twentieth edition of Morrison's *Feeds and Feeding*, a ton horse doing medium to heavy work every day of the year would require 104 bushels of ear corn and 7300 pounds (about 209 thirty-five-pound bales) of hay. Assuming that the hay is of grass and alfalfa, this much feed could be produced on less than two acres at today's yields. But not all draft horses would or should weigh a ton—1300 to 1800 pounds would be a realistic range. And very few indeed would work every day the year around. The above figures do not consider the horse's off-time subsistence on pasture alone or on a maintenance ration mostly of hay, and they do not consider his ability to utilize roughages such as cornstalks, now seldom used as feed. A more realistic accounting might be that of the *Draft Horse Journal*'s editorial, which states that a horse eats "the energy equivalent of 70 bushels" of corn.

One also notes this article's easy assumption that *all* of the thirty-one million needed people would be "workers" and not farmers. And that the mind that is "boggled" by the problems of "a movement of 26 million workers from city to farm" is apparently not boggled at all by the continuing and appalling problems of the recent movement of many more than that from farm to city. If there were any suspicion that such a reverse migration might be profitable to "agribusiness," we may be sure that there would be an unhesitating effort to bring it about. This is a kind of mind that is boggled only at its convenience.

The same is true of the "questions over feeding so many horses in this country while people abroad are starving." This serviceable charity is not at all troubled by work now being done at the University of Nebraska on the possibility of using grain alcohol as a motor fuel. It is morally questionable to feed grain to a work horse; but if the grain is to be consumed by engines to the profit of energy corporations and the machinery and automobile manufacturers, then the starving are forgotten. Nor do the people who attack the use of horses for farm work ever say a word against their use for racing, show competition, and other frivolous purposes.

"Horses," the *Draft Horse Journal* said, ". . . are no more anti-technological than legs on humans." They are simply a technological possibility that we have almost ceased to consider. We must learn to consider it again, for until we do we cannot complete the logic of Dr. Commoner's argument, nor can we answer the questions raised by the existence of, and the potential need for, such "marginal" lands as I described earlier. There are certain problems for which the use of horses is the appropriate solution—or for which we have so far found no more appropriate solution. There is also the possibility that a revival of the lapsed technology of horse-powered agriculture is necessary to complete our agricultural intelligence and judgment—to give us the diversity of choices required for the subsistence of intelligence and judgment.

The issue of economics merges finally into the much larger issue of health, just as the issue of the health of any one creature merges into that of the health of Creation. In the context of that issue, Dr. Commoner's "impractical standard" of near-perfect thermodynamic efficiency becomes not just thinkable but indispensable. It is no more impractical than the standard of perfect health, which we all apply to our bodies. We desire —our bodies desire—to be perfectly healthy. That is what we hope and strive for. And it is the way we understand our effort; without the ideal of perfect health, we could not know how healthy we are. To be three-quarters or seven-eighths healthy is not an ambition that ever occurs to us.

There is no point in saying that perfection of health, as of all else, is not attainable by humans. The point is that we must have the vision of perfection, we must strive for it, we must sense the possibility of approaching it, or we cannot live. Jesus enjoined his followers to be perfect—not, I think, because they could hope for perfection, but because perfection is the necessary standard. People cannot understand themselves, or live fully and humanly, without it. To reconcile ourselves to imperfection, to place great practical barriers in our own way, is brutish. It condemns us immediately to great suffering of the spirit and undoubtedly, in the long run, of the body as well. Common sense alone requires us to consider well any technology that might bring us nearer the vision of perfect health. To repudiate such technology on the ground that it is "old-fashioned" is madness.

It is hard to overestimate the importance of applying the correct standard to agricultural performance. I do not see how a stable, abundant, long-term agriculture can be built up and maintained by any standard less comprehensive than that of the perfect health of individual human bodies, of the community, and of the community's sources and supports in the natural world—whereas the standards of orthodox agriculture tend to be extremely simple and exclusive: productivity (as determined by "records" and by the equation between the number of eaters and the amount of food) and the financial prosperity of "agribusiness."

It is easy to say, as former Secretary Butz said in his own fatuous attack on the "anti-technological revolution," that "To return to the 'good old days' in agriculture, or indeed just to cling stubbornly to the farming methods of today, would be to condemn hundreds of millions of people to a lingering death by malnutrition and starvation in the years ahead." But that is simply the oldest—and the most profitable—cliché of the industrial revolution, supported only by a thoughtless obeisance to "progress." We must look beyond that to what is assumed. So far as I can make out, Mr. Butz's statement rests upon two main assumptions, both suspect: that the health of humans may be safely distinguished from the health of the rest of Creation and that there is no distinction between affluence and survival.

People who argue for ways of farming that are ecologically sound, says Mr. Butz, are "placing the needs of man second to the needs of all other creatures." Man, he says, is "as much a 'part of nature'" as the other creatures. But what he means is that human needs must be put *ahead* of the needs of all other creatures, as we see when he equates a hydroelectric or an irrigation dam with a beaver dam. His solution to the problem of hunger is therefore remarkably unencumbered by moral, cultural, or ecological considerations: "We'll turn to science and technology for the answer—we'll modify the environment." Thus, with a shrug, he sets agriculture free of ecology. But we are left with an awesome ecological question: How can humans, who are creatures, hope to survive in a world in which other creatures perish? Or how much can we "modify" the environment before we fatally "modify" ourselves? Here is Mr. Butz's answer: "The challenge to agriculture and science is to find the right application of technology to modify the environment in a way that will benefit both man and the rest of nature." One can only agree, pointing out, however, that the applications of technology so far advocated and defended by Mr. Butz have notoriously failed to do so and that his colleagues and constituents in the "agribusiness" system have so far failed notoriously even to consider the advisability of doing so.

The second assumption is, of course, closely related to the first. Mr. Butz begins with the term "survival" and a most dramatic issue he makes of it: "Backed in a corner with no job, no income, and an empty stomach churning from hunger, the most dedicated environmentalist will forget his fight for the seagull or the walrus. He will get down and scrap for survival like any other creature . . ." Of course he will—though he may even then remember that he and the sea gull and the walrus are all scrapping for survival in the same world and against the same abuses. But by the end of Mr. Butz's speech, without transition or warning, the term has changed: ". . . there can be little hope for mankind's continued *affluence* [my emphasis] unless we face up to the moral question of the need to limit our numbers. In the meantime, science and agriculture will have to buy the time for us to reach that solution." And with this shift of terms, "science and agriculture" have been nominated to do the work that can be done safely and adequately only by complex cultural changes leading to restraint of consumption and competent care of the earth. It is exactly this refusal to consider survival except as "continued affluence" that has brought our survival into doubt. There is, anyhow, only a fanciful connection between affluence and survival; we do not have to be as comfortable and extravagant as we are in order to survive. And there is no connection between affluence, as we understand it, and civilization. All that civilization requires is enough; it does not require extravagance. Until these distinctions are made, we cannot even begin to talk sensibly about the problem of hunger.

The fact is that Mr. Butz and his colleagues in the corporations and the universities do not know whether unorthodox technologies and methods will produce more food or less. The only information that they have, or that they acknowledge, is that which "proves" the efficacy of "agribusiness" technology. Where are the control plots which test the various organic systems of soil management? Where are the performance figures for present-day small farms using draft animals, small-scale machine technologies, and alternative energy sources? Where are the plots kept free of agricultural chemicals? If these exist, then they are the best-kept secrets of our time. But if they do not exist, whence comes the scientific authority of orthodox agriculture? Without appropriate controls, one has no proof; one does not, in any respectable sense, have an experiment.

HORSE-POWERED FARMS

Mr. Gavett, followed by Mr. Butz and others, bases an amazingly bitter attack against the use of horses upon a "projection." There was no need for so speculative a maneuver, for a number of horse-powered farms presently exist—not experiments or controls, but living examples, requiring only to be carefully observed. I have visited a number of farms powered either partially or exclusively by draft horses. I can offer only a few random figures having to do with these farms, and so what I have to say is offered as proof of nothing except their possibility. But the fact of their possibility suggests strongly that we ought to have thorough studies of their ecological and economic performance.

In the early spring of 1975 I visited three good Iowa farms, all of which made extensive use of horses. All three were farmed by older men, working for the most part alone, who farmed this way by conviction; who were thoughtful, indeed passionate, holdouts against the capital-intensive, highly mechanized farming of their neighbors; and who lived in the isolation of those who are "different." Of their financial condition, I can say only that from all visible signs they were better than solvent. Their homes were comfortable, their farm buildings well kept up, etc.

The first of these men farmed 120 acres. He had two teams, one of which was a young pair he was breaking for another horseman. He owned two thirty-year-old H Farmall tractors that he used mainly for the heavy work of plowing and disking; the rest of his work he did with the horses. Aside from the manure from his barn, he used no fertilizer. He did not use insecticides. Herbicides he used only selectively, for the control of thistles. In addition to the considerable saving of fuel, he mentioned two other benefits from his way of farming: he believed that he had less erosion than his neighbors and that his ground worked easier. His corn yielded an average of seventy bushels per acre.

The second of these farmers had 300 acres of excellent land surrounded by cash-grain farm "businesses" of the orthodox make. He did most of the farming with horses, keeping an ancient Farmall tractor to do only the heaviest field work and to provide stationary power. Except for plowing down "a little" fertilizer, he used no chemicals. His fields were fertilized with manure, and tilled in rotation from corn to beans to corn again to oats to hay. His corn yield ran to about seventy-five bushels to the acre. The economy of this farm was carefully diversified. Among his other enterprises, the farmer had a dairy herd of six cows, which he milked by hand.

The third farm was similar to the second in size and in the combination of horse power with an old tractor (a 1946 WC Allis-Chalmers) used for power-takeoff work and for the heaviest work in the field. This farmer said that he had "never used a pound of fertilizer." He owned twelve horses, one a Percheron stud. The income from this horse-breeding operation was paying *all* the operating expenses of the farm. I neglected to ask this farmer what his corn yield was. But I did ask him how his economic situation compared to that of his neighbors. He said that he couldn't say, but that they often called him over "to buy something that they ought to keep."

My visits to these farms involved long distance and short time, and so were necessarily far too hasty. And it was too early in the season to get a fair look at the condition of the fields. For those reasons my information is not nearly so complete as I now wish it were. I am able to fill out the impression somewhat by quoting a letter from an Iowa agriculture student who spent much more time on the second of these farms than I did. His visit, like mine, was before the growing season, and again the facts are scanty. But his description is much more detailed than mine, and a context is given in which the *meaning* of such a farm is made plain.

"At this time of the year 95% of the land has either been cut (soybeans) or plowed up (corn). The few guys who have any livestock at all have it on an enormous scale and they are among the few who have let fences remain. So the majority don't plow to the fence row; they plow to the culvert's edge. Row crop cultivation is done on such a large scale here that they must fall plow so that they can be timely in planting the vast acreages come spring. But what I saw on the south and west sides of fields in the culverts were snow drifts that were black to dark grey—each layer of snow has a layer of topsoil on top. You see this beside every field without exception that has been fall plowed. There's no protection from the wind in this flat country and if they don't get early freezes with heavy snow the land is vulnerable. There are many abandoned farm buildings with the ground cultivated within a few feet of them. I often saw good stock barns, the likes of which you rarely see in the southeast, with the south or east end cut out, and all they hold are large tractors and implements. You don't have to travel far to see whole square miles of land with no farm houses or outbuildings, plowed up north to south and east to west—right up to the culverts. And in the midst of this land, where farmers are no less dependent on Shell Oil Co. and John Deere than they are on the weather, stands _____'s place; honestly, to see it is to believe that it's an oasis in the midst of a desert. I knew from a mile and a half down the road that it was his place. His milking shorthorns were out gleaning corn. The fields were well fenced, the buildings being used for the purposes intended. His rotation is the old Iowa standard: 60 acres of corn (and some sorghum), 30 of oats, 30 of hay (clover), 30 of soybeans all on the home place. He has another 120 acres on a neighboring section. He keeps 40 shorthorns, five* of

*One had evidently been turned dry since my visit.

which he milks by hand. He fattens about twenty hogs, keeps 200 chickens by which he's able to sell eggs to his neighbors."

It will be observed that the use of horses is not just a means of doing work, a kind of power added to a farm from outside as petroleum or electricity is added to it. The use of horses is a means that *belongs* to the farm; it is a way of farming; it is, as Maurice Telleen points out, invariably accompanied or followed by a set of practices that belong together. If made to belong to the land by good care and good sense, horses tend to preserve its health. With horses come pastures and hay fields, because the horses must eat. And if one is going to grow forage for horses, then one finds it natural and economical to grow it also for other animals. From the growing of forage and the diversification of animal species, there follow naturally the principles of diversification and rotation of field crops. Having animals, one has manure, and so manure is used instead of commercial chemical fertilizers. And the use of manure, the conservation of humus, and the practices of rotation and diversification tend to work against diseases, insects, and weeds, and so one uses few or no pesticides. It is a way of farming that involves year-round use of the land by animals, plants, and the farm people—in contrast to the "corn, beans, and Florida" rotation of orthodox cash-grain farmers. Moreover, the farmer who farms with horses is not likely to be an expander. His way of farming tends to confine him to a limited acreage near home. He therefore concentrates his attention and, instead of getting more, takes good care of what he has—sows cover crops, guards against erosion, etc.

What we have here is a description of a permanent, settled, careful, largely independent agriculture that uses the land more efficiently, at least in the sense that it uses it more months in the year and more conservatively, than the orthodox agriculture. The defenders of the orthodoxy will immediately point out that the corn yields I have cited are extremely low and that we would run great risks should we reduce all yields to that level. This point must be taken seriously—not just by people on my side of the argument, but by all students and scholars of agriculture, for what is required is a definitive, scientifically sound answer. In the absence of such an answer, there are still a couple of points that need to be made.

First, it must be emphasized that all three of these farmers are older men whose children have left the farm and who are working for the most part alone. We must therefore consider that they may lack the energy, help, and motivation to push themselves and their fields toward maximum production. Second, these farms are survivals of an old way—a good way, when well followed, but not necessarily the best. The pressures of surviving, of keeping their inherited values intact in an increasingly alien atmosphere, have undoubtedly kept these farmers from being as innovative as they might have been in kinder circumstances. Particularly suggestive is the possibility of grafting the soil management methods of the more advanced organic farmers upon the traditional structures and skills of the old horse-powered farming.

But offsetting the smallness of these yields is their relative independence of economic and political conditions. Such yields are attainable on these farms year after year, *whatever* the availability of credit or of petroleum products—something that cannot be said of the much larger yields of orthodox farms, which depend absolutely on credit and on "purchased inputs" from the oil industries. The horse teams will go to the fields no matter what is happening on Wall Street or in the capitals of the Middle East. Seventy bushels of corn per acre is only half as good a yield as 140 bushels, true enough. But then it is infinitely preferable to no bushels at all.

THE AMISH

My final example of an exemplary marginal agriculture is that of the Amish. Nothing, I think, is more peculiarly characteristic of the agricultural orthodoxy—as of American society in general—than its inability to see the Amish for what they are. Oh, it *sees* them, all right. It sees them as quaint, picturesque, old-fashioned, backward, unprogressive, strange, extreme, different, perhaps slightly subversive. And that "sight" is perfect blindness. What is not seen is that the Amish are a community in the full sense of the word; they may well be the last surviving white community of any considerable size in this country. And for this there are reasons. It is especially the reasons that we do not want to see, for these reasons invalidate most of the assumptions and ambitions by which we proudly characterize ourselves as "modern."

My knowledge of the Amish, as of the other farmers I have discussed, is by no means thorough or detailed enough to satisfy the demands of strict scholarship. And I shall not pretend to be "objective" about them. I admire and respect them deeply, with few reservations; in many ways I envy them. In addition to reading several published accounts of Amish culture and agriculture, I am able to speak to some extent from experience. I have looked carefully at Amish farming in Iowa, Pennsylvania, Indiana, and Ohio, and these travels have involved some personal contacts.

What, then, are the reasons that the Amish have been able to survive as a community—or, it might be more correct to say, as a closely bound fellowship of many communities? I think that there are three primary reasons, from which spring many others.

First, the Amish communities are, at their center, religious. They are bound together not just by various worldly necessities, but by spiritual authority. Theirs is, moreover, a religion unusually attentive to its effects and obligations in this world. Whereas most contemporary sects of Christianity have tended to specialize in the interests of the spirit, leaving aside the issues of the use of the world, the Amish have not secularized their earthly life. They have not hesitated to see communal and agricultural implications in their religious principles, and these implications directly influence their behavior. The "goal" of Amish culture is not just the welfare of the spirit, but a larger harmony "among God, nature, family, and community."*

Second, the Amish have severely restricted the growth of institutions among themselves, and so they are not victimized, as we so frequently are, by organizations set up ostensibly to

---

*The Amish have two considerable problems, now, which trouble this ideal of harmony. They are having more children than, in present economic conditions, they can provide farms for, and so some of their young people are taking town jobs. And where coal underlies their land, some are permitting the stripminers to come in. Reclamation was better than usual on the Amish farms I saw that had been stripped, and the land was going back into pasture. Some Amishmen nevertheless feel the practice to be wrong.

"serve" them. Though they pay the required deferences to our institutions, they accept few of the benefits, and so remain, in perhaps the most important respects, free of them. They do not become dependent on them and so maintain their integrity. As far as I know, the only institutions in our sense that the Amish have started are their schools—and this, by our standards, for a strange reason: to *keep* the responsibility for educating their children and so, in consequence, to keep their children. Amish ministers and bishops are chosen by lot, after fasting and prayer (as Mathias was chosen), and so they do not have a professional, a paid, an economically dependent, or an ambitious clergy. Their religious services are held in barns or homes; their charities are not organized or abstract but are usually in direct response to observed needs. And so they do not have a church building or a building fund or church functionaries or administrators. There is little distinction between the church and its members.

There are, one may as well say, only two Amish institutions: the family and the community. And these institutions fulfill directly, humanly, simply, and quietly nearly all the functions that we have delegated to our obtrusive, inhuman, indifferent, clumsy, expensive institutions. Family and community serve as insurance, welfare, social security, public safety. Indeed, they serve as, and replace, government. The simple living together of relatives and neighbors makes unnecessary to them our obsession with "security."

Third, the Amish are the truest geniuses of technology, for they understand the necessity of limiting it, and they know *how* to limit it. They have refused to see "technological innovation as an end in itself." And so their "religiously enforced family and community values are safeguarded against the social costs of changes which in their estimation did more harm than good to the community as a whole." Whereas our society tends to conceive of community as a loose political-economic mechanism of mutually competing producers, suppliers, and consumers, the Amish think of "the community as a whole"—that is, as all of the people, or perhaps, considering the excellence both of their neighborliness and their husbandry, as all the people and their land together. If the community is whole, then it is healthy, at once earthly and holy. The wholeness or health of the community is their standard. And by this standard they have been required to limit their technology.

By living well without such "necessities" as automobiles, tractors, electrical power, and telephones, the Amish prove them unnecessary and so give the lie to our "economy." And by these restraints they have kept their health, for by them they have kept themselves at home and have, for the most part, kept their children at home. They have not the knowledge of experts, which is by definition a homeless or rootless knowledge —the knowledge, in Sir Albert Howard's words, of people who cannot "take their own advice before offering it to other people"—and which is, as such, dangerous. They do not use knowledge to prey upon one another.

The healthy results of these restraints are readily visible to anyone who so much as drives an automobile through such an Amish community as the one in Holmes County, Ohio. Unlike so much of the best farmland, which has become a kind of agricultural desert, the Amish landscape in Holmes County is vibrantly populated with both people and animals. Busy people are seen everywhere. All the houses are lived in. All the buildings are in use. Fences and buildings are in excellent repair. And there are signs of a thriving and thrifty home life: vegetable gardens, flower gardens, fruit trees, grape arbors, berry vines, beehives, bird houses. People are making careful, comely, dignified work of the essential tasks defined by modern values as "drudgery." And because they have thought of the well-being of all the people, all are busy. There is a use for everyone. The Amish do not have the abandoned children, cast-off old people, criminals, indigents, and vagrants whom we have "freed from drudgery."

And these people practice a way of farming capable of taking exquisite care of the land. In the fall of 1976 I stood on a hillside that had been used and cared for by three generations of Amish farmers. It was steep land of the sort more often than not worn out under the old American agriculture and simply unusable by the new. This hillside had been cropped in alternating strips of corn and sod. The corn crop, which was excellent, had been cut with a binder and shocked. The farmer and his sons had carried the bundles off the plowed ground—eight

rows up the hill, eight rows down—and shocked them on the
sod strips, so as to get the cover crop sowed as soon as possible.
By orthodox standards, this work was demeaning drudgery. By
the standard of the health of the field, it was simply necessary,
and so it had been done. When I was there the cover crop was
coming up to safeguard the ground over the winter. I looked
for marks of erosion. There were none. It is possible, I think, to
say that this is a Christian agriculture, formed upon the understanding that it is sinful for people to misuse or destroy what
they did not make. The Creation is a unique, irreplaceable gift,
therefore to be used with humility, respect, and skill.

And so, though Amish agriculture is not modern or progressive, it is by no means ignorant or unintelligent. By the
correct standard, it is much more sophisticated than orthodox
agriculture. The Amish were among the first to understand the
uses of rotation, manure, and legumes. They keep a balance
between livestock and crops. They benefit from exchanges
of labor and other forms of neighborliness. In lieu of massive consumption of fossil fuels and electricity, they make the
fullest possible use of energies available on the farm—of the
wind, of draft animals, and, of course, of their own bodies.
Their technological restraints are balanced, quite naturally, by
inventiveness. The Amish are good mechanics, and they have
displayed much ingenuity in, among other things, the adaptation of tractor implements for use with teams of two to eight
or more horses.

Another observer of Amish farming wrote a letter to the
editor of the *Draft Horse Journal*, who published excerpts in
the issue of Autumn 1976. The editor noted that "the writer of
this letter owns no horses, has no vested interest in the horse
business, that I know of." The following paragraphs are taken
from that letter.

"My farming consists of just under a hundred acres of rather
heavy, low lying land. At one time our family did something
or other on three farms, of which one has gotten completely
covered by houses and another has been sold to an Old Order
Amishman. We still call the third one home.

"When the one farm was sold to the Amishman I forgot to
tell him that one particular field was too heavy and low to grow
alfalfa. By experience, I knew it wouldn't work. For a couple

of years he had it in other crops, then he went to alfalfa. This embarrassed me because I knew I should have cautioned him on this. But he had a marvelous stand and a very heavy yield. It was puzzling. I puzzled over it for years but am now very persuaded of the why and wherefore.

"With our tractors we kept the soil rather permanently compacted because it was necessary to get on the land as soon as surface moisture conditions permitted. And the tire patterns pretty well rolled the entire area in the course of repeated passage. With his horses, this just didn't happen. And in the course of a couple of winters the deep frost had corrected my tractor compaction mistakes. Soil structure improved. Root penetration was facilitated. Water holding capacity as well as internal drainage both benefited, and the alfalfa flourished.

"The Amish will not argue the point because they don't think anyone is interested, but will say that a farm 'works' easier after a couple of years of horse farming. This compaction problem has to be the explanation. The resulting improved soil structure, allowing for better root penetration, is probably the reason they can get similar yields with less chemical fertilizer than their mechanized neighbors.

"But back to that heavy land alfalfa for a moment. The crop was disgustingly rank and lodged. Mowing it would be a problem. So I paid a visit, and . . . was flabbergasted to notice the farmer slowed down and mowed right through. . . . Then I realized that something very nice was being demonstrated; since the sickle drive was independent from ground travel (his horse drawn mower was equipped with an engine to drive the sickle . . .) this 'horse farmer' had sickle-cycle-to-ground-speed control that no tractor farmer could have unless he had a hydrostatic drive tractor! It gave him a control flexibility that I had never experienced and left me feeling sort of humble. I came away wondering which of us had the better technology . . . and I'm still not sure."

And then this writer addresses himself to the standard orthodox argument that we cannot feed draft horses without starving humans.

"Of course you have to feed draft animals. But this does not necessarily mean you'll have less to sell per acre farmed. In my observation, good horse farmers seem to have about as much to market per acre as the rest of us. Certainly they manage to nourish their animals very well, too. Closer examination would probably show the animals are at least partly nourished on what is wasted on fully mechanized farms. I'm not speaking just of corn fodder, either. Who else hand gathers the ears the picker missed now-a-days? . . . Does the man with the 4 row combine? And could he gather the shelling loss if he would? Hardly.

"Well, I'm not very impressed by the statistics that prove we'd starve if farming went back to animal power. In many sections of the country that is exactly what happens when a farm, or a group of farms, comes into the hands of an Amishman. The farming has gone from tractors back to horses on hundreds of farms in my part of the United States without any noticeable reduction in agricultural output. Any suggestion that our county produces less now than 20 years ago would seem outrageous to all the people I know."

In support of these impressions of the general good health of Amish agriculture, some more specific information is available in an article entitled "Agricultural Alternatives" in the March-April 1972 issue of *CBNS Notes*, published by the Center for the Biology of Natural Systems. I take the following quotations from that article:

"Amish attitudes toward fertility maintenance are amazingly varied. . . . Preliminary analysis of the data shows three distinct patterns. The traditional pattern consists of crop rotation which includes nitrogen-fixing legumes, heavy application of manure to at least 1/5 of the farm in any growing season and lime and rock phosphate to one of the fields in the rotation every year. With the aid of hybrid seed corn some farmers estimate their yields at 90 to 100 bushels per acre although some estimates fall as low as 70.

"A second pattern is the conventional with Amish modifications. The soil will be tested for its acidity and for its phosphorous and potash balance. A county agent or a fertilizer dealer will then make a recommendation for lime and fertilizer application in relation to specific cropping plans. This can include the use of anhydrous ammonia as a cheap source of nitrogen for corn. It is observed that Amish operators who follow this pattern 'factor in' the effects of their crop rotation and the availability of manure and thus apply fertilizer less heavily per acre. . . .

"The third pattern is the use of organic fertilizers. . . . Some of them tend to be costly in terms of additional yield per acre but a minority of the Amish farm operators are very enthusiastic users. The appeal is on the basis of a claim to a more nutritious quality of feed grain which in turn leads to healthier livestock, healthier soil and eventually healthier humans. Interestingly, several of the operators interviewed adopted a program of organic fertility maintenance after having been on a conventional program of commercial fertilizer for several years."

As for the effects of this agriculture, the article offers evidence which suggests that, ecologically and economically, the Amish methods are sounder than the orthodox. Water pollution from Amish fields was far less: "One of the comparative samples taken in March, 1971 showed concentrations of nitrate nitrogen of 12.1 ppm and 8.9 in the tile of conventional farms in Douglas County [Illinois]. The Amish tile had a concentration of 4.6 ppm (corn). A comparison in May showed a concentration of 26.6 ppm (corn) and 10.9 ppm (beans) in the conventional farms and 4.6 ppm in the tiles of the Amish farmer."

And the Amish, whose farms in 1965 averaged only 76.55 acres, were prospering financially during a time when many of the smaller orthodox farmers (with far larger holdings) were being "squeezed out": "Our best single indicator of economic viability is bank data which compares 88 Amish bank accounts in 1964 with those same accounts in 1971. During these years the Amish accounts showed an increase in net worth from $2,379,000 to $4,045,000."

Since the Amish are manifestly excellent farmers, and are so complexly successful in other ways, one wonders why they have been ignored by the officials and the scholars of agriculture —especially since their technology and methods are so well suited to land not even farmable by orthodox methods and to farmers not able to survive in the orthodox economy. I have been able to think of only two answers, aside from the conventional contempt for anything small: first, the Amish are a thrifty people, hence poor consumers of "purchased inputs" from the "agribusiness" industries; and, second, they are living disproof of some of the fundamental assumptions of the orthodoxy.

PRODUCTION AND REPRODUCTION

To these exemplary forms of unorthodox agriculture, we may add the new work in urban homesteading, aquaculture, solar green-houses, alternative energy sources, small technology, organic pest control, etc., by the New Alchemy Institute, the Farallones Institute, Rodale Press, and others, as well as the various farmers' and consumers' cooperatives that have been started in the past few years both as strategies of health and as protests against the "agribusiness" juggernaut. Together, these have restored a sense of possibility, both cultural and agricultural, that has been nearly obliterated by the ambitions of agriculture specialists and businessmen. They make possible a vision of an agriculture many times more versatile and diverse than the orthodox, hence many times more responsive to the demands of good husbandry, to local conditions, and to human needs.

For the orthodox obsession with production, profit, and expansion, this healthier agriculture would substitute a more complex consciousness, the terms of which would be ecological integrity, nutrition, technological appropriateness, social stability, skill, quality, thrift, diversity, decentralization, independence, usufruct. Or, put more simply, it would replace the concern for production with a concern for reproduction. Production, some would say, is the male principle in isolation from the female principle. Thus isolated, the male principle wants to exert itself absolutely; it wants to "do everything at once" —which is, of course, what doomsday will do. But reproduction, which is the male and the female principles in union, is nurturing, patient, resigned to the pace of seasons and lives, respectful of the nature of things. Production's tendency is to go "all out"; it always aims to set a new record. Reproduction is more conservative and more modest; its aim is not to happen once, but to happen again and again and again, and so it seeks a balance between saving and spending. At their best, farmers have always had this ancient purpose of reproduction. Without it, they make their art as sterile as mining.

There would, of course, be no need for a different vision of agriculture if the one we had were demonstrably working in the best long-term interests of the people and the land, or even if it were generally believed to be doing so. In fact, there are a great many people who do *not* believe that it is doing so, and their number is growing. And so the last agricultural margin remaining to be noticed is a political one: the people who feel that they are being victimized by orthodox agriculture and whose dissatisfaction is either ignored or held in contempt.

There are, first, many people—ex-farmers, heirs of farmers, and would-be farmers—who want to farm but are prevented from doing so by high land costs, taxes, inheritance taxes, and interest rates. And these economic barriers, which exclude the small operator, directly favor not just the survival, but also the expansion, of the big operator. This is not a necessary result of "the way things are." It is the calculated effect of a deliberate policy to allow the big to grow bigger at the expense of the small. In addition, there are many farmers of the same kinds who are presently farming, but whose survival is in doubt for the same reasons.

Second, there is a rapidly increasing number of consumers who wish to buy food that is nutritionally whole and uncontaminated by pesticides and other toxic chemical residues. And these people would prefer not to pay the exorbitant food prices required by long-distance transportation, processing, packaging, and advertising, all of which result from "agribusiness" control of food.

PUBLIC REMEDIES

And so we come to the question of what, in a public or governmental sense, ought to be done. Any criticism of an established way, if it is to be valid, must have as its standard not only a need, but a better way. It must show that a better way is desirable, and it must give examples to show that it is possible. I have produced the argument and the examples—not definitively, I am sure, but sufficiently to provide an agenda for the further work that is necessary.

It remains for me to suggest public changes that are necessary to bring the better way to realization. This is the most fearful part of my task, for what I have described at such length here is a big problem, and it is the overwhelming tendency of our time to assume that a big problem calls for a big solution.

I do not believe in the efficacy of big solutions. I believe that they not only tend to prolong and complicate the problems they are meant to solve, but that they cause new problems. On the other hand, if the solution is small, obvious, simple, and cheap, then it may quickly and permanently solve the immediate problem and many others as well. For example, if a city-dweller walks or rides a bicycle to work, he has found the simplest solution to his transportation problem—and at the same time he is reducing pollution, reducing the waste of natural resources, reducing the public expenditure for traffic control, saving his money, and improving his health. The same ramifying pattern of solutions attends all skills and strategies of economic independence: gardening, cooking, household maintenance, etc. To turn an agricultural problem over to the developers, promoters, and salesmen of industrial technology is not to ask for a solution; it is to ask for more industrial technology and for a bigger bureaucracy to handle the resulting problems of social upset, unemployment, ill health, urban sprawl, and overcrowding. Whatever their claims to "objectivity," these people will not examine the problem and apply the most fitting solution; they will reverse that procedure and define the problem to fit the solution in which their ambitions and their livelihoods have been invested. They are thriving on the problem and so can have little interest in solving it.

And so the first necessary public change is simply a withdrawal of confidence from the league of specialists, officials, and corporation executives who for at least a generation have had almost exclusive charge of the problem and who have enormously enriched and empowered themselves by making it worse.

Second, as a people, we must learn again to think of human energy, *our* energy, not as something to be saved, but as something to be used and to be enjoyed in use. We must understand that our strength is, first of all, strength of body, and that this strength cannot thrive except in useful, decent, satisfying, comely work. There is no such thing as a reservoir of bodily energy. By saving it—as our ideals of labor-saving and luxury bid us to do—we simply waste it, and waste much else along with it.

Third, we must see again, as I think the founders of our

government saw, that the most appropriate governmental powers are negative—those, that is, that protect the small and weak from the great and powerful, *not* those by which the government becomes the profligate, ineffectual parent of the small and weak after it has permitted the great and powerful to make them helpless. The governmental power that can be used most effectively to assure an equitable distribution of property, which alone can give some measure of strength and independence to ordinary citizens, is that of taxation. As our present economy clearly shows, the small can survive only if the great are restrained. And there is nothing undemocratic or anti-libertarian about restraining them. To assume that ordinary citizens can compete successfully with people of wealth and with corporations, as our government presently tends to do, is simply to abandon the ordinary citizens. Restraint by taxation is the smallest, most obvious, simplest, and cheapest answer. This is not my idea. It is Thomas Jefferson's. Writing to Reverend James Madison on October 28, 1785, Jefferson spoke of the desirability of freehold tenure of property. And then he said "Another means of silently lessening the inequality of property is to exempt all from taxation below a certain point, and to tax the higher portions of property in geometric progression as they rise. The earth is given as a common stock for man to labor and live on. If for the encouragement of industry [he means, of course, mainly agriculture] we allow it to be appropriated, we must take care that employment be provided to those excluded from the appropriation. If we do not, the fundamental right to labor the earth returns to the unemployed . . . it is not too soon to provide by every possible means that as few as possible shall be without a little portion of land. The small landholders are the most precious part of a state. . . ." It would, of course, be necessary to consider how much land in any region ought to constitute a living for a family.

Fourth, considering that the price of farmland has now been driven up by urban pressures and speculation until farmers often cannot afford to own it, low-interest loans ought to be made available to people wishing to buy family-size farms. This would probably need to be only a temporary or transitional measure.

Fifth, there should be a system of production and price controls that would tend to adjust production both to need and to the carrying capacities of farms. One purpose of this would be to curb the extreme fluctuations of supply, which work in the long run to the disadvantage of small producers. Another would be the elimination of the phenomenon of "harvest-time depressed prices"—which, in practice, means that the price of grain is low when it is in the hands of the wrong people (small farmers who cannot afford storage) and high when it is in the hands of the right people (big farmers and "agribusiness" corporations).

Sixth, there should be a program to promote local self-sufficiency in food. The cheapest, freshest food is that which is produced closest to home and is not delayed for processing. This should work toward the most direct dealing between farmers and merchants and farmers and consumers. Much might be done by the promotion of growers' and consumers' cooperatives.

Seventh, every town and city should be required to operate an organic-waste depot where sewage, garbage, waste paper, and the like would be composted and given or sold at cost to farmers. Every truck bringing a load of produce to town should go home with a load of compost. This would greatly improve the health of both the rivers and the fields and it would lower the cost of food.

Eighth, there should be a strenuous review of all sanitation laws governing the production of food, and those that are unnecessary should be eliminated. Sanitation laws have almost invariably worked against the small producer, destroying his markets or prohibitively increasing the cost of production. If we are as technologically adept as we claim to be, then it is inexcusable that we do not have, for instance, an acceptable, inexpensive technology for small dairies. And there is no reason, given the necessary collecting points, that we should not have markets for small quantities of other foods. If we are serious about increasing food production, then we must make room for the small producer. Moreover, decency and common sense require us to learn if it is necessary for cleanliness invariably to be expensive.

Ninth, we should encourage the greatest possible technological and genetic diversity, in conformation to local need, as opposed to the present dangerous uniformity in both categories. This diversity should be the primary goal of the land-grant schools. To this end, they should be *required*, as the Hatch Act instructs, "to assure agriculture a position in research equal to that of industry." These schools, and their professors individually, should be forbidden to accept work on assignment from any corporation or other outside interest that might wish to market any resulting product. (This, of course, would not apply to professors working on their own time outside the university.)

Tenth, to de-specialize the interests of the colleges of agriculture—that is, to shift their loyalty from "agribusiness" and industry back to the farmers—two other measures might be useful: (1) The faculties should be opened, on a part-time basis, to farmers, just as faculties of medicine and law are opened to doctors and lawyers; and (2) faculty members could be paid half their salary in cash and given the use of a boundary of college farmland the potential annual income from which would be equivalent to the other half. In both instances, the professor would be in a position to "take his own advice before offering it to other people." And much good might be expected from that. Professors might again become people of experience rather than experts. They might again be able to apply their learning to the small problems of ordinary people and to recommend means and methods not profitable to the suppliers of "purchased inputs."

Eleventh, we must address ourselves seriously, and not a little fearfully, to the problem of human scale. What is it? How do we stay within it? What sort of technology enhances our humanity? What sort reduces it? The reason is simply that we cannot live except within limits, and these limits are of many kinds: spatial, material, moral, spiritual. The world has room for many people who are content to live as humans, but only for a relative few intent upon living as giants or as gods.

Twelfth, having exploited "relativism" until, as a people, we have no deeply believed reasons for doing anything, we must now ask ourselves if there is not, after all, an absolute good by which we must measure ourselves and for which we must work. That absolute good, I think, is health—not in the merely hygienic sense of personal health, but the health, the wholeness, finally the holiness, of Creation, of which our personal health is only a share.

THE NECESSITY OF MARGINS

In Michigan in the fall of 1975, a fire-retarding chemical known as PBB was mistaken for a trace mineral and mixed into a large order of livestock feed. This feed was sent to four Michigan mills run by the Farm Bureau, and from there it went to farms and to the stock troughs. The resulting contamination of meat, milk, and eggs produced a disaster which is still continuing after three-and-a-half years and the limits of which are not known. The immediate and most noticeable result was a state program to destroy contaminated animals and food products. This did away with "about 1.5 million chickens, 29,000 head of cattle, 5,920 hogs, 1,470 sheep, 2,600 lb. of butter, 18,000 lb. of cheese, 34,000 lb. of dry milk products, and 5 million eggs." But many people were also affected, some seriously, and the long-term effects on human health are not known.

This was a tragedy—personal and economic, private and public—caused by one error that "may have been as simple as pulling the wrong lever." And we must recognize that, both in its carelessness and in its magnitude, this tragedy is characteristic of an agriculture, indeed of a culture, without margins. In a highly centralized and industrialized food-supply system there can be no small disaster. Whether it be a production "error" or a corn blight, the disaster is not foreseen until it exists; it is not recognized until it is widespread. By contrast, a highly diversified, small-farm agriculture combined with local marketing is literally crisscrossed with margins, and these margins work both to allow and encourage care and to contain damage.

But such an agriculture would do more than provide us with protective margins. In reducing industrial uniformity it would give us a new sense of our real unity, our common sharing in the good of health. It is a rule, apparently, that whatever is divided must compete. We have been wrong to believe that competition invariably results in the triumph of the best. Divided, body and soul, man and woman, producer and consumer, nature and technology, city and country are thrown into competition with one another. And none of these competitions is ever resolved in the triumph of one competitor, but only in the exhaustion of both.

For our healing we have on our side one great force: the power of Creation, with good care, with kindly use, to heal itself.

Afterword to the Third Edition

In *The Unsettling of America* I argue that industrial agriculture and the assumptions on which it rests are wrong, root and branch; I argue that this kind of agriculture grows out of the worst of human history and the worst of human nature. From my own point of view, the happiest fate of my labors would have been disproof. I would have been much relieved if somebody had proved me wrong, or if events had shown that I need not have worried. For this book certainly was written out of worry. It was written, in fact, out of the belief that we were living under the rule of an ideology that was destroying our land, our communities, and our culture—as we still are.

My argument, as I saw it twenty years ago, was addressed to leaders in the schools, the governments, and other places, who presumably were interested in argument as a way of approaching certain kinds of truths. The years since its publication have demonstrated, among other things, my naiveté. The argument set forth in this book, though it has been much and sometimes vehemently disagreed with, has never been answered, let alone disproved.

For this, surely the paramount reason is that events have continued to confirm my argument at every point. The enormous productivity of industrial agriculture cannot be denied, but neither can its enormous ecological, economic, and human costs, which are bound eventually to damage its productivity. This book's tragedy is that it is true.

Moreover, those who disagree with what I wrote are mainly those against whom I wrote it: adherents of the industrial program, which is too powerful, too rich, and too preoccupied with conquest to be diverted by anybody's mere argument. They simply are not obliged to care whether or not they may be wrong.

Another reason my book has received no vigorous counterargument, I fear, is that in centers of learning and power argument itself has become virtually obsolete, a lost art. Public discourse of all kinds now tends to pattern itself either upon the arts of advertisement and propaganda (that is, the arts of persuasion without argument, which lead to reasonless and even unconscious acquiescence) or upon the allegedly objective or value-free demonstrations of science.

When I was writing this book I still supposed, for example, that the land-grant universities, if confronted by an argument against their governing assumptions, would either have to produce a stronger counterargument or change their assumptions. That supposition may have been naive, but it was nonetheless one that I had every right to make, for the pursuit of truth by argument and counterargument is a major part of our cultural tradition from the Gospels and the Platonic dialogues to every county courthouse today.

The response to this book has shown, instead, that the universities are not interested in the pursuit of truth by argument. They are interested in preserving the conclusion of an old argument that for the most part they no longer bother to make: namely, that the world and all its creatures are machines. The organization of the modern university—and of modern intellectual life—rests upon this argument. Perhaps this line of thought began in metaphor, but now the likeness has become identity. It is assumed, as my friend Gene Logsdon puts it, that biology and mechanics are the same thing, and that other things don't matter. Here are some smaller assumptions that derive from and help to preserve the larger one:

1. If the world and all its creatures are machines, then the world and all its creatures are entirely comprehensible, manipulable, and controllable by humans.
2. The humans who have this power are experts.
3. Experts are made by education.
4. Education only happens in schools.
5. Experts are smarter than other people.
6. Thinking is best done by experts in offices and laboratories.
7. People who *do* work cannot be trusted to think about it.
8. People who work would prefer not to work.
9. Human workers are inefficient machines, encumbered by extraneous needs and desires, and they should be replaced by more efficient machines or by chemicals.
10. In general, the human machine is better at consumption than production.

11. A farm is or ought to be a factory in which plant and animal machines serve the economic machine in the most efficient way.

12. Efficiency has nothing to do with human or biological needs and desires.

13. Farm bankruptcy increases agricultural efficiency.

14. All farmers actually dislike farming and are secretly glad when they go bankrupt, because that gets them out of the sticks and into the bright lights where they have a chance to become experts.

15. Conventional agricultural science (like all conventional science) is disinterested and objective and serves no interest other than the advancement of human knowledge.

And so on.

Eventually this mechanistic line of thought brings us to the doctrine that whatever happens is inevitable. Actually, this stark determinism is altered in general use to a doctrine that is even more contemptible: every *bad* thing that happens is inevitable. For every good thing that happens there are mobs of claimers of credit. Every good and perfect gift comes from politicians, scientists, researchers, governments, and corporations. Evils, however, are inevitable; there is just no use in trying to choose against them. Thus all industrial comforts and labor-saving devices are the result only of human ingenuity and determination (not to mention the charity and altruism that have so conspicuously distinguished the industrial subspecies for the past two centuries), but the consequent pollution, land destruction, and social upheaval have been "inevitable."

Thus President Clinton (for whom I voted) could tell an audience of "farmers and agricultural organization leaders" in Billings, Montana, on June 1, 1995, that the American farm population now is "dramatically lower, obviously, than it was a generation ago. And that was inevitable because of the increasing productivity of agriculture." (See assumptions 9–13 above.)

That is to say that what happened happened because it had to happen. Thus the apologists for the ruin of agricultural lands, economies, and communities have shown always that they did nothing to stop it because there was nothing they could have done to stop it. (It's just progress, folks. Be glad your children won't suffer the drudgery and degradation of farm ownership.) The president also said that he wants to save the family farm, which is "alive and well" in Montana. He said he believes that we have "bottomed out in the shrinking of the farm sector." He said he wants to help young farmers. He spoke of the need to make American agriculture "competitive with people around the world." He praised our huge volume of agricultural exports. All of this had been said countless times before, and all of it sinks beneath that weighty adjective *inevitable*. If an utterly brainless and destructive agricultural economy has been inevitable for half a century, why should it now suddenly cease to be inevitable?

The president's remarks in Montana provide evidence enough that our national conversation about agriculture, at the "upper levels" of policy and research, is frozen solid. The people up there have not had a new or divergent thought in two generations. Furthermore, they do not wish to think a new or divergent thought; the old thoughts have suited their careers and their pocketbooks well enough. When threatening ideas or even threatening statistics or experimental evidence are introduced, those upper-level experts just freeze them into the ice and skate over the top of them.

But if the publication of *The Unsettling of America* and subsequent events have shown me that throwing a rock into a frozen river does not make a ripple, they have also shown that beneath the ice the waters are strongly flowing and stirred up and full of nutrients. Beneath the clichés of official science and policy, our national conversation about agriculture is more vigorous and exciting now than it has been since the 1930s.

This book has not had the happy fate of being proved wrong, but it has had the next-happiest fate of belonging to a growing effort to think again about the issues of American land use and to start the changes that are needed.

My book does not stand alone. It occurs in a lineage of works, influences, and exemplars that it acknowledges, and that I have more fully acknowledged in writings since. And it belongs to a company of present-day works and exemplars joined by a common commitment and a common aim: books by Marty Strange, Gene Logsdon, and Wes Jackson, among others; organizations such as the Land Institute, the Center for Rural Affairs, the Land Stewardship Project, Tilth, and the

E. F. Schumacher Society; the several conservation organizations; a rapidly increasing number of organizations interested in the local marketing of local products; and thousands of farmers and gardeners. I am much encouraged by the knowledge that if this book (and my other books) suddenly disappeared from print and from memory, its advocacy and its hope would continue undiminished.

The people to whom this book belongs are thinking about agriculture and other land-based enterprises in a way radically different from the way the people at the upper levels of policy and research are thinking. They think so differently, I believe, because their motives are different. Their thinking does not begin with a set of predetermining ideas but rather with particular places, people, needs, and desires. This book's friends and allies began to think and to work not because they had careers to make or ideologies to serve but because they loved certain places, people, possibilities, and ways that they could not indifferently see destroyed.

What we are working for, I think, is an authentic settlement and inhabitation of our country. We would like to see all human work lovingly adapted to the nature of the places where it is done and to the real needs of the people by whom and for whom it is done. We do not believe that any violence to places, to people, or to other creatures is "inevitable." We believe that the industrial ideology is wrong because it obscures and disrupts this necessary work of local adaptation or home making.

I do not believe that this effort will lead to perfection; the best agriculture and economics we can imagine will not return us to Eden; we will carry with us into whatever we do the weaknesses and limits inherent in our nature. But our imperfections argue more strongly than any hope of perfection for the adoption of ways and aims that can lead us beyond the selfishness that is institutionalized in the present economic system. To suggest that the health of places and communities might be the indispensable standard of economic behavior is finally to ask how a mere human, whose years are like the grass that is cut down in the evening, can justify on his or her own behalf the permanent destruction of anything.

Our effort to make something comely and enduring of our life on this earth will last as long as our species, I am confident of that; it is, after all, an ancient effort. I am confident that our present effort here in the United States can grow and accomplish much without the help of the upper levels of research and policy. It does worry me, however, that the people working in various ways to protect places and communities and ways of life now make up a sizable constituency that is virtually unclaimed and unrepresented. The dangers in this are obvious enough to anybody willing to look. Our government has shown considerable enthusiasm for "leveling the playing field" in the interest of international corporations. Its enthusiasm for leveling the playing field in the interest of local economies and local ecosystems remains to be demonstrated.

Wendell Berry
August 1995

Horse-Drawn Tools and
the Doctrine of Labor Saving

FIVE YEARS AGO, when we enlarged our farm from about twelve acres to about fifty, we saw that we had come to the limits of the equipment we had on hand: mainly a rotary tiller and a Gravely walking tractor; we had been borrowing a tractor and mower to clip our few acres of pasture. Now we would have perhaps twenty-five acres of pasture, three acres of hay, and the garden; and we would also be clearing some land and dragging the cut trees out for firewood. I thought for a while of buying a second-hand 8N Ford tractor, but decided finally to buy a team of horses instead.

I have several reasons for being glad that I did. One reason is that it started me thinking more particularly and carefully than before about the development of agricultural technology. I had learned to use a team when I was a boy, and then had learned to use the tractor equipment that replaced virtually all the horse and mule teams in this part of the country after World War II. Now I was turning around, as if in the middle of my own history, and taking up the old way again.

Buying and borrowing, I gathered up the equipment I needed to get started: wagon, manure spreader, mowing machine, disk, a one-row cultivating plow for the garden. Most of these machines had been sitting idle for years. I put them back into working shape, and started using them. That was 1973. In the years since, I have bought a number of other horse-drawn tools, for myself and other people. My own outfit now includes a breaking plow, a two-horse riding cultivator, and a grain drill.

As I have repaired these old machines and used them, I have seen how well designed and durable they are, and what good work they do. When the manufacturers modified them for use with tractors, they did not much improve either the machines or the quality of their work. (It is necessary, of course, to note some exceptions. Some horsemen, for instance, would argue that alfalfa sod is best plowed with a tractor. And one must also except such tools as hay conditioners and chisel plows that came after the development of horse-drawn tools had ceased. We do not know what innovations, refinements, and improvements would have come if it had continued.) At the peak of their development, the old horse tools were excellent. The coming of the tractor made it possible for a farmer to do more work, but not better. And there comes a point, as we know, when *more* begins to imply *worse*. The mechanization of farming passed that point long ago—probably, or so I will argue, when it passed from horse power to tractor power.

The increase of power has made it possible for one worker to crop an enormous acreage, but for this "efficiency" the country has paid a high price. From 1946 to 1976, because fewer people were needed, the farm population declined from thirty million to nine million; the rapid movement of these millions into the cities greatly aggravated that complex of problems which we now call the "urban crisis," and the land is suffering for want of the care of those absent families. The coming of a tool, then, can be a cultural event of great influence and power. Once that is understood, it is no longer possible to be simpleminded about technological progress. It is no longer possible to ask, What is a good tool? without asking at the same time, How *well* does it work? and, What is its influence?

One could say, as a rule of thumb, that a good tool is one that makes it possible to work faster *and* better than before. When companies quit making them, the horse-drawn tools fulfilled both requirements. Consider, for example, the International High Gear No. 9 mowing machine. This is a horse-drawn mower that certainly improved on everything that came before it, from the scythe to previous machines in the International line. Up to that point, to cut fast and to cut well were two aspects of the same problem. Past that point the speed of the work could be increased, but not the quality.

I own one of these mowers. I have used it in my hayfield at the same time that a neighbor mowed there with a tractor mower; I have gone from my own freshly cut hayfield into others just mowed by tractors; and I can say unhesitatingly that, though the tractors do faster work, they do not do it better. The same is substantially true, I think, of other tools: plows, cultivators, harrows, grain drills, seeders, spreaders, etc. Through the development of the standard horse-drawn equipment, quality and speed increased together; after that, the principal increase has been in speed.

Moreover, as the speed has increased, care has tended to decline. For this, one's eyes can furnish ample evidence. But we have it also by the testimony of the equipment manufacturers themselves. Here, for example, is a quote from the public relations paper of one of the largest companies: "Today we have multi-row planters that slap in a crop in a hurry, putting down seed, fertilizer, insecticide and herbicide in one quick swipe across the field."

But good work and good workmanship cannot be accomplished by "slaps" and "swipes." Such language seems to be derived from the he-man vocabulary of TV westerns, not from any known principles of good agriculture. What does the language of good agricultural workmanship sound like? Here is the voice of an old-time English farmworker and horseman, Harry Groom, as quoted in George Ewart Evans's *The Horse in the Furrow*: "It's all rush today. You hear a young chap say in the pub: 'I done thirty acres today.' But it ain't messed over, let alone done. You take the rolling, for instance. Two mile an hour is fast enough for a roll or a harrow. With a roll, the slower the better. If you roll fast, the clods are not broken up, they're just pressed in further. Speed is everything now; just jump on the tractor and way across the field as if it's a dirt-track. You see it when a farmer takes over a new farm: he goes in and plants straight-way, right out of the book. But if one of the old farmers took a new farm, and you walked round the land with him and asked him: 'What are you going to plant here and here?' he'd look at you some queer; because he wouldn't plant nothing much at first. He'd wait a bit and see what the land was like: he'd *prove* the land first. A good practical man would hold on for a few weeks, and get the feel of the land under his feet. He'd walk on it and feel it through his boots and see if it was in good heart, before he planted anything: he'd sow only when he knew what the land was fit for."

Granted that there is always plenty of room to disagree about farming methods, there is still no way to deny that in the first quotation we have a description of careless farming, and in the second a description of a way of farming as careful —as knowing, skillful, and loving—as any other kind of high workmanship. The difference between the two is simply that the second considers where and how the machine is used, whereas the first considers only the machine. The first is the point of view of a man high up in the air-conditioned cab of a tractor described as "a beast that eats acres." The second is that of a man who has worked close to the ground in the open air of the field, who has studied the condition of the ground as he drove over it, and who has cared and thought about it.

If we had tools thirty-five years ago that made it possible to do farm work both faster and better than before, then why did we choose to go ahead and make them no longer better, but just bigger and bigger and faster and faster? It was, I think, because we were already allowing the wrong people to give the wrong answers to questions raised by the improved horse-drawn machines. Those machines, like the ones that followed them, were *labor savers*. They may seem old-timey in comparison to today's "acre eaters," but when they came on the market they greatly increased the amount of work that one worker could do in a day. And so they confronted us with a critical question: How would we define labor saving?

We defined it, or allowed it to be defined for us by the corporations and the specialists, as if it involved no human considerations at all, as if the labor to be "saved" were not human labor. We decided, in the language of some experts, to look on technology as a "substitute for labor." Which means that we did not intend to "save" labor at all, but to *replace* it, and to *displace* the people who once supplied it. We never asked what should be done with the "saved" labor; we let the "labor market" take care of that. Nor did we ask the larger questions of what values we should place on people and their work and on the land. It appears that we abandoned ourselves unquestioningly to a course of technological evolution, which would value the development of machines far above the development of people.

And so it becomes clear that, by itself, my rule-of-thumb definition of a good tool (one that permits a worker to work both better and faster) does not go far enough. Even such a tool can cause bad results if its use is not directed by a benign and healthy social purpose. The coming of a tool, then, is not just a cultural event; it is also an historical crossroad—a point at which people must choose between two possibilities: to become more intensive or more extensive; to use the tool for quality or for quantity, for care or for speed.

In speaking of this as a choice, I am obviously assuming that the evolution of technology is *not* unquestionable or uncontrollable; that "progress" and the "labor market" do *not* represent anything so unyielding as natural law, but are aspects of an economy; and that any economy is in some sense a "managed" economy, managed by an intention to distribute the benefits of work, land, and materials in a certain way. (The present agricultural economy, for instance, is slanted to give the greater portion of these benefits to the "agribusiness" corporations. If this were not so, the recent farmers' strike would have been an "agribusiness" strike as well.) If those assumptions are correct, we are at liberty to do a little historical supposing, not meant, of course, to "change history" or "rewrite it," but to clarify somewhat this question of technological choice.

Suppose, then, that in 1945 we had valued the human life of farms and farm communities 1 percent more than we valued "economic growth" and technological progress. And suppose we had espoused the health of homes, farms, towns, and cities with anything like the resolve and energy with which we built the "military-industrial complex." Suppose, in other words, that we had really meant what, all that time, most of us and most of our leaders were saying, and that we had really tried to live by the traditional values to which we gave lip service.

Then, it seems to me, we might have accepted certain mechanical and economic limits. We might have used the improved horse-drawn tools, or even the small tractor equipment that followed, not to displace workers and decrease care and skill, but to intensify production, improve maintenance, increase care and skill, and widen the margins of leisure, pleasure, and community life. We might, in other words, by limiting technology to a human or a democratic scale, have been able to use the saved labor *in the same places where we saved it.*

It is important to remember that "labor" is a very crude, industrial term, fitted to the huge economic structures, the dehumanized technology, and the abstract social organization of urban-industrial society. In such circumstances, "labor" means little more than the sum of two human quantities, human energy plus human time, which we identify as "man-hours." But the nearer home we put "labor" to work, and the smaller and more familiar we make its circumstances, the more we enlarge and complicate and enhance its meaning. At work in a factory, workers are only workers, "units of production" expending "man-hours" at a task set for them by strangers. At work in their own communities, on their own farms or in their own households or shops, workers are *never* only workers, but rather persons, relatives, and neighbors. They work *for* those they work *among* and *with*. Moreover, workers tend to be independent in inverse proportion to the size of the circumstance in which they work. That is, the work of factory workers is ruled by the factory, whereas the work of housewives, small craftsmen, or small farmers is ruled by their own morality, skill, and intelligence. And so, when workers work independently and at home, the society as a whole may lose something in the way of organizational efficiency and economies of scale. But it begins to *gain* values not so readily quantifiable in the fulfilled humanity of the workers, who then bring to their work not just contracted quantities of "man-hours," but qualities such as independence, skill, intelligence, judgment, pride, respect, loyalty, love, reverence.

To put the matter in concrete terms, if the farm communities had been able to use the best horse-drawn tools to save labor in the true sense, then they might have used the saved time and energy, first of all, for leisure—something that technological progress has given to farmers. Second, they might have used it to improve their farms: to enrich the soil, prevent erosion, conserve water, put up better and more permanent fences and buildings; to practice forestry and its dependent crafts and economies; to plant orchards, vineyards, gardens of bush fruits; to plant market gardens; to improve pasture, breeding, husbandry, and the subsidiary enterprises of a local, small-herd livestock economy; to enlarge, diversify, and deepen the economies of households and homesteads. Third, they might have used it to expand and improve the specialized crafts necessary to the health and beauty of communities: carpentry, masonry, leatherwork, cabinetwork, metalwork, pottery, etc. Fourth, they might have used it to improve the homelife and the home instruction of children, thereby preventing the hardships and expenses now placed on schools, courts, and jails.

It is probable also that, if we *had* followed such a course, we would have averted or greatly ameliorated the present shortages of energy and employment. The cities would be much less crowded; the rates of crime and welfare dependency would be much lower; the standards of industrial production would probably be higher. And farmers might have avoided their present crippling dependence on money lenders.

I am aware that all this is exactly the sort of thinking that the technological determinists will dismiss as nostalgic or wishful. I mean it, however, not as a recommendation that we "return to the past," but as a criticism of the past; and my criticism is based on the assumption that we had in the past, and that we have now, a *choice* about how we should use technology and what we should use it for. As I understand it, this choice depends absolutely on our willingness to limit our desires as well as the scale and kind of technology we use to satisfy them. Without that willingness, there is no choice; we must simply abandon ourselves to whatever the technologists may discover to be possible.

The technological determinists, of course, do not accept that such a choice exists—undoubtedly because they resent the moral limits on their work that such a choice implies. They speak romantically of "man's destiny" to go on to bigger and more sophisticated machines. Or they take the opposite course and speak the tooth-and-claw language of Darwinism. Ex-secretary of agriculture Earl Butz speaks, for instance, of "Butz's Law of Economics" which is "Adapt or Die."

I am, I think, as enthusiastic about the principle of adaptation as Mr. Butz. We differ only on the question of what should be adapted. He believes that we should adapt to the machines, that humans should be forced to conform to technological conditions or standards. I believe that the machines should be adapted to us—to serve our *human* needs as our history, our heritage, and our most generous hopes have defined them.

Solving for Pattern

OUR DILEMMA in agriculture now is that the industrial methods that have so spectacularly solved some of the problems of food production have been accompanied by "side effects" so damaging as to threaten the survival of farming. Perhaps the best clue to the nature and the gravity of this dilemma is that it is not limited to agriculture. My immediate concern here is with the irony of agricultural methods that destroy, first, the health of the soil and, finally, the health of human communities. But I could just as easily be talking about sanitation systems that pollute, school systems that graduate illiterate students, medical cures that cause disease, or nuclear armaments that explode in the midst of the people they are meant to protect. This is a kind of surprise that is characteristic of our time: the cure proves incurable; security results in the evacuation of a neighborhood or a town. It is only when it is understood that our agricultural dilemma is characteristic not of our agriculture but of our time that we can begin to understand why these surprises happen, and to work out standards of judgment that may prevent them.

To the problems of farming, then, as to other problems of our time, there appear to be three kinds of solutions:

There is, first, the solution that causes a ramifying series of new problems, the only limiting criterion being, apparently, that the new problems should arise beyond the purview of the expertise that produced the solution—as, in agriculture, industrial solutions to the problem of production have invariably caused problems of maintenance, conservation, economics, community health, etc., etc.

If, for example, beef cattle are fed in large feed lots, within the boundaries of the feeding operation itself a certain factory-like order and efficiency can be achieved. But even within those boundaries that mechanical order immediately produces biological disorder, for we know that health problems and dependence on drugs will be greater among cattle so confined than among cattle on pasture.

And beyond those boundaries, the problems multiply. Pen feeding of cattle in large numbers involves, first, a manure-removal problem, which becomes at some point a health problem for the animals themselves, for the local watershed, and for the adjoining ecosystems and human communities. If the manure is disposed of without returning it to the soil that produced the feed, a serious problem of soil fertility is involved. But we know too that large concentrations of animals in feed lots in one place tend to be associated with, and to promote, large cash-grain monocultures in other places. These monocultures tend to be accompanied by a whole set of specifically agricultural problems: soil erosion, soil compaction, epidemic infestations of pests, weeds, and disease. But they are also accompanied by a set of agricultural-economic problems (dependence on purchased technology; dependence on purchased fuels, fertilizers, and poisons; dependence on credit)—and by a set of community problems, beginning with depopulation and the removal of sources, services, and markets to more and more distant towns. And these are, so to speak, only the first circle of the bad effects of a bad solution. With a little care, their branchings can be traced on into nature, into the life of the cities, and into the cultural and economic life of the nation.

The second kind of solution is that which immediately worsens the problem it is intended to solve, causing a hellish symbiosis in which problem and solution reciprocally enlarge one another in a sequence that, so far as its own logic is concerned, is limitless—as when the problem of soil compaction is "solved" by a bigger tractor, which further compacts the soil, which makes a need for a still bigger tractor, and so on and on. There is an identical symbiosis between coal-fired power plants and air conditioners. It is characteristic of such solutions that no one prospers by them but the suppliers of fuel and equipment.

These two kinds of solutions are obviously bad. They always serve one good at the expense of another or of several others, and I believe that if all their effects were ever to be accounted for they would be seen to involve, too frequently if not invariably, a net loss to nature, agriculture, and the human commonwealth.

Such solutions always involve a definition of the problem that is either false or so narrow as to be virtually false. To define an agricultural problem as if it were solely a problem of agriculture—or solely a problem of production or technology or economics—is simply to misunderstand the problem, either inadvertently or deliberately, either for profit or because of a prevalent fashion of thought. The whole problem must be solved, not just some handily identifiable and simplifiable aspect of it.

Both kinds of bad solutions leave their problems unsolved. Bigger tractors do not solve the problem of soil compaction any more than air conditioners solve the problem of air pollution. Nor does the large confinement-feeding operation solve the problem of food production; it is, rather, a way calculated to allow large-scale ambition and greed to profit from food production. The real problem of food production occurs within a complex, mutually influential relationship of soil, plants, animals, and people. A real solution to that problem will therefore be ecologically, agriculturally, and culturally healthful.

Perhaps it is not until health is set down as the aim that we come in sight of the third kind of solution: that which causes a ramifying series of solutions—as when meat animals are fed on the farm where the feed is raised, and where the feed is raised to be fed to the animals that are on the farm. Even so rudimentary a description implies a concern for pattern, for quality, which necessarily complicates the concern for production. The farmer has put plants and animals into a relationship of mutual dependence, and must perforce be concerned for balance or symmetry, a reciprocating connection in the pattern of the farm that is biological, not industrial, and that involves solutions to problems of fertility, soil husbandry, economics, sanitation—the whole complex of problems whose proper solutions add up to *health*: the health of the soil, of plants and animals, of farm and farmer, of farm family and farm community, all involved in the same internested, interlocking pattern —or pattern of patterns.

A bad solution is bad, then, because it acts destructively upon the larger patterns in which it is contained. It acts destructively upon those patterns, most likely, because it is formed in ignorance or disregard of them. A bad solution solves for a single purpose or goal, such as increased production. And it is typical of such solutions that they achieve stupendous increases in production at exorbitant biological and social costs.

A good solution is good because it is in harmony with those larger patterns—and this harmony will, I think, be found to have the nature of analogy. A bad solution acts within the larger pattern the way a disease or addiction acts within the body. A good solution acts within the larger pattern the way a healthy organ acts within the body. But it must at once be understood that a healthy organ does not—as the mechanistic or industrial mind would like to say—"give" health to the body, is not exploited for the body's health, but is *a part* of its health. The health of organ and organism is the same, just as the health of organism and ecosystem is the same. And these structures of organ, organism, and ecosystem—as John Todd has so ably understood—belong to a series of analogical integrities that begins with the organelle and ends with the biosphere.

It would be next to useless, of course, to talk about the possibility of good solutions if none existed in proof and in practice. A part of our work at *The New Farm* has been to locate and understand those farmers whose work is competently responsive to the requirements of health. Representative of these farmers, and among them remarkable for the thoroughness of his intelligence, is Earl F. Spencer, who has a 250-acre dairy farm near Palatine Bridge, New York.

Before 1972, Earl Spencer was following a "conventional" plan which would build his herd to 120 cows. According to this plan, he would eventually buy all the grain he fed, and he was already using as much as 30 tons per year of commercial fertilizer. But in 1972, when he had increased his herd to 70 cows, wet weather reduced his harvest by about half. The choice was clear: he had either to buy half his yearly feed supply, or sell half his herd.

He chose to sell half his herd—a very unconventional choice, which in itself required a lot of independent intelligence. But character and intelligence of an even more respectable order were involved in the next step, which was to understand that the initial decision implied a profound change in the pattern of the farm and of his life and assumptions as a farmer. With his herd now reduced by half, he saw that before the sale he had been overstocked, and had been abusing his land. On his 120 acres of tillable land, he had been growing 60 acres of corn and 60 of alfalfa. On most of his fields, he was growing corn three years in succession. The consequences of this he now saw as symptoms, and saw that they were serious: heavy dependence on purchased supplies, deteriorating soil structure, declining quantities of organic matter, increasing erosion, yield reductions despite continued large applications of fertilizer. In addition, because of his heavy feeding of concentrates, his cows were having serious digestive and other health problems.

He began to ask fundamental questions about the nature of the creatures and the land he was dealing with, and to ask if he could not bring about some sort of balance between their needs and his own. His conclusion was that "to be in balance with nature is to be successful." His farm, he says, had been going in a "dead run"; now he would slow it to a "walk."

From his crucial decision to reduce his herd, then, several other practical measures have followed:

1. A five-year plan (extended to eight years) to phase out entirely his use of purchased fertilizers.
2. A plan, involving construction of a concrete manure pit, to increase and improve his use of manure.
3. Better husbandry of cropland, more frequent rotation, better timing.
4. The gradual reduction of grain in the feed ration, and the concurrent increase of roughage—which has, to date, reduced the dependence on grain by half, from about 6000 pounds per cow to about 3000 pounds.
5. A breeding program which selects "for more efficient roughage conversion."

The most tangible results are that the costs of production have been "dramatically" reduced, and that per cow production has increased by 1500 to 2000 pounds. But the health of the whole farm has improved. There is a moral satisfaction in this, of which Earl Spencer is fully aware. But he is also aware that the satisfaction is not *purely* moral, for the good results are also practical and economic: "We have half the animals we had before and are feeding half as much grain to those remaining, so we now need to plant corn only two years in a row. Less corn means less plowing, less fuel for growing and harvesting, and less wear on the most expensive equipment." Veterinary bills have been reduced also. And in 1981, if the schedule holds, he will buy no commercial fertilizer at all.

From the work of Earl Spencer and other exemplary farmers, and from the understanding of destructive farming practices, it is possible to devise a set of critical standards for agriculture. I am aware that the list of standards which follows must be to some extent provisional, but am nevertheless confident that it will work to distinguish between healthy and unhealthy farms, as well as between the oversimplified minds that solve problems for some X such as profit or quantity of production, and those minds, sufficiently complex, that solve for health or quality or coherence of pattern. To me, the validity of these standards seems inherent in their general applicability. They will serve the making of sewer systems or households as readily as they will serve the making of farms:

1. A good solution accepts given limits, using so far as possible what is at hand. The farther-fetched the solution, the less it should be trusted. Granted that a farm can be too small, it is nevertheless true that enlarging scale is a deceptive solution; it solves one problem by acquiring another or several others.

2. A good solution accepts also the limitation of discipline. Agricultural problems should receive solutions that are agricultural, not technological or economic.

3. A good solution improves the balances, symmetries, or harmonies within a pattern—it is a qualitative solution—rather than enlarging or complicating some part of a pattern at the expense or in neglect of the rest.

4. A good solution solves more than one problem, and it does not make new problems. I am talking about health as opposed to almost any cure, coherence of pattern as opposed to almost any solution produced piecemeal or in isolation. The return of organic wastes to the soil may, at first glance, appear to be a good solution *per se*. But that is not invariably or necessarily true. It is true only if the wastes are returned to the right place at the right time in the pattern of the farm, if the

waste does not contain toxic materials, if the quantity is not too great, and if not too much energy or money is expended in transporting it.

5. A good solution will satisfy a whole range of criteria; it will be good in all respects. A farm that has found correct agricultural solutions to its problems will be fertile, productive, healthful, conservative, beautiful, pleasant to live on. This standard obviously must be qualified to the extent that the pattern of the life of a farm will be adversely affected by distortions in any of the larger patterns that contain it. It is hard, for instance, for the economy of a farm to maintain its health in a national industrial economy in which farm earnings are apt to be low and expenses high. But it is apparently true, even in such an economy, that the farmers most apt to survive are those who do not go too far out of agriculture into either industry or banking—and who, moreover, live like farmers, not like businessmen. This seems especially true for the smaller farmers.

6. A good solution embodies a clear distinction between biological order and mechanical order, between farming and industry. Farmers who fail to make this distinction are ideal customers of the equipment companies, but they often fail to understand that the real strength of a farm is in the soil.

7. Good solutions have wide margins, so that the failure of one solution does not imply the impossibility of another. Industrial agriculture tends to put its eggs into fewer and fewer baskets, and to make "going for broke" its only way of going. But to grow grain should not make it impossible to pasture livestock, and to have a lot of power should not make it impossible to use only a little.

8. A good solution always answers the question, How much is enough? Industrial solutions have always rested on the assumption that enough is all you can get. But that destroys agriculture, as it destroys nature and culture. The good health of a farm implies a limit of scale, because it implies a limit of attention, and because such a limit is invariably implied by any pattern. You destroy a square, for example, by enlarging one angle or lengthening one side. And in any sort of work there is a point past which more quantity necessarily implies less quality. In some kinds of industrial agriculture, such as cash grain farming, it is possible (to borrow an insight from Professor

Timothy Taylor) to think of technology as a substitute for skill. But even in such farming that possibility is illusory; the illusion can be maintained only so long as the consequences can be ignored. The illusion is much shorter lived when animals are included in the farm pattern, because the husbandry of animals is so insistently a human skill. A healthy farm incorporates a pattern that a single human mind can comprehend, make, maintain, vary in response to circumstances, and pay steady attention to. That this limit is obviously variable from one farmer and farm to another does not mean that it does not exist.

9. A good solution should be cheap, and it should not enrich one person by the distress or impoverishment of another. In agriculture, so-called "inputs" are, from a different point of view, outputs—*expenses*. In all things, I think, but especially in an agriculture struggling to survive in an industrial economy, any solution that calls for an expenditure to a manufacturer should be held in suspicion—not rejected necessarily, but *as a rule* mistrusted.

10. Good solutions exist only in proof, and are not to be expected from absentee owners or absentee experts. Problems must be solved in work and in place, with particular knowledge, fidelity, and care, by people who will suffer the consequences of their mistakes. There is no theoretical or ideal *practice*. Practical advice or direction from people who have no practice may have some value, but its value is questionable and is limited. The divisions of capital, management, and labor, characteristic of an industrial system, are therefore utterly alien to the health of farming—as they probably also are to the health of manufacturing. The good health of a farm depends on the farmer's mind; the good health of his mind has its dependence, and its proof, in physical work. The good farmer's mind and his body —his management and his labor—work together as intimately as his heart and his lungs. And the capital of a well-farmed farm by definition includes the farmer, mind and body both. Farmer and farm are one thing, an organism.

11. Once the farmer's mind, his body, and his farm are understood as a single organism, and once it is understood that the question of the endurance of this organism is a question about the sufficiency and integrity of a pattern, then the word *organic* can be usefully admitted into this series of standards.

It is a word that I have been defining all along, though I have not used it. An organic farm, properly speaking, is not one that uses certain methods and substances and avoids others; it is a farm whose structure is formed in imitation of the structure of a natural system; it has the integrity, the independence, and the benign dependence of an organism. Sir Albert Howard said that a good farm is an analogue of the forest which "manures itself." A farm that imports too much fertility, even as feed or manure, is in this sense as inorganic as a farm that exports too much or that imports chemical fertilizer.

12. The introduction of the term *organic* permits me to say more plainly and usefully some things that I have said or implied earlier. In an organism, what is good for one part is good for another. What is good for the mind is good for the body; what is good for the arm is good for the heart. We know that sometimes a part may be sacrificed for the whole; a life may be saved by the amputation of an arm. But we also know that such remedies are desperate, irreversible, and destructive; it is impossible to improve the body by amputation. And such remedies do not imply a safe logic. As *tendencies* they are fatal: you cannot save your arm by the sacrifice of your life.

Perhaps most of us who know local histories of agriculture know of fields that in hard times have been sacrificed to save a farm, and we know that though such a thing is possible it is dangerous. The danger is worse when topsoil is sacrificed for the sake of a crop. And if we understand the farm as an organism, we see that it is impossible to sacrifice the health of the soil to improve the health of plants, or to sacrifice the health of plants to improve the health of animals, or to sacrifice the health of animals to improve the health of people. In a biological pattern—as in the pattern of a community—the exploitive means and motives of industrial economics are immediately destructive and ultimately suicidal.

13. It is the nature of any organic pattern to be contained within a larger one. And so a good solution in one pattern preserves the integrity of the pattern that contains it. A good agricultural solution, for example, would not pollute or erode a watershed. What is good for the water is good for the ground, what is good for the ground is good for plants, what is good for plants is good for animals, what is good for

animals is good for people, what is good for people is good for the air, what is good for the air is good for the water. And vice versa.

14. But we must not forget that those human solutions that we may call organic are not natural. We are talking about organic *artifacts*, organic only by imitation or analogy. Our ability to make such artifacts depends on virtues that are specifically human: accurate memory, observation, insight, imagination, inventiveness, reverence, devotion, fidelity, restraint. Restraint —for us, now—above all: the ability to accept and live within limits; to resist changes that are merely novel or fashionable; to resist greed and pride; to resist the temptation to "solve" problems by ignoring them, accepting them as "trade-offs," or bequeathing them to posterity. A good solution, then, must be in harmony with good character, cultural value, and moral law.

Family Work

FOR THOSE of us who have wished to raise our food and our children at home, it is easy enough to state the ideal. Growing our own food, unlike buying it, is a complex activity, and it affects deeply the shape and value of our lives. We like the thought that the outdoor work that improves our health should produce food of excellent quality that, in turn, also improves and safeguards our health. We like no less the thought that the home production of food can improve the quality of family life. Not only do we intend to give our children better food than we can buy for them at the store, or than they will buy for themselves from vending machines or burger joints, we also know that growing and preparing food at home can provide family work—work for everybody. And by thus elaborating household chores and obligations, we hope to strengthen the bonds of interest, loyalty, affection, and cooperation that keep families together.

Forty years ago, for most of our people, whether they lived in the country or in town, this was less an ideal than a necessity, enforced both by tradition and by need. As is often so, it was only after family life and family work became (allegedly) unnecessary that we began to think of them as "ideals."

As ideals, they are threatened; as they have become (even allegedly) unnecessary, they have become by the same token less possible. I do not mean to imply that *I* think the ideal is any less valuable than it ever was, or that it is—in reality, in the long run—less necessary. Nor do I think that less possible means impossible.

I do think that the ideal is more difficult now than it was. We are trying to uphold it now mainly by will, without much help from necessity, and with no help at all from custom or public value. For most people now do seem to think that family life and family work are unnecessary, and this thought has been institutionalized in our economy and in our public values. Never before has private life been so preyed upon by public life. How can we preserve family life—if by that we mean, as I think we must, *home* life—when our attention is so forcibly drawn away from home?

We know the causes well enough.

Automobiles and several decades of supposedly cheap fuel have put longer and longer distances between home and work, household and daily needs.

TV and other media have learned to suggest with increasing subtlety and callousness—especially, and most wickedly, to children—that it is better to consume than to produce, to buy than to grow or to make, to "go out" than to stay home. If you have a TV, your children will be subjected almost from the cradle to an overwhelming insinuation that all worth experiencing is somewhere else and that all worth having must be bought. The purpose is blatantly to supplant the joy and beauty of health with cosmetics, clothes, cars, and ready-made desserts. There is clearly too narrow a limit on how much money can be made from health, but the profitability of disease—especially disease of spirit or character—has so far, for profiteers, no visible limit.

Another cause, and one that seems particularly regrettable, is public education. The idea that the public should be educated is altogether salutary, and since we insist on making this education compulsory we ought, in reason, to reconcile ourselves to the likelihood that it will be mainly poor. I am not nearly so much concerned about its quality as I am about its *length*. My impression is that the chief, if unadmitted, purpose of the school system is to keep children away from home as much as possible. Parents want their children kept out of their hair; education is merely a by-product, not overly prized. In many places, thanks to school consolidation, two hours or more of travel time have been added to the school day. For my own children the regular school day from the first grade—counting from the time they went to catch the bus until they came home —was nine hours. An extracurricular activity would lengthen the day to eleven hours or more. This is not education, but a form of incarceration. Why should anyone be surprised if, under these circumstances, children should become "disruptive" or even "ineducable"?

If public education is to have any meaning or value at all,

then public education *must* be supplemented by home education. I know this from my own experience as a college teacher.
What can you teach a student whose entire education has been
public, whose daily family life for twenty years has consisted of
four or five hours of TV, who has never read a book for pleasure or even *seen* a book so read; whose only work has been
schoolwork, who has never learned to perform any essential
task? Not much, so far as I could tell.

We can see clearly enough at least a couple of solutions.

We can get rid of the television set. As soon as we see that
the TV cord is a vacuum line, pumping life and meaning out of
the household, we can unplug it. What a grand and neglected
privilege it is to be shed of the glibness, the gleeful idiocy, the
idiotic gravity, the unctuous or lubricious greed of those public
faces and voices!

And we can try to make our homes centers of attention and
interest. Getting rid of the TV, we understand, is not just a
practical act, but also a symbolical one: we thus turn our backs
on the invitation to consume; we shut out the racket of consumption. The ensuing silence is an invitation to our homes,
to our own places and lives, to come into being. And we begin
to recognize a truth disguised or denied by TV and all that
it speaks and stands for: no life and no place is destitute; all
have possibilities of productivity and pleasure, rest and work,
solitude and conviviality that belong particularly to themselves.
These possibilities exist everywhere, in the country or in the
city, it makes no difference. All that is necessary is the time and
the inner quietness to look for them, the sense to recognize
them, and the grace to welcome them. They are now most
often lived out in home gardens and kitchens, libraries, and
workrooms. But they are beginning to be worked out, too, in
little parks, in vacant lots, in neighborhood streets. Where we
live is also a place where our interest and our effort can be. But
they can't be there by the means and modes of consumption.
If we consume nothing but what we buy, we are living in "the
economy," in "television land," not at home. It is productivity
that rights the balance, and brings us home. Any way at all of
joining and using the air and light and weather of your own
place—even if it is only a window box, even if it is only an

opened window—is a making and a having that you cannot get from TV or government or school.

That local productivity, however small, is a gift. If we are parents we cannot help but see it as a gift to our children—and the *best* of gifts. How will it be received?

Well, not ideally. Sometimes it will be received gratefully enough. But sometimes indifferently, and sometimes resentfully.

According to my observation, one of the likeliest results of a wholesome diet of home-raised, home-cooked food is a heightened relish for cokes and hot dogs. And if you "deprive" your children of TV at home, they are going to watch it with something like rapture away from home. And obligations, jobs, and chores at home will almost certainly cause your child to wish, sometimes at least, to be somewhere else, watching TV.

Because, of course, parents are not the only ones raising their children. They are being raised also by their schools and by their friends and by the parents of their friends. Some of this outside raising is good, some is not. It is, anyhow, unavoidable.

What this means, I think, is about what it has always meant. Children, no matter how nurtured at home, must be risked to the world. And parenthood is not an exact science, but a vexed privilege and a blessed trial, absolutely necessary and not altogether possible.

If your children spurn your healthful meals in favor of those concocted by some reincarnation of Col. Sanders, Long John Silver, or the Royal Family of Burger; if they flee from books to a friend's house to watch TV, if your old-fashioned notions and ways embarrass them in front of their friends—does that mean you are a failure?

It may. And what parent has not considered that possibility? I know, at least, that I have considered it—and have wailed and gnashed my teeth, found fault, laid blame, preached and ranted. In weaker moments, I have even blamed myself.

But I have thought, too, that the term of human judgment is longer than parenthood, that the upbringing we give our children is not just for their childhood but for all their lives. And it is surely the *duty* of the older generation to be embarrassingly old-fashioned, for the claims of the "newness" of any younger generation are mostly frivolous. The young are born to the human condition more than to their time, and they face mainly the same trials and obligations as their elders have faced.

The real failure is to give in. If we make our house a household instead of a motel, provide healthy nourishment for mind and body, enforce moral distinctions and restraints, teach essential skills and disciplines and require their use, there is no certainty that we are providing our children a "better life" that they will embrace wholeheartedly during childhood. But we are providing them a choice that they may make intelligently as adults.

A Few Words for Motherhood

IT IS the season of motherhood again, and we are preoccupied with the pregnant and the unborn. When birth is imminent, especially with a ewe or a mare, we are at the barn the last thing before we go to bed, at least once in the middle of the night, and well before daylight in the morning. It is a sort of joke here that we have almost never had anything born in the middle of the night. And yet somebody must get up and go out anyway. With motherhood, you don't argue probabilities.

I set the alarm, but always wake up before it goes off. Some part of the mind is given to the barn, these times, and you can't put it to sleep. For a few minutes after I wake up, I lie there wondering where I will get the will and the energy to drag myself out of bed again. Anxiety takes care of that: maybe the ewe has started into labor, and is in trouble. But it isn't just anxiety. It is curiosity too, and the eagerness for new life that goes with motherhood. I want to see what nature and breeding and care and the passage of time have led to. If I open the barn door and hear a little bleat coming out of the darkness, I will be glad to be awake. My liking for that always returns with a force that surprises me.

These are bad times for motherhood—a kind of biological drudgery, some say, using up women who could do better things. Thoreau may have been the first to assert that people should not belong to farm animals, but the idea is now established doctrine with many farmers—and it has received amendments to the effect that people should not belong to children, or to each other. But we all have to belong to something, if only to the idea that we should not belong to anything. We all have to be used up by something. And though I will never be a mother, I am glad to be used up by motherhood and what it leads to, just as—most of the time—I gladly belong to my wife, my children, and several head of cattle, sheep, and horses. What better way to be used up? How else to be a farmer?

There are good arguments against female animals that need help in giving birth; I know what they are, and have gone over them many times. And yet—if the ordeal is not too painful or too long, and if it succeeds—I always wind up a little grateful to the ones that need help. Then I get to take part, get to go through the process another time, and I invariably come away from it feeling instructed and awed and pleased.

My wife and son and I find the heifer in a far corner of the field. In maybe two hours of labor she has managed to give birth to one small foot. We know how it has been with her. Time and again she has lain down and heaved at her burden, and got up and turned and smelled the ground. She is a heifer —how does she know that something is supposed to *be* there?

It takes some doing even for the three of us to get her into the barn. Her orders are to be alone, and she does all in her power to obey. But finally we shut the door behind her and get her into a stall. She isn't wild; once she is confined it isn't even necessary to tie her. I wash in a bucket of icy water and soap my right hand and forearm. She is quiet now. And so are we humans—worried, and excited too, for if there is a chance for failure here, there is also a chance for success.

I loop a bale string onto the calf's exposed foot, knot the string short around a stick which my son then holds. I press my hand gently into the birth canal until I find the second foot and then, a little further on, a nose. I loop a string around the second foot, fasten on another stick for a handhold. And then we pull. The heifer stands and pulls against us for a few seconds, then gives up and goes down. We brace ourselves the best we can into our work, pulling as the heifer pushes. Finally the head comes, and then, more easily, the rest.

We clear the calf's nose, help him to breathe, and then, because the heifer has not yet stood up, we lay him on the bedding in front of her. And what always seems to me the miracle of it begins. She has never calved before. If she ever saw another cow calve, she paid little attention. She has, as we humans say, no education and no experience. And yet she recognizes the calf as her own, and knows what to do for it. Some heifers don't, but most do, as this one does. Even before she gets up, she begins to lick it about the nose and face with loud, vigorous swipes of her tongue. And all the while she utters a kind of moan, meant to comfort, encourage, and reassure—or so I understand it.

How does she know so much? How did all this come about? Instinct. Evolution. I know those words. I understand the logic of the survival of the fittest: good mothering instincts have survived because bad mothers lost their calves: the good traits triumphed, the bad perished. But how come some are fit in the first place? What prepared in the mind of the first cow or ewe or mare—or, for that matter, in the mind of the first human mother—this intricate, careful, passionate welcome to the newborn? I don't know. I don't think anybody does. I distrust any mortal who claims to know. We call these animals dumb brutes, and so far as we can tell they are more or less dumb, and there are certainly times when those of us who live with them will seem to find evidence that they are plenty stupid. And yet, they are indisputably allied with intelligence more articulate and more refined than is to be found in any obstetrics textbook. What is one to make of it? Here is a dumb brute lying in dung and straw, licking her calf, and as always I am feeling honored to be associated with her.

The heifer has stood up now, and the calf is trying to stand, wobbling up onto its hind feet and knees, only to be knocked over by an exuberant caress of its mother's tongue. We have involved ourselves too much in this story by now to leave before the end, but we have our chores to finish too, and so to hasten things I lend a hand.

I help the calf onto his feet and maneuver him over to the heifer's flank. I am not supposed to be there, but her calf is, and so she accepts, or at least permits, my help. In these situations it sometimes seems to me that animals know that help is needed, and that they accept it with some kind of understanding. The thought moves me, but I am never sure, any more than I am sure what the cow means by the low moans she makes as the calf at last begins to nurse. To me, they sound like praise and encouragement—but how would I know?

Always when I hear that little smacking as the calf takes hold of the tit and swallows its first milk, I feel a pressure of laughter under my ribs. I am not sure what that means either. It certainly affirms more than the saved money value of the calf and the continued availability of beef. We all three feel it. We look at each other and grin with relief and satisfaction. Life is on its legs again, and we exult.

A Talent for Necessity

IN THE DAYS when the Southdown ram was king of the sheep pastures and the show-ring, Henry Besuden of Vinewood Farm in Clark County, Kentucky, was perhaps the premier breeder and showman of Southdown sheep in the United States. The list of his winnings at major shows would be too long to put down here, but the character of his achievement can be indicated by his success in showing carload lots of fat lambs in the Chicago International Livestock Exposition. Starting in 1946, he sent eighteen carloads to the International, and won the competition twelve times. "I had 'em fat," he says, remembering. "I had 'em good." Such was the esteem and demand for his stock among fellow breeders that in 1954 he sold a yearling ram for $1200, then a record price for a Southdown.

One would imagine that such accomplishments must have rested on the very best of Bluegrass farmland. But the truth, nearly opposite to that, is much more interesting. "If I'd inherited good land," Henry Besuden says, "I'd probably have been just another Bluegrass farmer."

What he inherited, in fact, was 632 acres of rolling land, fairly steep in places, thin soiled even originally, and by the time he got it, worn out, "corned to death." His grandfather would rent the land out to corn, two hundred acres at a time, and not even get up to see where it would be planted—even though "it was understood to be the rule that renters ruined land." By the time Henry Besuden was eight years old both his mother and father were dead, and the land was farmed by tenants under the trusteeship of a Cincinnati bank. When the farm came to him in 1927, it was heavily encumbered by debt and covered with gullies, some of which were deep enough to hide a standing man.

And so Mr. Besuden began his life as a farmer with the odds against him. But his predicament became his education and, finally, his triumph. "I was lucky," he told Grant Cannon of *The Farm Quarterly* in 1951. "I found that I had some talent for doing the things I *had* to do. I *had* to improve the farm or starve to death; and I *had* to go into the sheep business because sheep were the only animals that could have lived off the farm."

Now seventy-six years old and not in the best of health, Mr. Besuden has not owned a sheep for several years, but he speaks of them with exact remembrance and exacting intelligence; he is one of the best talkers I have had the luck to listen to. How did he get started with sheep? "I was told they'd eat weeds and briars," he says, looking sideways through pipesmoke to see if I get the connection, for the connection between sheep and land is the critical one for him. The history of his sheep and the history of his farm are one history, and it is his own.

Having only talent and necessity—and unusual energy and determination—Mr. Besuden set about the restoration of his ravaged fields. There was no Soil Conservation Service then, but a young man in his predicament was bound to get plenty of advice. To check erosion he first tried building rock dams across the gullies. That wasn't satisfactory; the dams did catch some dirt, but then the fields were marred by half buried rock walls that interfered with work. He tried huge windrows of weeds and brush to the same purpose, but that was not satisfactory either.

Some of the worst gullies he eventually had to fill with a bulldozer. But his main erosion-stopping tool turned out, strangely enough, to be the plow; the tool that in the wrong hands had nearly ruined the farm, in the right hands healed it. Starting at the edge of a gulley he would run a backfurrow up one side and down the other, continuing to plow until he had completed a sizable land. And then he would start at the gulley again, turning the furrows inward as before. He repeated this process until what had been a ditch had become a saucer, so that the runoff, rather than concentrating its force in an abrasive torrent, would be shallowly dispersed over as wide an area as possible. This, as he knew, had been the method of the renters to prepare the gullied land for yet another crop of corn. For them, it had been a temporary remedy; he made it a permanent one.

Nowadays Kentucky fescue 31 would be the grass to sow on such places, but fescue was not available then. Mr. Besuden used small grains, timothy, sweet clover, Korean lespedeza. He used mulches, and he did not overlook the usefulness of what he knew for certain would grow on his land—weeds: "Briars are a good thing for a little hollow." In places he planted thickets of black locust—a native leguminous tree that would serve four purposes: hold the land, encourage grass to grow, provide shade for livestock, and produce posts. But his highest praise is given to the sweet clover which he calls "the best land builder I've ever run into. It'll open up clay, and throw a lot of nitrogen into the ground." The grass would come then, and the real healing would start.

Once the land was in grass, his policy generally was to leave it in grass. Only the best-laying, least vulnerable land was broken for tobacco, the region's major money crop then as now. Even today, I noticed, he sees that his fields are plowed very conservatively. The plowlands are small and carefully placed, leaving out thin places and waterways.

The basic work of restoration continued for twenty-three years. By 1950 the scars were grassed over, and the land was supporting one of the great Southdown flocks of the time. But it was not healed. What was there is gone, and Henry Besuden knows that it will be a long time building back. "'Tain't in good shape, yet," he told an interviewer in 1978.

And so if Mr. Besuden built a reputation as one of the best of livestock showmen, the focus of his interest was nevertheless not the show-ring but the farm. It would be true, it seems, to say that he became a master sheepman and shepherd as one of the ways of becoming a master farmer. For this reason, his standards of quality were never frivolous or freakish, as show-ring standards have sometimes been accused of being, but insistently practical. He never forgot that the purpose of a sheep is to produce a living for the farmer and to put good meat on the table: "When they asked me, 'What do you consider a perfect lamb?' I said, 'One a farmer can make money on!' The foundation has to be the commercial flock." And he wrote in praise of the Southdown ram that "he paid his rent."

But it was perhaps even more characteristic of him to write in 1945 that "one very important thing is that sheep are land builders," and to plead for their continued inclusion in farm livestock programs. He had seen the handwriting on the wall: the new emphasis on row cropping and "production" which in the years after World War II would radically alter the balance of crops and animals on farms, and which, as he feared, would help to destroy the sheep business in his own state. (In 1947, Mr. Besuden's county of Clark had twenty-four breeding flocks of Southdowns, and 30,000 head of grade ewes. That is more than remain now in the whole state of Kentucky.) What he called for instead—and events are rapidly proving him right —was "a long-time program of land building" by which he meant a way of farming based on grass and forage crops, which would build up and maintain reserves of fertility. And in that kind of farming, he was prepared to insist, because he knew, sheep would have an important place.

"I think," he wrote in his series of columns, "Sheep Sense," published in *The Sheepman* in 1945 and 1946, "the fertilizing effect of sheep on the farm has never received the attention it deserves. As one who has had to farm poor land where the least amount of fertilizer shows up plainly, I have noticed that on land often thought too poor for cattle the sheep do well and in time benefit the crops and grass to such an extent that other stock can then be carried. I have seldom seen sheep bed down for the night on anything but high land, and their droppings are evenly scattered on the pasture while grazing, so that no vegetation is killed."

What he wanted was "a way of farming compatible with nature"; this was the constant theme of his work, and he followed it faithfully, both in his pleasure in the lives and events of nature, and in his practical solutions to the problems of farming and soil husbandry. He was never too busy to appreciate, and to praise, the spiritual by-products, as he called them, of farm life. Nor was he too busy to attend to the smallest needs of his land. At one time, for example, he built "two small houses on skids," each of which would hold twenty-five bales of hay. These could be pulled to places where the soil was thin, where the hay would be fed out, and then moved on to other such places. (In the spring they could be used to raise chickens.)

"It's good to have Nature working for you," he says. "She works for a minimum wage." But in reading his "Sheep Sense" columns, one realizes that he not only did not separate the spiritual from the practical, but insisted that they cannot be separated: "This thing of soil conservation involves more than laying out a few terraces and diversion ditches and sowing to grass and legumes, it also involves the heart of the man managing the land. If he loves his soil he will save it." Once, he says, he thought of numbering his fields, but decided against it—"That didn't seem fair to them"—for each has its own character and potential.

As a rule, he would have 400 head of ewes in two flocks—a flock of registered Southdowns and a flock of "Western" commercial ewes. After lambing, he would be running something in the neighborhood of 1000 head. To handle so many sheep on a diversified farm required a great deal of care, and Mr. Besuden's system of management, worked out with thorough understanding and attention to detail, is worth the interest and reflection of any raiser of livestock.

It was a system intended, first of all, to get the maximum use of forage. This rested on what he understands to be a sound principle of livestock farming and soil conservation, but it was forced upon him by the poor quality of his land. He had to keep row cropping to a minimum, and if that meant buying grain, then he would buy it. But he did not buy much. He usually fed, he told me, one pound of corn per ewe per day for sixty days. But in "Sheep Sense" for December 1945, he wrote: "One-half pound grain with three pounds legume hay should do the job, starting with the hay and adding the grain later." He creep-fed his early lambs, but took them off grain as soon as pasture was available. In "Sheep Sense," March 1946, he stated flatly that "creep-feeding after good grass arrives does not pay."

Grain, then, he considered not a diet, but a supplement, almost an emergency ration, to assure health and growth in the flock during the time when he had no pasture. It must be remembered that he was talking about a kind of sheep bred to make efficient use of pasture and hay, and that the market then favored that kind. In the decades following World War II, cheap energy and cheap grain allowed interest to shift to the larger breeds of sheep and larger slaughter lambs that must be grain fed. But now with the cost of energy rising, pushing up the cost of grain, and the human consumption of grain rising with the increase of population, Henry Besuden's sentence of a generation ago resounds with good sense: "Due to the shortage of grain throughout the world, the sheep farmer needs to study the possibilities of grass fattening."

Those, anyhow, were the possibilities that *he* was studying. And the management of pasture, the management of sheep *on* pasture, was his art.

In the fall he would select certain pastures close to the barn to be used for late grazing. This is what is now called "stock-piling"—which, he points out, is only a new word for old common sense. It was sometimes possible, in favorable years, to keep the ewes on grass all through December, feeding "very little hay" and "a small amount of grain." Sometimes he sowed rye early to provide late fall pasture and so extend the grazing season.

His ewes were bred to lamb in January and February. He fed good clover or alfalfa hay, and from about the middle of January to about the middle of March he gave the ewes their sixty daily rations of grain. In mid March the grain feeding ended, and ewes and lambs went out on early pasture of rye which had been sown as a cover crop on the last year's tobacco patches. "A sack of Balboa rye sown in the early fall," he wrote, "is worth several sacks of feed fed in the spring and is much cheaper." From the rye they went to the clover fields where tobacco had grown two years before. From the clover they were moved onto the grass pastures. The market lambs were sold straight off the pastures, at eighty to eighty-five pounds, starting the first of May.

After fescue became available, Mr. Besuden made extensive use of it in his pastures. But he feels that this grass, though an excellent land conserver, is not nutritious or palatable enough to make the best sheep pasture, and so he took pains to diversify his fescue stands with timothy and legumes. His favorite pasture legume is Korean lespedeza, though he joins in the fairly common complaint that it is less vigorous and productive now than it used to be. He has also used red clover, alsike, ladino, and birdsfoot trefoil. He says that he had trouble getting his ewes with lamb in the first heat when they were bred on clover pastures, but that he never had this trouble on lespedeza.

His pastures were regularly reseeded to legumes, usually in March, the sheep tramping in the seed, and he found this method of "renovation" to be as good as any. The pastures were clipped twice during the growing season, sometimes oftener, to keep the growth vigorous and uniform.

The key to efficient management of sheep on pasture is paying attention, and it was important to Mr. Besuden that he should be on horseback among his sheep in the early mornings. The sheep would be out of the shade then, grazing, and he could study their condition and the condition of the field. He speaks of the "bloom" of a pasture, referring to a certain freshness of appearance made by new, tender growth sprigging up through the old. When that bloom is gone, he thinks, the sheep should be moved. The move from a stale pasture to a fresh one can lengthen the grazing time by as much as two hours a day. He believes also that lambs do best when the flock is not too large. That is because sheep tend to bunch together when grazing, the least vigorous lambs coming last and having to feed on grass mouthed over and rejected by the others. He saw to it that his pastures were amply provided with shade, and he knew that the shade needed to be well placed: "I think the best lamb-growing pastures I have are the ones where the shade is close to the water. I have seen times during July and August when sheep would not leave the shade and go to water if the shade and water happened to be at opposite ends of a large field."

The crisis of the shepherd's year, of course, is lambing time. That is the time that the year's work stands or falls by. And because it usually takes place in cold weather, the success of lambing is almost as dependent on the shepherd's facilities as on his knowledge. The lambing barn at Vinewood is an instructive embodiment of Mr. Besuden's understanding of his work and his gift for order. He gives a good description of it himself in one of his columns:

"Practically all the lambing here at Vinewood in recent years has been in a barn especially made for the purpose, shiplap (tongue groove) boxing with a low loft and a window in each bent. The east end of the barn [away from the prevailing winds] is rarely ever closed, a gate being used. Often in extremely cold weather the temperature can be raised fifteen or twenty degrees by the heat from the sheep. Some thirty feet out in the front and extending the width of the barn [is] a heavy layer

of rock. . . . This prevents the muddy place that often appears at the barn door and . . . pulls at the sheep as they walk
through it, causing slipped lambs. Also at the entrance . . . a
locust post is half embedded across the door. This serves as a
protection in case of dogs trying to dig under the door or gate
and helps to hold the bedding in the barn as the sheep go out.
Any kind of a sill that is too high or causes the heavy-in-lamb
ewes to jump or strain to cross is too risky."

The barn is admirably laid out, with pens, chutes, and gates
to permit the feeding, handling, sorting, and loading of a large
number of sheep with the least trouble. There were lambing
pens for forty ewes. There was also a small room with pens that
could be heated by a stove. Above each pen was a red wooden
"button" that could be turned down to indicate that a ewe was
near to lambing or for any other reason in need of close attention. These were used when Mr. Besuden had an experienced
helper to share the nighttime duty with him. "They saved a lot
of cold midnight talk," he says.

But experienced help was not always available, and then he
would have to work through the days and nights of lambing
alone. Staying awake would get to be a problem. Sometimes,
sitting beside one of the pens, waiting for a ewe to lamb, he
would tie a string from one of her hind legs to his wrist. When
her labor pains came and she began to shift around, she would
tug the string and he would wake up and tend to her.

And so the talent for what he "had to do" was in large measure the ability to bear the good outcome in mind: to envi-
sion, in spite of rocks and gullies, the good health of the fields;
to foresee in the pregnant ewes and the advancing seasons a
good crop of lambs. And it was the ability to carry in his head
for nearly half a century the ideal character and pattern of the
Southdown, and to measure his animals relentlessly against it—
an ability, rare enough, that marked him as a master stockman.

He told me a story that suggests very well the distinction
and the effect of that ability. On one of his trips to the International he competed against a western sheepman who had
selected his carload of fifty fat lambs out of ten thousand head.

After the Vinewood carload had won the class, this gentleman came up and asked: "How many did you pick yours from,
Mr. Besuden?"

"About seventy-five."

"Well," the western breeder said, "I guess it's better to have the right seventy-five than the wrong ten thousand."

But the ability to recognize the right seventy-five is worthless by itself. Just as necessary is the ability to do the work and to pay attention. To pay attention, above all—that is another of the persistent themes of Mr. Besuden's talk and of his life. He is convinced that paying attention pays, and this sets him apart from the mechanized "modern" farmers who are pushed to accept more responsibility than they can properly meet, and to work at freeway speeds. He wrote in his column of the importance of "little things done on time." He said that they paid, but he knew that people did them for more than pay.

He told me also about a farmer who wouldn't scrape the manure off his shoes until he came to a spot that was bare of grass. "That's what I mean," he said. "You have to keep it on your mind."

Seven Amish Farms

IN TYPICAL Midwestern farming country the distances between inhabited houses are stretching out as bigger farmers buy out their smaller neighbors in order to "stay in." The signs of this "movement" and its consequent specialization are everywhere: good houses standing empty, going to ruin; good stock barns going to ruin; pasture fences fallen down or gone; machines too large for available doorways left in the weather; windbreaks and woodlots gone down before the bulldozers; small schoolhouses and churches deserted or filled with grain.

In the latter part of March this country shows little life. Field after field lies under the dead stalks of last year's corn and soybeans, or lies broken for the next crop; one may drive many miles between fields that are either sodded or planted in winter grain. If the weather is wet, the country will seem virtually deserted. If the ground is dry enough to support their wheels, there will be tractors at work, huge machines with glassed cabs, rolling into the distances of fields larger than whole farms used to be, as solitary as seaborne ships.

The difference between such country and the Amish farmlands in northeast Indiana seems almost as great as that between a desert and an oasis. And it is the *same* difference. In the Amish country there is a great deal more life: more natural life, more agricultural life, more human life. Because the farms are small—most of them containing well under a hundred acres—the Amish neighborhoods are more thickly populated than most rural areas, and you see more people at work. And because the Amish are diversified farmers, their plowed croplands are interspersed with pastures and hayfields and often with woodlots. It is a varied, interesting, healthy looking farm country, pleasant to drive through. When we were there, on the twentieth and twenty-first of last March, the spring plowing had just started, and so you could still see everywhere the annual covering of stable manure on the fields, and the teams of Belgians or Percherons still coming out from the barns with loaded spreaders.

Our host, those days, was William J. Yoder, a widely respected breeder of Belgian horses, an able farmer and carpenter, and a most generous and enjoyable companion. He is a vigorous man, strenuously involved in the work of his farm and in the life of his family and community. From the look of him and the look of his place, you know that he has not just done a lot of work in his time, but has done it well, learned from it, mastered the necessary disciplines. He speaks with heavy stress on certain words—the emphasis of conviction, but also of pleasure, for he enjoys the talk that goes on among people interested in horses and in farming. But unlike many people who enjoy talking, he speaks with care. Bill was born in this community, has lived there all his life, and he has grandchildren who will probably live there all their lives. He belongs there, then, root and branch, and he knows the history and the quality of many of the farms. On the two days, we visited farms belonging to Bill himself, four of his sons, and two of his sons-in-law.

The Amish farms tend to divide up between established ones, which are prosperous looking and well maintained, and run-down, abused, or neglected ones, on which young farmers are getting started. Young Amish farmers *are* still getting started, in spite of inflation, speculators' prices, and usurious interest rates. My impression is that the proportion of young farmers buying farms is significantly greater among the Amish than among conventional farmers.

Bill Yoder's own eighty-acre farm is among the established ones. I had been there in the fall of 1975 and had not forgotten its aspect of cleanness and good order, its well-kept white buildings, neat lawns, and garden plots. Bill has owned the place for twenty-six years. Before he bought it, it had been rented and row cropped, with the usual result: it was nearly played out. "The buildings," he says, "were nothing," and there were no fences. The first year, the place produced five loads (maybe five tons) of hay, "and that was mostly sorrel." The only healthy plants on it were the spurts of grass and clover that grew out of the previous year's manure piles. The corn crop that first year "might have been thirty bushels an acre," all nubbins. The sandy soil blew in every strong wind, and when he plowed the fields his horses' feet sank into "quicksand potholes" that the share uncovered.

The remedy has been a set of farming practices traditional among the Amish since the seventeenth century: diversification, rotation of crops, use of manure, seeding of legumes. These practices began when the Anabaptist sects were disfranchised in their European homelands and forced to the use of poor soil. We saw them still working to restore farmed-out soils in Indiana. One thing these practices do is build humus in the soil, and humus does several things: increases fertility, improves soil structure, improves both water-holding capacity and drainage. "No humus, you're in trouble," Bill says.

After his rotations were established and the land had begun to be properly manured, the potholes disappeared, and the soil quit blowing. "There's something in it now—there's some substance there." Now the farm produces abundant crops of corn, oats, wheat, and alfalfa. Oats now yield 90–100 bushels per acre. The corn averages 100–125 bushels per acre, and the ears are long, thick, and well filled.

Bill's rotation begins and ends with alfalfa. Every fall he puts in a new seeding of alfalfa with his wheat; every spring he plows down an old stand of alfalfa, "no matter how good it is." From alfalfa he goes to corn for two years, planting thirty acres, twenty-five for ear corn and five for silage. After the second year of corn, he sows oats in the spring, wheat and alfalfa in the fall. In the fourth year the wheat is harvested; the alfalfa then comes on and remains through the fifth and sixth years. Two cuttings of alfalfa are taken each year. After curing in the field, the hay is hauled to the barn, chopped, and blown into the loft. The third cutting is pastured.

Unlike cow manure, which is heavy and chunky, horse manure is light and breaks up well coming out of the spreader; it interferes less with the growth of small seedlings and is less likely to be picked up by a hay rake. On Bill's place, horse manure is used on the fall seedings of wheat and alfalfa, on the young alfalfa after the wheat harvest, and both years on the established alfalfa stands. The cow manure goes on the corn ground both years. He usually has about 350 eighty-bushel spreader loads of manure, and each year he covers the whole farm—cropland, hayland, and pasture.

With such an abundance of manure there obviously is no *dependence* on chemical fertilizers, but Bill uses some as a "starter" on his corn and oats. On corn he applies 125 pounds of nitrogen in the row. On oats he uses 200–250 pounds of 16-16-16, 20-20-20, or 24-24-24. He routinely spreads two tons of lime to the acre on the ground being prepared for wheat.

His out-of-pocket costs per acre of corn last year were as follows:

Seed (planted at a rate of seven acres
    per bushel). . . . . $ 7.00
Fertilizer 7.75
Herbicide (custom applied, first year only) 16.40

That comes to a total of $31.15 per acre—or, if the corn makes only a hundred bushels per acre, a little over $0.31 per bushel. In the second year his per acre cost is $14.75, less than $0.15 per bushel, bringing the two-year average to $22.95 per acre or about $0.23 per bushel.

The herbicide is used because, extra horses being on the farm during the winter, Bill has to buy eighty to a hundred tons of hay, and in that way brings in weed seed. He had no weed problem until he started buying hay. Even though he uses the herbicide, he still cultivates his corn three times.

His cost per acre of oats came to $33.00 ($12.00 for seed and $21.00 for fertilizer)—or, at ninety bushels per acre, about $0.37 per bushel.

Of Bill's eighty acres, sixty-two are tillable. He has ten acres of permanent pasture, and seven or eight of woodland, which produced the lumber for all the building he has done on the place. In addition, for $500 a year he rents an adjoining eighty acres of "hill and woods pasture" which provides summer grazing for twenty heifers; and on another neighboring farm he rents varying amounts of cropland.

All the field work is done with horses, and this, of course, comes virtually free—a by-product of the horse-breeding enterprise. Bill has an ancient Model D John Deere tractor that he uses for belt power.

At the time of our visit, there were twenty-two head of horses on the place. But that number was unusually low, for Bill aims to keep "around thirty head." He has a band of excellent brood mares and three stallions, plus young stock of assorted ages. Since October 1 of last year, he had sold eighteen head of registered Belgian horses. In the winters he operates a "urine line," collecting "pregnant mare urine," which is sold to a pharmaceutical company for the extraction of various hormones. For this purpose he boards a good many mares belonging to neighbors; that is why he must buy the extra hay that causes his weed problem. (Horses are so numerous on this farm because they are one of its money-making enterprises. If horses were used only for work on this farm, four good geldings would be enough.)

One bad result of the dramatic rise in draft horse prices over the last eight or ten years is that it has tended to focus attention on such characteristics as size and color to the neglect of less obvious qualities such as good feet. To me, foot quality seems a critical issue. A good horse with bad feet is good for nothing but decoration, and at sales and shows there are far too many flawed feet disguised by plastic wood and black shoe polish. And so I was pleased to see that every horse on Bill Yoder's place had sound, strong-walled, correctly shaped feet. They were good horses all around, but their other qualities were well founded; they stood on good feet, and this speaks of the thoroughness of his judgment and also of his honesty.

Though he is a master horseman, and the draft horse business is more lucrative now than ever in its history, Bill does not specialize in horses, and that is perhaps the clearest indication of his integrity as a farmer. Whatever may be the dependability of the horse economy, on this farm it rests upon a diversified agricultural economy that is sound.

He was milking five Holstein cows; he had fifteen Holstein heifers that he had raised to sell; and he had just marketed thirty finished hogs, which is the number that he usually has on fence. All the animals had been well wintered—Bill quotes his father approvingly: "Well wintered is half summered"—and were in excellent condition. Another saying of his father's that Bill likes to quote—"Keep the horses on the side of the fence the feed is on"—has obviously been obeyed here. The feeding is careful, the feed is good, and it is abundant. Though it was almost spring, there were ample surpluses in the hayloft and in the corn cribs.

Other signs of the farm's good health were three sizable garden plots, and newly pruned grapevines and raspberry canes. The gardener of the family is Mrs. Yoder. Though most of the children are now gone from home, Bill says that she still grows as much garden stuff as she ever did.

All seven of the Yoders' sons live in the community. Floyd, the youngest, is still at home. Harley has a house on nearly three acres, works in town, and returns in the afternoons to his own shop where he works as a farrier. Henry, who also works in town, lives with Harley and his wife. The other four sons are now settled on farms that they are in the process of paying for. Richard has eighty acres, Orla eighty, Mel fifty-seven, and Wilbur eighty. Two sons-in-law also living in the community are Perry Bontrager, who owns ninety-five acres, and Ervin Mast, who owns sixty-five. Counting Bill's eighty acres, the seven families are living on 537 acres. Of the seven farms, only Mel's is entirely tillable, the acreages in woods or permanent pasture varying from five to twenty-six.

These young men have all taken over run-down farms, on which they are establishing rotations and soil husbandry practices that, being traditional, more or less resemble Bill's. It seemed generally agreed that after three years of this treatment the land would grow corn, as Perry Bontrager said, "like anywhere else."

These are good farmers, capable of the intelligent planning, sound judgment, and hard work that good farming requires. Abused land heals and flourishes in their care. None of them expressed a wish to own more land; all, I believe, feel that what they have will be enough—when it is paid for. The big problems are high land prices and high interest rates, the latter apparently being the worst.

The answer, for Bill's sons so far, has been town work. All of them, after leaving home, have worked for Redman Industries, a manufacturer of mobile homes in Topeka. They do piecework, starting at seven in the morning and quitting at two in the afternoon, using the rest of the day for farming or other work. This, Bill thinks, is now "the only way" to get started farming. Even so, there is "a lot of debt" in the community —"more than ever."

With a start in factory work, with family help, with government and bank loans, with extraordinary industry and perseverance, with highly developed farming skills, it is still possible for young Amish families to own a small farm that will eventually support them. But there is more strain in that effort now than there used to be, and more than there should be. When the burden of usurious interest becomes too great, these young men are finding it necessary to make temporary returns to their town jobs.

The only one who spoke of his income was Mel, who owns fifty-seven acres, which, he says, *will be* enough. He and his family milk six Holsteins. He had nine mares on the urine line last winter, seven of which belonged to him. And he had twelve brood sows. Last year his gross income was $43,000. Of this, $12,000 came from hogs, $7,000 from his milk cows, the rest from his horses and the sale of his wheat. After his production costs, but *before* payment of interest, he netted $22,000. In order to cope with the interest payments, Mel was preparing to return to work in town.

These little Amish farms thus become the measure both of "conventional" American agriculture and of the cultural meaning of the national industrial economy.

To begin with, these farms give the lie direct to that false god of "agribusiness": the so-called economy of scale. The small farm is not an anachronism, is not unproductive, is not unprofitable. Among the Amish, it is still thriving, and is still the economic foundation of what John A. Hostetler (in *Amish Society*, third edition) rightly calls "a healthy culture." Though they do not produce the "record-breaking yields" so touted by the "agribusiness" establishment, these farms are nevertheless highly productive. And if they are not likely to make their owners rich (never an Amish goal), they can certainly be said to be sufficiently profitable. The economy of scale has helped corporations and banks, not farmers and farm communities. It has been an economy of dispossession and waste—plutocratic, if not in aim, then certainly in result.

What these Amish farms suggest, on the contrary, is that in farming there is inevitably a scale that is suitable both to the productive capacity of the land and to the abilities of the farmer; and that agricultural problems are to be properly solved, not in expansion, but in management, diversity, balance, order, responsible maintenance, good character, and in the sensible limitation of investment and overhead. (Bill makes a careful distinction between "healthy" and "unhealthy" debt, a "healthy debt" being "one you can hope to pay off in a reasonable way.")

Most significant, perhaps, is that while conventional agriculture, blindly following the tendency of any industry to exhaust its sources, has made soil erosion a national catastrophe, these Amish farms conserve the land and improve it in use.

And what is one to think of a national economy that drives such obviously able and valuable farmers to factory work? What value does such an economy impose upon thrift, effort, skill, good husbandry, family and community health?

In spite of the unrelenting destructiveness of the larger economy, the Amish—as Hostetler points out with acknowledged surprise and respect—have almost doubled in population in the last twenty years. The doubling of a population is, of course, no significant achievement. What is significant is that these agricultural communities have doubled their population *and yet remained agricultural communities* during a time when conventional farmers have failed by the millions. This alone would seem to call for a careful look at Amish ways of farming. That those ways have, during the same time, been ignored by the colleges and the agencies of agriculture must rank as a prime intellectual wonder.

Amish farming has been so ignored, I think, because it involves a complicated structure that is at once biological and cultural, rather than industrial or economic. I suspect that anyone who might attempt an accounting of the economy of an Amish farm would soon find himself dealing with virtually unaccountable values, expenses, and benefits. He would be dealing with biological forces and processes not always measurable, with spiritual and community values not quantifiable; at certain points he would be dealing with mysteries—and he would be finding that these unaccountables and inscrutables have results, among others, that are economic. Hardly an appropriate study for the "science" of agricultural economics.

The economy of conventional agriculture or "agribusiness"

is remarkable for the simplicity of its arithmetic. It involves a manipulation of quantities that are all entirely accountable. List your costs (land, equipment, fuel, fertilizer, pesticides, herbicides, wages), add them up, subtract them from your earnings, or subtract your earnings from them, and you have the result.

Suppose, on the other hand, that you have an eighty-acre farm that is not a "food factory" but your home, your given portion of Creation which you are morally and spiritually obliged "to dress and to keep." Suppose you farm, not for wealth, but to maintain the integrity and the practical supports of your family and community. Suppose that, the farm being small enough, you farm it with family work and work exchanged with neighbors. Suppose you have six Belgian brood mares that you use for field work. Suppose that you also have milk cows and hogs, and that you raise a variety of grain and hay crops in rotation. What happens to your accounting then?

To start with, several of the costs of conventional farming are greatly diminished or done away with. Equipment, fertilizer, chemicals all cost much less. Fuel becomes feed, but you have the mares and are feeding them anyway; the work ration for a brood mare is not a lot more costly than a maintenance ration. And the horses, like the rest of the livestock, are making manure. Figure that in, and figure, if you can, the value of the difference between manure and chemical fertilizer. You can probably get an estimate of the value of the nitrogen fixed by your alfalfa, but how will you quantify the value to the soil of its residues and deep roots? Try to compute the value of humus in the soil—in improved drainage, improved drought resistance, improved tilth, improved health. Wages, if you pay your children, will still be among your costs. But compute the difference between paying your children and paying "labor." Work exchanged with neighbors can be reduced to "man-hours" and assigned a dollar value. But compute the difference between a neighbor and "labor." Compute the value of a family or a community to any one of its members. We may, as we must, grant that among the values of family and community there is economic value—but what is it?

In the Louisville *Courier-Journal* of April 5, 1981, the Mobil Oil Corporation ran an advertisement which was yet another celebration of "scientific agriculture." American farming, the

Mobil people are of course happy to say, "requires *more petroleum products than almost any other industry.* A gallon of gasoline to produce a single bushel of corn, for example. . . ." This, they say, enables "each American farmer to feed sixty-seven people." And they say that this is "a-maizing."

Well, it certainly is! And the chances are good that an agriculture totally dependent on the petroleum industry is not yet as amazing as it is going to be. But one thing that is already sufficiently amazing is that a bushel of corn produced by the burning of one gallon of gasoline has already cost more than *six times* as much as a bushel of corn grown by Bill Yoder. How does Bill Yoder escape what may justly be called the petroleum tax on agriculture? He does so by a series of substitutions: of horses for tractors, of feed for fuel, of manure for fertilizer, of sound agricultural methods and patterns for the exploitive methods and patterns of industry. But he has done more than that—or, rather, he and his people and their tradition have done more. They have substituted themselves, their families, and their communities for petroleum. The Amish use little petroleum—and need little—because they have those other things.

I do not think that we can make sense of Amish farming until we see it, until we become willing to see it, as belonging essentially to the Amish practice of Christianity, which instructs that one's neighbors are to be loved as oneself. To farmers who give priority to the maintenance of their community, the economy of scale (that is, the economy of *large* scale, of "growth") can make no sense, for it requires the ruination and displacement of neighbors. A farm cannot be increased except by the decrease of a neighborhood. What the interest of the community proposes is invariably an economy of *proper* scale. A whole set of agricultural proprieties must be observed: of farm size, of methods, of tools, of energy sources, of plant and animal species. Community interest also requires charity, neighborliness, the care and instruction of the young, respect for the old; thus it assures its integrity and survival. Above all, it requires good stewardship of the land, for the community, as the Amish have always understood, is no better than its land. "If treated violently or exploited selfishly," John Hostetler writes, the land "will yield poorly." There could be no better statement of the meaning of the *practice* and the practicality of charity. Except to the insane narrow-mindedness of industrial economics, selfishness does not pay.

The Amish have steadfastly subordinated economic value to the values of religion and community. What is too readily overlooked by a secular, exploitive society is that their ways of doing this are not "empty gestures" and are not "backward." In the first place, these ways have kept the communities intact through many varieties of hard times. In the second place, they conserve the land. In the third place, they yield economic benefits. The community, the religious fellowship, has many kinds of value, and among them is economic value. It is the result of the practice of neighborliness, and of the practice of stewardship. What moved me most, what I liked best, in those days we spent with Bill Yoder was the sense of the continuity of the community in his dealings with his children and in their dealings with their children.

Bill has helped his sons financially so far as he has been able. He has helped them with his work. He has helped them by sharing what he has—lending a stallion, say, at breeding time, or lending a team. And he helps them by buying good pieces of equipment that come up for sale. "If he ever gets any money," he says of one of the boys, for whom he has bought an implement, "he'll pay me for it. If he don't, he'll just use it." He has been their teacher, and he remains their advisor. But he does not stand before them as a domineering patriarch or "authority figure." He seems to speak, rather, as a representative of family and community experience. In their respect for him, his sons respect their tradition. They are glad for his help, advice, and example, but there is nothing servile in this. It seems to be given and taken in a kind of familial friendship, respect going both ways.

Everywhere we went, when school was not in session, the children were at the barns, helping with the work, watching, listening, learning to farm in the way it is best learned. Wilbur told us that his eleven-year-old son had cultivated twenty-three acres of corn last year with a team and a riding cultivator. That reminded Bill of the way he taught Wilbur to do the same job.

Wilbur was little then, and he loved to sit in his father's lap and drive the team while Bill worked the cultivator. If Wilbur could drive, Bill thought, he could do the rest of it. So he got off and shortened the stirrups so the boy could reach them with his feet. Wilbur started the team, and within a few steps began plowing up the corn.

"Whoa!" he said.

And Bill, who was walking behind him, said, "Come up!"

And it went that way for a little bit:

"Whoa!"

"Come up!"

And then Wilbur started to cry, and Bill said:

"Don't cry! Go ahead!"

The Gift of Good Land

"Dream not of other Worlds . . ."
Paradise Lost VIII, 175

M Y PURPOSE here is double. I want, first, to attempt a
Biblical argument for ecological and agricultural responsibility. Second, I want to examine some of the practical implications of such an argument. I am prompted to the first of these tasks partly because of its importance in our unresolved conflict about how we should use the world. That those who affirm the divinity of the Creator should come to the rescue of His creature is a logical consistency of great potential force.

The second task is obviously related to the first, but my motive here is somewhat more personal. I wish to deal directly at last with my own long held belief that Christianity, as usually presented by its organizations, is not *earthly* enough—that a valid spiritual life, in this world, must have a practice and a practicality—it must have a material result. (I am well aware that in this belief I am not alone.) What I shall be working toward is some sort of practical understanding of what Arthur O. Lovejoy called the "this-worldly" aspect of Biblical thought. I want to see if there is not at least implicit in the Judeo-Christian heritage a doctrine such as that the Buddhists call "right livelihood" or "right occupation."

Some of the reluctance to make a forthright Biblical argument against the industrial rape of the natural world seems to come from the suspicion that this rape originates with the Bible, that Christianity cannot cure what, in effect, it has caused. Judging from conversations I have had, the best known spokesman for this view is Professor Lynn White, Jr., whose essay, "The Historical Roots of Our Ecologic Crisis" has been widely published.

Professor White asserts that it is a "Christian axiom that nature has no reason for existence save to serve man." He seems to base his argument on one Biblical passage, Genesis 1:28, in which Adam and Eve are instructed to "subdue" the earth. "Man," says Professor White, "named all the animals, thus

519

establishing his dominance over them." There is no doubt that Adam's superiority over the rest of Creation was represented, if not established, by this act of naming; he *was* given dominance. But that this dominance was meant to be tyrannical, or that "subdue" meant to destroy, is by no means a necessary inference. Indeed, it might be argued that the correct understanding of this "dominance" is given in Genesis 2:15, which says that Adam and Eve were put into the Garden "to dress it and to keep it."

But these early verses of Genesis can give us only limited help. The instruction in Genesis 1:28 was, after all, given to Adam and Eve in the time of their innocence, and it seems certain that the word "subdue" would have had a different intent and sense for them at that time than it could have for them, or for us, after the Fall.

It is tempting to quarrel at length with various statements in Professor White's essay, but he has made that unnecessary by giving us two sentences that define both his problem and my task. He writes, first, that "God planned all of this [the Creation] explicitly for man's benefit and rule: no item in the physical creation had any purpose save to serve man's purposes." And a few sentences later he says: "Christianity . . . insisted that it is God's will that man exploit nature for his *proper* ends" [My emphasis].

It is certainly possible that there might be a critical difference between "man's purposes" and "man's *proper* ends." And one's belief or disbelief in that difference, and one's seriousness about the issue of propriety, will tell a great deal about one's understanding of the Judeo-Christian tradition.

I do not mean to imply that I see no involvement between that tradition and the abuse of nature. I know very well that Christians have not only been often indifferent to such abuse, but have often condoned it and often perpetrated it. That is not the issue. The issue is whether or not the Bible explicitly or implicitly defines a *proper* human use of Creation or the natural world. Proper use, as opposed to improper use, or abuse, is a matter of great complexity, and to find it adequately treated it is necessary to turn to a more complex story than that of Adam and Eve.

The story of the giving of the Promised Land to the Israelites is more serviceable than the story of the giving of the

Garden of Eden, because the Promised Land is a divine gift to a *fallen* people. For that reason the giving is more problematical, and the receiving is more conditional and more difficult. In the Bible's long working out of the understanding of this gift, we may find the beginning—and, by implication, the end—of the definition of an ecological discipline.

The effort to make sense of this story involves considerable difficulty because the tribes of Israel, though they see the Promised Land as a gift to them from God, are also obliged to take it by force from its established inhabitants. And so a lot of the "divine sanction" by which they act sounds like the sort of rationalization that invariably accompanies nationalistic aggression and theft. It is impossible to ignore the similarities to the westward movement of the American frontier. The Israelites were following their own doctrine of "manifest destiny," which for them, as for us, disallowed any human standing to their opponents. In Canaan, as in America, the conquerors acted upon the broadest possible definition of idolatry and the narrowest possible definition of justice. They conquered with the same ferocity and with the same genocidal intent.

But for all these similarities, there is a significant difference. Whereas the greed and violence of the American frontier produced an ethic of greed and violence that justified American industrialization, the ferocity of the conquest of Canaan was accompanied from the beginning by the working out of an ethical system antithetical to it—and antithetical, for that matter, to the American conquest with which I have compared it. The difficulty but also the wonder of the story of the Promised Land is that, there, the primordial and still continuing dark story of human rapaciousness begins to be accompanied by a vein of light which, however improbably and uncertainly, still accompanies us. This light originates in the idea of the land as a gift—not a free or a deserved gift, but a gift given upon certain rigorous conditions.

It is a gift because the people who are to possess it did not create it. It is accompanied by careful warnings and demonstrations of the folly of saying that "My power and the might of mine hand hath gotten me this wealth" (Deuteronomy 8:17). Thus, deeply implicated in the very definition of this gift is a specific warning against *hubris* which is the great ecological sin, just as it is the great sin of politics. People are not gods.

They must not act like gods or assume godly authority. If they do, terrible retributions are in store. In this warning we have the root of the idea of propriety, of *proper* human purposes and ends. We must not use the world as though we created it ourselves.

The Promised Land is not a permanent gift. It is "given," but only for a time, and only for so long as it is properly used. It is stated unequivocally, and repeated again and again, that "the heaven and the heaven of heavens is the Lord's thy God, the earth also, with all that therein is" (Deuteronomy 10:14). What is given is not ownership, but a sort of tenancy, the right of habitation and use: "The land shall not be sold forever: for the land is mine; for ye are strangers and sojourners with me" (Leviticus 25:23).

In token of His landlordship, God required a sabbath for the land, which was to be left fallow every seventh year; and a sabbath of sabbaths every fiftieth year, a "year of jubilee," during which not only would the fields lie fallow, but the land would be returned to its original owners, as if to free it of the taint of trade and the conceit of human ownership. But beyond their agricultural and social intent, these sabbaths ritualize an observance of the limits of "my power and the might of mine hand"—the limits of human control. Looking at their fallowed fields, the people are to be reminded that the land is theirs only by gift; it exists in its own right, and does not begin or end with any human purpose.

The Promised Land, moreover, is "a land which the Lord thy God careth for: the eyes of the Lord thy God are always upon it" (Deuteronomy 11:12). And this care promises a re-possession by the true landlord, and a fulfillment not in the power of its human inhabitants: "as truly as I live, all the earth shall be filled with the glory of the Lord" (Numbers 14:21)—a promise recalled by St. Paul in Romans 8:21: "the creature [the Creation] itself also shall be delivered from the bond-age of corruption into the glorious liberty of the children of God."

Finally, and most difficult, the good land is not given as a reward. It is made clear that the people chosen for this gift do not deserve it, for they are "a stiff-necked people" who have been wicked and faithless. To such a people such a gift can be given only as a moral predicament: having failed to deserve it beforehand, they must prove worthy of it afterwards; they must use it well, or they will not continue long in it.

How are they to prove worthy?

First of all, they must be faithful, grateful, and humble; they must remember that the land is a gift: "When thou hast eaten and art full, then thou shalt bless the Lord thy God for the good land which he hath given thee." (Deuteronomy 8:10).

Second, they must be neighborly. They must be just, kind to one another, generous to strangers, honest in trading, etc. These are social virtues, but, as they invariably do, they have ecological and agricultural implications. For the land is described as an "inheritance"; the community is understood to exist not just in space, but also in time. One lives in the neighborhood, not just of those who now live "next door," but of the dead who have bequeathed the land to the living, and of the unborn to whom the living will in turn bequeath it. But we can have no direct behavioral connection to those who are not yet alive. The only neighborly thing we can do for them is to preserve their inheritance: we must take care, among other things, of the land, which is never a possession, but an inheritance to the living, as it will be to the unborn.

And so the third thing the possessors of the land must do to be worthy of it is to practice good husbandry. The story of the Promised Land has a good deal to say on this subject, and yet its account is rather fragmentary. We must depend heavily on implication. For sake of brevity, let us consider just two verses (Deuteronomy 22:6–7):

> If a bird's nest chance to be before thee in the way in any tree, or on the ground, whether they be young ones, or eggs, and the dam sitting upon the young, or upon the eggs, thou shalt not take the dam with the young:
> But thou shalt in any wise let the dam go, and take the young to thee; that it may be well with thee, and that thou mayest prolong thy days.

This, obviously, is a perfect paradigm of ecological and agricultural discipline, in which the idea of inheritance is necessarily paramount. The inflexible rule is that the source must be preserved. You may take the young, but you must save the breeding stock. You may eat the harvest, but you must save seed, and you must preserve the fertility of the fields.

What we are talking about is an elaborate understanding of charity. It is so elaborate because of the perception, implicit here, explicit in the New Testament, that charity by its nature cannot be selective—that it is, so to speak, out of human control. It cannot be selective because between any two humans, or any two creatures, all Creation exists as a bond. Charity cannot be just human, any more than it can be just Jewish or just Samaritan. Once begun, wherever it begins, it cannot stop until it includes all Creation, for all creatures are parts of a whole upon which each is dependent, and it is a contradiction to love your neighbor and despise the great inheritance on which his life depends. Charity even for one person does not make sense except in terms of an effort to love all Creation in response to the Creator's love for it.

And how is this charity answerable to "man's purposes"? It is not, any more than the Creation itself is. Professor White's contention that the Bible proposes any such thing is, so far as I can see, simply wrong. It is not allowable to love the Creation according to the purposes one has for it, any more than it is allowable to love one's neighbor in order to borrow his tools. The wild ass and the unicorn are said in the Book of Job (39:5–12) to be "free," precisely in the sense that they are not subject or serviceable to human purposes. The same point —though it is not the main point of that passage—is made in the Sermon on the Mount in reference to "the fowls of the air" and "the lilies of the field." Faced with this problem in Book VIII of *Paradise Lost*, Milton scrupulously observes the same reticence. Adam asks about "celestial Motions," and Raphael refuses to explain, making the ultimate mysteriousness of Creation a test of intellectual propriety and humility:

> . . . for the Heav'n's wide Circuit, let it speak
> The Maker's high magnificence, who built
> So spacious, and his Line stretcht out so far;
> That Man may know he dwells not in his own;
> An Edifice too large for him to fill,
> Lodg'd in a small partition, and the rest
> Ordain'd for uses to his Lord best known.
>
> *(lines 100–106)*

The Creator's love for the Creation is mysterious precisely because it does not conform to human purposes. The wild ass and the wild lilies are loved by God for their own sake and yet they are part of a pattern that we must love because it includes us. This is a pattern that humans can understand well enough to respect and preserve, though they cannot "control" it or hope to understand it completely. The mysterious and the practical, the Heavenly and the earthly, are thus joined. Charity is a theological virtue and is prompted, no doubt, by a theological emotion, but it is also a practical virtue because it must be practiced. The requirements of this complex charity cannot be fulfilled by smiling in abstract beneficence on our neighbors and on the scenery. It must come to acts, which must come from skills. Real charity calls for the study of agriculture, soil husbandry, engineering, architecture, mining, manufacturing, transportation, the making of monuments and pictures, songs and stories. It calls not just for skills but for the study and criticism of skills, because in all of them a choice must be made: they can be used either charitably or uncharitably.

How can you love your neighbor if you don't know how to build or mend a fence, how to keep your filth out of his water supply and your poison out of his air; or if you do not produce anything and so have nothing to offer, or do not take care of yourself and so become a burden? How can you be a neighbor without *applying* principle—without bringing virtue to a practical issue? How will you practice virtue without skill?

The ability to be good is not the ability to do nothing. It is not negative or passive. It is the ability to do something well —to do good work for good reasons. In order to be good you have to know how—and this knowing is vast, complex, humble and humbling; it is of the mind and of the hands, of neither alone.

The divine mandate to use the world justly and charitably, then, defines every person's moral predicament as that of a steward. But this predicament is hopeless and meaningless unless it produces an appropriate discipline: stewardship. And stewardship is hopeless and meaningless unless it involves long-term courage, perseverance, devotion, and skill. This skill is not to be confused with any accomplishment or grace of spirit or of intellect. It has to do with everyday proprieties in the practical use and care of created things—with "right livelihood."

If "the earth is the Lord's" and we are His stewards, then obviously some livelihoods are "right" and some are not. Is

there, for instance, any such thing as a Christian strip mine? A
Christian atomic bomb? A Christian nuclear power plant or radioactive waste dump? What might be the design of a Christian
transportation or sewer system? Does not Christianity imply
limitations on the scale of technology, architecture, and land
holding? Is it Christian to profit or otherwise benefit from violence? Is there not, in Christian ethics, an implied requirement
of practical separation from a destructive or wasteful economy?
Do not Christian values require the enactment of a distinction
between an organization and a community?

It is impossible to understand, much less to answer, such
questions except in reference to issues of practical skill, because they all have to do with distinctions between kinds of
action. These questions, moreover, are intransigently personal,
for they ask, ultimately, how each livelihood and each life will
be taken from the world, and what each will cost in terms
of the livelihoods and lives of others. Organizations and even
communities cannot hope to answer such questions until individuals have begun to answer them.

But here we must acknowledge one inadequacy of Judeo-
Christian tradition. At least in its most prominent and best
known examples, this tradition does not provide us with a precise enough understanding of the commonplace issues of livelihood. There are two reasons for this.

One is the "otherworldly philosophy" that, according to
Lovejoy, "has, in one form or another, been the dominant official philosophy of the larger part of civilized mankind through
most of its history. . . . The greater number of the subtler
speculative minds and of the great religious teachers have . . .
been engaged in weaning man's thought or his affections, or
both, from . . . Nature." The connection here is plain.

The second reason is that the Judeo-Christian tradition as we
have it in its art and literature, including the Bible, is so strongly
heroic. The poets and storytellers in this tradition have tended
to be interested in the extraordinary actions of "great men"—
actions unique in grandeur, such as may occur only once in the
history of the world. These extraordinary actions do indeed
bear a universal significance, but they cannot very well serve as
examples of ordinary behavior. Ordinary behavior belongs to
a different dramatic mode, a different understanding of action,

even a different understanding of virtue. The drama of heroism raises above all the issue of physical and moral courage: Does the hero have, in extreme circumstances, the courage to obey —to perform the task, the sacrifice, the resistance, the pilgrimage that he is called on to perform? The drama of ordinary or daily behavior also raises the issue of courage, but it raises at the same time the issue of skill; and, because ordinary behavior lasts so much longer than heroic action, it raises in a more complex and difficult way the issue of perseverance. It may, in some ways, be easier to be Samson than to be a good husband or wife day after day for fifty years.

These heroic works are meant to be (among other things) instructive and inspiring to ordinary people in ordinary life, and they are, grandly and deeply so. But there are two issues that they are prohibited by their nature from raising: the issue of life-long devotion and perseverance in unheroic tasks, and the issue of good workmanship or "right livelihood."

It can be argued, I believe, that until fairly recently there was simply no need for attention to such issues, for there existed yeoman or peasant or artisan classes: these were the people who did the work of feeding and clothing and housing, and who were responsible for the necessary skills, disciplines, and restraints. As long as those earth-keeping classes and their traditions were strong, there was at least the hope that the world would be well used. But probably the most revolutionary accomplishment of the industrial revolution was to destroy the traditional livelihoods and so break down the cultural lineage of those classes.

The industrial revolution has held in contempt not only the "obsolete skills" of those classes, but the concern for quality, for responsible workmanship and good work, that supported their skills. For the principle of good work it substituted a secularized version of the heroic tradition: the ambition to be a "pioneer" of science or technology, to make a "breakthrough" that will "save the world" from some "crisis" (which now is usually the result of some previous "breakthrough").

The best example we have of this kind of hero, I am afraid, is the fallen Satan of *Paradise Lost*—Milton undoubtedly having observed in his time the prototypes of industrial heroism. This is a hero who instigates and influences the actions of others,

but does not act himself. His heroism is of the mind only—
escaped as far as possible, not only from divine rule, from its
place in the order of creation or the Chain of Being, but also
from the influence of material creation:

> A mind not to be chang'd by Place or Time.
> The mind is its own place, and in itself
> Can make a Heav'n of Hell, a Hell of Heav'n.
>
> *(Book I, lines 253–255)*

This would-be heroism is guilty of two evils that are prerequi-
site to its very identity: *hubris* and abstraction. The industrial
hero supposes that "mine own *mind* hath saved me"—and
moreover that it may save the world. Implicit in this is the assumption that one's mind is one's own, and that it may choose
its own place in the order of things; one usurps divine authority, and thus, in classic style, becomes the author of results that
one can neither foresee nor control.

And because this mind is understood only as a cause, its
primary works are necessarily abstract. We should remind
ourselves that materialism in the sense of the love of material
things is not in itself an evil. As C. S. Lewis pointed out, God
too loves material things; He invented them. The Devil's work
is abstraction—not the love of material things, but the love
of their quantities—which, of course, is why "David's heart
smote him after that he had numbered the people" (II Samuel
24:10). It is not the lover of material things but the abstrac-
tionist who defends long-term damage for short-term gain,
or who calculates the "acceptability" of industrial damage to
ecological or human health, or who counts dead bodies on the
battlefield. The true lover of material things does not think in
this way, but is answerable instead to the paradox of the parable
of the lost sheep: that each is more precious than all.

But perhaps we cannot understand this secular heroic mind
until we understand its opposite: the mind obedient and in
place. And for that we can look again at Raphael's warning in
Book VIII of *Paradise Lost*:

> . . . apt the Mind or Fancy is to rove
> Uncheckt, and of her roving is no end;
> Till warn'd, or by experience taught, she learn

> That not to know at large of things remote
> From use, obscure and subtle, but to know
> That which before us lies in daily life,
> Is the prime Wisdom; what is more, is fume,
> Or emptiness, or fond impertinence,
> And renders us in things that most concern
> Unpractic'd, unprepar'd, and still to seek.
> Therefore from this high pitch let us descend
> A lower flight, and speak of things at hand
> Useful . . .
>
> *(lines 188–200)*

In its immediate sense this is a warning against thought that is theoretical or speculative (and therefore abstract), but in its broader sense it is a warning against disobedience—the eating of the forbidden fruit, an act of *hubris*, which Satan justifies by a compellingly reasonable theory and which Eve undertakes as a speculation.

A typical example of the conduct of industrial heroism is to be found in the present rush of experts to "solve the problem of world hunger"—which is rarely defined except as a "world problem" known, in industrial heroic jargon, as "the world food problematique." As is characteristic of industrial heroism, the professed intention here is entirely salutary: nobody should starve. The trouble is that "world hunger" is not a problem that can be solved by a "world solution." Except in a very limited sense, it is not an industrial problem, and industrial attempts to solve it—such as the "Green Revolution" and "Food for Peace"—have often had grotesque and destructive results. "The problem of world hunger" cannot be solved until it is understood and dealt with by local people as a multitude of local problems of ecology, agriculture, and culture.

The most necessary thing in agriculture, for instance, is not to invent new technologies or methods, not to achieve "breakthroughs," but to determine what tools and methods are appropriate to specific people, places, and needs, and to apply them correctly. Application (which the heroic approach ignores) is the crux, because no two farms or farmers are alike; no two fields are alike. Just the changing shape or topography of the land makes for differences of the most formidable kind.

Abstractions never cross these boundaries without either ceasing to be abstractions or doing damage. And prefabricated industrial methods and technologies *are* abstractions. The bigger and more expensive, the more heroic, they are, the harder they are to apply considerately and conservingly.

Application is the most important work, but also the most modest, complex, difficult, and long—and so it goes against the grain of industrial heroism. It destroys forever the notions that the world can be thought of (by humans) as a whole and that humans can "save" it as a whole—notions we can well do without, for they prevent us from understanding our problems and from growing up.

To use knowledge and tools in a particular place with good long-term results is not heroic. It is not a grand action visible for a long distance or a long time. It is a small action, but more complex and difficult, more skillful and responsible, more whole and enduring, than most grand actions. It comes of a willingness to devote oneself to work that perhaps only the eye of Heaven will see in its full intricacy and excellence. Perhaps the real work, like real prayer and real charity, must be done in secret.

The great study of stewardship, then, is "to know / That which before us lies in daily life" and to be practiced and prepared "in things that most concern." The angel is talking about good work, which is to talk about skill. In the loss of skill we lose stewardship; in losing stewardship we lose fellowship; we become outcasts from the great neighborhood of Creation. It is possible—as our experience in *this* good land shows—to exile ourselves from Creation, and to ally ourselves with the principle of destruction—which is, ultimately, the principle of nonentity. It is to be willing *in general* for beings to not-be. And once we have allied ourselves with that principle, we are foolish to think that we can control the results. The "regulation" of abominations is a modern governmental exercise that never succeeds. If we are willing to pollute the air—to harm the elegant creature known as the atmosphere—by that token we are willing to harm all creatures that breathe, ourselves and our children among them. There is no begging off or "trading off." You cannot affirm the power plant and condemn the smokestack, or affirm the smoke and condemn the cough.

That is not to suggest that we can live harmlessly, or strictly at our own expense; we depend upon other creatures and survive by their deaths. To live, we must daily break the body and shed the blood of Creation. When we do this knowingly, lovingly, skillfully, reverently, it is a sacrament. When we do it ignorantly, greedily, clumsily, destructively, it is a desecration. In such desecration we condemn ourselves to spiritual and moral loneliness, and others to want.

FROM
STANDING BY WORDS
(1983)

Standing by Words

"He said, and stood . . ."
Paradise Regained, IV, 561.

Two epidemic illnesses of our time—upon both of which virtual industries of cures have been founded—are the disintegration of communities and the disintegration of persons. That these two are related (that private loneliness, for instance, will necessarily accompany public confusion) is clear enough. And I take for granted that most people have explored in themselves and their surroundings some of the intricacies of the practical causes and effects; most of us, for example, have understood that the results are usually bad when people act in social or moral isolation, and also when, because of such isolation, they fail to act.

What seems not so well understood, because not so much examined, is the relation between these disintegrations and the disintegration of language. My impression is that we have seen, for perhaps a hundred and fifty years, a gradual increase in language that is either meaningless or destructive of meaning. And I believe that this increasing unreliability of language parallels the increasing disintegration, over the same period, of persons and communities.

My concern is for the *accountability* of language—hence, for the accountability of the users of language. To deal with this matter I will use a pair of economic concepts: *internal accounting*, which considers costs and benefits in reference only to the interest of the money-making enterprise itself; and *external accounting*, which considers the costs and benefits to the "larger community." By altering the application of these terms a little, any statement may be said to account well or poorly for what is going on inside the speaker, or outside him, or both.

It will be found, I believe, that the accounting will be poor —incomprehensible or unreliable—if it attempts to be purely internal or purely external. One of the primary obligations of language is to connect and balance the two kinds of accounting.

535

And so, in trying to understand the degeneracy of language, it is necessary to examine, not one kind of unaccountability, but two complementary kinds. There is language that is diminished by subjectivity, which ends in meaninglessness. But that kind of language rarely exists alone (or so I believe), but is accompanied, in a complex relationship of both cause and effect, by a language diminished by objectivity, or so-called objectivity (inordinate or irresponsible ambition), which ends in confusion.

My standpoint here is defined by the assumption that no statement is complete or comprehensible in itself, that in order for a statement to be complete and comprehensible three conditions are required:

1. It must designate its object precisely.

2. Its speaker must stand by it: must believe it, be accountable for it, be willing to act on it.

3. This relation of speaker, word, and object must be conventional; the community must know what it is.

These are still the common assumptions of private conversations. In our ordinary dealings with each other, we take for granted that we cannot understand what is said if we cannot assume the accountability of the speaker, the accuracy of his speech, and mutual agreement on the structures of language and the meanings of words. We assume, in short, that language is communal, and that its purpose is to tell the truth.

That these common assumptions are becoming increasingly uncommon, particularly in the discourse of specialists of various sorts, is readily evident to anyone looking for evidence. How far they have passed from favor among specialists of language, to use the handiest example, is probably implicit in the existence of such specialists; one could hardly become a language specialist (a "scientist" of language) so long as one adhered to the old assumptions.

But the influence of these specialists is, of course, not confined to the boundaries of their specializations. They write textbooks for people who are not specialists of language, but who are apt to become specialists of other kinds. The general purpose of at least some of these specialists, and its conformability to the purposes of specialists of other kinds, is readily suggested by a couple of recently published textbooks for freshman English.

One of these, *The Contemporary Writer*, by W. Ross Winterowd, contains a chapter on language, the main purpose of which is to convince the student of the illegitimate tyranny of any kind of prescriptive grammar and of the absurdity of judging language "on the basis of extra-linguistic considerations." This chapter proposes four rules that completely overturn all the old common assumptions:

1. "Languages apparently do not become better or worse in any sense. They simply change."

2. "Language is arbitrary."

3. "Rightness and wrongness are determined . . . by the purpose for which the language is being used, by the audience at which it is directed, and by the situation in which the use is taking place."

4. ". . . a grammar of a language is a description of that language, nothing more and nothing less."

And these rules have a pair of corollaries that Mr. Winterowd states plainly. One is that "you [the freshman student] have a more or less complete mastery of the English language. . . ." The other is that art—specifically, here, the literary art—is "the highest expression of the human need to play, of the desire to escape from the world of reality into the world of fantasy."

The second of these texts, *Rhetoric: Discovery and Change*, by Richard E. Young, Alton L. Becker, and Kenneth L. Pike, takes the standardless functionalism of Mr. Winterowd's understanding of language and applies it to the use of language. "The ethical dimension of the art of rhetoric," these authors say, is in "the attempt to reduce another's sense of threat in the effort to reach the goal of cooperation and mutual benefit. . . ." They distinguish between evaluative writing and descriptive writing, preferring the latter because evaluative writing tends to cause people "to become defensive," whereas "a description . . . does not make judgments. . . ." When, however, a writer "must make judgments, he can make them in a way that minimizes the reader's sense of threat." Among other things, "he can acknowledge the personal element in his judgment. . . . There is a subtle but important difference between saying 'I don't like it' and 'It's bad.'"

The authors equate evaluation—functionally, at least—with dogmatism: "The problem with dogmatism is that, like evaluation, it forces the reader to take sides." And finally they recommend a variety of writing which they call "provisional" because it "focuses on the process of enquiry itself and acknowledges the tentative nature of conclusions. . . . Provisional writing implies that more than one reasonable conclusion is possible."

The first of these books attempts to make the study of language an "objective" science by eliminating from that study all extra-linguistic values and the issue of quality. Mr. Winterowd asserts that "the language grows according to its own dynamics." He does not say, apparently because he does not believe, that its dynamics include the influence of the best practice. There is no "best." Anyone who speaks English is a "master" of the language. And the writers once acknowledged as masters of English are removed from "the world of reality" to the "world of fantasy," where they lose their force within the dynamics of the growth of language. Their works are reduced to the feckless status of "experiences": "we are much more interested in the imaginative statement of the message . . . than we are in the message. . . ." Mr. Winterowd's linguistic "science" thus views language as an organism that has evolved without reference to habitat. Its growth has been "arbitrary," without any principle of selectivity.

Against Mr. Winterowd's definition of literature, it will be instructive to place a definition by Gary Snyder, who says of poetry that it is "a tool, a net or trap to catch and present; a sharp edge; a medicine, or the little awl that unties knots." It will be quickly observed that this sentence enormously complicates Mr. Winterowd's simplistic statement-message dichotomy. What Mr. Winterowd means by "message" is an "idea" written in the dullest possible prose. His book is glib, and glibness is an inescapable doom of language without standards. One of the great practical uses of literary disciplines, of course, is to resist glibness—to slow language down and make it thoughtful. This accounts for the influence of verse, in its formal aspect, within the dynamics of the growth of language: verse checks the merely impulsive flow of speech, subjects it to another pulse, to measure, to extra-linguistic considerations; by inducing the hesitations of difficulty, it admits into language the influence of the Muse and of musing.

The three authors of the second book attempt to found an ethics of rhetoric on the idea expressed in one of Mr. Winterowd's rules: "Rightness and wrongness are determined" by purpose, audience, and situation. This idea apparently derives from, though it significantly reduces, the ancient artistic concern for propriety or decorum. A part of this concern was indeed the fittingness of the work to its occasion: that is, one would not write an elegy in the meter of a drinking song—though that is putting it too plainly, for the sense of occasion exercised an influence both broad and subtle on form, diction, syntax, small points of grammar and prosody—everything. But occasion, as I understand it, was invariably second in importance to the subject. It is only the modern specialist who departs from this. The specialist poet, for instance, degrades the subject to "subject matter" or raw material, so that the subject exists for the poem's sake, is *subjected* to the poem, in the same way as industrial specialists see trees or ore-bearing rocks as raw material subjected to their manufactured end-products. Quantity thus begins to dominate the work of the specialist poet at its source. Like an industrialist, he is interested in the subjects of the world for the sake of what they can be made to produce. He mines his experience for subject matter. The first aim of the propriety of the old poets, by contrast, was to make the language true to its subject—to see that it told the truth. That is why they invoked the Muse. The truth the poet chose as his subject was perceived as *superior* to his powers—and, by clear implication, to his occasion and purpose. But the aim of truth-telling is not stated in either of these textbooks. The second, in fact, makes an "ethical" aim of avoiding the issue, for, as the authors say, coining a formidable truth: "Truth has become increasingly elusive and men are driven to embrace conflicting ideologies."

This sort of talk about language, it seems to me, is fundamentally impractical. It does not propose as an outcome any fidelity between words and speakers or words and things or words and acts. It leads instead to muteness and paralysis. So

far as I can tell, it is unlikely that one could speak at all, in even
the most casual conversation, without some informing sense of
what would be best to say—that is, without some sort of *standard*. And I do not believe that it is possible to act on the basis of a "tentative" or "provisional" conclusion. We may know
that we are forming a conclusion on the basis of provisional or
insufficient knowledge—that is a part of what we understand
as the tragedy of our condition. But we must act, nevertheless, on the basis of *final* conclusions, because we know that
actions, occurring in time, are irrevocable. That is another part
of our tragedy. People who make a conventional agreement
that all conclusions are provisional—a convention almost invariably implied by academic uses of the word "objectivity"
—characteristically talk but do not act. Or they do not act deliberately, though time and materiality carry them into action
of a sort, willy-nilly.

And there are times, according to the only reliable ethics we
have, when one is required to tell the truth, whatever the urgings of purpose, audience, and situation. Ethics requires this
because, in the terms of the practical realities of our lives, the
truth is safer than falsehood. To ignore this is simply to put language at the service of purpose—*any* purpose. It is, in terms of
the most urgent realities of our own time, to abet a dangerous
confusion between public responsibility and public relations.
Remote as these theories of language are from practical contexts, they are nevertheless serviceable to expedient practices.

In affirming that there is a necessary and indispensable connection between language and truth, and therefore between
language and deeds, I have certain precedents in mind. I begin with the Christian idea of the Incarnate Word, the Word
entering the world as flesh, and inevitably therefore as action
—which leads logically enough to the insistence in the Epistle
of James that faith without works is dead:

> For if any be a hearer of the word, and not a doer, he is like
>   unto a man beholding his natural face in a glass:
> For he beholdeth himself, and goeth his way, and
>   straightway forgetteth what manner of man he was.

I also have in mind the Confucian insistence on sincerity (precision) and on fidelity between speaker and word as essentials of political health: "Honesty is the treasure of states." I have returned to Ezra Pound's observation that Confucius "collected *The Odes* to keep his followers from abstract discussion. That is, *The Odes* give particular instances. They do not lead to exaggerations of dogma."

And I have remembered Thoreau's sentence: "Where shall we look for standard English, but to the words of a standard man?"

The idea of standing by one's word, of words precisely designating things, of deeds faithful to words, is probably native to our understanding. Indeed, it seems doubtful that we could understand anything without that idea.

But in order to discover what makes language that can be understood, stood by, and acted on, it is necessary to return to my borrowed concepts of internal and external accounting. And it will be useful to add two further precedents.

In *Mind and Nature*, Gregory Bateson writes that "'things' . . . can only enter the world of communication and meaning by their names, their qualities and their attributes (i.e., by reports of their internal and external relations and interactions)."

And Gary Snyder, in a remarkably practical or practicable or practice-able definition of where he takes his stand, makes the poet responsible for "possibilities opening both inward and outward."

There can be little doubt, I think, that any accounting that is *purely* internal will be incomprehensible. If the connection between inward and outward is broken—if, for instance, the experience of a single human does not resonate within the common experience of humanity—then language fails. In *The Family Reunion*, Harry says: "I talk in general terms / Because the particular has no language." But he speaks, too, in despair, having no hope that his general terms can communicate the particular burden of his experience. We readily identify this loneliness of personal experience as "modern." Many poems of our century have this loneliness, this failure of speech, as a subject; many more exhibit it as a symptom.

But it begins at least as far back as Shelley, in such lines as these from "Stanzas Written in Dejection, Near Naples":

> Alas! I have nor hope nor health,
>   Nor peace within nor calm around,
> Nor that content surpassing wealth
>
> The sage in meditation found,
>   And walked with inward glory crowned—
> Nor fame, nor power, nor love, nor leisure.
> . . . .
> I could lie down like a tired child,
>   And weep away the life of care
>   Which I have borne and must bear,
> Till death like sleep might steal on me . . .

This too is an example of particular experience concealing itself in "general terms"—though here the failure, if it was suspected, is not acknowledged. The generality of the language does not objectify it, but seals it in its subjectivity. In reading this—as, I think, in reading a great many poems of our own time—we sooner or later realize that we are reading a "complaint" that we do not credit or understand. If we fail to realize this, it is because we have departed from the text of the poem, summoning particularities of our own experience in support of Shelley's general assertions. The fact remains that Shelley's poem doesn't tell us what he is complaining about; his lines fail to "create the object [here, the experience] which they contemplate." The poem has forsaken its story. This failure is implicitly conceded by the editors of *The Norton Anthology of English Literature*, who felt it necessary to provide the following footnote:

> Shelley's first wife, Harriet, had drowned herself; Clara, his baby daughter by Mary Shelley, had just died; and Shelley himself was plagued by ill health, pain, financial worries, and the sense that he had failed as a poet.

But I think the poem itself calls attention to the failure by its easy descent into self-pity, finally asserting that "I am one / Whom men love not. . . ." Language that becomes too subjective lacks currency, to use another economic metaphor; it will not pass. Self-pity, like self-praise, will not pass. The powers of language are used illegitimately, to impose, rather than to elicit, the desired response.

Shelley is not writing gibberish here. It is possible to imagine that someone who does not dislike this poem may see in it a certain beauty. But it is the sickly beauty of generalized emotionalism. For once precision is abandoned as a linguistic or literary virtue, vague generalization is one of the two remaining possibilities, gibberish being the second. And without precise accounting, leading to responsible action, there is no escape from the Shelleyan rhythm of exaltation and despair—ideal passions culminating in real disasters.

It is true, in a sense, that "the particular has no language"— that at least in public writing, and in speech passing between strangers, there may only be degrees of generalization. But there are, I think, two kinds of precision that are particular and particularizing. There is, first, the precision in the speech of people who share the same knowledge of place and history and work. This is the precision of direct reference or designation. It sounds like this: "How about letting me borrow your tall jack?" Or: "The old hollow beech blew down last night." Or, beginning a story, "Do you remember that time . . . ?" I would call this community speech. Its words have the power of pointing to things visible either to eyesight or to memory. Where it is not much corrupted by public or media speech, this community speech is wonderfully vital. Because it so often works designatively it *has* to be precise, and its precisions are formed by persistent testing against its objects.

This community speech, unconsciously taught and learned, in which words live in the presence of their objects, is the very root and foundation of language. It is the source, the unconscious inheritance that is carried, both with and without schooling, into consciousness—but never *all* the way, and so it remains rich, mysterious, and enlivening. Cut off from this source, language becomes a paltry work of conscious purpose, at the service and the mercy of expedient aims. Theories such as those underlying the two textbooks I have discussed seem to be attempts to detach language from its source in communal experience, by making it arbitrary in origin and provisional in use. And this may be a "realistic" way of "accepting" the degradation of community life. The task, I think, is hopeless, and it shows the extremes of futility that academic specialization can lead to. If one wishes to promote the life of language, one must promote the life of the community—a discipline many times more trying, difficult, and long than that of linguistics, but having at least the virtue of hopefulness. It escapes the despair always implicit in specializations: the cultivation of discrete parts without respect or responsibility for the whole.

The other sort of precision—the sort available to public speech or writing as well as to community speech—is a precision that comes of tension either between a statement and a prepared context or, within a single statement, between more or less conflicting feelings or ideas. Shelley's complaint is incomprehensible not just because it is set in "general terms," but because the generalities are too simple. One doesn't credit the emotion of the poem because it is too purely mournful. We are—conventionally, maybe, but also properly—unprepared to believe without overpowering evidence that things are *all* bad. Self-pity may deal in such absolutes of feeling, but we don't deal with other people in the manner of self-pity.

Another general complaint about mortality is given in Act V, Scene 2, of *King Lear*, when Edgar says to Gloucester: "Men must endure / Their going hence, even as their coming hither." Out of context this statement is even more general than Shelley's. It is, unlike Shelley's, deeply moving because it is tensely poised within a narrative context that makes it precise. We know exactly, for instance, what is meant by that "must": a responsible performance is required until death. But the complaint is followed immediately by a statement of another kind, forcing the speech of the play back into its action: "Ripeness is all. Come on."

Almost the same thing is done in a single line of Robert Herrick, in the tension between the complaint of mortality and the jaunty metric:

Out of the world he must, who once comes in . . .

Here the very statement of inevitable death sings its acceptability. How would you divide, there, the "statement of the message" from "the message"?

And see how the tension between contradictory thoughts particularizes the feeling in these three lines by John Dryden:

> Old as I am, for Ladies Love unfit,
> The Pow'r of Beauty I remember yet,
> Which once inflam'd my Soul, and still inspires my Wit.

These last three examples immediately receive our belief and sympathy because they satisfy our sense of the complexity, the cross-graining, of real experience. In them, an inward possibility is made to open outward. Internal accounting has made itself externally accountable.

Shelley's poem, on the other hand, exemplifies the solitude of inward experience that continues with us, both in and out of poetry. I don't pretend to understand all the causes and effects of this, but I will offer the opinion that one of its chief causes is a simplistic idea of "freedom," which also continues with us, and is also to be found in Shelley. At the end of Act III of *Prometheus Unbound*, we are given this vision of a liberated humanity:

> The loathsome mask has fallen, the man remains
> Sceptreless, free, uncircumscribed, but man
> Equal, unclassed, tribeless, and nationless,
> Exempt from awe, worship, degree, the king
> Over himself . . .

This passage, like the one from the "Stanzas Written in Dejection," is vague enough, and for the same reason; as the first hastened to emotional absolutes, this hastens to an absolute idea. It is less a vision of a free man than a vision of a definition of a free man. But Shelley apparently did not notice that this headlong scramble of adjectives, though it may produce one of the possible definitions of a free man, also defines a lonely one, unattached and displaced. This free man is described as loving, and love is an emotion highly esteemed by Shelley. But it is, like his misery, a "free" emotion, detached and absolute. In this same passage of *Prometheus Unbound*, he calls it "the nepenthe, love"—love forgetful, or inducing forgetfulness, of grief or pain.

Shelley thought himself, particularly in *Prometheus Unbound*, a follower of Milton—an assumption based on a misunderstanding of *Paradise Lost*. And so it is instructive in two ways to set beside Shelley's definition of freedom this one by Milton:

> To be free is precisely the same thing as to be pious, wise, just, and temperate, careful of one's own, abstinent from what is another's, and thence, in fine, magnanimous and brave.

And Milton's definition, like the lines previously quoted from Shakespeare, Herrick, and Dryden, derives its precision from tension: he defines freedom in terms of responsibilities. And it is only this tension that can suggest the possibility of *living* (for any length of time) in freedom—just as it is the tension between love and pain that suggests the possibility of carrying love into acts. Shelley's freedom, defined in terms of freedom, gives us only this from a "Chorus of Spirits" in Act IV, Scene 1, of *Prometheus Unbound*:

> Our task is done,
> We are free to dive, or soar, or run;
> Beyond and around,
> Or within the bound
> Which clips the world with darkness round.
>
> We'll pass the eyes
> Of the starry skies
> Into the hoar deep to colonize . . .

Which, as we will see, has more in common with the technological romanticism of Buckminster Fuller than with anything in Milton.

In supposed opposition to this remote subjectivity of internal accounting, our age has developed a stance or state of mind which it calls "objective," and which produces a kind of accounting supposed to be external—that is, free from personal biases and considerations. This objective mentality, within the safe confines of its various specialized disciplines, operates with great precision and confidence. It follows tested and trusted procedures and uses a professional language that an outsider must assume to be a very exact code. When this language is used by its accustomed speakers on their accustomed ground, even when one does not understand it, it clearly voices the implication of a marvelously precise control over objective reality. It is only when it is overheard in confrontation with failure that this implication falters, and the adequacy of this sort of language comes into doubt.

The transcribed conversations of the members of the Nuclear Regulatory Commission during the crisis of Three Mile Island provide a valuable exhibit of the limitations of one of these objective languages. At one point, for example, the commissioners received a call from Roger Mattson, Nuclear Reactor Regulation chief of systems safety. He said, among other things, the following:

> That bubble will be 5,000 cubic feet. The available volume in the upper head and the candy canes, that's the hot legs, is on the order of 2,000 cubic feet total. I get 3,000 excess cubic feet of noncondensibles. I've got a horse race. . . . We have got every systems engineer we can find . . . thinking the problem: how the hell do we get the noncondensibles out of there, do we win the horse race or do we lose the horse race.

At another time the commissioners were working to "engineer a press release," of which "The focus . . . has to be reassuring. . . ." Commissioner Ahearne apparently felt that it was a bit *too* reassuring, and he would have liked to *suggest* the possibility of a bad outcome, apparently a meltdown. He said:

> I think it would be technically a lot better if you said —something about there's a possibility—it's small, but it could lead to serious problems.

And, a few sentences later, Commissioner Kennedy told him:

> Well I understand what you're saying. . . . You could put a little sentence in right there . . . to say, were this —in the unlikely event that this occurred, increased temperatures would result and possible further fuel damage.

What is remarkable, and frightening, about this language is its inability to admit what it is talking about. Because these specialists have routinely eliminated themselves, as such and

as representative human beings, from consideration, according
to the prescribed "objectivity" of their discipline, they cannot
bring themselves to acknowledge to each other, much less to
the public, that their problem involves an extreme danger to a
lot of people. Their subject, as bearers of a public trust, is this
danger, and it can be nothing else. It is a technical problem
least of all. And yet when their language approaches this subject, it either diminishes it, or dissolves into confusions of both
syntax and purpose. Mr. Mattson speaks clearly and coherently
enough so long as numbers and the jargon of "candy canes"
and "hot legs" are adequate to his purpose. But as soon as he
tries to communicate his sense of the human urgency of the
problem, his language collapses into a kind of rant around the
metaphor of "a horse race," a metaphor that works, not to reveal, but to obscure his meaning. And the two commissioners,
struggling with their obligation to inform the public of the
possibility of a disaster, find themselves virtually languageless
—without the necessary words and with only the shambles of
a syntax. They cannot say what they are talking about. And
so their obligation to *inform* becomes a tongue-tied—and
therefore surely futile—effort to *reassure*. Public responsibility
becomes public relations, apparently, for want of a language
adequately responsive to its subject.

So inept is the speech of these commissioners that we must
deliberately remind ourselves that they are not stupid and are
probably not amoral. They are highly trained, intelligent, worried men, whose understanding of language is by now to a
considerable extent a public one. They are atomic scientists
whose criteria of language are identical to those of at least
some linguistic scientists. They determine the correctness of
their statement to the press exactly according to Mr. Winterowd's rule: by their purpose, audience, and situation. Their
language is governed by the ethical aim prescribed by the three
authors of *Rhetoric: Discovery and Change*: they wish above all
to speak in such a way as to "reduce another's sense of threat."
But the result was not "cooperation and mutual benefit"; it
was incoherence and dishonesty, leading to public suspicion,
distrust, and fear. It is beneficial, surely, to "reduce another's
sense of threat" only if there is no threat.

The commissioners speak a language that is diminished by

inordinate ambition: the taking of more power than can be responsibly or beneficently held. It is perhaps a law of human nature that such ambition always produces a confusion of tongues:

> And they said, Go to, let us build us a city and a tower, whose top may reach unto heaven; and let us make us a name, lest we be scattered
> . . . .
> And the Lord said . . . now nothing will be restrained from them, which they have imagined to do.
> Go to, let us go down, and there confound their language, that they may not understand one another's speech.

The professed aim is to bring people together—usually for the implicit, though unstated, purpose of subjecting them to some public power or project. Why else would rulers seek to "unify" people? The idea is to cause them to speak the same language—meaning either that they will agree with the government or be quiet, as in communist and fascist states, or that they will politely ignore their disagreements or disagree "provisionally," as in American universities. But the result—though power may survive for a while in spite of it—is confusion and dispersal. Real language, real discourse are destroyed. People lose understanding of each other, are divided and scattered. Speech of whatever kind begins to resemble the speech of drunkenness or madness.

What this dialogue of the Nuclear Regulatory Commissioners causes one to suspect—and I believe the suspicion is confirmed by every other such exhibit I have seen—is that there is simply no such thing as an accounting that is *purely* external. The notion that external accounting can be accomplished by "objectivity" is an illusion. Apparently the only way to free the accounting of what is internal to people, or subjective, is to make it internal to (that is subject to) some other entity or structure just as limiting, or more so—as the commissioners attempted to deal with a possible public catastrophe in terms either of nuclear technology or of public relations. The only thing really externalized by such accounting is a bad result that one does not wish to pay for.

And so external accounting, alone, is only another form of internal accounting. The only difference is that this "objective" accounting does pretty effectively rule out personal considerations *of a certain kind*. (It does *not* rule out the personal desire for wealth, power, or for intellectual certainty.) Otherwise, the talk of the commissioners and the lines from "Stanzas Written in Dejection" are equally and similarly incomprehensible. The languages of both are obviously troubled, we recognize the words, and learn something about the occasions, but we cannot learn from the language itself exactly what the trouble is. The commissioners' language cannot define the problem of their public responsibility, and Shelley's does not develop what I suppose should be called the narrative context of his emotion, which therefore remains incommunicable.

Moreover, these two sorts of accounting, so long as they remain discrete, both work to keep the problem abstract, all in the mind. They are both, in different ways, internal to the mind. The real occasions of the problems are not admitted into consideration. In Shelley's poem, this may be caused by a despairing acceptance of loneliness. In the Nuclear Regulatory Commission deliberations it is caused, I think, by fear; the commissioners take refuge in the impersonality of technological procedures. They cannot bear to acknowledge considerations and feelings that might break the insulating spell of their "objective" dispassion.

Or, to put it another way, their language and their way of thought make it possible for them to think of the crisis only as a technical event or problem. Even a meltdown is fairly understandable and predictable within the terms of their expertise. What is unthinkable is the evacuation of a massively populated region. It is the disorder, confusion, and uncertainty of that exodus that they cannot face. In dealing with the unstudied failure mode, the commissioners' minds do not have to leave their meeting room. It is an *internal* problem. The other, the human, possibility, if they were really to deal with it, would send them shouting into the streets. Even worse, perhaps, from the point of view of their discipline, it would force them to face the absurdity of the idea of "emergency planning"—the idea, in other words, of a controlled catastrophe. They would have to admit, against all the claims of professional standing and "job security," that the only way to control the danger of a nuclear power plant is not to build it. That is to say, if they had a language strong and fine enough to consider *all* the considerations, it would tend to force them out of the confines of "objective" thought and into action, out of solitude into community.

It is the *purity* of objective thought that finally seduces and destroys it. The same thing happens, it seems to me, to the subjective mind. For certain emotions, especially the extremely subjective ones of self-pity and self-love, isolation holds a strong enticement: it offers to keep them pure and neat, aloof from the disorderliness and the mundane obligations of the human common ground.

The only way, so far as I can see, to achieve an accounting that is verifiably and reliably external is to admit the internal, the personal, as an appropriate, necessary consideration. If the Nuclear Regulatory Commissioners, for example, had spoken a good common English, instead of the languages of their specialization and of public relations, then they might have spoken of their personal anxiety and bewilderment, and so brought into consideration what they had in common with the people whose health and lives they were responsible for. They might, in short, have sympathized openly with those people—and so have understood the probably unbearable burden of their public trusteeship.

To be bound within the confines of either the internal or the external way of accounting is to be diseased. To hold the two in balance is to validate both kinds, and to have health. I am not using these terms "disease" and "health" according to any clinical definitions, but am speaking simply from my own observation that when my awareness of how I feel overpowers my awareness of where I am and who is there with me, I am sick, diseased. This can be appropriately extended to say that if what I think obscures my sense of whereabouts and company, I am diseased. And the converse is also true: I am diseased if I become so aware of my surroundings that my own inward life is obscured, as if I should so fix upon the value of some mineral in the ground as to forget that the world is God's work and my home.

But still another example is necessary, and other terms.

*

In an article entitled "The Evolution and Future of American Animal Agriculture," G. W. Salisbury and R. G. Hart consider the transformation of American agriculture "from an art form into a science." The difference, they say, is that the art of agriculture is concerned "only" with the "how . . . of farming," whereas the science is interested in the "whys."

As an example—or, as they say, a "reference index"—of this change, the authors use the modern history of milk production: the effort, from their point of view entirely successful, "to change the dairy cow from the family companion animal she became after domestication and through all of man's subsequent history into an appropriate manufacturing unit of the twentieth century for the efficient transformation of unprocessed feed into food for man."

The authors produce "two observations" about this change, and these constitute their entire justification of it:

> First, the total cow population was reduced in the period 1944 through 1975 by 67 percent, but second, the yield per cow during the same period increased by 60 percent. In practical terms, the research that yielded such dramatic gains produced a savings for the American public as a whole of approximately 50 billion pounds of total digestible nitrogen per year in the production of a relatively constant level of milk.

The authors proceed to work this out in dollar values, and to say that the quantity of saved dollars finally "gets to the point that people simply do not believe it." And later they say that, in making this change, "The major disciplines were genetics, reproduction, and nutrition."

This is obviously a prime example of internal accounting in the economic sense. The external account is not fully renderable; the context of the accounting is vast, some quantities are not known, and some of the costs are not quantifiable. However, there can be no question that the externalized costs are large. The net gain is not, as these authors imply, identical with the gross. And the industrialization of milk production is a part of a much larger enterprise that may finally produce a highly visible, if not entirely computable, net loss.

At least two further observations are necessary:

1. The period, 1944–1975, also saw a drastic decrease in the number of dairies, by reason of the very change cited by Salisbury and Hart. The smaller—invariably the smaller—dairies were forced out because of the comparative "inefficiency" of their "manufacturing units." Their failure was part of a major population shift, which seriously disrupted the life both of the country communities and of the cities, broke down traditional community forms, and so on.

2. The industrialization of agriculture, of which the industrialization of milk production is a part, has caused serious problems that even agricultural specialists are beginning to recognize: soil erosion, soil compaction, chemical poisoning and pollution, energy shortages, several kinds of money troubles, obliteration of plant and animal species, disruption of soil biology.

The human, the agricultural, and the ecological costs are all obviously great. Some of them have begun to force their way into the accounts and are straining the economy. Others are, and are likely to remain, external to all ledgers.

The passages I have quoted from Professors Salisbury and Hart provide a very neat demonstration of the shift from a balanced internal-external accounting (the dairy cow as "family companion animal") to a so-called "objective" accounting (the dairy cow as "appropriate manufacturing unit of the twentieth century"), which is, in fact, internal to an extremely limited definition of agricultural progress.

The discarded language, oddly phrased though it is, comes close to a kind of accountability: the internal (family) and the external (cow) are joined by a moral connection (companionship). A proof of its accountability is that this statement can be the basis of moral behavior: "Be good to the cow, for she is our companion."

The preferred phrase—"appropriate manufacturing unit of the twentieth century"—has nothing of this accountability. One can say, of course: "Be good to the cow, for she is productive (or expensive)." But that could be said of a machine; it takes no account of the cow as a living, much less a fellow, creature. But the phrase is equally unaccountable as language. "Appropriate" to what? Though the authors write

"appropriate . . . of the twentieth century," they may mean
"appropriate . . . *to* the twentieth century." But are there
no families and no needs for companionship with animals in
the twentieth century? Or perhaps they mean "appropriate
. . . for the efficient transformation of unprocessed feed into
food for man." But the problem remains. Who is this "man"?
Someone, perhaps, who needs no companionship with family
or animals? We are constrained to suppose so, for "objectivity"
has apparently eliminated "family" and "companion" as terms
subject to personal bias—perhaps as "merely sentimental."
By the terms of this "objective" accounting, then, "man" is
a creature who needs to eat, and who is for some unspecified
reason more important than a cow. But for a reader who considers himself a "man" by any broader definition, this language
is virtually meaningless. Because the terms of personal bias
(that is, the terms of *value*) have been eliminated, the terms of
judgment ("appropriate" and "efficient") mean nothing. The
authors' conditions would be just as well satisfied if the man
produced the milk and the cow bought it, or if a machine produced it and a machine bought it.

Sense, and the possibility of sense, break down here because
too much that clearly belongs in has been left out. Like the
Nuclear Regulatory Commissioners, Salisbury and Hart have
eliminated themselves as representative human beings, and
they go on to eliminate the cow as a representative animal—
all "interests" are thus removed from the computation. They
scrupulously pluck out the representative or symbolic terms
in order to achieve a pristinely "objective" accounting of the
performance of a "unit." And so we are astonished to discover,
at the end of this process, that they have complacently allowed
the dollar to stand as representative of *all* value. What announced itself as a statement about animal agriculture has become, by way of several obscure changes of subject, a crudely
simplified statement about industrial economics. This is not, in
any respectable sense, language, or thought, or even computation. Like the textbooks I have discussed, and like the dialogue
of the Nuclear Regulatory Commission, it is a pretentious and
dangerous deception, forgiveable only insofar as it may involve
self-deception.

*

If we are to begin to make a reliable account of it, this recent history of milk production must be seen as occurring within a system of nested systems: the individual human within the family within the community within agriculture within nature:

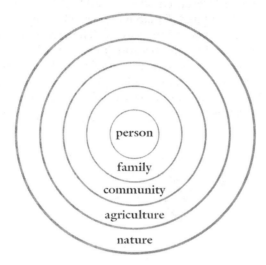

So long as the smaller systems are enclosed within the larger, and so long as all are connected by complex patterns of inter-dependency, as we know they are, then whatever affects one system will affect the others.

It seems that this system of systems is safe so long as each system is controlled by the next larger one. If at any point the hierarchy is reversed, and the smaller begins to control the larger, then the destruction of the entire system of systems begins. This system of systems is perhaps an updated, ecological version of the Great Chain of Being. That is, it may bring us back to a hierarchical structure not too different from the one that underlies *Paradise Lost*—a theory of the form of Creation which is at the same time a moral form, and which is violated by the "disobedience" or *hubris* of attempting to rise and take power above one's proper place.

But the sketch I have made of the system of systems is much too crude, for the connections between systems, insofar as this is a human structure, are not "given" or unconscious or automatic, but involve disciplines. Persons are joined to families, families to communities, etc., by disciplines that must be deliberately made, remembered, taught, learned, and practiced.

The system of systems begins to disintegrate when the hierarchy is reversed because that begins the disintegration of the connecting disciplines. Disciplines, typically, degenerate into professions, professions into careers. The accounting of Salisbury and Hart is defective because it upsets the hierarchies and so, perhaps unwittingly, fails to consider all the necessary considerations. Their "reference index" does occur within a system of systems—but a drastically abbreviated one, which involves a serious distortion:

Two things are wrong with this. First, too much has been left out; the claims of family, community, and nature are all ignored. Second, the outer circle is too much within the interest of the inner. The dairyman is not *necessarily* under the control of simple greed—but this structure supplies no hint of a reason why he should not be.

The system of systems, as I first described it, involves three different kinds of interests:

1. The ontogenetic. This is self-interest and is at the center.
2. The phylogenetic. This is the interest that we would call

"humanistic." It reaches through family and community and into agriculture. But it does not reach far enough into agriculture because, by its own terms, it cannot.

3. The ecogenetic. This is the interest of the whole "household" in which life is lived. (I don't know whether I invented this term or not. If I did, I apologize.)

These terms give us another way to characterize the flaw in the accounting of Salisbury and Hart. Their abbreviated system of systems fails either to assemble enough facts, to account fully for the meaning of the facts, or to provide any standard of judgment, because the ontogenetic interest is both internal and external to it.

The system of systems, as I first sketched it, has this vulnerability: that the higher interests can be controlled or exploited by the lower interests simply by leaving things out—a procedure just as available to ignorance as to the highest cunning of "applied science." And given even the most generous motives, ignorance is always going to be involved.

There is no reliable standard for behavior anywhere within the system of systems except truth. Lesser standards produce destruction—as, for example, the standards of public relations make gibberish of language.

The trouble, obviously, is that we do not know much of the truth. In particular, we know comparatively little ecogenetic truth.

And so yet another term has to be introduced. The system of systems has to be controlled from above and outside. There has to be a religious interest of some kind above the ecogenetic. It will be sufficient to my purpose to say simply that the system of systems is enclosed within mystery, in which some truth can be known, but never all truth.

Neither the known truth nor the mystery is internal to any system. And here, however paradoxical it may seem, we begin to see a possibility of reliable accounting and of responsible behavior. The appropriateness of words or deeds can be determined only in reference to the whole "household" in which they occur. But this whole, as such, cannot enter into the accounting. (If it could, then the only necessary language would be mathematics, and the only necessary discipline would be military.) It can only come in as mystery: a factor of X which stands not for the unknown but the unknowable. This is an X that cannot be solved—which may be thought a disadvantage by some; its advantage is that, once it has been let into the account, it cannot easily be ignored. You cannot leave anything out of mystery, because by definition everything is always in it.

The practical use of religion, then, is to keep the accounting in as large a context as possible—to see, in fact, that the account is never "closed." Religion forces the accountant to reckon with mystery—the unsolvable X that keeps the debit and credit or cost and benefit columns open so that no "profit" can ever be safely declared. It forces the accounting outside of every enclosure that it might be internal to. Practically, this X means that all "answers" must be worked out within a limit of humility and restraint, so that the initiative to act would always imply a knowing acceptance of accountability for the results. The establishment and maintenance of this limit seems to me the ultimate empirical problem—the real "frontier" of science, or at least of the definition of the possibility of a *moral* science. It would place science under the rule of the old concern for propriety, correct proportion, proper scale—from which, in modern times, even the arts have been "liberated." That is, it would return to all work, artistic or scientific, the possibility of an external standard of quality. The quality of work or of a made thing would be determined by how conservingly it fitted into the system of systems. Judgment could then begin to articulate what is already obvious: that some work preserves the household of life, and some work destroys it. And thus a real liberation could take place: life and work could go free of those "professional standards" (and professional languages) that are invariably destructive of quality, because they always work as sheep's clothing for various kinds of ontogenetic motives. It is because of these professional standards that the industries and governments, while *talking* of the "betterment of the human condition," can *act* to enrich and empower themselves.

The connections within the system of systems are *practical* connections. The practicality consists in the realization that—despite the blandishments of the various short-circuited "professional" languages—you cannot speak or act in your own best interest without espousing and serving a higher interest.

It is not knowledge that enforces this realization, but the humbling awareness of the insufficiency of knowledge, of mystery.

Applying, then, the standard of ecogenetic health to the work of Salisbury and Hart, we get a third way to describe its failure: it makes a principle of replacing the complex concern for quality ("how") with the drastically simplifying concern for quantity. Thus motive is entirely "liberated" from method: any way is good so long as it increases per unit production. But everything except production is diminished in the process, and Salisbury and Hart do not have a way of accounting for, hence no ability even to recognize, these diminishments. All that has been diminished could have been protected by a lively interest in the question of "how," which Salisbury and Hart—like the interests they are accounting for—rule out at the beginning. They were nevertheless working under what I take to be a rule: When you subtract quality from quantity, the gross result is not a net gain.

And so a reliable account is personal at the beginning and religious at the end. This does not mean that a reliable account includes the whole system of systems, for no account can do that. It does mean that the account is made in precise reference to the system of systems—which is another way of saying that it is made in respect for it. Without this respect for the larger structures, the accounting shrinks into the confines of some smaller structure and becomes specialized, partial, and destructive.

It is this sort of external accounting that deals with connections and thus inevitably raises the issue of quality. Which, I take it, is always the same as the issue of propriety: how appropriate is the tool to the work, the work to the need, the need to other needs and the needs of others, and to the health of the household or community of all creatures?

And this kind of accounting gives us the great structures of poetry—as in Homer, Dante, and Milton. It is these great structures, I think, that carry us into the sense of being, in Gary Snyder's phrase, "at one with each other." They teach us to imagine the life that is divided from us by difference or enmity: as Homer imagined the "enemy" hero, Hector; as Dante, on his pilgrimage to Heaven, imagined the damned; as

Milton, in his awed study of the meaning of obedience, epitomized sympathetically in his Satan the disobedient personality. And as, now, ecological insight proposes again a poetry with the power to imagine the lives of animals and plants and streams and stones. And this imagining is eminently proprietous, fitting to the claims and privileges of the great household.

Unlike the problems of quantity, the problems of propriety are never "solved," but are ceaselessly challenging and interesting. This is the antidote to the romance of big technological solutions. Life would be interesting—there would be exciting work to do—even if there were no nuclear power plants or "agri-industries" or space adventures. The elaborations of elegance are at least as fascinating, and more various, more democratic, more healthy, more practical—though less glamorous —than elaborations of power.

Without this ultimate reference to the system of systems, and this ultimate concern for quality, any rendering of account falls into the service of a kind of tyranny: it accompanies, and in one way or another invariably enables, the taking of power, from people first and last, but also from all other created things.

In this degenerative accounting, language is almost without the power of designation because it is used conscientiously to refer to nothing in particular. Attention rests upon percentages, categories, abstract functions. The reference drifts inevitably toward the merely provisional. It is not language that the user will very likely be required to stand by or to act on, for it does not define any personal ground for standing or acting. Its only practical utility is to support with "expert opinion" a vast, impersonal technological action already begun. And it works directly against the conventionality, the community life, of language, for it holds in contempt, not only all particular grounds of private fidelity and action, but the common ground of human experience, memory, and understanding from which language rises and on which meaning is shaped. It is a tyrannical language: tyrannese.

Do people come consciously to such language and to such implicit purpose? I hope not. I do not think so. It seems likely to me that, first, a certain kind of confusion must occur. It is a confusion about the human place in the universe, and it has been produced by diligent "educational" labors. This confusion is almost invariably founded on some romantic proposition about the "high destiny of man" or "unlimited human horizons." For an example I turn to R. Buckminster Fuller, here defending the "cosmic realism" of space colonization:

> Conceptualizing realistically about humans as passengers on board 8,000-mile diameter Spaceship Earth traveling around the Sun at 60,000 miles an hour while flying formation with the Moon, which formation involves the 365 revolutions per each Sun circuit, and recalling that humans have always been born naked, helpless and ignorant though superbly equipped cerebrally, and endowed with hunger, thirst, curiosity and procreative instincts, it has been logical for humans to employ their minds' progressive discoveries of the cosmic principles governing all physical interattractions, interactions, reactions and intertransformings, and to use those principles in progressively organizing, to humanity's increasing advantage, the complex of cosmic principles interacting locally to produce their initial environment which most probably was that of a verdant south seas coral atoll—built by the coral on a volcano risen from ocean bottom ergo unoccupied by any animals, having only fish and birds as well as fruits, nuts and coconut milk.

That is a single sentence. I call attention not only to the vagueness and oversimplification of its generalities, but, more important, to the weakness of its grammar and the shapelessness and aimlessness of its syntax. The subject is an "it" of very tentative reference, buried in the middle of the sentence—an "it," moreover, that cannot possibly be the subject of the two complicated participial constructions that precede it. The sentence, then, begins with a dangling modifier half its length. On the other end, it peters out in a description of the biology of a coral atoll, the pertinence of which is never articulated. In general, the sentence is a labyrinth of syntactical confusions impossible to map. When we reflect that "sentence" means, literally, "a way of thinking" (Latin: *sententia*) and that it comes from the Latin *sentire*, to feel, we realize that the concepts of sentence and sentence structure are not merely grammatical or merely academic—not negligible in any sense. A sentence is both the opportunity and the limit of thought—what we have to think with, and what we have to think in. It is, moreover, a *feelable* thought, a thought that impresses its sense not just on our understanding, but on our hearing, our sense of rhythm and proportion. It is a pattern of felt sense.

A sentence that is completely shapeless is therefore a loss of thought, an act of self-abandonment to incoherence. And indeed Mr. Fuller shows himself here a man who conceives a sentence, not as a pattern of thought apprehensible to sense, but merely as a clot of abstract concepts. In such a syntactical clot, words and concepts will necessarily tend to function abstractly rather than referentially. It is the statement of a man for whom words have replaced things, and who has therefore ceased to think particularly about any thing.

The idea buried in all those words is, so far as I can tell, a simple one: humans are born earth-bound, ignorant, and vulnerable, but intelligent; by their intelligence they lift themselves up from their primitive origin and move to fulfill their destiny. As we learn in later sentences, this destiny is universal in scope, limitless ("ever larger"), and humans are to approach it by larger and larger technology. The end is not stated, obviously, because it is not envisioned, because it is not envisionable. It seems to me that the aimlessness, the limitlessness, of Mr. Fuller's idea produces the aimlessness and shapelessness of his sentence.

By contrast, consider another view of the human place in the universe, not a simple view or simply stated, but nevertheless comely, orderly, and clear:

> There wanted yet the Master work, the end
> Of all yet done: A Creature who not prone
> And Brute as other Creatures, but endu'd
> With Sanctity of Reason, might erect
> His Stature, and upright with Front serene
> Govern the rest, self-knowing, and from thence
> Magnanimous to correspond with Heav'n,
> But grateful to acknowledge whence his good
> Descends, thither with heart and voice and eyes
> Directed in Devotion to adore
> And worship God Supreme . . .

These lines of Milton immediately suggest what is wrong, first, with Mr. Fuller's sentence, and then with the examples of tyrannese that preceded it. They all assume that the human prerogative is unlimited, that we *must* do whatever we have the power to do. Specifically, what is lacking is the idea that humans have a place in Creation and that this place is limited by responsibility on the one hand and by humility on the other —or, in Milton's terms, by magnanimity and devotion. Without this precision of definition, this setting of bounds or ends to thought, we cannot mean, or say what we mean, or mean what we say; we cannot stand by our words because we cannot utter words that can be stood by; we cannot speak of our own actions as persons, or even as communities, but only of the actions of percentages, large organizations, concepts, historical trends, or the impersonal "forces" of destiny or evolution.

Or let us consider another pair of statements. The first, again from Buckminster Fuller, following the one just quoted, elaborates his theme of technological destiny:

> First the humans developed fish catching and carving tools, then rafts, dug-out canoes and paddles and then sailing outrigger canoes.
>
> Reaching the greater islands and the mainland they developed animal skin, grass and leafwoven clothing and skin tents. They gradually entered safely into geographical areas where they would previously have perished. Slowly they learned to tame, then breed, cows, bullocks, water buffalo, horses and elephants. Next they developed oxen, then horse-drawn vehicles, then horseless vehicles, then ships of the sky. Then employing rocketry and packaging up the essential life-supporting environmental constituents of the biosphere they made sorties away from their mothership Earth and finally ferried over to their Sun orbiting-companion, the Moon.

The other is from William Faulkner's story, "The Bear." Isaac McCaslin is speaking of his relinquishment of the ownership of land:

> He created the earth, made it and looked at it and said it was all right, and then He made man. He made the earth first and peopled it with dumb creatures, and then

He created man to be His overseer on the earth and to hold suzerainty over the earth and the animals on it in His name, not to hold for himself and his descendants inviolable title forever, generation after generation, to the oblongs and squares of the earth, but to hold the earth mutual and intact in the communal anonymity of brotherhood, and all the fee He asked was pity and humility and sufferance and endurance and the sweat of his face for bread.

The only continuity recognized by Mr. Fuller is that of technological development, which is in fact not a continuity at all, for, as he sees it, it does not proceed by building on the past but by outmoding and replacing it. And if any other human concern accompanied the development from canoe to spaceship, it is either not manifest to Mr. Fuller, or he does not think it important enough to mention.

The passage from Faulkner, on the other hand, cannot be understood except in terms of the historical and cultural continuity that produced it. It awakens our memory of Genesis and *Paradise Lost*, as *Paradise Lost* awakens our memory of Genesis. In each of these the human place in Creation is described as a moral circumstance, and this circumstance is understood each time, it seems to me, with a deeper sense of crisis, as history has proved humanity more and more the exploiter and destroyer of Creation rather than its devout suzerain or steward. Milton knew of the conquests of Africa and the Americas, the brutality of which had outraged the humane minds of Europe and provided occasion to raise again the question of the human place in Creation; the devils of *Paradise Lost* are, among other things, conquistadores. (They are also the most expedient of politicians and technologists: ". . . by strength / They measure all. . . .") Faulkner's Isaac McCaslin, a white Mississippian of our own time, speaks not just with Milton's passion as a moral witness, but with the anguish of a man who inherits directly the guilt of the conqueror, the history of expropriation, despoliation, and slavery.

It is of the greatest importance to see how steadfastly this thrust of tradition, from Genesis to Milton to Faulkner, works toward the definition of personal place and condition, responsibility and action. And one feels the potency of this tradition to reach past the negativity of Isaac McCaslin's too simple relinquishment toward the definition of the atoning and renewing work that each person must do. Mr. Fuller's vision, by contrast, proposes that we have ahead of us only the next technological "breakthrough"—which, now that we have "progressed" to the scale of spaceships, is not work for persons or communities but for governments and corporations. We have in these two statements an open conflict between unlimited technology and traditional value. It is foolish to think that these are compatible. Value and technology can meet only on the ground of restraint.

The technological determinists have tyrannical attitudes, and speak tyrannese, at least partly because their assumptions cannot produce a moral or a responsible definition of the human place in Creation. Because they assume that the human place is any place, they are necessarily confused about where they belong.

Where does this confusion come from? I think it comes from the specialization and abstraction of intellect, separating it from responsibility and humility, magnanimity and devotion, and thus giving it an importance that, in the order of things and in its own nature, it does not and cannot have. The specialized intellectual assumes, in other words, that intelligence is all in the mind. For illustration, I turn again to *Paradise Lost*, where Satan, fallen, boasts in "heroic" defiance that he has

> A mind not to be chang'd by Place or Time.
> The mind is its own place, and in itself
> Can make a Heav'n of Hell, a Hell of Heav'n,
> What matter where, if I be still the same . . .

I do not know where one could find a better motto for the modernist or technological experiment, which assumes that we can fulfill a high human destiny anywhere, any way, so long as we can keep up the momentum of innovation; that the mind is "its own place" even within ecological degradation, pollution, poverty, hatred, and violence.

What we know, on the contrary, is that in any culture that could be called healthy or sane we find a much richer, larger concept of intelligence. We find, first, some way of acknowledging in action the existence of "higher intelligence." And we find that the human mind, in such a culture, is invariably strongly *placed*, in reference to other minds in the community and in cultural memory and tradition, and in reference to earthly localities and landmarks. Intelligence survives both by internal coherence and external pattern; it is both inside and outside the mind. People are born both with and into intelligence. What is thought refers precisely to what is thought about. It is this outside intelligence that we are now ignoring and consequently destroying.

As industrial technology advances and enlarges, and in the process assumes greater social, economic, and political force, it carries people away from where they belong by history, culture, deeds, association, and affection. And it destroys the landmarks by which they might return. Often it destroys the nature or the character of the places they have left. The very possibility of a practical connection between thought and the world is thus destroyed. Culture is driven into the mind, where it cannot be preserved. Displaced memory, for instance, is hard to keep in mind, harder to hand down. The little that survives is attenuated—without practical force. That is why the Jews, in Babylon, wept when they remembered Zion. The mere memory of a place cannot preserve it, nor apart from the place itself can it long survive in the mind. "How shall we sing the Lord's song in a strange land?"

The enlargement of industrial technology is thus analogous to war. It continually requires the movement of knowledge and responsibility away from home. It thrives upon the disintegration of homes, the subjugation of homelands. It requires that people cease to cooperate directly to fulfill local needs from local sources and begin instead to deal with each other always across the rift that divides producer and consumer, and always competitively. The idea of the independence of individual farms, shops, communities, and households is anathema to industrial technologists. The rush to nuclear energy and the growth of the space colony idea are powered by the industrial will to cut off the possibility of a small-scale energy technology—which is to say the possibility of small-scale personal and community acts. The corporate producers and their sycophants in the universities and the government will do virtually anything (or so they have obliged us to assume) to keep people from acquiring necessities in any way except by *buying* them.

Industrial technology and its aspirations enlarge along a line described by changes of verb tense: I need this tool; I will need this tool; I would need this tool. The conditional verb rests by nature upon *ifs*. The ifs of technological rationalization (*if* there were sufficient demand, money, knowledge, energy, power) act as wedges between history and futurity, inside and outside, value and desire, and ultimately between people and the earth and between one person and another.

By such shifts in the tenses of thought (as sometimes also by the substitution of the indefinite for the definite article) it is possible to impair or destroy the power of language to designate, to shift the focus of reference from what is outside the mind to what is inside it. And thus what already exists is devalued, subjugated, or destroyed for the sake of what *might* exist. The modern cult of planners and "futurologists" has thus achieved a startling resemblance to Swift's "academy of projectors":

> In these colleges, the professors contrive new rules and methods of agriculture and building, and new instruments and tools for all trades and manufactures; whereby, as they undertake, one man shall do the work of ten: a palace may be built in a week, of materials so durable, as to last for ever without repairing. All the fruits of the earth shall come to maturity, at whatever season we think fit to chuse, and encrease an hundred fold more than they do at present, with innumerable other happy proposals. The only inconvenience is, that none of these projects are yet brought to perfection; and, in the mean time, the whole country lies miserably waste. . . .

People who are willing to follow technology wherever it leads are necessarily willing to follow it away from home, off the earth, and outside the sphere of human definition, meaning, and responsibility. One has to suppose that this would be all right if they did it only for themselves and if they accepted the terms of their technological romanticism absolutely—that is, if they would depart absolutely from all that they propose to supersede, never to return. But past a certain scale, as C. S. Lewis wrote, the person who makes a technological choice does not choose for himself alone, but for others; past a certain scale, he chooses for *all* others. Past a certain scale, if the break with the past is great enough, he chooses for the past, and if the effects are lasting enough he chooses for the future. He makes, then, a choice that can neither be chosen against nor unchosen. Past a certain scale, there is no dissent from a technological choice.

People speaking out of this technological willingness cannot speak precisely, for what they are talking about does not yet exist. They cannot mean what they say because their words are avowedly speculative. They cannot stand by their words because they are talking about, if not *in*, the future, where they are not standing and cannot stand until long after they have spoken. All the grand and perfect dreams of the technologists are happening in the future, but nobody is there.

What can turn us from this deserted future, back into the sphere of our being, the great dance that joins us to our home, to each other and to other creatures, to the dead and the unborn? I think it is love. I am perforce aware how baldly and embarrassingly that word now lies on the page—for we have learned at once to overuse it, abuse it, and hold it in suspicion. But I do not mean any kind of abstract love, which is probably a contradiction in terms, but particular love for particular things, places, creatures, and people, requiring stands and acts, showing its successes or failures in practical or tangible effects. And it implies a responsibility just as particular, not grim or merely dutiful, but rising out of generosity. I think that this sort of love defines the effective range of human intelligence, the range within which its works can be dependably beneficent. Only the action that is moved by love for the good at hand has the hope of being responsible and generous. Desire for the future produces words that cannot be stood by. But love makes language exact, because one loves only what one knows. One cannot love the future or anything in it, for nothing is known there. And one cannot unselfishly make a future for someone else. Love for the future is self-love—love for the present self, projected and magnified into the future, and it is an irremediable loneliness.

Because love is not abstract, it does not lead to trends or

percentages or general behavior. It leads, on the contrary, to the perception that there is no such thing as general behavior. There is no abstract action. Love proposes the work of settled households and communities, whose innovations come about in response to immediate needs and immediate conditions, as opposed to the work of governments and corporations, whose innovations are produced out of the implicitly limitless desire for future power or profit. This difference is the unacknowledged cultural break in Mr. Fuller's evolutionary series: oxen, horse-drawn vehicles, horseless vehicles, ships of the sky. Between horse-drawn vehicles and horseless vehicles, human life disconnected itself from local sources; energy started to flow away from home. A biological limit was overrun, and with it the deepest human propriety.

Or, to shift the terms, love defines the difference between the "global village" which is a technological and a totalitarian ideal, directly suited to the purposes of centralized governments and corporations, and the Taoist village-as-globe, where the people live frugally and at peace, pleased with the good qualities of necessary things, so satisfied where they are that they live and die without visiting the next village, though they can hear its dogs bark and its roosters crow.

We might conjecture and argue a long time about the meaning and even the habitability of such a village. But one thing, I think, is certain: it would not be a linguistic no-man's-land in which words and things, words and deeds, words and people failed to stand in reliable connection or fidelity to one another. People and other creatures would be known by their names and histories, not by their numbers or percentages. History would be handed down in songs and stories, not reduced to evolutionary or technological trends. Generalizations would exist, of course, but they would be distilled from experience, not "projected" from statistics. They would sound, says Lao Tzu, this way:

> "Alert as a winter-farer on an icy stream,"
> "Wary as a man in ambush,"
> "Considerate as a welcome guest,"
> "Selfless as melting ice,"
> "Green as an uncut tree,"
> "Open as a valley . . ."

I come, in conclusion, to the difference between "projecting" the future and making a promise. The "projecting" of "futurologists" *uses* the future as the safest possible context for whatever is desired; it binds one only to selfish interest. But making a promise binds one *to someone else's future.* If the promise is serious enough, one is brought to it by love, and in awe and fear. Fear, awe, and love bind us to no selfish aims, but to each other. And they enforce a speech more exact, more clarifying, and more binding than any speech that can be used to sell or advocate some "future." For when we promise in love and awe and fear there is a certain kind of mobility that we give up. We give up the romanticism of progress, that is always shifting its terms to fit its occasions. We are speaking where we stand, and we shall stand afterwards in the presence of what we have said.

Poetry and Marriage: The Use of Old Forms

THE MEANING of marriage begins in the giving of words. We cannot join ourselves to one another without giving our word. And this must be an unconditional giving, for in joining ourselves to one another we join ourselves to the unknown. We can join one another *only* by joining the unknown. We must not be misled by the procedures of experimental thought: in life, in the world, we are never given two known results to choose between, but only *one* result that we choose without knowing what it is.

Marriage rests upon the immutable *givens* that compose it: words, bodies, characters, histories, places. Some wishes cannot succeed; some victories cannot be won; some loneliness is incorrigible. But there is relief and freedom in knowing what is real; these givens come to us out of the perennial reality of the world, like the terrain we live on. One does not care for this ground to make it a different place, or to make it perfect, but to make it inhabitable and to make it better. To flee from its realities is only to arrive at them unprepared.

Because the condition of marriage is worldly and its meaning communal, no one party to it can be solely in charge. What you alone think it ought to be, it is not going to be. Where you alone think you want it to go, it is not going to go. It is going where the two of you—and marriage, time, life, history, and the world—will take it. You do not know the road; you have committed your life to a way.

In marriage as in poetry, the given word implies the acceptance of a form that is never entirely of one's own making. When understood seriously enough, a form is a way of accepting and of living within the limits of creaturely life. We live only one life, and die only one death. A marriage cannot include everybody, because the reach of responsibility is short. A poem cannot be about everything, for the reach of attention and insight is short.

There are two aspects to these forms. The first is the *way* of making or acting or doing, which is to some extent technical. That is to say that definitions—settings of limits—are

involved. The names *poetry* and *marriage* are given only to
certain things, not to anything or to everything. Poetry is
made of words; it is expected to keep a certain fidelity to everyday speech and a certain fidelity to music; if it is unspeakable or unmusical, it is not poetry. Marriage is the mutual
promise of a man and a woman to live together, to love and
help each other, in mutual fidelity, until death. It is understood that these definitions cannot be altered to suit convenience or circumstance, any more than we can call a rabbit a
squirrel because we preferred to see a squirrel. Poetry of the
traditionally formed sort, for instance, does not propose that
its difficulties should be solved by skipping or forcing a rhyme
or by mutilating syntax or by writing prose. Marriage does
not invite one to solve one's quarrel with one's wife by marrying a more compliant woman. Certain limits, in short, are
prescribed—imposed *before* the beginning.

The second aspect of these forms is an opening, a generosity, toward possibility. The forms acknowledge that good is
possible; they hope for it, await it, and prepare its welcome
—though they dare not *require* it. These two aspects are inseparable. To forsake the way is to forsake the possibility. To give
up the form is to abandon the hope.

A certain awesome futurity, then, is the inescapable condition of word-giving—as it is, in fact, of all speech—for we
speak into no future that we know, much less into one that we
desire, but into one that is unknown. But that it is unknown
requires us to be generous toward it, and requires our generosity to be full and unconditional. The unknown is the mercy
and it may be the redemption of the known. The given word
may come to appear to be wrong, or wrongly given. But the
unknown still lies ahead of it, and so who is finally to say? If
time has apparently proved it wrong, more time may prove it
right. As growth has called it into question, further growth
may reaffirm it.

These forms are artificial; if they exist they have to be made.
Sexual love is natural, but marriage is not. The impulse to sing
is natural, but language and the forms of song are not.

These forms are also initially arbitrary, because at the outset
they can always be argued against. Until the wedding vows
are said, the argument that one might find a better spouse has

standing because there is no argument or evidence that can be produced against it; statistical probability would seem to support it: given the great number of theoretically possible choices, one *might* make a better choice. The vows answer that argument simply by cloture: the marriage now exists beyond all possibility of objection. A vow, Beatrice says, in *Paradiso* V, "is never canceled save by being kept . . . ," and must not be changed or broken on one's own judgment, though her stricture does not apply to vows that are foolish or perverse like those of Jephthah and Agamemnon. (She is acknowledging, let me emphasize, that some vows *ought* to be broken. Undoubtedly, some marriages are wrong, some divorces right. But it must also be understood, I think, that the possibility of breaking a vow can tell us nothing of what is meant by making and keeping one. Divorce is the contradiction of marriage, not one of its proposed results.)

Similarly, before the poem, there is no necessity governing the choice of form. Why *The Faerie Queene* should have been written in Spenserian stanzas is an unanswerable and probably a pointless question. All we can do is acknowledge that it *is* so written and go on to questions that are answerable.

The forms, then, are arbitrary *before* they are entered upon. Afterward, they have the same undoubtable existence and standing as the forms of an elm or a river. A poet such as Spenser evidently entered upon the form of a poem as solemnly as he entered upon any other cultural form—that of public service, say, or that of marriage. He understood it both as enablement and as constraint, and he meant not to break it, for in keeping the form he did not merely obey an arbitrarily imposed technical requirement but maintained his place in his cultural lineage, as both inheritor and bequeather, which saved him from loneliness and enabled him to mean—as witness his filial apostrophe to Chaucer in *The Faerie Queen*:

> through infusion sweete
> Of thine owne spirit, which doth in me survive
> I follow here the footing of thy feete,
> That with thy meaning so I may the rather meete.

(Though my concern here is with *making*, I should point out that the reader of a poem also participates in its form and in the community it makes, precisely as Spenser says: by following the footing of the poet's feet.)

Arbitrary in the choosing, these forms, once chosen and kept, are not arbitrary, but become inseparable from our definition as human beings. But the decision to break them is again arbitrary, for it cannot be based on any sufficient argument or evidence. And this decision does not return us to the state (perfectly inhabitable and respectable) that we were in before we chose the form but throws us into a state of formlessness, which is painful and dangerous, and which we can escape only by a return to form. The choice of a form is rightly solemn, both because of the gravity of the responsibility to keep it and because of the danger of failing to keep it. This choice does not present itself in any way that we may safely take less than seriously. To have a life or a place or a poem that is formless —into which anything at all may, or may not, enter—is to be condemned, at best, to bewilderment.

It is often assumed, as if under the influence of the promises of advertisements, that need or desire, ambition or inspiration may proceed straight to satisfaction. But this is false, contrary to the nature of form and the nature of discipline, as it is to common experience. It may sometimes happen by chance, but it does not dependably happen by chance. When it happens by luck, it will generally be found to be the luck of the well prepared. These impulses dependably come to fruition only by encountering the resistances of form, by being balked, baffled, forced to turn back and start again. They come to fruition by error and correction. Form is the means by which error is recognized and the means by which correctness is recognized.

There are, it seems, two Muses: the Muse of Inspiration, who gives us inarticulate visions and desires, and the Muse of Realization, who returns again and again to say, "It is yet more difficult than you thought." This is the muse of form.

The first muse is the one mainly listened to in a cheap-energy civilization, in which "economic health" depends on the assumption that everything desirable lies within easy reach of anyone. To hear the second muse one must move outside the cheap-energy enclosure. It is the willingness to hear the second muse that keeps us cheerful in our work. To hear only the first is to live in the bitterness of disappointment.

It is true that any form can be applied with a stupid rigidity. It can be falsely exclusive, consigning all that it cannot contain "to cold oblivion," as Shelley wrongly believed marriage was supposed to do. A set verse form can, of course, be used like a cookie cutter or a shovel, including and excluding arbitrarily by its own rule. But a set form can be used also to summon into a poem, or into a life, its unforeseen belongings, and thus is not rigid but freeing—an invocation to unknown possibility.

Form, crudely or stupidly used, may indeed be inimical to freedom. Well used, it may be the means of earning freedom, the price of admission or permission, the enablement to be free. But the connection may be even closer, more active and interesting, than that; it may be that form, strictly kept, *enforces* freedom. The form can be fulfilled only by a kind of abandonment to hope and to possibility, to unexpected gifts. The argument for freedom is not an argument against form. Form, like topsoil (which is intricately formal), empowers time to do good.

Properly used, a verse form, like a marriage, creates impasses, which the will and present understanding can solve only arbitrarily and superficially. These halts and difficulties do not ask for immediate remedy; we fail them by making emergencies of them. They ask, rather, for patience, forbearance, inspiration —the gifts and graces of time, circumstance, and faith. They are, perhaps, the true occasions of the poem: occasions for surpassing what we know or have reason to expect. They are points of growth, like the axils of leaves. Writing in a set form, rightly understood, is anything but force and predetermination. One puts down the first line of the pattern *in trust* that life and language are abundant enough to complete it. Rightly understood, a set form prescribes its restraint to the poet, not to the subject.

Marriage too is an attempt to rhyme, to bring two different lives—within the one life of their troth and household —periodically into agreement or consent. The two lives stray apart necessarily, and by consent come together again: to "feel together," to "be of the same mind." Difficult virtues are again necessary. And failure, permanent failure, is possible. But it is this possibility of failure, together with the formal bounds, that turns us back from fantasy, wishful thinking, and self-pity into the real terms and occasions of our lives.

It may be, then, that form serves us best when it works as an obstruction to baffle us and deflect our intended course. It may be that when we no longer know what to do we have come to our real work and that when we no longer know which way to go we have begun our real journey. The mind that is not baffled is not employed. The impeded stream is the one that sings.

In this way the keeping of the form instructs us. We had been prepared to learn what we had the poor power to expect. But fidelity to the form has driven us beyond expectation. The world, the truth, is more abounding, more delightful, more demanding than we thought. What appeared for a time perhaps to be mere dutifulness, that dried skull, suddenly breaks open in sweetness—and we are not where we thought we were, nowhere that we could have expected to be. It was expectation that would have kept us where we were.

In *Taking the Path of Zen*, Robert Aitken Roshi says: "It is not unusual to find true resonance with a so-called advanced koan in just a single dokusan, though often more time is necessary, and sometimes one gets stuck, and must stay there for a while." That necessity to "stay there for a while" is the gist of the meaning of form. Forms join us to time, to the consequences and fruitions of our own passing. The Zen student, the poet, the husband, the wife—none knows with certainty what he or she is staying for, but all know the likelihood that they will be staying "a while": to find out what they are staying for. And it is the faith of all of these disciplines that they will not stay to find that they should not have stayed.

That faith has nothing to do with what is usually called optimism. As the traditional marriage ceremony insists, not everything that we stay to find out will make us happy. The faith, rather, is that by staying, and only by staying, we will learn something of the truth, that the truth is good to know, and that it is always both different and larger than we thought.

The exploiter and the Shelleyan romantic (who are often the same person) are always in flight from consequence, the troubles of duration. The religious disciple, the husband and wife, the poet, like the true husbandman, accept the duration and effort, even the struggle, of formal commitment. They must come prepared to stay; if they mean to stay they will have to work, and they must learn the difference between good work and bad—which is to say that the capability of good work must be handed down from old to young.

The filial piety of Spenser, following the footing of the feet of Chaucer, has its opposite in the character of Michael in Eliot's *The Elder Statesman*, who longs to be fatherless in a present "freed" from the past, so that his name will be, as we say, "just a word":

> I simply want to lead a life of my own,
> According to my own ideas of good and bad,
> Of right and wrong. I want to go far away
> To some country where no one has heard the name of
>   Claverton . . .

We will find that dream of escape in Shelley, who might be surprised—though *we* should not be—to discover that what Michael wants to do with his freedom is

>   to go abroad
>   As a partner in some interesting business.
>   . . . .
>   Import and export,
>   With opportunity of profits both ways.

Part of the nature of a form seems to be that it is communal —that it can be bequeathed and inherited, that it can be taught, not as an instance (a relic), but as a way still usable. Both its validity and its availability depend upon our common understanding that we humans are all fundamentally alike.

Forms are broken, usually, on the authority of the opposing principle that we are all fundamentally or essentially different. Each individual, each experience, each life is assumed to be unique—hence, each individual should be "free" to express or fulfill his or her unique self in a way appropriately unique.

Both the communal and the individual emphases can be carried to extremes, and the extremity of each is loneliness. One can be lonely in the totalitarian crowd, in which no difference is perceived or tolerated; and one can be lonely in the difference or uniqueness of individuality in which community is repudiated.

The whole range of possibilities can be exemplified within language itself. It is possible to speak a language so com-monized by generality or jargon or slang that one's own mind and life virtually disappear into it. And it is possible to speak a language made so personal by contrivance, affectation, or slovenliness that one makes no sense. Between the two are the Confucian principles, dear to Pound, of fidelity ("the man . . . standing by his word") and sincerity (precise speech: words that can be stood by).

This word-keeping, standing by one's word, is a double fi-delity: to the community and to oneself. It is the willingness of difference both to represent itself and to account for itself. The individual is thereby at once free and a member. To break one's word in order to be "free" of it, on the other hand, is to make and enforce a damning equation between freedom and loneliness.

The freedom that depends upon or results in the breaking of words and the breaking of forms comes, I think, from a faith in the individual intelligence, in "genius," as opposed to a faith in the community or in culture. Belief in culture calls for the same disciplines as belief in marriage. It calls, indeed, for *more* patience and *more* faith, for it requires dedication to work and hope of more than a lifetime. This work, as I under-stand it, consists of the accumulation of local knowledge *in place*, generation after generation, children learning the visions and failures, stories and songs, names, ways, and skills of their elders, so that the costs of individual trial-and-error learning can be lived with and repaid, and the community thus enabled to preserve both itself and its natural place and neighborhood.

I do not mean that the rebellions of genius are necessarily bad; they may be both inevitable and indispensable. But they *can* be bad; they are dangerous, and it is prudent to under-stand what their dangers are. We may be properly suspicious of them when they are carried out in the name of "personal freedom" or "personal fulfillment," and when their program is a general rebellion against the imperfect past.

"Freedom" and "fulfillment" are often coupled together in the idea that one must be free in order to fulfill oneself. Be-hind the divorce epidemic—as behind the epidemics of vene-real disease and accidental parenthood among the young—is the unappeasable demon of "sexual fulfillment." Behind the epidemic of poetic illiteracy is the conviction that one must free oneself from poetic tradition (by not learning it) in order to fulfill one's ambition. The question here is whether we want "fulfill" to mean more than "satisfy"—as it must if we take seriously even its literal sense: to fill full. Sexual desire is an appetite, and so, in a different way, is ambition; they can be fulfilled only temporarily and soon have to be refulfilled. Though we would like it to be, "sexual fulfillment" is really no more exalted an idea than "hunger fulfillment." Ambition fulfillment, as we still understand when talking of politics or war, is a dangerous enterprise—for a mental appetite may be larger and more rapacious than a physical one, may require to be refulfilled more frequently and is less likely to be twice appeased by the same refulfillment.

If we mean "fulfill" in its larger sense   to discharge or put to rest, as when one fulfills an obligation or a promise—then we see that what we fulfill are not appetites but forms. Marriage is a form of sexual love which allows its fulfillment in both senses: in satisfaction and in responsibility for its consequences, and it sets a term to this responsibility—"until death"—at which it may be said to be fulfilled. The form of a particular poem similarly allows a valid fulfillment of poetic ambition. And the particularity of these fulfillments, working as they do to define complex obligations to discipline, to community, to tradition, to forebears and successors, mitigates the dangers of self-renewing appetites. The seriousness of this formality can hardly be overstated, for in the fulfillment of form is death—as the marriage vows instruct us. Fulfillment "bears it out even to the edge of doom." Past fulfillment, desire is at an end; there is no need to return. Great cost is obviously involved. But to stop short of such fulfillment is to make love "time's fool," and to fix a value upon a worth unknown.

Outrage and rebellion against the past are undoubtedly human necessities, but they are limited necessities, and they probably should be limited to youth. Things are obviously wrong with the past; young people have the clarity to see them and the energy to rebel against them. But as a general principle, such rebellion is destructive, for it keeps us from seeing that the past, unsatisfactory as it is, is the source of nearly all our good. Maturity sees that the past is not to be rejected, destroyed, or replaced, but rather that it is to be judged and corrected, that the work of judgment and correction is endless, and that it necessarily involves one's *own* past.

The industrial economy has made a general principle of the youthful antipathy to the past, and the modern world abounds with heralds of "a better future" and with debunkers happy to point out that Yeats was "silly like us" or that Thomas Jefferson may have had a Negro slave as a mistress—and so we are disencumbered of the burden of great lives, set free to be as cynical or desperate as we please. Cultural forms, it is held, should change apace to keep up with technology. Sexual discipline should be replaced by the chemicals, devices, and procedures of "birth control," and poetry must hasten to accept the influence of typewriter or computer.

It can be better argued that cultural forms ought to change by analogy with biological forms. I assume that they *do* change in that way, and by the same necessity to respond to changes of circumstance. It is necessary, nevertheless, to recognize a difference in kinds of cultural change: there is change by necessity, or adaptation; and there is contrived change, or novelty. The first is the work of species or communities or lineages of descent, occurring usually by slow increments over a long time. The second is the work of individual minds, and it happens, or is intended to happen, by fiat. Individual attempts to change cultural form—as to make a new kind of marriage or family or community—are nearly always shallow or foolish and are frequently totalitarian. The assumption that it can easily be otherwise comes from the faith in genius.

To adopt a communal form with the idea of changing or discarding it according to individual judgment is hopeless, the despair and death of meaning. To keep the form is an act of faith in possibility, not of the form, but of the life that is given to it; the form is a question addressed to life and time, which only life and time can answer.

Individual genius, then, goes astray when it proposes to do the work of community. We rightly follow its promptings, on the other hand, when it can point out correctly that *we* have gone astray—when forms have become rigid or empty, when we have forgot their use or their meaning. We then follow our genius or our geniuses back to reverence, to truth, or to nature. This alternation is one of the long rhythms of our history.

But the faith in genius and the rebellions of genius, at the times when these are necessary, should lead to the renewal of forms, not to their destruction. No individual can justifiably destroy anything of communal value on behalf of the community. Though individual geniuses have often enough assumed otherwise, there is no reason to grant them special privileges or exemptions. No artistic or scientific genius is justified in abusing nature or culture.

The analogy I have been working with here is most readily apparent if we think of marriage and poetic forms as *set* forms—that is, forms that in a sense *precede* the content, that are in a sense *prescriptive*. These set forms are indispensable, I believe, because they accommodate and serve that part of our life which is cyclic, drawing minds and lives back repeatedly through the same patterns, as each year moves through the same four seasons in the same order.

My remaining problem is to see how so-called free verse may fit into my scheme. It *has* to be fitted in if I am to respect my scheme, and if I acknowledge, as I certainly do, that much free verse is poetry. There is some danger of becoming cute or precious in carrying this analogy out to such length, and yet I am working on the assumption that the analogy is valid.

One analogue of free verse, I think, must be courtship, a way of accommodating our minds to something new, of finding out what it is that we are doing or about to do. The encounter between the English language and the subjects and objects of a new continent strange to it and to its poetic tradition seems to have required this of a succession of American poets from Whitman's time until now. Who can help feeling in the early *Leaves of Grass* a kind of falling in love? The lines must reach out impulsively, become capacious and tensile, to include in their full stance and particularity the images of American experience:

> The negro that drives the long dray of the stone-yard,
>   steady and tall he stands pois'd on one leg on the
>   string-piece . . .

Never mind that these loving inventories have occurred in

poetry before. This is new, this confrontation with a continent needing to be realized, and we grant Whitman his liberty and his exultation; we feel ourselves free and exultant with him; we willingly forgive the absurdities that occasionally jeopardize his exuberance—his ostentatious French phrases, for instance, or his industrial optimism.

But if Whitman's work is the prime example of the freeing of verse, it is an example of something else too, for at its best Whitman's line, as we will see if we try to shorten or lengthen it, is a form. He set his line free only to make it into a *kind* of line that we recognize anywhere we see it—a new power, a new music, added to poetry, which can be learned and used. Theoretically, I suppose, any line can be written a different way, but I don't think that we are tempted to imagine this line as anything but what it is:

Seasons pursuing each other the plougher ploughs, the
     mower mows, and the winter-grain falls in the ground . . .

And such newness does not destroy the old set forms, but renews them in renewing our understanding of what a line of verse is, our sense of its properties of duration and coherence, beginning and end. The term "organic," when applied to free or "open" poetic forms, should alert us to the nature of all form, traditional or new, for the organic forms of nature, like the ballad stanza or the stanza of Spenser, are principles that are repeatable and recognizable through a series of variations. This recognizability within difference suggests the proper relation of abstraction and particularity—and suggests, moreover, that the right function of abstraction is to give appropriate clarity and distinction to the particular. The form, recognizable from verse to verse, shapes and measures what is said, makes it musical.

So it is with Whitman's line. So it is, I believe, with the line of every writer of free verse worth reading. At some point the poet ceases to be an experimenter or innovator and begins a life's work. *Leaves of Grass* answers for itself the questions that stood before *The Divine Comedy* or *The Canterbury Tales*: What can be done in this way? What *must* be done in this way? Once a way is chosen with enough seriousness, the analogy shifts from courtship to marriage.

The work of poetic form is coherence, joining things that need to be joined, as marriage joins them—in words by which a man or a woman can stand, words confirmable in acts. Forms join, and this is why forms tend to be analogues of each other and to resonate with each other. Forms join the diverse things that they contain; they join their contents to their context; they join us to themselves; they join us to each other; they join writers and readers; they join the generations together, the young and the old, the living and the dead. Thus, for a couple, marriage is an entrance into a timeless community. So, for a poet (or a reader), is the mastery of poetic form. Joining the form, we join all that the form has joined.

Getting Along with Nature

THE DEFENDERS of nature and wilderness—like their enemies the defenders of the industrial economy—sometimes sound as if the natural and the human were two separate estates, radically different and radically divided. The defenders of nature and wilderness sometimes seem to feel that they must oppose any human encroachment whatsoever, just as the industrialists often apparently feel that they must make the human encroachment absolute or, as they say, "complete the conquest of nature." But there is danger in this opposition, and it can be best dealt with by realizing that these pure and separate categories are pure ideas and do not otherwise exist.

Pure nature, anyhow, is not good for humans to live in, and humans do not want to live in it—or not for very long. Any exposure to the elements that lasts more than a few hours will remind us of the desirability of the basic human amenities: clothing, shelter, cooked food, the company of kinfolk and friends—perhaps even of hot baths and music and books.

It is equally true that a condition that is *purely* human is not good for people to live in, and people do not want to live for very long in it. Obviously, the more artificial a human environment becomes, the more the word "natural" becomes a term of value. It can be argued, indeed, that the conservation movement, as we know it today, is largely a product of the industrial revolution. The people who want clean air, clear streams, and wild forests, prairies, and deserts are the people who no longer have them.

People cannot live apart from nature; that is the first principle of the conservationists. And yet, people cannot live in nature without changing it. But this is true of *all* creatures; they depend upon nature, and they change it. What we call nature is, in a sense, the sum of the changes made by all the various creatures and natural forces in their intricate actions and influences upon each other and upon their places. Because of the woodpeckers, nature is different from what it would be without them. It is different also because of the borers and ants that live in tree trunks, and because of the bacteria that live in the soil under the trees. The making of these differences is the making of the world.

Some of the changes made by wild creatures we would call beneficent: beavers are famous for making ponds that turn into fertile meadows; trees and prairie grasses build soil. But sometimes, too, we would call natural changes destructive. According to early witnesses, for instance, large areas around Kentucky salt licks were severely trampled and eroded by the great herds of hoofed animals that gathered there. The buffalo "streets" through hilly country were so hollowed out by hoof-wear and erosion that they remain visible almost two centuries after the disappearance of the buffalo. And so it can hardly be expected that humans would not change nature. Humans, like all other creatures, must make a difference; otherwise, they cannot live. But unlike other creatures, humans must make a choice as to the kind and scale of the difference they make. If they choose to make too small a difference, they diminish their humanity. If they choose to make too great a difference, they diminish nature, and narrow their subsequent choices; ultimately, they diminish or destroy themselves. Nature, then, is not only our source but also our limit and measure. Or, as the poet Edmund Spenser put it almost four hundred years ago, Nature, who is the "greatest goddesse," acts as a sort of earthly lieutenant of God, and Spenser represents her as both a mother and judge. Her jurisdiction is over the relations between the creatures; she deals "Right to all . . . indifferently," for she is "the equall mother" of all "And knittest each to each, as brother unto brother." Thus, in Spenser, the natural principles of fecundity and order are pointedly linked with the principle of justice, which we may be a little surprised to see that he attributes also to nature. And yet in his insistence on an "indifferent" natural justice, resting on the "brotherhood" of *all* creatures, not just of humans, Spenser would now be said to be on sound ecological footing.

In nature we know that wild creatures sometimes exhaust their vital sources and suffer the natural remedy: drastic population reductions. If lynxes eat too many snowshoe rabbits —which they are said to do repeatedly—then the lynxes starve down to the carrying capacity of their habitat. It is the carrying capacity of the lynx's habitat, not the carrying capacity of the lynx's stomach, that determines the prosperity of lynxes. Similarly, if humans use up too much soil—which they have often done and are doing—then they will starve down to the carrying capacity of *their* habitat. This is nature's "indifferent" justice. As Spenser saw in the sixteenth century, and as we must learn to see now, there is no appeal from this justice. In the hereafter, the Lord may forgive our wrongs against nature, but on earth, so far as we know, He does not overturn her decisions.

One of the differences between humans and lynxes is that humans can see that the principle of balance operates between lynxes and snowshoe rabbits, as between humans and topsoil; another difference, we hope, is that humans have the sense to act on their understanding. We can see, too, that a stable balance is preferable to a balance that tilts back and forth like a seesaw, dumping a surplus of creatures alternately from either end. To say this is to renew the question of whether or not the human relationship with nature is necessarily an adversary relationship, and it is to suggest that the answer is not simple.

But in dealing with this question and in trying to do justice to the presumed complexity of the answer, we are up against an American convention of simple opposition to nature that is deeply established both in our minds and in our ways. We have opposed the primeval forests of the East and the primeval prairies and deserts of the West, we have opposed man-eating beasts and crop-eating insects, sheep-eating coyotes and chicken-eating hawks. In our lawns and gardens and fields, we oppose what we call weeds. And yet more and more of us are beginning to see that this opposition is ultimately destructive even of ourselves, that it does not explain many things that need explaining—in short, that it is untrue.

If our proper relation to nature is not opposition, then what is it? This question becomes complicated and difficult for us because none of us, as I have said, wants to live in a "pure" primeval forest or in a "pure" primeval prairie; we do not want to be eaten by grizzly bears; if we are gardeners, we have a legitimate quarrel with weeds; if, in Kentucky, we are trying to improve our pastures, we are likely to be enemies of the nodding thistle. But, do what we will, we remain under the spell of the primeval forests and prairies that we have cut down and broken; we turn repeatedly and with love to the thought of them and to their surviving remnants. We find ourselves attracted to the grizzly bears, too, and know that they and other great, dangerous animals remain alive in our imaginations as they have been all through human time. Though we cut down the nodding thistles, we acknowledge their beauty and are glad to think that there must be some place where they belong. (They may, in fact, not always be out of place in pastures; if, as seems evident, overgrazing makes an ideal seedbed for these plants, then we must understand them as a part of nature's strategy to protect the ground against abuse by animals.) Even the ugliest garden weeds earn affection from us when we consider how faithfully they perform an indispensable duty in covering the bare ground and in building humus. The weeds, too, are involved in the business of fertility.

We know, then, that the conflict between the human and the natural estates really exists and that it is to some extent necessary. But we are learning, or relearning, something else, too, that frightens us: namely, that this conflict often occurs at the expense of *both* estates. It is not only possible but altogether probable that by diminishing nature we diminish ourselves, and vice versa.

The conflict comes to light most suggestively, perhaps, when advocates for the two sides throw themselves into absolute conflict where no absolute difference can exist. An example of this is the battle between defenders of coyotes and defenders of sheep, in which the coyote-defenders may find it easy to forget that the sheep ranchers are human beings with some authentic complaints against coyotes, and the sheep-defenders find it easy to sound as if they advocate the total eradication of both coyotes and conservationists. Such conflicts—like the old one between hawk-defenders and chicken-defenders—tend to occur between people who use nature indirectly and people who use it directly. It is a dangerous mistake, I think, for either side to pursue such a quarrel on the assumption that victory would be a desirable result.

The fact is that people need both coyotes and sheep, need a world in which both kinds of life are possible. Outside the heat of conflict, conservationists probably know that a sheep is one of the best devices for making coarse foliage humanly edible and that wool is ecologically better than the synthetic fibers, just as most shepherds will be aware that wild nature is of value to them and not lacking in interest and pleasure.

The usefulness of coyotes is, of course, much harder to define than the usefulness of sheep. Coyote fur is not a likely substitute for wool, and, except as a last resort, most people don't want to eat coyotes. The difficulty lies in the difference between what is ours and what is nature's: What is ours is ours because it is directly useful. Coyotes are useful *indirectly*, as part of the health of nature, from which we and our sheep alike must live and take our health. The fact, moreover, may be that sheep and coyotes need each other, at least in the sense that neither would prosper in a place totally unfit for the other.

This sort of conflict, then, does not suggest the possibility of victory so much as it suggests the possibility of a compromise —some kind of peace, even an alliance, between the domestic and the wild. We know that such an alliance is necessary. Most conservationists now take for granted that humans thrive best in ecological health and that the test or sign of this health is the survival of a diversity of wild creatures. We know, too, that we cannot imagine ourselves apart from those necessary survivals of our own wildness that we call our instincts. And we know that we cannot have a healthy agriculture apart from the teeming wilderness in the topsoil, in which worms, bacteria, and other wild creatures are carrying on the fundamental work of decomposition, humus making, water storage, and drainage. "In wildness is the preservation of the world," as Thoreau said, may be a spiritual truth, but it is also a practical fact.

On the other hand, we must not fail to consider the opposite proposition—that, so long at least as humans are in the world, in human culture is the preservation of wildness—which is equally, and more demandingly, true. If wildness is to survive, then *we* must preserve it. We must preserve it by public act, by law, by institutionalizing wildernesses in some places. But such preservation is probably not enough. I have heard Wes Jackson of the Land Institute say, rightly I think, that if we cannot preserve our farmland, we cannot preserve the wilderness. That said, it becomes obvious that if we cannot preserve our cities, we cannot preserve the wilderness. This can be demonstrated practically by saying that the same attitudes that destroy wildness in the topsoil will finally destroy it everywhere; or by saying that if *everyone* has to go to a designated public wilderness for the necessary contact with wildness, then our parks will be no more natural than our cities.

But I am trying to say something more fundamental than that. What I am aiming at—because a lot of evidence seems to point this way—is the probability that nature and human culture, wildness and domesticity, are not opposed but are interdependent. Authentic experience of either will reveal the need of one for the other. In fact, examples from both past and present prove that a human economy and wildness can exist together not only in compatibility but to their mutual benefit.

One of the best examples I have come upon recently is the story of two Sonora Desert oases in Gary Nabhan's book, *The Desert Smells Like Rain.* The first of these oases, A'al Waipia, in Arizona, is dying because the park service, intending to preserve the natural integrity of the place as a bird sanctuary for tourists, removed the Papago Indians who had lived and farmed there. The place was naturally purer after the Indians were gone, but the oasis also began to shrink as the irrigation ditches silted up. As Mr. Nabhan puts it, "an odd thing is happening to their 'natural' bird sanctuary. They are losing the heterogeneity of the habitat, and with it, the birds. The old trees are dying. . . . These riparian trees are essential for the breeding habitat of certain birds. Summer annual seed plants are conspicuously absent. . . . Without the soil disturbance associated with plowing and flood irrigation, these natural foods for birds and rodents no longer germinate."

The other oasis, Ki:towak, in old Mexico, still thrives because a Papago village is still there, still farming. The village's oldest man, Luis Nolia, is the caretaker of the oasis, cleaning the springs and ditches, farming, planting trees: "Luis . . . blesses the oasis," Mr. Nabhan says, "for his work keeps it healthy." An ornithologist who accompanied Mr. Nabhan found twice as many species of birds at the farmed oasis as he found at the bird sanctuary, a fact that Mr. Nabhan's Papago friend, Remedio, explained in this way: "That's because those birds, they come where the people are. When the people live and work in a place, and plant their seeds and water their trees, the birds go live with them. They like those places, there's plenty to eat and that's when we are friends to them."

Another example, from my own experience, is suggestive in a somewhat different way. At the end of July 1981, while I was using a team of horses to mow a small triangular hillside pasture that is bordered on two sides by trees, I was suddenly aware of wings close below me. It was a young red-tailed hawk, who flew up into a walnut tree. I mowed on to the turn and stopped the team. The hawk then glided to the ground not twenty feet away. I got off the mower, stood and watched, even spoke, and the hawk showed no fear. I could see every feather distinctly, claw and beak and eye, the creamy down of the breast. Only when I took a step toward him, separating myself from the team and mower, did he fly. While I mowed three or four rounds, he stayed near, perched in trees or standing erect and watchful on the ground. Once, when I stopped to watch him, he was clearly watching me, stooping to see under the leaves that screened me from him. Again, when I could not find him, I stooped, saying to myself, "This is what he did to look at me," and as I did so I saw him looking at me.

Why had he come? To catch mice? Had he seen me scare one out of the grass? Or was it curiosity?

A human, of course, cannot speak with authority of the motives of hawks. I am aware of the possibility of explaining the episode merely by the hawk's youth and inexperience. And yet it does not happen often or dependably that one is approached so closely by a hawk of any age. I feel safe in making a couple of assumptions. The first is that the hawk came because of the conjunction of the small pasture and its wooded borders, of open hunting ground and the security of trees. This is the phenomenon of edge or margin that we know to be one of the powerful attractions of a diversified landscape, both to wildlife and to humans. The human eye itself seems drawn to such margins, hungering for the difference made in the countryside by a hedgy fencerow, a stream, or a grove of trees. And we know that these margins are biologically rich, the meeting of two kinds of habitat. But another difference also is important here: the difference between a large pasture and a small one, or, to use Wes Jackson's terms, the difference between a field and a patch. The pasture I was mowing was a patch—small, intimate, nowhere distant from its edges.

My second assumption is that the hawk was emboldened

to come so near because, though he obviously recognized me as a man, I was there with the team of horses, with whom he familiarly and confidently shared the world.

I am saying, in other words, that this little visit between the hawk and me happened because the kind and scale of my farm, my way of farming, and my technology *allowed* it to happen. If I had been driving a tractor in a hundred-acre cornfield, it would not have happened.

In some circles I would certainly be asked if one can or should be serious about such an encounter, if it has any value. And though I cannot produce any hard evidence, I would unhesitatingly answer yes. Such encounters involve another margin —the one between domesticity and wildness—that attracts us irresistibly; they are among the best rewards of outdoor work and among the reasons for loving to farm. When the scale of farming grows so great and obtrusive as to forbid them, the *life* of farming is impoverished.

But perhaps we do find hard evidence of a sort when we consider that *all* of us—the hawk, the horses, and I—were there for our benefit and, to some extent, for our *mutual* benefit: The horses live from the pasture and maintain it with their work, grazing, and manure; the team and I together furnish hunting ground to the hawk; the hawk serves us by controlling the field-mouse population.

These meetings of the human and the natural estates, the domestic and the wild, occur invisibly, of course, in any well-farmed field. The wilderness of a healthy soil, too complex for human comprehension, can yet be husbanded, can benefit from human care, and can deliver incalculable benefits in return. Mutuality of interest and reward is a possibility that can reach to any city backyard, garden, and park, but in any place under human dominance—which is, now, virtually everyplace —it is a possibility that is *both* natural and cultural. If humans want wildness to be possible, then they have to make it possible. If balance is the ruling principle and a stable balance the goal, then, for humans, attaining this goal requires a consciously chosen and deliberately made partnership with nature.

In other words, we can be true to nature only by being true to human nature—to our animal nature as well as to cultural patterns and restraints that keep us from acting like animals. When humans act like animals, they become the most dangerous of animals to themselves and other humans, and this is because of another critical difference between humans and animals: Whereas animals are usually restrained by the limits of physical appetites, humans have mental appetites that can be far more gross and capacious than physical ones. Only humans squander and hoard, murder and pillage because of notions.

The work by which good human and natural possibilities are preserved is complex and difficult, and it probably cannot be accomplished by raw intelligence and information. It requires knowledge, skills, and restraints, some of which must come from our past. In the hurry of technological progress, we have replaced some tools and methods that worked with some that do not work. But we also need culture-borne instructions about who or what humans are and how and on what assumptions they should act. The Chain of Being, for instance—which gave humans a place between animals and angels in the order of Creation—is an old idea that has not been replaced by any adequate new one. It was simply rejected, and the lack of it leaves us without a definition.

Lacking that ancient definition, or any such definition, we do not know at what point to restrain or deny ourselves. We do not know how ambitious to be, what or how much we may safely desire, when or where to stop. I knew a barber once who refused to give a discount to a bald client, explaining that his artistry consisted, not in the cutting off, but in the knowing when to stop. He spoke, I think, as a true artist and a true human. The lack of such knowledge is extremely dangerous in and to an individual. But ignorance of when to stop is a modern epidemic; it is the basis of "industrial progress" and "economic growth." The most obvious practical result of this ignorance is a critical disproportion of scale between the scale of human enterprises and their sources in nature.

The scale of the energy industry, for example, is too big, as is the scale of the transportation industry. The scale of agriculture, from a technological or economic point of view, is too big, but from a demographic point of view, the scale is too small. When there are enough people on the land to use it but not enough to husband it, then the wildness of the soil that we call fertility begins to diminish, and the soil itself begins to flee from us in water and wind.

If the human economy is to be fitted into the natural economy in such a way that both may thrive, the human economy must be built to proper scale. It is possible to talk at great length about the difference between proper and improper scale. It may be enough to say here that that difference is *suggested* by the difference between amplified and unamplified music in the countryside, or the difference between the sound of a motorboat and the sound of oarlocks. A proper human sound, we may say, is one that allows other sounds to be heard. A properly scaled human economy or technology allows a diversity of other creatures to thrive.

"The proper scale," a friend wrote to me, "confers freedom and simplicity . . . and doubtless leads to long life and health." I think that it also confers joy. The renewal of our partnership with nature, the rejoining of our works to their proper places in the natural order, reshaped to their proper scale, implies the reenjoyment both of nature and of human domesticity. Though our task will be difficult, we will greatly mistake its nature if we see it as grim, or if we suppose that it must always be necessary to suffer at work in order to enjoy ourselves in places specializing in "recreation."

Once we grant the possibility of a proper human scale, we see that we have made a radical change of assumptions and values. We realize that we are less interested in technological "breakthroughs" than in technological elegance. Of a new tool or method we will no longer ask: Is it fast? Is it powerful? Is it a labor saver? How many workers will it replace? We will ask instead: Can we (and our children) afford it? Is it fitting to our real needs? Is it becoming to us? Is it unhealthy or ugly? And though we may keep a certain interest in innovation and in what we may become, we will renew our interest in what we have been, realizing that conservationists must necessarily conserve *both* inheritances, the natural and the cultural.

To argue the necessity of wildness to, and in, the human economy is by no means to argue against the necessity of wilderness. The survival of wilderness—of places that we do not change, where we allow the existence even of creatures we perceive as dangerous—is necessary. Our sanity probably requires it. Whether we go to those places or not, we need to know that they exist. And I would argue that we do not need just the great public wildernesses, but millions of small private or semi-private ones. Every farm should have one; wildernesses can occupy corners of factory grounds and city lots—places where nature is given a free hand, where no human work is done, where people go only as guests. These places function, I think, whether we intend them to or not, as sacred groves—places we respect and leave alone, not because we understand well what goes on there, but because we do not.

We go to wilderness places to be restored, to be instructed in the natural economies of fertility and healing, to admire what we cannot make. Sometimes, as we find to our surprise, we go to be chastened or corrected. And we go in order to return with renewed knowledge by which to judge the health of our human economy and our dwelling places. As we return from our visits to the wilderness, it is sometimes possible to imagine a series of fitting and decent transitions from wild nature to the human community and its supports: from forest to wood-lot to the "two-story agriculture" of tree crops and pasture to orchard to meadow to grainfield to garden to household to neighborhood to village to city—so that even when we reached the city we would not be entirely beyond the influence of the nature of that place.

What I have been implying is that I think there is a bad reason to go to the wilderness. We must not go there to escape the ugliness and the dangers of the present human economy. We must not let ourselves feel that to go there is to escape. In the first place, such an escape is now illusory. In the second place, if, even as conservationists, we see the human and the natural economies as necessarily opposite or opposed, we subscribe to the very opposition that threatens to destroy them both. The wild and the domestic now often seem isolated values, estranged from one another. And yet these are not exclusive polarities like good and evil. There can be continuity between them, and there must be.

What we find, if we weight the balance too much in favor of the domestic, is that we involve ourselves in dangers both personal and public. Not the least of these dangers is dependence on distant sources of money and materials. Farmers are in deep trouble now because they have become too dependent on corporations and banks. They have been using methods and species that enforce this dependence. But such a dependence is not safe, either for farmers or for agriculture. It is not safe for urban consumers. Ultimately, as we are beginning to see, it is not safe for banks and corporations—which, though they have evidently not thought so, are dependent upon farmers. Our farms are endangered because—like the interstate highways or modern hospitals or modern universities—they cannot be inexpensively used. To be usable at all they require great expense.

When the human estate becomes so precarious, our only recourse is to move it back toward the estate of nature. We undoubtedly need better plant and animal species than nature provided us. But we are beginning to see that they can be too much better—too dependent on us and on "the economy," too expensive. In farm animals, for instance, we want good commercial quality, but we can see that the ability to produce meat or milk can actually be a threat to the farmer and to the animal if not accompanied by qualities we would call natural: thriftiness, hardiness, physical vigor, resistance to disease and parasites, ability to breed and give birth without assistance, strong mothering instincts. These natural qualities decrease care, work, and worry; they also decrease the costs of production. They save feed and time; they make diseases and cures exceptional rather than routine.

We need crop and forage species of high productive ability also, but we do not need species that will not produce at all without expensive fertilizers and chemicals. Contrary to the premise of agribusiness advertisements and of most expert advice, farmers do not thrive by production or by "skimming" a large "cash flow." They cannot solve their problems merely by increasing production or income. They thrive, like all other creatures, according to the difference between their income and their expenses.

One of the strangest characteristics of the industrial economy is the ability to increase production again and again without ever noticing—or without acknowledging—the *costs* of production. That one Holstein cow should produce 50,000 pounds of milk in a year may appear to be marvelous—a miracle of modern science. But what if her productivity is dependent upon the consumption of a huge amount of grain (about a bushel a day), and therefore upon the availability of cheap petroleum? What if she is too valuable (and too delicate) to be allowed outdoors in the rain? What if the proliferation of her kind will again drastically reduce the number of dairy farms and farmers? Or, to use a more obvious example, can we afford a bushel of grain at a cost of five to twenty bushels of topsoil lost to erosion?

"It is good to have Nature working for you," said Henry Besuden, the dean of American Southdown breeders. "She works for a minimum wage." That is true. She works at times for almost nothing, requiring only that we respect her work and give her a chance, as when she maintains—indeed, improves—the fertility and productivity of a pasture by the natural succession of clover and grass or when she improves a clay soil for us by means of the roots of a grass sod. She works for us by preserving health or wholeness, which for all our ingenuity we cannot make. If we fail to respect her health, she deals out her justice by withdrawing her protection against disease—which we *can* make, and do.

To make this continuity between the natural and the human, we have only two sources of instruction: nature herself and our cultural tradition. If we listen only to the apologists for the industrial economy, who respect neither nature nor culture, we get the idea that it is somehow our goodness that makes us so destructive: The air is unfit to breathe, the water is unfit to drink, the soil is washing away, the cities are violent and the countryside neglected, all because we are intelligent, enterprising, industrious, and generous, concerned only to feed the hungry and to "make a better future for our children." Respect for nature causes us to doubt this, and our cultural tradition confirms and illuminates our doubt: No good thing is destroyed by goodness; good things are destroyed by wickedness. We may identify that insight as Biblical, but it is taken for granted by both the Greek and the Biblical lineages of our culture, from Homer and Moses to William Blake. Since the start of the industrial revolution, there have been voices urging that this inheritance may be safely replaced by intelligence, information, energy, and money. No idea, I believe, could be more dangerous.

Two Economies

S OME TIME AGO, in a conversation with Wes Jackson in which we were laboring to define the causes of the modern ruination of farmland, we finally got around to the money economy. I said that an economy based on energy would be more benign because it would be more comprehensive.

Wes would not agree. "An energy economy still wouldn't be comprehensive enough."

"Well," I said, "then what kind of economy *would* be comprehensive enough?"

He hesitated a moment, and then, grinning, said, "The Kingdom of God."

I assume that Wes used the term because he found it, at that point in our conversation, indispensable; I assume so because, in my pondering over its occurrence at that point, I have found it indispensable myself. For the thing that troubles us about the industrial economy is exactly that it is not comprehensive enough, that, moreover, it tends to destroy what it does not comprehend, and that it is *dependent* upon much that it does not comprehend. In attempting to criticize such an economy, we naturally pose against it an economy that does not leave anything out, and we can say without presuming too much that the first principle of the Kingdom of God is that it includes everything; in it, the fall of every sparrow is a significant event. We are in it whether we know it or not and whether we wish to be or not. Another principle, both ecological and traditional, is that everything in the Kingdom of God is joined both to it and to everything else that is in it; that is to say, the Kingdom of God is orderly. A third principle is that humans do not and can never know either all the creatures that the Kingdom of God contains or the whole pattern or order by which it contains them.

The suitability of the Kingdom of God as, so to speak, a place name is partly owing to the fact that it still means pretty much what it has always meant. Because, I think, of the embarrassment that the phrase has increasingly caused among the educated, it has not been much tainted or tampered with by the disinterested processes of academic thought; it is a phrase that comes to us with its cultural strings still attached. To say that we live in the Kingdom of God is both to suggest the difficulty of our condition and to imply a fairly complete set of culture-borne instructions for living in it. These instructions are not always explicitly ecological, but it can be argued that they are always implicitly so, for all of them rest ultimately on the assumptions that I have given as the second and third principles of the Kingdom of God: that we live within order and that this order is both greater and more intricate than we can know. The difficulty of our predicament, then, is made clear if we add a fourth principle: Though we cannot produce a complete or even an adequate description of this order, severe penalties are in store for us if we presume upon it or violate it.

I am not dealing, of course, with perceptions that are only Biblical. The ancient Greeks, according to Aubrey de Sélincourt, saw "a continuing moral pattern in the vicissitudes of human fortune," a pattern "formed from the belief that men, as men, are subject to certain limitations imposed by a Power —call it Fate or God—which they cannot fully comprehend, and that any attempt to transcend those limitations is met by inevitable punishment." The Greek name for the pride that attempts to transcend human limitations was *hubris*, and hubris was the cause of what the Greeks understood as tragedy.

Nearly the same sense of *necessary* human limitation is implied in the Old Testament's repeated remonstrances against too great a human confidence in the power of "mine own hand." Gideon's army against the Midianites, for example, was reduced from thirty-two thousand to three hundred expressly to prevent the Israelites from saying, "Mine own hand hath saved me." A similar purpose was served by the institution of the Sabbath, when, by not working, the Israelites were meant to see the limited efficacy of their work and thus to understand their true dependence.

Though I hope that my insistence on the usefulness of the term, the Kingdom of God, will be understood, I must acknowledge that the term is local, in the sense that it is fully available only to those whose languages are involved in Western or Biblical tradition. A person of Eastern heritage, for example, might speak of the totality of all creation, visible and invisible, as "the Tao." I am well aware also that many people would not willingly use either term, or any such term. For these reasons, I do not want to make a statement that is specially or exclusively Biblical, and so I would like now to introduce a more culturally neutral term for that economy that I have been calling the Kingdom of God. Sometimes, in thinking about it, I have called it the Great Economy, which is the name I am going to make do with here—though I will remain under the personal necessity of Biblical reference. And that, I think, must be one of my points: We can name it whatever we wish, but we cannot define it except by way of a religious tradition. The Great Economy, like the Tao or the Kingdom of God, is both known and unknown, visible and invisible, comprehensible and mysterious. It is, thus, the ultimate condition of our experience and of the practical questions rising from our experience, and it imposes on our consideration of those questions an extremity of seriousness and an extremity of humility.

I am assuming that the Great Economy, whatever we may name it, is indeed—and in ways that are, to some extent, practical—an economy: It includes principles and patterns by which values or powers or necessities are parceled out and exchanged. But if the Great Economy comprehends humans and thus cannot be fully comprehended by them, then it is also not an economy in which humans can participate directly. What this suggests, in fact, is that humans can live in the Great Economy only with great uneasiness, subject to powers and laws that they can understand only in part. There is no human accounting for the Great Economy. This obviously is a description of the circumstance of religion, the circumstance that *causes* religion. De Sélincourt states the problem succinctly: "Religion in every age is concerned with the vast and fluctuant regions of experience which knowledge cannot penetrate, the regions which a man knows, or feels, to stretch away beyond the narrow, closed circle of what he can *manage* by the use of his wits."

If there is no denying our dependence on the Great Economy, there is also no denying our need for a little economy—a narrow circle within which things are manageable by the use of our wits. I don't think Wes Jackson was denying this need when he invoked the Kingdom of God as the complete economy; rather, he was, I think, insisting upon a priority that is both proper and practical. If he had a text in mind, it must have been the sixth chapter of Matthew, in which, after speaking of God's care for nature, the fowls of the air and the lilies of the field, Jesus says: "Therefore take no thought, saying, What shall we eat? or, What shall we drink? or, Wherewithal shall we be clothed? . . . But seek ye first the kingdom of God, and his righteousness; and all these things shall be added unto you."

There is an attitude that sees in this text a denial of the value of *any* economy of this world, but this attitude makes the text useless and meaningless to humans who must live in this world. These verses make usable sense only if we read them as a statement of considerable practical import about the real nature of worldly economy. If this passage meant for us to seek *only* the Kingdom of God, it would have the odd result of making good people not only feckless but also dependent upon bad people busy with quite other seekings. It says, rather, to seek the Kingdom of God *first*; that is, it gives an obviously necessary priority to the Great Economy over any little economy made within it. The passage also clearly includes nature within the Great Economy, and it affirms the goodness, indeed the sanctity, of natural creatures.

The fowls of the air and the lilies of the field live within the Great Economy entirely by nature, whereas humans, though entirely dependent upon it, must live in it partly by artifice. The birds can live in the Great Economy only as birds, the flowers only as flowers, the humans only as humans. The humans, unlike the wild creatures, may choose not to live in it —or, rather, since no creature can escape it, they may choose to *act* as if they do not, or they may choose to try to live in it on their own terms. If humans choose to live in the Great Economy on *its* terms, then they must live in harmony with it, maintaining it in trust and learning to consider the lives of the wild creatures.

Certain economic restrictions are clearly implied, and these restrictions have mainly to do with the economics of futurity. We know from other passages in the Gospels that a certain preparedness or provisioning for the future is required of us.

It may be that such preparedness is part of our obligation to today, and for *that* reason we need "take no thought for the morrow." But it is clear that such preparations can be carried too far, that we can provide too much for the future. The sin of "a certain rich man" in the twelfth chapter of Luke is that he has "much goods laid up for many years" and thus believes that he can "eat, drink, and be merry." The offense seems to be that he has stored up too much and in the process has belittled the future, for he has reduced it to the size of his own hopes and expectations. He is prepared for a future in which he will be prosperous, not for one in which he will be dead. We know from our own experience that it is possible to live in the present in such a way as to diminish the future practically as well as spiritually. By laying up "much goods" in the present—and, in the process, *using up* such goods as topsoil, fossil fuel, and fossil water—we incur a debt to the future that we cannot repay. That is, we diminish the future by deeds that we call "use" but that the future will call "theft." We may say, then, that we seek the Kingdom of God, in part, by our economic behavior, and we fail to find it if that behavior is wrong.

If we read Matthew 6:24–34 as a teaching that is *both* practical and spiritual, as I think we must, then we must see it as prescribing the terms of a kind of little economy or human economy. Since I am deriving it here from a Christian text, we could call it a Christian economy. But we need not call it that. A Buddhist might look at the working principles of the economy I am talking about and call it a Buddhist economy. E. F. Schumacher, in fact, says that the aim of "Buddhist economics" is "to obtain the maximum of well-being with the minimum of consumption," which I think is partly the sense of Matthew 6:24–34. Or we could call this economy (from Matthew 6:28) a "considerate" economy or, simply, a good economy. Whatever the name, the human economy, if it is to be a good economy, must fit harmoniously within and must correspond to the Great Economy; in certain important ways, it must be an analogue of the Great Economy.

A fifth principle of the Great Economy that must now be added to the previous four is that *we* cannot foresee an end to it: The same basic stuff is going to be shifting from one form to another, so far as we know, forever. From a human point of view, this is a rather heartless endurance. As cynics sometimes point out, conservation is always working, for what is lost or wasted in one place always turns up someplace else. Thus, soil erosion in Iowa involves no loss because the soil is conserved in the Gulf of Mexico. Such people like to point out that soil erosion is as "natural" as birdsong. And so it is, though these people neglect to observe that soil conservation is also natural, and that, before the advent of farming, nature alone worked effectively to keep Iowa topsoil in Iowa. But to say that soil erosion is natural is only a way of saying that there are some things that the Great Economy cannot do for humans. Only a little economy, only a good human economy, can define for us the value of keeping the topsoil where it is.

A good human economy, that is, defines and values human goods, and, like the Great Economy, it conserves and protects its goods. It proposes to endure. Like the Great Economy, a good human economy does not propose for itself a term to be set by humans. That termlessness, with all its implied human limits and restraints, is a human good.

The difference between the Great Economy and any human economy is pretty much the difference between the goose that laid the golden egg and the golden egg. For the goose to have value as a layer of golden eggs, she must be a live goose and therefore joined to the life cycle, which means that she is joined to all manner of things, patterns, and processes that sooner or later surpass human comprehension. The golden egg, on the other hand, can be fully valued by humans according to kind, weight, and measure—but it will not hatch, and it cannot be eaten. To make the value of the egg *fully* accountable, then, we must make it "golden," must remove it from life. But if in our valuation of it, we wish to consider its relation to the goose, we have to undertake a different kind of accounting, more exacting if less exact. That is, if we wish to value the egg in such a way as to preserve the goose that laid it, we find that we must behave, not scientifically, but humanely; we must understand ourselves as humans as fully as our traditional knowledge of ourselves permits. We participate in our little human economy to a considerable extent, that is, by factual knowledge, calculation, and manipulation; our participation in the Great Economy also requires those things, but requires as well humility, sympathy, forbearance, generosity, imagination.

Another critical difference, implicit in the foregoing, is that, though a human economy can evaluate, distribute, use, and preserve things of value, it cannot make value. Value can originate only in the Great Economy. It is true enough that humans can add value to natural things: We may transform trees into boards, and transform boards into chairs, adding value at each transformation. In a good human economy, these transformations would be made by good work, which would be properly valued and the workers properly rewarded. But a good human economy would recognize at the same time that it was dealing all along with materials and powers that it did not make. It did not make trees, and it did not make the intelligence and talents of the human workers. What the humans have added at every step is artificial, made by art, and though the value of art is critical to human life, it is a secondary value.

When humans presume to originate value, they make value that is first abstract and then false, tyrannical, and destructive of real value. Money value, for instance, can be said to be true only when it justly and stably represents the value of necessary goods, such as clothing, food, and shelter, which originate ultimately in the Great Economy. Humans can originate money value in the abstract, but only by inflation and usury, which falsify the value of necessary things and damage their natural and human sources. Inflation and usury and the damages that follow can be understood, perhaps, as retributions for the presumption that humans can make value.

We may say, then, that a human economy originates, manages, and distributes secondary or added values but that, if it is to last long, it must also manage in such a way as to make continuously available those values that are primary or given, the secondary values having mainly to do with husbandry and trusteeship. A little economy is obliged to receive them gratefully and to use them in such a way as not to diminish them. We might make a long list of things that we would have to describe as primary values, which come directly into the little economy from the Great, but the one I want to talk about, because it is the one with which we have the most intimate working relationship, is the topsoil.

We cannot speak of topsoil, indeed we cannot know what it is, without acknowledging at the outset that we cannot make it. We can care for it (or not), we can even, as we say, "build" it, but we can do so only by assenting to, preserving, and perhaps collaborating in its own processes. To those processes themselves we have nothing to contribute. We cannot make topsoil, and we cannot make any substitute for it; we cannot do what it does. It is apparently impossible to make an adequate description of topsoil in the sort of language that we have come to call "scientific." For, although any soil sample can be reduced to its inert quantities, a handful of the real thing has life in it; it is full of living creatures. And if we try to describe the behavior of that life we will see that it is doing something that, if we are not careful, we will call "unearthly": It is making life out of death. Not so very long ago, had we known about it what we know now, we would probably have called it "miraculous." In a time when death is looked upon with almost universal enmity, it is hard to believe that the land we live on and the lives we live are the gifts of death. Yet that is so, and it is the topsoil that makes it so. In fact, in talking about topsoil, it is hard to avoid the language of religion. When, in "This Compost," Whitman says, "The resurrection of the wheat appears with pale visage out of its graves," he is speaking in the Christian tradition, and yet he is describing what happens, with language that is entirely accurate and appropriate. And when at last he says of the earth that "It gives such divine materials to men," we feel that the propriety of the words comes not from convention but from the actuality of the uncanny transformation that his poem has required us to imagine, as if in obedience to the summons to "consider the lilies of the field."

Even in its functions that may seem, to mechanists, to be mechanical, the topsoil behaves complexly and wonderfully. A healthy topsoil, for instance, has at once the ability to hold water and to drain well. When we speak of the health of a watershed, these abilities are what we are talking about, and the word "health," which we do use in speaking of watersheds, warns us that we are not speaking merely of mechanics. A healthy soil is made by the life dying into it and by the life living in it, and to its double ability to drain and retain water we are complexly indebted, for it not only gives us good crops but also erosion control as well as *both* flood control and a constant water supply.

Obviously, topsoil, not energy or money, is the critical quantity in agriculture. And topsoil *is* a quantity; we need it in quantities. We now need more of it than we have; we need to help it to make more of itself. But it is a most peculiar quantity, for it is inseparable from quality. Topsoil is by definition *good* soil, and it can be preserved in human use only by good care. When humans see it as a mere quantity, they tend to make it that; they destroy the life in it, and they begin to measure in inches and feet and tons how much of it they have "lost."

When we see the topsoil as the foundation of that household of living creatures and their nonliving supports that we now call an "ecosystem" but which some of us understand better as a "neighborhood," we find ourselves in debt for other benefits that baffle our mechanical logic and defy our measures. For example, one of the principles of an ecosystem is that diversity increases capacity—or, to put it another way, that complications of form or pattern can increase greatly within quantitative limits. I suppose that this may be true only up to a point, but I suppose also that that point is far beyond the human capacity to understand or diagram the pattern.

On a farm put together on a sound ecological pattern, the same principle holds. Henry Besuden, the great farmer and shepherd of Clark County, Kentucky, compares the small sheep flock to the two spoons of sugar that can be added to a brimful cup of coffee, which then becomes "more palatable [but] doesn't run over. You can stock your farm to the limit with other livestock and still add a small flock of sheep." He says this, characteristically, after rejecting the efforts of sheep specialists to get beyond "the natural physical limits of the ewe" by breeding out of season in order to get three lamb crops in two years or by striving for "litters" of lambs rather than nature's optimum of twins. Rather than chafe at "natural physical limits," he would turn to nature's elegant way of enriching herself *within* her physical limits by diversification, by complication of pattern. Rather than strain the productive capacity of the ewe, he would, without strain, enlarge the productive capacity of the farm—a healthier, safer, and cheaper procedure. Like many of the better traditional farmers, Henry Besuden is suspicious of "the measure of land in length and width," for he would be mindful as well of "the depth and quality."

A small flock of ewes, fitted properly into a farm's pattern, virtually disappears into the farm and does it good, just as it virtually disappears into the time and energy economy of a farm family and does it good. And, properly fitted into the farm's pattern, the small flock virtually disappears from the debit side of the farm's accounts but shows up plainly on the credit side. This "disappearance" is possible, not to the extent that the farm is a human artifact, a belonging of the human economy, but to the extent that it remains, by its obedience to natural principle, a belonging of the Great Economy.

A little economy may be said to be good insofar as it perceives the excellence of these benefits and husbands and preserves them. It is by holding up this standard of goodness that we can best see what is wrong with the industrial economy. For the industrial economy does not see itself as a little economy; it sees itself as the *only* economy. It makes itself thus exclusive by the simple expedient of valuing only what it can use—that is, only what it can regard as "raw material" to be transformed mechanically into something else. What it cannot use, it characteristically describes as "useless," "worthless," "random," or "wild," and gives it some such name as "chaos," "disorder," or "waste"—and thus ruins it or cheapens it in preparation for eventual use. That western deserts or eastern mountains were once perceived as "useless" made it easy to dignify them by the "use" of strip mining. Once we acknowledge the existence of the Great Economy, however, we are astonished and frightened to see how much modern enterprise is the work of hubris, occurring outside the human boundary established by ancient tradition. The industrial economy is based on invasion and pillage of the Great Economy.

The weakness of the industrial economy is clearly revealed when it imposes its terms upon agriculture, for its terms cannot define those natural principles that are most vital to the life and longevity of farms. Even if the industrial economists could afford to do so, they could not describe the dependence of agriculture upon nature. If asked to consider the lilies of the field or told that the wheat is resurrected out of its graves, the agricultural industrialist would reply that "my engineer's mind inclines less toward the poetic and philosophical, and more toward the practical and possible," unable even to suspect that such a division of mind induces blindness to possibilities of the utmost practical concern.

That good topsoil both drains and retains water, that diversity increases capacity, are facts similarly alien to industrial logic. Industrialists see retention and drainage as different and opposite functions, and they would promote one at the expense of the other, just as, diversity being inimical to industrial procedure, they would commit themselves to the forlorn expedient of enlarging capacity by increasing area. They are thus encumbered by dependence on mechanical solutions that can work only by isolating and oversimplifying problems. Industrialists are condemned to proceed by devices. To facilitate water retention, they must resort to a specialized water-holding device such as a terrace or a dam; to facilitate drainage, they must use drain tile, or a ditch, or a "subsoiler." It is possible, I know, to argue that this analysis is too general and to produce exceptions, but I do not think it deniable that the discipline of soil conservation is now principally that of the engineer, not that of the farmer or soil husband—that it is now a matter of digging in the earth, not of enriching it.

I do not mean to say that the devices of engineering are always inappropriate; they have their place, not least in the restoration of land abused by the devices of engineering. My point is that, to facilitate both water retention and drainage in the same place, we must improve the soil, which is not a mechanical device but, among other things, a graveyard, a place of resurrection, and a community of living creatures. Devices may sometimes help, but only up to a point, for soil is improved by what humans do not do as well as by what they do. The proprieties of soil husbandry require acts that are much more complex than industrial acts, for these acts are conditioned by the ability *not* to act, by forbearance or self-restraint, sympathy or generosity. The industrial act is simply prescribed by thought, but the act of soil building is also *limited* by thought. We build soil by knowing what to do but also by knowing what not to do and by knowing when to stop. Both kinds of knowledge are necessary because invariably, at some point, the reach of human comprehension becomes too short, and at that point the work of the human economy must end in absolute deference to the working of the Great Economy. This, I take it, is the practical significance of the idea of the Sabbath.

To push our work beyond that point, invading the Great Economy, is to become guilty of hubris, of presuming to be greater than we are. We cannot do what the topsoil does, any more than we can do what God does or what a swallow does. We can fly, but only as humans—very crudely, noisily, and clumsily. We can dispose of corpses and garbage, but we cannot, by our devices, turn them into fertility and new life. And we are discovering, to our great uneasiness, that we cannot dispose at all of some of our so-called wastes that are toxic or radioactive. We can appropriate and in some fashion use godly powers, but we cannot use them safely, and we cannot control the results. That is to say that the human condition remains for us what it was for Homer and the authors of the Bible. Now that we have brought such enormous powers to our aid (we hope), it seems more necessary than ever to observe how inexorably the human condition still contains us. We only do what humans can do, and our machines, however they may appear to enlarge our possibilities, are invariably infected with our limitations. Sometimes, in enlarging our possibilities, they narrow our limits and leave us more powerful but less content, less safe, and less free. The mechanical means by which we propose to escape the human condition only extend it; thinking to transcend our definition as fallen creatures, we have only colonized more and more territory eastward of Eden.

II

Like the rich man of the parable, the industrialist thinks to escape the persistent obligations of the human condition by means of "much goods laid up for many years"—by means, in other words, of quantities: resources, supplies, stockpiles, funds, reserves. But this is a grossly oversimplifying dream and, thus, a dangerous one. All the great natural goods that empower agriculture, some of which I have discussed, have to do with quantities, but they have to do also with qualities, and

they involve principles that are not static but active; they have
to do with formal processes. The topsoil exists as such because
it is ceaselessly transforming death into life, ceaselessly supply-
ing food and water to all that lives in it and from it; otherwise,
"All flesh shall perish together, and man shall turn again unto
dust." If we are to live well on and from our land, we must
live by faith in the ceaselessness of these processes and by faith
in our own willingness and ability to collaborate with them.
Christ's prayer for "daily bread" is an affirmation of such faith,
just as it is a repudiation of faith in "much goods laid up." Our
life and livelihood are the gift of the topsoil and of our willingness and ability to care for it, to grow good wheat, to make
good bread; they do not derive from stockpiles of raw materials
or accumulations of purchasing power.

The industrial economy can define potentiality, even the potentiality of the living topsoil, only as a *fund*, and thus it must
accept impoverishment as the inescapable condition of abundance. The invariable mode of its relation both to nature and
to human culture is that of mining: withdrawal from a limited
fund until that fund is exhausted. It removes natural fertility
and human workmanship from the land, just as it removes
nourishment and human workmanship from bread. Thus the
land is reduced to abstract marketable quantities of length and
width, and bread to merchandise that is high in money value
but low in food value. "Our bread," Guy Davenport once said
to me, "is more obscene than our movies."

But the industrial use of *any* "resource" implies its exhaustion. It is for this reason that the industrial economy has been
accompanied by an ever-increasing hurry of research and ex-
ploration, the motive of which is not "free enterprise" or "the
spirit of free inquiry," as industrial scientists and apologists
would have us believe, but the desperation that naturally and
logically accompanies gluttony.

One of the favorite words of the industrial economy is "control": We want "to keep things under control"; we wish (or
so we say) to "control" inflation and erosion; we have a discipline known as "crowd control"; we believe in "controlled
growth" and "controlled development," in "traffic control"
and "self-control." But, because we are always setting out to
control something that we refuse to limit, we have made control a permanent and a helpless enterprise. If we will not limit

causes, there can be no controlling of effects. What is to be the fate of self-control in an economy that encourages and rewards unlimited selfishness?

More than anything else, we would like to "control the forces of nature," refusing at the same time to impose any limit on human nature. We assume that such control and such freedom are our "rights," which seems to ensure that our means of control (of nature and of all else that we see as alien) will be violent. It is startling to recognize the extent to which the industrial economy depends upon controlled explosions—in mines, in weapons, in the cylinders of engines, in the economic pattern known as "boom and bust." This dependence is the result of a progress that can be argued for, but those who argue for it must recognize that, in all these means, good ends are served by a destructive principle, an association that is difficult to control if it is not limited; moreover, they must recognize that our failure to limit this association has raised the specter of uncontrollable explosion. Nuclear holocaust, if it comes, will be the final detonation of an explosive economy.

An explosive economy, then, is not only an economy that is dependent upon explosions but also one that sets no limits on itself. Any little economy that sees itself as unlimited is obviously self-blinded. It does not see its real relation of dependence and obligation to the Great Economy; in fact, it does not see that there *is* a Great Economy. Instead, it calls the Great Economy "raw material" or "natural resources" or "nature" and proceeds with the business of putting it "under control."

But "control" is a word more than ordinarily revealing here, for its root meaning is to roll against, in the sense of a little wheel turning in opposition. The principle of control, then, involves necessarily the principle of division: One thing may turn against another thing only by being divided from it. This mechanical division and turning in opposition William Blake understood as evil, and he spoke of "Satanic wheels" and "Satanic mills": "wheel without wheel, with cogs tyrannic / Moving by compulsion each other." By "wheel without wheel," Blake meant wheel outside of wheel, one wheel communicating motion to the other in the manner of two cogwheels, the point being that one wheel can turn another wheel outside itself only in a direction opposite to its own. This, I suppose, is acceptable enough as a mechanism. It becomes "Satanic" when it becomes a ruling metaphor and is used to describe and to organize fundamental relationships. Against the Satanic "wheel without wheel," Blake set the wheels of Eden, which "Wheel within wheel in freedom revolve, in harmony and peace." This is the "wheel in the middle of a wheel" of Ezekiel's vision, and it is an image of harmony. That the relation of these wheels is not mechanical we know from Ezekiel 1:21: "the spirit of the living creature was in the wheels." The wheels of opposition oppose the spirit of the living creature.

What had happened, as Blake saw accurately and feared justifiably, was a fundamental shift in the relation of humankind to the rest of creation. Sometime between, say, Pope's verses on the Chain of Being in *An Essay on Man* and Blake's "London," the dominant minds had begun to see the human race, not as a part or a member of Creation, but as outside it and opposed to it. The industrial revolution was only a part of this change, but it is true that, when the wheels of the industrial revolution began to revolve, they turned against nature, which became the name for all of Creation thought to be below humanity, as well as, incidentally, against all once thought to be above humanity. Perhaps this would have been safe enough if nature—that is, if all the rest of Creation—had been, as proposed, passively subject to human purpose.

Of course, it never has been. As Blake foresaw, and as we now know, what we turn against must turn against us. Blake's image of the cogwheels turning in relentless opposition is terrifyingly apt, for in our vaunted war against nature, nature fights back. The earth may answer our pinches and pokes "only with spring," as e. e. cummings said, but if we pinch and poke too much, she can answer also with flood or drouth, with catastrophic soil erosion, with plague and famine. Many of the occurrences that we call "acts of God" or "accidents of nature" are simply forthright natural responses to human provocations. Not always; I do not mean to imply here that, by living in harmony with nature, we can be free of floods and storms and drouths and earthquakes and volcanic eruptions; I am only pointing out, as many others have done, that, by living in opposition to nature, we can *cause* natural calamities of which we would otherwise be free.

*

The problem seems to be that a human economy cannot prescribe the terms of its own success. In a time when we wish to
believe that humans are the sole authors of the truth, that truth
is relative, and that value judgments are all subjective, it is hard
to say that a human economy can be wrong, and yet we have
good, sound, practical reasons for saying so. It is indeed possible for a human economy to be wrong—not relatively wrong,
in the sense of being "out of adjustment," or unfair according
to some human definition of fairness, or weak according to the
definition of its own purposes—but wrong absolutely and according to practical measures. Of course, if we see the human
economy as the *only* economy, we will see its errors as political
failures, and we will continue to talk about "recovery." It is
only when we think of the little human economy in relation to
the Great Economy that we begin to understand our errors for
what they are and to see the qualitative meanings of our quantitative measures. If we see the industrial economy in terms of
the Great Economy, then we begin to see industrial wastes and
losses not as "trade-offs" or "necessary risks" but as costs that,
like all costs, are chargeable to somebody, sometime.

That we can prescribe the terms of our own success, that
we can live outside or in ignorance of the Great Economy
are the greatest errors. They condemn us to a life without a
standard, wavering in inescapable bewilderment from paltry
self-satisfaction to paltry self-dissatisfaction. But since we have
no place to live but in the Great Economy, whether or not
we know that and act accordingly is the critical question, not
about economy merely, but about human life itself.

It is possible to make a little economy, such as our present
one, that is so short-sighted and in which accounting is of so
short a term as to give the impression that vices are necessary
and practically justifiable. When we make our economy a little
wheel turning in opposition to what we call "nature," then we
set up competitiveness as the ruling principle in our explanation of reality and in our understanding of economy; we make
of it, willy-nilly, a virtue. But competitiveness, as a ruling principle and a virtue, imposes a logic that is extremely difficult,
perhaps impossible, to control. That logic explains why our
cars and our clothes are shoddily made, why our "wastes" are

toxic, and why our "defensive" weapons are suicidal; it explains why it is so difficult for us to draw a line between "free enterprise" and crime. If our economic ideal is maximum profit with minimum responsibility, why should we be surprised to find our corporations so frequently in court and robbery on the increase? Why should we be surprised to find that medicine has become an exploitive industry, profitable in direct proportion to its hurry and its mechanical indifference? People who pay for shoddy products or careless services and people who are robbed outright are equally victims of theft, the only difference being that the robbers outright are not guilty of fraud.

If, on the other hand, we see ourselves as living within the Great Economy, under the necessity of making our little human economy within it, according to its terms, the smaller wheel turning in sympathy with the greater, receiving its being and its motion from it, then we see that the traditional virtues are necessary and are practically justifiable. Then, because in the Great Economy *all* transactions count and the account is never "closed," the ideal changes. We see that we cannot *afford* maximum profit or power with minimum responsibility because, in the Great Economy, the loser's losses finally afflict the winner. Now the ideal must be "the maximum of well-being with the minimum of consumption," which both defines and requires neighborly love. Competitiveness cannot be the ruling principle, for the Great Economy is not a "side" that we can join nor are there such "sides" within it. Thus, it is not the "sum of its parts" but a *membership* of parts inextricably joined to each other, indebted to each other, receiving significance and worth from each other and from the whole. One is obliged to "consider the lilies of the field," not because they are lilies or because they are exemplary, but because they are fellow members and because, as fellow members, we and the lilies are in certain critical ways alike.

To say that within the Great Economy the virtues are necessary and practically justifiable is at once to remove them from that specialized, sanctimonious, condescending practice of virtuousness that is humorless, pointless, and intolerable to its beneficiaries. For a human, the good choice in the Great Economy is to see its membership as a neighborhood and oneself as a neighbor within it. I am sure that virtues count in a neighborhood—to "love thy neighbor as thyself" requires the help of all seven of them—but I am equally sure that in a neighborhood the virtues cannot be practiced as such. Temperance has no appearance or action of its own, nor does justice, prudence, fortitude, faith, hope, or charity. They can only be employed on occasions. "He who would do good to another," William Blake said, "must do it in Minute Particulars." To help each other, that is, we must go beyond the coldhearted charity of the "general good" and get down to work where we are:

Labour well the Minute Particulars, attend to the Little-ones,
And those who are in misery cannot remain so long
If we do but our duty: labour well the teeming Earth.

It is the Great Economy, not any little economy, that invests minute particulars with high and final importance. In the Great Economy, each part stands for the whole and is joined to it; the whole is present in the part and is its health. The industrial economy, by contrast, is always striving and failing to make fragments (pieces that *it* has broken) *add up* to an ever-fugitive wholeness.

Work that is authentically placed and understood within the Great Economy moves virtue toward virtuosity—that is, toward skill or technical competence. There is no use in helping our neighbors with their work if we do not know how to work. When the virtues are rightly practiced within the Great Economy, we do not call them virtues; we call them good farming, good forestry, good carpentry, good husbandry, good weaving and sewing, good homemaking, good parenthood, good neighborhood, and so on. The general principles are submerged in the particularities of their engagement with the world. Lao Tzu saw the appearance of the virtues as such, in the abstract, as indicative of their loss:

When people lost sight of the way to live
Came codes of love and honesty . . .
When differences weakened family ties
Came benevolent fathers and dutiful sons;
And when lands were disrupted and misgoverned
Came ministers commended as loyal.

And these lines might be read as an elaboration of the warning against the *appearances* of goodness at the beginning of the sixth chapter of Matthew.

The work of the small economy, when it is understandingly placed within the Great Economy, minutely particularizes the virtues and carries principle into practice; to the extent that it does so, it escapes specialization. The industrial economy requires the extreme specialization of work—the separation of work from its results—because it subsists upon divisions of interest and must deny the fundamental kinships of producer and consumer; seller and buyer; owner and worker; worker, work, and product; parent material and product; nature and artifice; thoughts, words, and deeds. Divided from those kinships, specialized artists and scientists identify themselves as "observers" or "objective observers"—that is, as outsiders without responsibility or involvement. But the industrialized arts and sciences are false, their division is a lie, for there is no specialization of results.

There is no "outside" to the Great Economy, no escape into either specialization or generality, no "time off." Even insignificance is no escape, for in the membership of the Great Economy everything signifies; whatever we do counts. If we do not serve what coheres and endures, we serve what disintegrates and destroys. We can *presume* that we are outside the membership that includes us, but that presumption only damages the membership—and ourselves, of course, along with it.

In the industrial economy, the arts and the sciences are specialized "professions," each having its own language, speaking to none of the others. But the Great Economy proposes arts and sciences of membership: ways of doing and ways of knowing that cannot be divided from each other or within themselves and that speak the common language of the communities where they are practiced.

The Loss of the University

THE PREDICAMENT of literature within the university is not fundamentally different from the predicament of any other discipline, which is not fundamentally different from the predicament of language. That is, the various disciplines have ceased to speak to each other; they have become too specialized, and this overspecialization, this separation, of the disciplines has been enabled and enforced by the specialization of their languages. As a result, the modern university has grown, not according to any unifying principle, like an expanding universe, but according to the principle of miscellaneous accretion, like a furniture storage business.

I assume that there is a degree of specialization that is unavoidable because concentration involves a narrowing of attention; we can only do one thing at a time. I assume further that there is a degree of specialization that is desirable because good work depends upon sustained practice. If we want the best work to be done in teaching or writing or stone masonry or farming, then we must arrange for that work to be done by proven master workers, people who are prepared for the work by long and excellent practice.

But to assume that there is a degree of specialization that is proper is at the same time to assume that there is a degree that is improper. The impropriety begins, I think, when the various kinds of workers come to be divided and cease to speak to one another. In this division they become makers of *parts* of things. This is the impropriety of industrial organization, of which Eric Gill wrote, "Skill in making . . . degenerates into mere dexterity, i.e. skill in doing, when the workman . . . ceases to be concerned for the thing made or . . . has no longer any responsibility for the thing made and has therefore lost the knowledge of what it is that he is making. . . . The factory hand can only know what he is *doing*. What is being made is no concern of his." Part of the problem in universities now (or part of the cause of the problem) is this loss of concern for the thing made and, back of that, I think, the loss of agreement on what the thing is that is being made.

The thing being made in a university is humanity. Given the current influence of universities, this is merely inevitable. But what universities, at least the public-supported ones, are *mandated* to make or to help to make is human beings in the fullest sense of those words—not just trained workers or knowledgeable citizens but responsible heirs and members of human culture. If the proper work of the university is only to equip people to fulfill private ambitions, then how do we justify public support? If it is only to prepare citizens to fulfill public responsibilities, then how do we justify the teaching of arts and sciences? The common denominator has to be larger than either career preparation or preparation for citizenship. Underlying the idea of a university—the bringing together, the combining into one, of all the disciplines—is the idea that good work and good citizenship are the inevitable by-products of the making of a good—that is, a fully developed—human being. This, as I understand it, is the definition of the name *university*.

In order to be concerned for the thing made, in order even to know what it is making, the university as a whole must speak the same language as all of its students and all of its graduates. There must, in other words, be a common tongue. Without a common tongue, a university not only loses concern for the thing made; it loses its own unity. Furthermore, when the departments of a university become so specialized that they can speak neither to each other nor to the students and graduates of other departments, then that university is displaced. As an institution, it no longer knows where it is, and therefore it cannot know either its responsibilities to its place or the effects of its irresponsibility. This too often is the practical meaning of "academic freedom": The teacher feels free to teach and learn, make and think, without concern for the thing made.

For example, it is still perfectly acceptable in land-grant universities for agricultural researchers to apply themselves to the development of more productive dairy cows without considering at all the fact that this development necessarily involves the failure of many thousands of dairies and dairy farmers—that it has already done so and will inevitably continue to do so. The researcher feels at liberty to justify such work merely on the basis of the ratio between the "production unit" and the volume of production. And such work is permitted to continue, I suspect, because it is reported in language that is unreadable and probably unintelligible to nearly everybody in the university, to nearly everybody who milks cows, and to nearly everybody who drinks milk. That a modern university might provide a forum in which such researchers might be required to defend their work before colleagues in, say, philosophy or history or literature is, at present, not likely, nor is it likely, at present, that the departments of philosophy, history, or literature could produce many colleagues able or willing to be interested in the ethics of agricultural research.

Language is at the heart of the problem. To profess, after all, is "to confess before"—to confess, I assume, before all who live within the neighborhood or under the influence of the confessor. But to confess before one's neighbors and clients in a language that few of them can understand is not to confess at all. The specialized professional language is thus not merely a contradiction in terms; it is a cheat and a hiding place; it may, indeed, be an ambush. At the very root of the idea of profession and professorship is the imperative to speak plainly in the common tongue.

That the common tongue should become the exclusive specialty of a department in a university is therefore a tragedy, and not just for the university and its worldly place; it is a tragedy for the common tongue. It means that the common tongue, so far as the university is concerned, *ceases* to be the common tongue; it becomes merely one tongue within a confusion of tongues. Our language and literature cease to be seen as occurring in the world, and begin to be seen as occurring within their university department and within themselves. Literature ceases to be the meeting ground of all readers of the common tongue and becomes only the occasion of a deafening clatter *about* literature. Teachers and students read the great songs and stories to learn *about* them, not to learn *from* them. The *texts* are tracked as by the passing of an army of ants, but the power of songs and stories to affect life is still little acknowledged, apparently because it is little felt.

The specialist approach, of course, is partly justifiable; in both speech and literature, language does occur within itself. It echoes within itself, reverberating endlessly like a voice echoing within a cave, and speaking in answer to its echo, and the answer again echoing. It must do this; its nature, in part, is to do this.

But its nature also is to turn outward to the world, to strike its worldly objects cleanly and cease to echo—to achieve a kind of rest and silence in them. The professionalization of language and of language study makes the cave inescapable; one strives without rest in the interior clamor.

The silence in which words return to their objects, touch them, and come to rest is not the silence of the plugged ear. It is the world's silence, such as occurs after the first hard freeze of autumn, when the weeks-long singing of the crickets is suddenly stopped, and when, by a blessedly recurring accident, all machine noises have stopped for the moment, too. It is a silence that must be prepared for and waited for; it requires a silence of one's own.

The reverberations of language within itself are, finally, mere noise, no better or worse than the noise of accumulated facts that grate aimlessly against each other in think tanks and other hollow places. Facts, like words, are not things but verbal tokens or signs of things that finally must be carried back to the things they stand for to be verified. This carrying back is not specialist work but an act generally human, though only properly humbled and quieted humans can do it. It is an act that at once enlarges and shapes, frees and limits us.

It is necessary, for example, that the word *tree* evoke memories that are both personal and cultural. In order to understand fully what a tree is, we must remember much of our experience with trees and much that we have heard and read about them. We destroy those memories by reducing trees to facts, by thinking of *tree* as a mere word, or by treating our memory of trees as "cultural history." When we call a tree a tree, we are not isolated among words and facts but are at once in the company of the tree itself and surrounded by ancestral voices calling out to us all that trees have been and meant. This is simply the condition of being human in this world, and there is nothing that art and science can do about it, except get used to it. But, of course, only specialized "professional" arts and sciences would propose or wish to do something about it.

This necessity for words and facts to return to their objects in the world describes one of the boundaries of a university, one of the boundaries of book learning anywhere, and it describes the need for humility, restraint, exacting discipline, and high standards within that boundary.

Beside every effort of making, which is necessarily narrow, there must be an effort of judgment, of criticism, which must be as broad as possible. That is, every made thing must be submitted to these questions: What is the quality of this thing as a human artifact, as an addition to the world of made and of created things? How suitable is it to the needs of human and natural neighborhoods?

It must, of course, sooner or later be submitted as well to the special question: How good is this poem or this farm or this hospital as such? For it to have a human value, it obviously must be well made; it must meet the specialized, technical criteria; it must be *good* as such. But the question of its quality as such is not interesting—in the long run it is probably not even askable—unless we ask it under the rule of the more general questions. If we are disposed to judge apart from the larger questions, if we judge, as well as make, as specialists, then a good forger has as valid a claim to our respect as a good artist.

These two problems, how to make and how to judge, are the business of education. But education has tended increasingly to ignore the doubleness of its obligation. It has concerned itself more and more exclusively with the problem of how to make, narrowing the issue of judgment virtually to the terms of the made thing itself. But the thing made by education now is not a fully developed human being; it is a specialist, a careerist, a graduate. In industrial education, the thing *finally* made is of no concern to the makers.

In some instances, this is because the specialized "fields" have grown so complicated within themselves that the curriculum leaves no time for the broad and basic studies that would inform judgment. In other instances, one feels that there is a potentially embarrassing conflict between judgment broadly informed and the specialized career for which the student is being prepared; teachers of advertising techniques, for example, could ill afford for their students to realize that they are learning the arts of lying and seduction. In all instances, this narrowing is justified by the improbable assumption that young students, before they know anything else, know what they need to learn.

If the disintegration of the university begins in its specialist ideology, it is enforced by a commercial compulsion to satisfy the customer. Since the student is now so much a free agent in determining his or her education, the department administrators and the faculty members must necessarily be preoccupied with the problem of how to keep enrollments up. Something obviously must be done to keep the classes filled; otherwise, the students will wander off to more attractive courses or to courses more directly useful to their proposed careers. Under such circumstances it is inevitable that requirements will be lightened, standards lowered, grades inflated, and instruction narrowed to the supposed requirements of some supposed career opportunity.

Dr. Johnson told Mrs. Thrale that his cousin, Cornelius Ford, "advised him to study the Principles of every thing, that a general Acquaintance with Life might be the Consequence of his Enquiries—Learn said he the leading Precognita of all things . . . grasp the Trunk hard only, and you will shake all the Branches." The soundness of this advice seems indisputable, and the metaphor entirely apt. From the trunk it is possible to "branch out." One can begin with a trunk and develop a single branch or any number of branches; although it may be possible to begin with a branch and develop a trunk, that is neither so probable nor so promising. The modern university, at any rate, more and more resembles a loose collection of lopped branches waving about randomly in the air. "Modern knowledge is departmentalized," H. J. Massingham wrote in 1943, "while the essence of culture is initiation into wholeness, so that all the divisions of knowledge are considered as the branches of one tree, the Tree of Life whose roots went deep into earth and whose top was in heaven."

This Tree, for many hundreds of years, seems to have come almost naturally to mind when we have sought to describe the form of knowledge. In Western tradition, it is at least as old as Genesis, and the form it gives us for all that we know is organic, unified, comprehensive, connective—and moral. The tree, at the beginning, was two trees: the tree of life and the tree of the knowledge of good and evil. Later, in our understanding of them, the two trees seem to have become one, or each seems to stand for the other—for in the world after the Fall, how can the two be separated? To know life is to know good and evil; to prepare young people for life is to prepare them to know the difference between good and evil. If we represent knowledge as a tree, we know that things that are divided are yet connected. We know that to observe the divisions and ignore the connections is to destroy the tree. The history of modern education may be the history of the loss of this image, and of its replacement by the pattern of the industrial machine, which subsists upon division—and by industrial economics ("publish or perish"), which is meaningless apart from division.

The need for broadly informed human judgment nevertheless remains, and this need requires inescapably an education that is broad and basic. In the face of this need, which is *both* private and public, "career preparation" is an improper use of public money, since "career preparation" serves merely private ends; it is also a waste of the student's time, since "career preparation" is best and most properly acquired in apprenticeships under the supervision of employers. The proper subject for a school, for example, is how to speak and write well, not how to be a "public speaker" or a "broadcaster" or a "creative writer" or a "technical writer" or a journalist or a practitioner of "business English." If one can speak and write well, then, given the need, one can make a speech or write an article or a story or a business letter. If one cannot speak or write well, then the tricks of a trade will be no help.

The work that should, and that can, unify a university is that of deciding what a student should be required to learn—what studies, that is, constitute the trunk of the tree of a person's education. "Career preparation," which has given so much practical support to academic specialization (and so many rewards to academic specialists), seems to have destroyed interest in this question. But the question exists and the failure to answer it (or even to ask it) imposes severe penalties on teachers, students, and the public alike. The penalties imposed on students and graduates by their failure to get a broad, basic education are, I think, obvious enough. The public penalties are also obvious if we consider, for instance, the number of certified expert speakers and writers who do not speak or write well, who do not know that they speak or write poorly, and who apparently do not care whether or not they speak or write honestly.

The penalties that this failure imposes on teachers are not so obvious, mainly, I suppose, because so far the penalties have been obscured by rewards. The penalties for teachers are the same as those for students and the public, plus one more: The failure to decide what students should be required to learn keeps the teacher from functioning as, and perhaps from becoming, a responsible adult.

There is no one to teach young people but older people, and so the older people must do it. That they do not know enough to do it, that they have never been smart enough or experienced enough or good enough to do it, does not matter. They must do it because there is no one else to do it. This is simply the elemental trial—some would say the elemental tragedy—of human life: the necessity to proceed on the basis merely of the knowledge that is available, the necessity to postpone until too late the question of the sufficiency and the truth of that knowledge.

There is, then, an inescapable component of trial and error in human education; some things that are taught will be wrong because fallible humans are the teachers. But the reason for education, its constant effort and discipline, is surely to reduce the young person's dependence on trial and error as far as possible. For it *can* be reduced. One should not have to learn everything, or the basic things, by trial and error. A child should not have to learn the danger of heat by falling into the fire. A student should not have to learn the penalties of illiteracy by being illiterate or the value of a good education by the "object lesson" of a poor one.

Teachers, moreover, are not providing "career preparation" so much as they are "preparing young people for life." This statement is not the result of educational doctrine; it is simply the fact of the matter. To prepare young people for life, teachers must dispense knowledge and enlighten ignorance, just as supposed. But ignorance is not only the affliction that teaching seeks to cure; it is also the condition, the predicament, in which teaching is done, for teachers do not know the life or the lives for which their students are being prepared.

This condition gives the lie to the claims for "career preparation," since students may not *have* the careers for which they have been prepared: The "job market" may be overfilled; the requirements for this or that career may change; the student may change, or the world may. The teacher, preparing the student for a life necessarily unknown to them both, has no excusable choice but to help the student to "grasp the Trunk."

Yet the arguments for "career preparation" continue to be made and to grow in ambition. On August 23, 1983, for example, the Associated Press announced that "the head of the Texas school board wants to require sixth-graders to choose career 'tracks' that will point them toward jobs." Thus, twelve-year-old children would be "free to choose" the kind of life they wish to live. They would even be free to change "career tracks," though, according to the article, such a change would involve the penalty of a delayed graduation.

But these are free choices granted to children not prepared or ready to make them. The idea, in reality, is to impose adult choices on children, and these "choices" mask the most vicious sort of economic determinism. This idea of education as "career track" diminishes everything it touches: education, teaching, childhood, the future. And such a thing could not be contemplated for sixth-graders, obviously, if it had not already been instituted in the undergraduate programs of colleges and universities.

To require or expect or even allow young people to choose courses of study and careers that they do not yet know anything about is not, as is claimed, a grant of freedom. It is a severe limitation upon freedom. It means, in practice, that when the student has finished school and is faced then, appropriately, with the need to choose a career, he or she is prepared to choose only one. At that point, the student stands in need of a freedom of choice uselessly granted years before and forfeited in that grant.

The responsibility to decide what to teach the young is an adult responsibility. When adults transfer this responsibility to the young, whether they do it by indifference or as a grant of freedom, they trap themselves in a kind of childishness. In that failure to accept responsibility, the teacher's own learning and character are disemployed, and, in the contemporary industrialized education system, they are easily replaced by bureaucratic and methodological procedures, "job market" specifications, and tests graded by machines.

This question of what all young people should be expected to learn is now little discussed. The reason, apparently, is the tacit belief that now, with the demands of specialization so numerous and varied, such a question would be extremely hard, if not impossible, to answer. And yet this question appears to be as much within the reach of reason and common sense as any other. It cannot be denied, to begin with, that all the disciplines rest upon the knowledge of letters and the knowledge of numbers. Some rest more on letters than numbers, some more on numbers than letters, but it is surely true to say that people without knowledge of both letters and numbers are not prepared to learn much else. From there, one can proceed confidently to say that history, literature, philosophy, and foreign languages rest principally on the knowledge of letters and carry it forward, and that biology, chemistry, and physics rest on the knowledge of numbers and carry it forward. This provides us with a description of a probably adequate "core curriculum"—one that would prepare a student well both to choose a direction of further study and to go in that direction. An equally obvious need, then, is to eliminate from the curriculum all courses without content—that is, all courses in methodologies and technologies that could, and should, be learned in apprenticeships.

Besides the innate human imperfections already mentioned, other painful problems are involved in expecting and requiring students to choose the course of their own education. These problems have to do mainly with the diversity of gifts and abilities: that is, some people are not talented in some kinds of work or study; some, moreover, who are poor in one discipline may be excellent in another. Why should such people be forced into situations in which they must see themselves as poor workers or as failures?

The question is not a comfortable one, and I do not believe that it can or should be comfortably answered. There is pain in the requirement to risk failure and pain in the failure that may result from that requirement. But failure is a possibility; in varying degrees for all of us, it is inescapable. The argument for removing the possibility of failure from schoolwork is therefore necessarily specious. The wrong is not in subjecting students to the possibility of failure or in calling their failures failures; the wrong is in the teacher's inability to see that failure in school is not necessarily synonymous with and does not necessarily lead to failure in the world. The wrong is in the failure to see or respect the boundaries between the school and the world. When those are not understood and respected, then the school, the school career, the diploma are all surrounded by such a spurious and modish dignity that failure in school *is* failure in the world. It is for this reason that it is so easy to give education a money value and to sell it to consumers in job lots.

It is a fact that some people with able minds do not fit well into schools and are not properly valued by schoolish standards and tests. If such people fail in a school, their failure should be so called; a school's worth and integrity depend upon its willingness to call things by their right names. But, by the same token, a failure in school is no more than that; it does not necessarily imply or cause failure in the world, any more than it implies or causes stupidity. It is not rare for the judgment of the world to overturn the judgment of schools. There are other tests for human abilities than those given in schools, and there are some that cannot be given in schools. My own life has happened to acquaint me with several people who did not attend high school but who have been more knowledgeable in their "field" and who have had better things to say about matters of general importance than most of the doctors of philosophy I have known. This is not an "anti-intellectual" statement; it is a statement of what I take to be fact, and it means only that the uses of schools are limited—another fact, which schools prepare us to learn by surprise.

Another necessary consideration is that low expectations and standards in universities encourage the lowering of expectations and standards in the high schools and elementary schools. If the universities raise their expectations and standards, the high schools and elementary schools will raise theirs; they will have to. On the other hand, if the universities teach high school courses because the students are not prepared for university courses, then they simply relieve the high schools of their duty and in the process make themselves unable to do their own duty. Once the school stoops to meet the student, the standards of judgment begin to topple at all levels. As standards are lowered—as they cease to be the measure of the students and come to be measured by them—it becomes manifestly less possible for students to fail. But for the same reason it becomes less possible for them to learn and for teachers to teach.

The question, then, is what is to determine the pattern of education. Shall we shape a university education according to the previous schooling of the students, which we suppose has made them unfit to meet high expectations and standards, and to the supposed needs of students in some future still dark to us all? Or shall we shape it according to the nature and demands of the "leading Precognita of all things"—that is, according to the essential subjects of study? If we shape education to fit the students, then we clearly can maintain no standards; we will lose the subjects and eventually will lose the students as well. If we shape it to the subjects, then we will save both the subjects and the students. The inescapable purpose of education must be to preserve and pass on the essential human means—the thoughts and words and works and ways and standards and hopes without which we are not human. To preserve these things and to pass them on is to prepare students for life.

That such work cannot be done without high standards ought not to have to be said. There are necessarily increasing degrees of complexity in the studies as students rise through the grades and the years, and yet the standards remain the same. The first-graders, that is, must read and write in simple sentences, but they read and write, even so, in the language of the King James Bible, of Shakespeare and Johnson, of Thoreau, Whitman, Dickinson, and Twain. The grade-schooler and the graduate student must study the same American history, and there is no excuse for falsifying it in order to make it elementary.

Moreover, if standards are to be upheld, they cannot be specialized, professionalized, or departmented. Only common standards can be upheld—standards that are held and upheld in common by the whole community. When, in a university, for instance, English composition is made the responsibility exclusively of the English department, or of the subdepartment of freshman English, then the quality of the work in composition courses declines and the standards decline. This happens necessarily and for an obvious reason: If students' writing is graded according to form and quality in composition class but according only to "content" in, say, history class and if in other classes students are not required to write at all, then the message to the students is clear: namely, that the form and quality of their writing matters only in composition class, which is to say that it matters very little indeed. High standards of composition can be upheld only if they are upheld everywhere in the university.

Not only must the standards be held and upheld in common but they must also be applied fairly—that is, there must be no conditions with respect to persons or groups. There must be no discrimination for or against any person for any reason. The quality of the individual performer is the issue, not the category of the performer. The aim is to recognize, reward, and promote good work. Special pleading for "disadvantaged" groups—whether disadvantaged by history, economics, or education—can only make it increasingly difficult for members of that group to do good work and have it recognized.

If the university faculties have failed to answer the question of the internal placement of the knowledges of the arts and sciences with respect to each other and to the university as a whole, they have, it seems to me, also failed to ask the question of the external placement of these knowledges with respect to truth and to the world. This, of course, is a dangerous question, and I raise it with appropriate fear. The danger is that such questions should be *settled* by any institution whatever; these questions are the proper business of the people in the institutions, not of the institutions as such. I am arguing here against the specialist absorption in career and procedure that destroys what I take to be the indispensable interest in the question of the truth of what is taught and learned, as well as the equally indispensable interest in the fate and the use of knowledge in the world.

I would be frightened to hear that some university had

suddenly taken a lively interest in the question of what is true and was in the process of answering it, perhaps by a faculty vote. But I am equally frightened by the fashionable lack of interest in the question among university teachers individually. I am more frightened when this disinterest, under the alias of "objectivity," is given the status of a public virtue.

Objectivity, in practice, means that one studies or teaches one's subject *as such*, without concern for its relation to other subjects or to the world—that is, without concern for its truth. If one is concerned, if one cares, about the truth or falsity of anything, one cannot be objective: one is glad if it is true and sorry if it is false; one believes it if it is judged to be true and disbelieves it if it is judged to be false. Moreover, the truth or falsity of some things cannot be objectively demonstrated, but must be determined by feeling and appearance, intuition and experience. And this work of judgment cannot take place at all with respect to one thing or one subject alone. The issue of truth rises out of the comparison of one thing with another, out of the study of the relations and influences between one thing and another and between one thing and many others.

Thus, if teachers aspire to the academic virtue of objectivity, they must teach as if their subject has nothing to do with anything beyond itself. The teacher of literature, for example, must propose the study of poems as relics left by people who, unlike our highly favored modern selves, believed in things not subject to measurable proof; religious poetry, that is, may be taught as having to do with matters once believed but not believable. The poetry is to be learned *about*; to learn *from* it would be an embarrassing betrayal of objectivity.

That this is more than a matter of classroom technique is made sufficiently evident in the current fracas over the teaching of the Bible in public schools. Judge Jackson Kiser of the federal district court in Bristol, Virginia, recently ruled that it would be constitutional to teach the Bible to public school students if the course is offered as an elective and "taught in an objective manner with no attempt made to indoctrinate the children as to either the truth or falsity of the biblical materials." James J. Kilpatrick, who discussed this ruling approvingly in one of his columns, suggested that the Bible might be taught "as Shakespeare is taught" and that this would be good because "the Bible is a rich lode of allusion, example and quotation." He warned that "The line that divides propaganda from instruction is a wavering line drawn on shifting sands," and he concluded by asserting that "Whatever else the Bible may be, the Bible is in fact literature. The trick is to teach it that way."

The interesting question here is not whether young English-speakers should know the Bible—they obviously should—but whether a book that so directly offers itself to our belief or disbelief can be taught "as literature." It clearly cannot be so taught except by ignoring "whatever else [it] may be," which is a very substantial part of it. The question, then, is whether it can be adequately or usefully taught as something less than it is. The fact is that the writers of the Bible did not think that they were writing what Judge Kiser and Mr. Kilpatrick call "literature." They thought they were writing the truth, which they expected to be believed by some and disbelieved by others. It is conceivable that the Bible could be well taught by a teacher who believes that it is true, by a teacher who believes that it is untrue, or by a teacher who believes that it is partly true. That it could be well taught by a teacher uninterested in the question of its truth is not conceivable. That a lively interest in the Bible could be maintained through several generations of teachers uninterested in the question of its truth is also not conceivable.

Obviously, this issue of the Bible in the public schools cannot be resolved by federal court decisions that prescribe teaching methods. It can only be settled in terms of the freedom of teachers to teach as they believe and in terms of the relation of teachers and schools to their local communities. It may be that in this controversy we are seeing the breakdown of the public school system, as an inevitable consequence of the breakdown of local communities. It is hard to believe that this can be remedied in courts of law.

My point, anyhow, is that we could not consider teaching the Bible "as literature" if we were not already teaching literature "as literature"—as if we do not care, as if it does not matter, whether or not it is true. The causes of this are undoubtedly numerous, but prominent among them is a kind of shame among teachers of literature and other "humanities" that their truths are not objectively provable as are the truths of science. There is now an embarrassment about any statement that depends for confirmation upon experience or imagination or feeling or faith, and this embarrassment has produced an overwhelming impulse to treat such statements merely as artifacts, cultural relics, bits of historical evidence, or things of "aesthetic value." We will study, record, analyze, criticize, and appreciate. But we will not believe; we will not, in the full sense, know.

The result is a stance of "critical objectivity" that causes many teachers, historians, and critics of literature to sound—not like mathematicians or chemists: their methodology does not permit that yet—but like ethologists, students of the behavior of a species to which they do not belong, in whose history and fate they have no part, their aim being, not to know anything for themselves, but to "advance knowledge." This may be said to work, as a textual mechanics, but it is not an approach by which one may know any great work of literature. That route is simply closed to people interested in what "they" thought "then"; it is closed to people who think that "Dante's world" or "Shakespeare's world" is far removed and completely alienated from "our world"; and it is closed to the viewers of poetic devices, emotional effects, and esthetic values.

The great distraction behind the modern fate of literature, I think, is expressed in Coleridge's statement that his endeavor in *Lyrical Ballads* was "to transfer from our inward nature . . . a semblance of truth sufficient to procure for these shadows of imagination that willing suspension of disbelief for the moment, which constitutes poetic faith." That is a sentence full of quakes and tremors. Is our inward nature true only by semblance? What is the difference, in a work of art, between truth and "semblance of truth"? What must be the result of separating "poetic faith" from faith of any other kind and then of making "poetic faith" dependent upon will?

The gist of the problem is in that adjective *willing*, which implies the superiority of the believer to what is believed. This implication, I am convinced, is simply untrue. Belief precedes will. One either believes or one does not, and, if one believes, then one willingly believes. If one disbelieves, even unwillingly, all the will in the world cannot make one believe. Belief is involuntary, as is the Ancient Mariner's recognition of the beauty and sanctity of the water snakes:

> A spring of love gushed from my heart,
> And I blessed them unaware . . .

This involuntary belief is the only approach to the great writings. One may, assuredly, not believe, and we must, of course, grant unbelievers the right to read and comment as unbelievers, for disbelief is a legitimate response, because it is a possible one. We must be aware of the possibility that belief may be false, and of the need to awaken from false belief; "one need not step into belief as into an abyss." But we must be aware also that to disbelieve is to remain, in an important sense, outside the work. When we are *in* the work, we are long past the possibility of any debate with ourselves about whether or not to be willing to believe. When we are *in* the work, we simply *know* that great Odysseus has come home, that Dante is in the presence of the celestial rose, that Cordelia, though her father carries her in his arms, is dead. If we know these things, we are apt to know too that Mary Magdalene mistook the risen Christ for the gardener—and are thus eligible to be taken lightly by objective scholars, and to be corrected by a federal judge.

We and these works meet in imagination; by imagination we know their truth. In imagination, there is no specifically or exclusively "poetic faith," just as there is no faith that is specifically or exclusively religious. Belief is the same wherever it happens, and its terms are invariably set by the imagination. One believes, that is, because one *sees*, not because one is informed. That is why, four hundred years after Copernicus, we still say, "The sun is rising."

When we read the ballad of Sir Patrick Spens we know that the knight and his men have drowned because "Thair hats they swam aboone," not because we have confirmed the event by the study of historical documents. And if our assent is forced also by the ballad of Thomas Rhymer, far stranger than that of Sir Patrick, what are we to say? Must we go, believing, into the poem, and then return from it in disbelief because we find the story in no official record, have read of no such thing in the newspaper, and know nothing like it in our own experience? Or must we live with the poem, with our awareness of its power over us, as a piece of evidence that reality may be larger than we thought?

"Does that mean," I am asked, "that it's not possible for us to read Homer properly because we don't believe in the Greek gods?" I can only answer that I suspect that a proper reading of Homer will *result* in some manner of belief in his gods. How else explain their survival in the works of Christian writers into our own time? This survival has its apotheosis, it seems to me, in C. S. Lewis's novel, *That Hideous Strength*, at the end of which the Greek planetary deities reappear on earth as angels. Lewis wrote as a Christian who had read Homer, but he had read, obviously, as a man whose imagination was not encumbered with any such clinical apparatus as the willing suspension of disbelief. As such a reader, though he was a Christian, his reading had told him that the pagan gods retained a certain authority and commanded a certain assent. Like many of his forebears in English literary tradition, he yearned toward them. Their triumphant return, at the end of *That Hideous Strength*, as members of the heavenly hierarchy of Christianity, is not altogether a surprise. It is a profound resolution, not only in the novel itself, but in the history of English literature. One hears the ghosts of Spenser and Milton sighing with relief.

Questions of the authenticity of imaginings invite answers, and yet may remain unanswered. For the imagination is not always subject to immediate proof or demonstration. It is often subject only to the slow and partial authentication of experience. It is subject, that is, to a practical, though not an exact, validation, and it is subject to correction. For a work of imagination to endure through time, it must prove valid, and it must survive correction. It is correctable by experience, by critical judgment, and by further works of imagination.

To say that a work of imagination is subject to correction is, of course, to imply that there is no "world of imagination" as distinct from or opposed to the "real world." The imagination is *in* the world, is at work in it, is necessary to it, and is correctable by it. This correcting of imagination by experience is inescapable, necessary, and endless, as is the correcting of experience by imagination. This is the great general work of criticism to which we all are called. It is not literary criticism any more than it is historical or agricultural or biological criticism, but it must nevertheless be a fundamental part of the work of literary criticism, as it must be of criticisms of all other kinds. One of the most profound of human needs is for the truth of imagination to prove itself in every life and place in the world, and for the truth of the world's lives and places to be proved in imagination.

This need takes us as far as possible from the argument for works of imagination, human artifacts, as special cases, privileged somehow to offer themselves to the world on their own terms. It is this argument and the consequent abandonment of the general criticism that have permitted the universities to organize themselves on the industrial principle, as if faculties and students and all that they might teach and learn are no more than parts of a machine, the purpose of which they have, in general, not bothered to define, much less to question. And largely through the agency of the universities, this principle and this metaphor now dominate our relation to nature and to one another.

If, for the sake of its own health, a university must be interested in the question of the truth of what it teaches, then, for the sake of the world's health, it must be interested in the fate of that truth and the uses made of it in the world. It must want to know where its graduates live, where they work, and what they do. Do they return home with their knowledge to enhance and protect the life of their neighborhoods? Do they join the "upwardly mobile" professional force now exploiting and destroying local communities, both human and natural, all over the country? Has the work of the university, over the last generation, increased or decreased literacy and knowledge of the classics? Has it increased or decreased the general understanding of the sciences? Has it increased or decreased pollution and soil erosion? Has it increased or decreased the ability and the willingness of public servants to tell the truth? Such questions are not, of course, precisely answerable. Questions about influence never are. But they are askable, and the asking, should we choose to ask, would be a unifying and a shaping force.

Preserving Wildness

THE ARGUMENT over the proper relation of humanity to nature is becoming, as the sixties used to say, polarized. And the result, as before, is bad talk on both sides. At one extreme are those who sound as if they are entirely in favor of nature; they assume that there is no necessary disjuncture or difference between the human estate and the estate of nature, that human good is in some simple way the same as natural good. They believe, at least in principle, that the biosphere is an egalitarian system, in which all creatures, including humans, are equal in value and have an equal right to live and flourish. These people tend to stand aloof from the issue of the proper human use of nature. Indeed, they have begun to use "stewardship" (meaning the responsible use of nature) as a term of denigration.

At the other extreme are the nature conquerors, who have no patience with an old-fashioned outdoor farm, let alone a wilderness. These people divide all reality into two parts: human good, which they define as profit, comfort, and security; and everything else, which they understand as a stockpile of "natural resources" or "raw materials," which will sooner or later be transformed into human good. The aims of these militant tinkerers invariably manage to be at once unimpeachable and suspect. They wish earnestly, for example, to solve what they call "the problem of hunger"—if it can be done glamorously, comfortably, and profitably. They believe that the ability to do something is the reason to do it. According to a recent press release from the University of Illinois College of Agriculture, researchers there are looking forward to "food production without either farmers or farms." (This is perhaps the first explicit acknowledgment of the program that has been implicit in the work of the land-grant universities for forty or fifty years.)

If I had to choose, I would join the nature extremists against the technology extremists, but this choice seems poor, even assuming that it is possible. I would prefer to stay in the middle, not to avoid taking sides, but because I think the middle *is* a side, as well as the real location of the problem.

*

The middle, of course, is always rather roomy and bewildering territory, and so I should state plainly the assumptions that define the ground on which I intend to stand:

1. We live in a wilderness, in which we and our works occupy a tiny space and play a tiny part. We exist under its dispensation and by its tolerance.

2. This wilderness, the universe, is *somewhat* hospitable to us, but it is also absolutely dangerous to us (it is going to kill us, sooner or later), and we are absolutely dependent upon it.

3. That we depend upon what we are endangered by is a problem not solvable by "problem solving." It does not have what the nature romantic or the technocrat would regard as a solution. We are not going back to the Garden of Eden, nor are we going to manufacture an Industrial Paradise.

4. There does exist a possibility that we can live more or less in harmony with our native wilderness; I am betting my life that such a harmony is possible. But I do not believe that it can be achieved simply or easily or that it can ever be perfect, and I am certain that it can never be made, once and for all, but is the forever unfinished lifework of our species.

5. It is not possible (at least, not for very long) for humans to intend their own good specifically or exclusively. We cannot intend our good, in the long run, without intending the good of our place—which means, ultimately, the good of the world.

6. To use or not to use nature is not a choice that is available to us; we can live only at the expense of other lives. Our choice has rather to do with how and how much to use. This is not a choice that can be decided satisfactorily in principle or in theory; it is a choice intransigently impractical. That is, it must be worked out in local practice because, by necessity, the practice will vary somewhat from one locality to another. There is, thus, no *practical* way that we can intend the good of the world; practice can only be local.

7. If there is no escape from the human use of nature, then human good cannot be simply synonymous with natural good.

What these assumptions describe, of course, is the human predicament. It is a spiritual predicament, for it requires us to be properly humble and grateful; time and again, it asks us to be still and wait. But it is also a practical problem, for it requires us to *do* things.

In going to work on this problem it is a mistake to proceed on the basis of an assumed division or divisibility between nature and humanity, or wildness and domesticity. But it is also a mistake to assume that there is no difference between the natural and the human. If these things could be divided, our life would be far simpler and easier than it is, just as it would be if they were not different. Our problem, exactly, is that the human and the natural are indivisible, and yet are different.

The indivisibility of wildness and domesticity, even within the fabric of human life itself, is easy enough to demonstrate. Our bodily life, to begin at the nearest place, is half wild. Perhaps it is more than half wild, for it is dependent upon reflexes, instincts, and appetites that we do not cause or intend and that we cannot, or had better not, stop. We live, partly, because we are domestic creatures—that is, we participate in our human economy to the extent that we "make a living"; we are able, with variable success, to discipline our appetites and instincts in order to produce this artifact, this human living. And yet it is equally true that we breathe and our hearts beat and we survive as a species because we are wild.

The same is true of a healthy human economy as it branches upward out of the soil. The topsoil, to the extent that it is fertile, is wild; it is a dark wilderness, ultimately unknowable, teeming with wildlife. A forest or a crop, no matter how intentionally husbanded by human foresters or farmers, will be found to be healthy precisely to the extent that it is wild—able to collaborate with earth, air, light, and water in the way common to plants before humans walked the earth. We know from experience that we can increase our domestic demands upon plants so far that we force them into kinds of failure that wild plants do not experience.

Breeders of domestic animals, likewise, know that, when a breeding program is too much governed by human intention, by economic considerations, or by fashion, uselessness is the result. Size or productivity, for instance, will be gained at the cost of health, vigor, or reproductive ability. In other words, so-called domestic animals must remain half wild, or more than half, because they are creatures of nature. Humans are intelligent enough to select for a type of creature; they are not intelligent enough to *make* a creature. Their efforts to make an entirely domestic animal, like their efforts to make an entirely domestic human, are doomed to failure because they do not have and undoubtedly are never going to have the full set of production standards for the making of creatures. From a human point of view, then, creature making is wild. The effort to make plants, animals, and humans ever more governable by human intentions is continuing with more determination and more violence than ever, but that does not mean that it is nearer to success. It means only that we are increasing the violence and the magnitude of the expectable reactions.

To be divided against nature, against wildness, then, is a human disaster because it is to be divided against ourselves. It confines our identity as creatures entirely within the bounds of our own understanding, which is invariably a mistake because it is invariably reductive. It reduces our largeness, our mystery, to a petty and sickly comprehensibility.

But to say that we are not divided and not dividable from nature is not to say that there is no difference between us and the other creatures. Human nature partakes of nature, participates in it, is dependent on it, and yet is different from it. We feel the difference as discomfort or difficulty or danger. Nature is not easy to live with. It is hard to have rain on your cut hay, or floodwater over your cropland, or coyotes in your sheep; it is hard when nature does not respect your intentions, and she never does exactly respect them. Moreover, such problems belong to all of us, to the human lot. Humans who do not experience them are exempt only because they are paying (or underpaying) other humans such as farmers to deal with nature on their behalf. Further, it is not just agriculture-dependent humanity that has had to put up with natural dangers and frustrations; these have been the lot of hunting and gathering societies also, and the wild creatures do not always live comfortably or easily with nature either.

But humans differ most from other creatures in the extent to which they must be *made* what they are—that is, in the extent to which they are artifacts of their culture. It is true that what we might as well call culture does go into the making of some birds and animals, but this teaching is so much less than the teaching that makes a human as to be almost a different thing. To take a creature who is biologically a human and to make him or her fully human is a task that requires many years (some of us sometimes fear that it requires more than a lifetime), and this long effort of human making is necessary, I think, because of our power. In the hierarchy of power among the earth's creatures, we are at the top, and we have been growing stronger for a long time. We are now, to ourselves, incomprehensibly powerful, capable of doing more damage than floods, storms, volcanoes, and earthquakes. And so it is more important than ever that we should have cultures capable of making us into humans—creatures capable of prudence, justice, fortitude, temperance, and the other virtues. For our history reveals that, stripped of the restraints, disciplines, and ameliorations of culture, humans are not "natural," not "thinking animals" or "naked apes," but monsters—indiscriminate and insatiable killers and destroyers. We differ from other creatures, partly, in our susceptibility to monstrosity. It is perhaps for this reason that, in the wake of the great wars of our century, we have seen poets such as T. S. Eliot, Ezra Pound, and David Jones making an effort to reweave the tattered garment of culture and to reestablish the cultural tasks, which are, as Pound put it, "To know the histories / to know good from evil / And know whom to trust." And we see, if we follow Pound a little further, that the recovery of culture involves, leads to, or is the recovery of nature:

> the trees rise
> and there is a wide sward between them
> . . . myrrh and olibanum on the altar stone
> giving perfume,
>   and where was nothing
> now is furry assemblage
>   and in the boughs now are voices . . .

In the recovery of culture *and* nature is the knowledge of how to farm well, how to preserve, harvest, and replenish the forests, how to make, build, and use, return and restore. In this *double* recovery, which is the recovery of our humanity, is the hope that the domestic and the wild can exist together in lasting harmony.

This doubleness of allegiance and responsibility, difficult as it always is, confusing as it sometimes is, apparently is inescapable. A culture that does not measure itself by nature, by an understanding of its debts to nature, becomes destructive of nature and thus of itself. A culture that does not measure itself by its own best work and the best work of other cultures (the determination of which is its unending task) becomes destructive of itself and thus of nature.

Harmony is one phase, the good phase, of the inescapable dialogue between culture and nature. In this phase, humans consciously and conscientiously ask of their work: Is this good for us? Is this good for our place? And the questioning and answering in this phase is minutely particular: It can occur only with reference to particular artifacts, events, places, ecosystems, and neighborhoods. When the cultural side of the dialogue becomes too theoretical or abstract, the other phase, the bad one, begins. Then the conscious, responsible questions are not asked; acts begin to be committed and things to be made on their own terms for their own sakes, culture deteriorates, and nature retaliates.

The awareness that we are slowly growing into now is that the earthly wildness that we are so complexly dependent upon is at our mercy. It has become, in a sense, our artifact because it can only survive by a human understanding and forbearance that we now must make. The only thing we have to preserve nature with is culture; the only thing we have to preserve wildness with is domesticity.

To me, this means simply that we are not safe in assuming that we can preserve wildness by making wilderness preserves. Those of us who see that wildness and wilderness need to be preserved are going to have to understand the dependence of these things upon our domestic economy and our domestic behavior. If we do not have an economy capable of valuing in particular terms the durable good of localities and communities, then we are not going to be able to preserve anything. We are going to have to see that, if we want our forests to last, then we must make wood products that last, for our forests are more threatened by shoddy workmanship than by clear-cutting or by fire. Good workmanship—that is, careful, considerate, and loving work—requires us to think considerately of the whole process, natural and cultural, involved in the making of wooden artifacts, because the good worker does not share the industrial contempt for "raw material." The good worker loves the board before it becomes a table, loves the tree before it yields the board, loves the forest before it gives up the tree. The good worker understands that a badly made artifact is both an insult to its user and a danger to its source. We could say, then, that good forestry begins with the respectful husbanding of the forest that we call stewardship and ends with well-made tables and chairs and houses, just as good agriculture begins with stewardship of the fields and ends with good meals.

In other words, conservation is going to prove increasingly futile and increasingly meaningless if its proscriptions are not answered positively by an economy that rewards and enforces good use. I would call this a loving economy, for it would strive to place a proper value on all the materials of the world, in all their metamorphoses from soil and water, air and light to the finished goods of our towns and households, and I think that the only effective motive for this would be a particularizing love for local things, rising out of local knowledge and local allegiance.

Our present economy, by contrast, does not account for affection at all, which is to say that it does not account for value. It is simply a description of the career of money as it preys upon both nature and human society. Apparently because our age is so manifestly unconcerned for the life of the spirit, many people conclude that it places an undue value on material things. But that cannot be so, for people who valued material things would take care of them and would care for the sources of them. We could argue that an age that *properly* valued and cared for material things would be an age properly spiritual. In my part of the country, the Shakers, "unworldly" as they were, were the true materialists, for they truly valued materials.

And they valued them in the only way that such things *can* be valued in practice: by good workmanship, both elegant and sound. The so-called materialism of our own time is, by contrast, at once indifferent to spiritual concerns and insatiably destructive of the material world. And I would call our economy, not materialistic, but abstract, intent upon the subversion of both spirit and matter by abstractions of value and of power. In such an economy, it is impossible to value anything that one *has*. What one has (house or job, spouse or car) is only valuable insofar as it can be exchanged for what one believes that one wants—a limitless economic process based upon boundless dissatisfaction.

Now that the practical processes of industrial civilization have become so threatening to humanity and to nature, it is easy for us, or for some of us, to see that practicality needs to be made subject to spiritual values and spiritual measures. But we must not forget that it is also necessary for spirituality to be responsive to practical questions. For human beings the spiritual and the practical are, and should be, inseparable. Alone, practicality becomes dangerous; spirituality, alone, becomes feeble and pointless. Alone, either becomes dull. Each is the other's discipline, in a sense, and in good work the two are joined.

"The dignity of toil is undermined when its necessity is gone," Kathleen Raine says, and she is right. It is an insight that we dare not ignore, and I would emphasize that it applies to *all* toil. What is not needed is frivolous. Everything depends on our right relation to necessity—and therefore on our right definition of necessity. In defining our necessity, we must be careful to discount the subsidies, the unrepaid borrowings, from nature that have so far sustained industrial civilization: the "cheap" fossil fuels and ores; the forests that have been cut down and not replanted; the virgin soils of much of the world, whose fertility has not been replenished.

And so, though I am trying to unspecialize the idea and the job of preserving wildness, I am not against wilderness preservation. I am only pointing out, as the Reagan administration has done, that the wildernesses we are trying to preserve are standing squarely in the way of our present economy, and that the wildernesses cannot survive if our economy does not change.

The reason to preserve wilderness is that we need it. We need wilderness of all kinds, large and small, public and private. We need to go now and again into places where our work is disallowed, where our hopes and plans have no standing. We need to come into the presence of the unqualified and mysterious formality of Creation. And I would agree with Edward Abbey that we need as well some tracts of what he calls "absolute wilderness," which "through general agreement none of us enters at all."

We need wilderness also because wildness—nature—is one of our indispensable studies. We need to understand it as our source and preserver, as an essential measure of our history and behavior, and as the ultimate definer of our possibilities. There are, I think, three questions that must be asked with respect to a human economy in any given place:

1. What is here?
2. What will nature permit us to do here?
3. What will nature help us to do here?

The second and third questions are obviously the ones that would define agendas of practical research and of work. If we do not work with and within natural tolerances, then we will not be permitted to work for long. It is plain enough, for example, that if we use soil fertility faster than nature can replenish it, we are proposing an end that we do not desire. And to ignore the possibility of help from nature makes farming, for example, too expensive for farmers—as we are seeing. It may make life too expensive for humans.

But the second and third questions are ruled by the first. They cannot be answered—they cannot intelligently be asked —until the first has been answered. And yet the first question has not been answered, or asked, so far as I know, in the whole history of the American economy. All the great changes, from the Indian wars and the opening of agricultural frontiers to the inauguration of genetic engineering, have been made without a backward look and in ignorance of whereabouts. Our response to the forest and the prairie that covered our present fields was to get them out of the way as soon as possible. And the obstructive human populations of Indians and "inefficient" or small farmers have been dealt with in the same spirit. We have never known what we were doing because we have never known what we were *un*doing. We cannot know what we are doing until we know what nature would be doing if we were doing nothing. And that is why we need small native wildernesses widely dispersed over the countryside as well as large ones in spectacular places.

However, to say that wilderness and wildness are indispensable to us, indivisible from us, is not to say that we can find sufficient standards for our life and work in nature. To suggest that, for humans, there is a simple equation between "natural" and "good" is to fall prey immediately to the cynics who love to point out that, after all, "everything is natural." They are, of course, correct. Nature provides bountifully for her children, but, as we would now say, she is also extremely permissive. If her children want to destroy one another entirely or to commit suicide, that is all right with her. There is nothing, after all, more natural than the extinction of species; the extinction of *all* species, we must assume, would also be perfectly natural.

Clearly, if we want to argue for the existence of the world as we know it, we will have to find some way of qualifying and supplementing this relentless criterion of "natural." Perhaps we can do so only by a reaffirmation of a lesser kind of naturalness —that of self-interest. Certainly human self-interest has much wickedness to answer for, and we are living in just fear of it; nevertheless, we must take care not to condemn it absolutely. After all, we value this passing work of nature that we call "the natural world," with its graceful plenty of animals and plants, precisely because *we* need it and love it and want it for a home.

We are creatures obviously subordinate to nature, dependent upon a wild world that we did not make. And yet we are joined to that larger nature by our own nature, a part of which is our self-interest. A common complaint nowadays is that humans think the world is "anthropocentric," or human-centered. I understand the complaint; the assumptions of so-called anthropocentrism often result in gross and dangerous insubordination. And yet I don't know how the human species can avoid some version of self-centeredness; I don't know how any species can. An earthworm, I think, is living in an earthworm-centered world; the thrush who eats the earthworm is living in a thrush-centered world; the hawk who eats the thrush is living in a hawk-centered world. Each creature, that is, does what is necessary in its own behalf, and is domestic in its own *domus* or home.

Humans differ from earthworms, thrushes, and hawks in their capacity to do more—in modern times, a great deal more —in their own behalf than is necessary. Moreover, the vast majority of humans in the industrial nations are guilty of this extravagance. One of the oldest human arguments is over the question of how much is necessary. How much must humans do in their own behalf in order to be fully human? The number and variety of the answers ought to notify us that we never have known for sure, and yet we have the disquieting suspicion that, almost always, the honest answer has been "less."

We have no way to work at this question, it seems to me, except by perceiving that, in order to have the world, we must share it, both with each other and with other creatures, which is immediately complicated by the further perception that, in order to live in the world, we must use it somewhat at the expense of other creatures. We must acknowledge both the centrality and the limits of our self-interest. One can hardly imagine a tougher situation.

But in the recognition of the difficulty of our situation is a kind of relief, for it makes us give up the hope that a solution can be found in a simple preference for humanity over nature or nature over humanity. The only solutions we have ahead of us will need to be worked for and worked out. They will have to be practical solutions, resulting in good local practice. There is work to do that can be done.

As we undertake this work, perhaps the greatest immediate danger lies in our dislike of ourselves as a species. This is an understandable dislike—we are justly afraid of ourselves—but we are nevertheless obliged to think and act out of a proper self-interest and a genuine self-respect as human beings. Otherwise, we will allow our dislike and fear of ourselves to justify further abuses of one another and the world. We must come to terms with the fact that it is not natural to be disloyal to one's own kind.

For these reasons, there is great danger in the perception that "there are too many people," whatever truth may be in it, for this is a premise from which it is too likely that somebody, sooner or later, will proceed to a determination of *who* are the surplus. If we conclude that there are too many, it is hard to avoid the further conclusion that there are some we do not need. But how many do we need, and which ones? Which ones, now apparently unnecessary, may turn out later to be indispensable? We do not know; it is a part of our mystery, our wildness, that we do not know.

I would argue that, at least for us in the United States, the conclusion that "there are too many people" is premature, not because I know that there are *not* too many people, but because I do not think we are prepared to come to such a conclusion. I grant that questions about population size need to be asked, but they are not the *first* questions that need to be asked.

The "population problem," initially, should be examined as a problem, not of quantity, but of pattern. Before we conclude that we have too many people, we must ask if we have people who are misused, people who are misplaced, or people who are abusing the places they have. The facts of most immediate importance may be, not how many we are, but where we are and what we are doing. At any rate, the attempt to solve our problems by reducing our numbers may be a distraction from the overriding population statistic of our time: that *one* human with a nuclear bomb and the will to use it is 100 percent too many. I would argue that it is not human fecundity that is overcrowding the world so much as technological multipliers of the power of individual humans. The worst disease of the world now is probably the ideology of technological heroism, according to which more and more people willingly cause large-scale effects that they do not foresee and that they cannot control. This is the ideology of the professional class of the industrial nations—a class whose allegiance to communities and places has been dissolved by their economic motives and by their educations. These are people who will go anywhere and jeopardize anything in order to assure the success of their careers.

We may or may not have room for more people, but it is certain that we do not have more room for technological heroics. We do not need any more thousand-dollar solutions to ten-dollar problems or million-dollar solutions to thousand-dollar problems—or multibillion-dollar solutions where there was never a problem at all. We have no way to compute the inhabitability of our places; we cannot weigh or measure the pleasures we take in them; we cannot say how many dollars domestic tranquillity is worth. And yet we must now learn to bear in mind the memory of communities destroyed, disfigured, or made desolate by technological events, as well as the memory of families dispossessed, displaced, and impoverished by "labor-saving" machines. The issue of human obsolescence may be more urgent for us now than the issue of human population.

The population issue thus leads directly to the issue of proportion and scale. What is the proper amount of power for a human to use? What are the proper limits of human enterprise? How may these proprieties be determined? Such questions may seem inordinately difficult, but that is because we have gone too long without asking them. One of the fundamental assumptions of industrial economics has been that such questions are outmoded and that we need never ask them again. The failure of that assumption now requires us to reconsider the claims of wildness and to renew our understanding of the old ideas of propriety and harmony.

When we propose that humans should learn to behave properly with respect to nature so as to place their domestic economy harmoniously upon and within the sustaining and surrounding wilderness, then we make possible a sort of landscape criticism. Then we can see that it is not primarily the number of people inhabiting a landscape that determines the propriety of the ratio and the relation between human domesticity and wildness, but it is the way the people divide the landscape and use it. We can see that it is the landscape of monoculture in which both nature and humanity are most at risk. We feel the human fragility of the huge one-class housing development, just as we feel the natural fragility of the huge one-crop field.

Looking at the monocultures of industrial civilization, we yearn with a kind of homesickness for the humanness and the naturalness of a highly diversified, multipurpose landscape, democratically divided, with many margins. The margins are of the utmost importance. They are the divisions between holdings, as well as between kinds of work and kinds of land. These margins—lanes, streamsides, wooded fencerows, and the like—are always freeholds of wildness, where limits are set on human intention. Such places are hospitable to the wild lives of plants and animals and to the wild play of human children. They enact, within the bounds of human domesticity itself, a human courtesy toward the wild that is one of the best safeguards of designated tracts of true wilderness. This is the landscape of harmony, safer far for life of all kinds than the landscape of monoculture. And we should not neglect to notice that, whereas the monocultural landscape is totalitarian in tendency, the landscape of harmony is democratic and free.

A Good Farmer of the Old School

A T THE 1982 Draft Horse Sale in Columbus, Ohio, Maury Telleen summoned me over to the group of horsemen with whom he was talking: "Come here," he said, "I want you to hear this." One of those horsemen was Lancie Clippinger, and what Maury wanted me to hear was the story of Lancie's corn crop of the year before.

The story, which Lancie obligingly told again, was as interesting to me as Maury had expected it to be. Lancie, that year, had planted forty acres of corn; he had also bred forty gilts that he had raised so that their pigs would be ready to feed when the corn would be ripe. The gilts produced 360 pigs, an average of nine per head. When the corn was ready for harvest, Lancie divided off a strip of the field with an electric fence and turned in the 360 shoats. After the shoats had fed on that strip for a while, Lancie opened a new strip for them. He then picked the strip where they had just fed. In that way, he fattened his 360 shoats and also harvested all the corn he needed for his other stock.

The shoats brought $40,000. Lancie's expenses had been for seed corn, 275 pounds of fertilizer per acre, and one quart per acre of herbicide. He did not say what the total costs amounted to, but it was clear enough that his net income from the forty acres of corn had been high, in a year when the corn itself would have brought perhaps two dollars a bushel.

At the end of the story, I remember, Lancie and Maury had a conversation that went about like this:

"Do you farrow your sows in a farrowing house?"

"No."

"Oh, you do it in huts, then?"

"No, I have a field I turn them out in. It has plenty of shade and water. And I see them every day."

Here was an intelligent man, obviously, who knew the value of doing his own thinking and paying attention, who understood clearly that the profit is in the difference between costs and earnings, and who proceeded directly to minimize his costs. In a time when hog farmers often spend many thousands of dollars on highly specialized housing and equipment, Lancie's "hog operation" consisted almost entirely of hogs. His principal outlays otherwise were for the farm itself and for fencing. But what struck me most, I think, was the way he had employed nature and the hogs themselves to his own advantage. The bred sows needed plenty of shade, water, and room for exercise; Lancie provided those things, and nature did the rest. He also supplied his own care and attention, which came free; they did not have to be purchased at an inflated cost from an industrial supplier. And then, instead of harvesting his corn mechanically, hauling it, storing it, grinding it, and hauling it to his shoats, he let the shoats harvest and grind it for themselves. He had the use of the whole hog, whereas in a "confinement operation," the hog's feet, teeth, and eyes have virtually no use and produce no profit.

At the next Columbus Sale, I hunted Lancie up, and again we spent a long time talking. We talked about draft horses, of course, but also about milk cows and dairying. And that part of our conversation interested me about as much as the hog story had the year before. What so impressed me was Lancie's belief that there is a limit to the number of cows that a dairy farmer can manage well; he thought the maximum number to be about twenty-five: "If a fellow milks twenty-five cows, he'll *see* them all." If he milks more than that, Lancie said, even though he may touch them all, he will not *see* them all. As in Lancie's account of his corn crop and the 360 shoats, the emphasis here was on the importance of seeing, of paying attention. That this is important economically, he made clear in something he said to me later: "You can take care of twenty or twenty-five cows and do it right. More, you're overlooking things that cost you money." It is necessary, Lancie thinks, to limit the scale of operation, not only in dairying, but in all other enterprises on the farm because proper scale permits a correct balance between work and care. The distinction he was making, it seemed to me, was between work, as it has been understood traditionally on the farm, and processing, as it is understood in industry.

Those two conversations stayed in my mind, proving useful many times in my effort to understand the troubles developing in our agricultural economy. I knew that Lancie Clippinger was one of the best farmers of the old school, and I promised myself that I would visit him at his farm, which I was finally able to do in October 1985.

The farm is on somewhat rolling land, surrounded by wood-lots and brushy fencerows, so that it has a little of the feeling of a large forest clearing. There are 175 acres, of which about 135 are cropped; the rest is in permanent pasture and woods. Although conveniently close to the state road, the farm is at the end of a lane, set off to itself. It is pretty and quiet, a pleas-ant place to live and to farm, as well as to visit. Lancie and his wife, Verna Bell, bought the place and moved there in the fall of 1971.

When my wife and I drove into the yard, Kathy, one of Lan-cie's granddaughters, who had evidently been watching for us, came out of the house to meet us. She took us out through the barn lot to a granary where Lancie, his son Keith, and Sherri, another granddaughter, were sacking some oats. We waited, talking with Kathy, while they finished the job, and then we went with Lancie and Keith to look at the horses.

Lancie keeps only geldings, buying them at sales as wean-lings, raising and breaking them, selling them, and then re-placing them with new colts. When we were there, he had nine head: a pair of black Percherons, a handsome crossbred bay with black mane and tail, and six Belgians. Though he prefers Percherons, he does not specialize; at the sales, his only aim is to buy "colts that look like they'll grow into good big horses." He wants them big because the big ones bring the best prices, but, like nearly all draft horse people who use their horses, he would rather have smaller ones—fifteen hundred pounds or so—if he were keeping them only to work.

The horses he led out for us were in prime condition, and he had been right about them: They had, sure enough, grown into good big ones. These horses may be destined for pulling contests and show hitches, but while they are at Lancie's they put in a lot of time at farm work—they work their way through school, you might say. Like so many farmers of his time, Lancie once made the change from horses to tractors, but with him this did not last long. He was without horses "for a little while" in the seventies, and after that he began to use them again. Now he uses the horses for "just about everything" except cut-ting and baling his hay and picking his corn. Last spring he used his big tractor only two days. The last time he went to use it, it wouldn't start, and he left it sitting in the shed; it was still sitting there at the time of our visit.

Part of the justification for the return to the use of horses is economic. When he was doing all his work with tractors, Lancie's fuel bill was $6,000 a year; now it is about $2,000. Since the horses themselves are a profit-making enterprise on this farm, the $4,000 they save on fuel is money in the bank. But the economic reason is not the only one: "Pleasure," Lancie says, "is a big part of it." At the year's end, his bank account will show a difference that the horses have made, but day by day his reason for working them is that he *likes* to.

He does not need nine horses in order to do his farming. He has so many because he needs to keep replacements on hand for the horses he sells. He aims, he says, to sell "two or three or four horses every year." To farm his 175 acres, he needs only four good geldings, although he would probably like to keep five, in case he needed a spare. With four horses on his grain drill, he can plant fifteen or twenty acres in a day. He uses four horses also on an eight-foot tandem disk and a springtooth harrow, and he can plant twelve or fifteen acres of corn a day "and not half try."

In plowing, he goes by the old rule of thumb that you can plow an acre per horse per day, provided the horses are in hard condition. "If you start at seven in the morning and stay there the way you ought to," he says, "you can plow three acres a day with three horses." That is what he does, and he does it with a walking plow because, he says, it is easier to walk than to ride. That, of course, is hardly a popular opinion, and Lancie is amused by the surprise it sometimes causes.

One spring, he says, after he had started plowing, he ordered some lime. When the trucker brought the first load, he stopped by the house to ask where to spread it. Mrs. Clippinger told him that Lancie was plowing, and pointed out to the field where Lancie could be seen walking in the furrow behind his plow and team. The trucker was astonished: "Even the *Amish* ride!"

In 1936, Lancie remembers, he plowed a hundred acres, sixty of them in sod, with two horses, Bob and Joe. Together, that team weighed about thirty-five hundred pounds. They were blacks. Lancie had been logging with them before he started plowing, and they were in good shape, ready to go. They plowed two acres a day, six days a week, for nearly nine weeks. It is the sort of thing, one guesses, that could have been done only because all the conditions were right: a strong young man, a tough team, a good season. "Looked like, back then, there wasn't any bad weather," Lancie says, laughing. "You could work all the time."

This farmer's extensive use of live horsepower is possible because his farm is the right size for it and because a sensible rotation of crops both reduces the acreage to be plowed each year and distributes the other field work so that not too much needs to be done at any one time. Of the farm's 135 arable acres, approximately fifty-five will be in corn, forty in oats, and forty in alfalfa. Each of the crops will be grown on the same land two years in order to avoid buying alfalfa seed every year.

The two-year-old alfalfa, turned under, supplies enough nitrogen for the first year of corn. In the second year, the corn crop receives a little commercial nitrogen. The routine application of fertilizer on the corn is 275 pounds per acre of 10-10-20, drilled into the row with the planter. The oats are fertilized at the same rate as the corn, while the alfalfa field, because Lancie sells quite a bit of hay, receives 600 pounds per acre of 3-14-42 in two applications every year. The land is limed at a rate of two tons per acre every time it is plowed. Otherwise, for fertilization Lancie depends on manure from his cattle and horses. "That's what counts," he says. It counts because it pays but does not cost. He usually has enough manure to cover his corn ground every year.

This system of management has not only maintained the productive capacity of the farm but has greatly improved it. Fourteen years ago, when Lancie began on it, the place was farmed out. The previous farmer had plowed it all and planted it all in corn year after year. When the farm sold in the fall of 1971, the corn crop, which was still standing, was bought by a neighboring farmer, who found it not worth picking. Lancie plowed it under the next spring. In order to have a corn crop that first year, he used 900 pounds of fertilizer to the acre—300 pounds of nitrogen and 600 of "straight analysis." After that, when his rotations and other restorative practices had been established, he went to his present rate of 275 pounds of 10-10-20. The resulting rates of production speak well for good care: The corn has made 150 bushels per acre, Lancie says, "for a long time"; this year his oats made 109 bushels per acre, and he also harvested 11,000 fifty-pound bales of alfalfa hay from a forty-acre field (a per-acre yield of about seven tons) and sold 4,800 bales for $12,000.

In addition to seed and fertilizer, Lancie purchases some insecticide and herbicide. This year his alfalfa was sprayed once for weevils, and he used a half-pint of 2-4-D per acre on his corn. The 2-4-D, he says, would not have been necessary if he had cultivated four times instead of twice. Using the chemical saved two cultivations that would have interfered with hay harvest.

What is most significant about Lancie's management of his crops is that it gives his farm a degree of independence that is unusual in these times. The farm, first of all, is ordered and used according to its own nature and carrying capacity, not according to the dictates of farm policy, expert advice, or fluctuations of the economy. The possibility of solving one's economic problems by production alone is not, in Lancie's opinion, a good possibility. If you are losing money on the corn you produce, he points out, the more you produce the more you lose. That so many farmers continue to compensate for low grain prices by increasing production, at great cost to their farms and to themselves, is a sort of wonder to him. "The cheaper it is, the more they plow," he says. "I don't know what they mean." His own farm, by contrast, grows approximately the same acreages of the same crops every year, not because that is what the economy supposedly demands, but because that is what the land can produce at the least cost for the longest time.

Since the farm itself is so much the source of its own fertility and operating energy, Lancie's use of purchased supplies can be minimal, selective, and nonaddictive. Because his cropping pattern and system of management are sound, Lancie can buy these things to suit his convenience. His total expense for 2-4-D for his corn this year, for example, was fifty-six dollars —a very small price to pay in order to have his hands and his mind free at haying time. The point, I think, is that he had a choice: He could choose to do what made the most sense. A further point is that he can quit using chemicals and purchased fertilizer if it ever makes economic sense to do so. As a farmer, he is not addicted to these things.

The conventional industrial farmer, on the other hand, is too often the prisoner of his own technology and methods and has no choice but to continue to do as he has done, whatever the disadvantages. A farmer who has no fences cannot turn hogs in to harvest his corn when prices are low. A farmer who has invested heavily in a farrowing house and all the equipment that goes with it is stuck with that investment. If, for some reason, it ceases to be profitable for him to produce feeder pigs, he still has the farrowing house, which is good for little else, and perhaps a debt on it as well. Thus, mental paralysis and economic slavery can be instituted on a farm by the farmer's technological choices.

One of the main results of Lancie Clippinger's independence is versatility, enabling him to take advantage quickly of opportunities as they appear. Because he has invested in no expensive specialized equipment, he can change his ways to suit his wishes or his circumstances. That he did well raising and finishing shoats one year does not mean that he must continue to raise them. Last year, for instance, he thought there was money to be made on skinny sows. He bought sixty-two at $100 a head, turned them into his cornfield, and, while they ate, he picked. "We all worked together," he says. The sows did a nearly perfect job of gleaning the field, and they brought $200 a head when he sold them.

There is a direct economic payoff in this freedom of choice: It pays to be able to choose to substitute a team of horses for a tractor, or manure for fertilizer, or cultivation for herbicides. When you cultivate a field of corn, as Lancie says, "you're selling your labor"; in other words, you ensure a relation between production and consumption that is proper because it makes sound economic sense. If the farmer does not achieve that proper relation on his farm, he will be a victim. When Lancie prepares his ground with plow and harrow and cultivates his crop instead of buying chemicals, he is a producer, not a consumer; he is selling his labor, not buying an expensive substitute for labor. Moreover, when he does this with a team of horses instead of buying fuel, he is selling his team's labor, not paying for an expensive substitute. When he uses his own corn, oats, and hay to replace petroleum, he is selling those feeds for a far higher return than he could get on the market. He and his horses are functioning, in effect, as solar converters, making usable and profitable the free sunlight that falls onto the farm. They are producing at home the energy, weed control, and fertility that other farmers are going broke trying to pay for.

The industrial farmer consumes more than he produces and is a captive consumer of the suppliers who have prospered by the ruination of such farmers. So far as the national economy is concerned, this kind of farmer exists only to provide cheap food and to enrich the agribusiness corporations, at his own expense.

Sometimes Lancie's intelligent methods and his habit of paying attention yield unexpected dividends. The year after he hogged down the forty acres of corn with the 360 shoats, the field was covered with an excellent stand of alsike clover. "It was pretty," Lancie says, but he didn't know where it came from. He asked around in the neighborhood and discovered that the field had been in alsike seventeen years before. The seed had lain in the ground all that time, waiting for conditions to be right, and somehow the hogs had made them right. Thus, that year's very profitable corn harvest, which had been so well planned, resulted in a valuable gift that nobody had planned—or could have planned. There is no recipe, so far as I know, for making such a thing happen. Obviously, though, a certain eligibility is required. It happened on Lancie's farm undoubtedly because he is the kind of farmer he is. If he had been plowing the whole farm every year and planting it all in corn, as his predecessor had, such a thing would not have happened.

It is care, obviously, that makes the difference. The farm gives gifts because it is given a chance to do so; it is not overcropped or overused. One of Lancie's kindnesses to his farm is his regular rotation of his crops; another is his keeping of livestock, which gives him not only the advantages I have already described but also permits him to make appropriate use of land not suited to row cropping. Like many farms in the allegedly flat corn belt, Lancie's farm includes some land that should be kept permanently grassed, and on his farm, unlike many, it *is* kept permanently grassed. He can afford this because he can make good use of it that way, without damaging it, for these thirty or so acres give him five hundred bales of bluegrass hay early in the year and, after that, months of pasture, at the cost only of a second clipping. The crop on that land does not need to be planted or cultivated, and it is harvested by the animals; it is therefore the cheapest feed on the place.

Lancie Clippinger is as much in the business of growing crops and making money on them as any other farmer. But he is also in the business of making sense—making sense, that is, for himself, not for the oil, chemical, and equipment companies, or for the banks. He is taking his own advice, and his advice comes from his experience and the experience of farmers like him, not from experts who are not farmers. For those reasons, Lancie Clippinger is doing all right. He is farming well and earning a living by it in a time when many farmers are farming poorly and making money for everybody but themselves.

"I don't know what they mean," he says. "You'd think some in the bunch would use their heads a *little* bit."

WHAT ARE PEOPLE FOR?
(1990)

Damage

I

I have a steep wooded hillside that I wanted to be able to pasture occasionally, but it had no permanent water supply.

About halfway to the top of the slope there is a narrow bench, on which I thought I could make a small pond. I hired a man with a bulldozer to dig one. He cleared away the trees and then formed the pond, cutting into the hill on the upper side, piling the loosened dirt in a curving earthwork on the lower.

The pond appeared to be a success. Before the bulldozer quit work, water had already begun to seep in. Soon there was enough to support a few head of stock. To heal the exposed ground, I fertilized it and sowed it with grass and clover.

We had an extremely wet fall and winter, with the usual freezing and thawing. The ground grew heavy with water, and soft. The earthwork slumped; a large slice of the woods floor on the upper side slipped down into the pond.

The trouble was the familiar one: too much power, too little knowledge. The fault was mine.

I *was* careful to get expert advice. But this only exemplifies what I already knew. No expert knows everything about every place, not even everything about any place. If one's knowledge of one's whereabouts is insufficient, if one's judgment is unsound, then expert advice is of little use.

II

In general, I have used my farm carefully. It could be said, I think, that I have improved it more than I have damaged it.

My aim has been to go against its history and to repair the damage of other people. But now a part of its damage is my own.

The pond was a modest piece of work, and so the damage is not extensive. In the course of time and nature it will heal.

And yet there *is* damage—to my place, and to me. I have carried out, before my own eyes and against my intention, a part of the modern tragedy: I have made a lasting flaw in the face of the earth, for no lasting good.

Until that wound in the hillside, my place, is healed, there will be something impaired in my mind. My peace is damaged. I will not be able to forget it.

III

It used to be that I could think of art as a refuge from such troubles. From the imperfections of life, one could take refuge in the perfections of art. One could read a good poem—or better, write one.

Art was what was truly permanent, therefore what truly mattered. The rest was "but a spume that plays / Upon a ghostly paradigm of things."

I am no longer able to think that way. That is because I now live in my subject. My subject is my place in the world, and I live in my place.

There is a sense in which I no longer "go to work." If I live in my place, which is my subject, then I am "at" my work even when I am not working. It is "my" work because I cannot escape it.

If I live in my subject, then writing about it cannot "free" me of it or "get it out of my system." When I am finished writing, I can only return to what I have been writing about.

While I have been writing about it, time will have changed it. Over longer stretches of time, *I* will change it. Ultimately, it will be changed by what I write, inasmuch as I, who change my subject, am changed by what I write about it.

If I have damaged my subject, then I have damaged my art. What aspired to be whole has met damage face to face, and has come away wounded. And so it loses interest both in the anesthetic and in the purely esthetic.

It accepts the clarification of pain, and concerns itself with healing. It cultivates the scar that is the course of time and nature over damage: the landmark and mindmark that is the notation of a limit.

To lose the scar of knowledge is to renew the wound.

An art that heals and protects its subject is a geography of scars.

IV

"You never know what is enough unless you know what is more than enough."

I used to think of Blake's sentence as a justification of youthful excess. By now I know that it describes the peculiar condemnation of our species. When the road of excess has reached the palace of wisdom it is a healed wound, a long scar.

Culture preserves the map and the records of past journeys so that no generation will permanently destroy the route.

The more local and settled the culture, the better it stays put, the less the damage. It is the foreigner whose road of excess leads to a desert.

Blake gives the just proportion or control in another proverb: "No bird soars too high, if he soars with his own wings." Only when our acts are empowered with more than bodily strength do we need to think of limits.

It was no thought or word that called culture into being, but a tool or a weapon. After the stone axe we needed song and story to remember innocence, to record effect—and so to describe the limits, to say what can be done without damage.

The use only of our bodies for work or love or pleasure, or even for combat, sets us free again in the wilderness, and we exult.

But a man with a machine and inadequate culture—such as I was when I made my pond—is a pestilence. He shakes more than he can hold.

Wallace Stegner and the Great Community

I N THE SPRING of 1958 I received notice that I had been given a Stegner Fellowship for the next school year at Stanford University. I assumed, properly enough, that I was going to be dealing with people my own age who would know a great deal more than I did. Clearly, I needed to prepare myself.

My wife and new daughter and I were staying in Lexington, Kentucky, that summer, and I decided that I would read the works of Wallace Stegner. The only thing I had read by him up to then was *Field Guide to the Western Birds*, published by Ballantine in a collection called *New Short Novels 2*. In those days I assumed that anything that interested me had already interested everybody else, and so I was surprised to find that the university library yielded only his first novel, *Remembering Laughter*, and his second book of stories, *The City of the Living*. I found a copy of the first short story collection, *The Women on the Wall*, in a used book store.

My reading of those three books is still clear and immediate to me, perhaps because my uneasiness about going to Stanford had sharpened my wits. I thought *Remembering Laughter* a perfect little novel, clean and swift and assured, and I can still feel the weight of the disaster in it. As the would-be author of a first novel myself, I envied it and was intimidated by it. I saw plainly that this man, when he was perhaps my age, had known how to write, and that he had known how much better than I did.

Because I had no ambition to write short stories, I read *The Women on the Wall* and *The City of the Living* more dispassionately. By then I had read a good many short stories. Besides the ones scattered in textbooks and anthologies, I had read Hemingway's in the Modern Library collection and Faulkner's *Collected Stories*. But I had read those mainly, I think, as a reader—without asking how they had been made. *The Women on the Wall* and *The City of the Living* were the first collections of stories by one writer that I read asking that question, seeing how an able workman made use of a form.

667

And so Wallace Stegner became my teacher before I ever laid eyes on him, and he was already teaching me in a way that I have come to see as characteristic of him: by bestowing a kindness that implied an expectation, and by setting an example.

It has been twenty-seven years since 1958, and I am no longer certain what I expected him to be. All I am sure of is that I did not expect him to be as he was when he took over the writing seminar at the start of the winter quarter. Perhaps, since he taught at a great university and had written many books, I expected him to be magisterial. Perhaps I expected him to display the indisputable field marks of Literary Genius. Or perhaps I expected a dogmatist of some sort, for I remember that he had a reputation with some former writing students still around as a stickler for grammar—as if *grammar* had anything to do with *art*.

The man himself, it turned out, was something altogether different, and a great deal better. When I was asked to write about him as a teacher, I had the feeling that I was expected to tell anecdotes of memorable classroom performances. But that is not what I have to tell. As a teacher, as I knew him, he had none of the performer in him. When he spoke to the class, the class felt spoken to. You did not feel that he was glancing at himself.

What struck me first about him was the way he looked. He was, and still is, an extraordinarily handsome man. And unlike many writers and professors, he dressed with care and looked good in his clothes. I don't mean that he was a dandy; he was as far from that as he was from being a performer. His was a neatness, one felt, that had to do with respect, for himself and for others.

But just as striking, just as unexpected, was the way he understated his role as teacher. I think that he is naturally a reticent man, not given to self-revelation or self-advertisement, and for years I assumed that his reticence in the classroom was merely a part of his character. But as I have come to know him better, have read him more, and have thought more about him, I have changed my mind. I am sure that character had something to do with it, but it seems to me now that his stance and manner in the classroom were the results of an accurate and generous understanding of his situation as teacher and of ours as students. The best explanation may be in these sentences from his essay "The Book and the Great Community":

> Thought is neither instant nor noisy. . . . It thrives best in solitude, in quiet, and in the company of the past, the great community of recorded human experience. That recorded experience is essential whether one hopes to reassert some aspect of it, or attack it.

The community here is that of "recorded human experience," not the Pantheon of Great Writers. It is immense and diverse, more like the Library of Congress than the Harvard Five-Foot Shelf. But it does include the great writers. It is bewildering both in its amplitude and in the eminence of some of its members. A teacher leading his students to the entrance to that community, as would-be contributors to it, must know that both he and they are coming into the possibility of error. The teacher may make mistakes about the students; the students may make much more serious ones about themselves. He is leading them, moreover, to a community, not to some singular stump or rostrum from which he will declare the Truth.

I think, then, that when Mr. Stegner sat with us at the long table in the Jones Room of the Stanford Library, he felt that a certain modesty and a certain discretion were in order. He did not speak as though he confused any utterance of his with the earthquake that one afternoon gently shook us while we talked.

There must have been about twenty of us in the class, and among its members were Ernest Gaines, Ken Kesey, and Nancy Packer. Nancy, I think, was the oldest of us. She had read a great deal, talked well, and was useful to us all. But as a group we had a good deal to say, and Mr. Stegner would let us say it. He would read a piece of somebody's work and then sit back, sometimes with a cigar, listening attentively while we had our say.

One sunny afternoon he read us a piece of work of his own —a chapter, I think, from *All the Little Live Things*—and then graciously paid attention to all we said to him about it. I remember that I made an extensive comment myself, and I am tempted to wish I could remember what I said. Probably I should be glad to have forgotten. I would be glad, anyhow, to know that he has forgotten.

He had his say too, of course. He commended generously where commendation was due. He was a good encourager when encouragement was needed. And, more important, he was a willing dealer with problems, even little problems, even —yes—problems of grammar. That is, he gave good practical help.

He did not adopt the pose or use the tactics of the authoritarian teacher, and yet one felt his authority. I am not talking, of course, about institutional authority, which must be asserted to be felt, and which he did not assert. We felt in him, rather, the authority of authentic membership in the great community, of one who had thought and worked in solitude, in quiet, in the company of the past—an authority that would be destroyed in being asserted.

It was this implicit authority in him that most impressed me during my time at Stanford. That and an expectation implicit in his dealings with me, that perhaps was not conscious with him, but which I felt keenly. At one time in the midst of my fellowship year I thought (mistakenly, it turned out) that I had finished the novel I was working on, and I could not see clearly what I ought to do next. For a while, paralyzed in my confusion, I was not able to write anything. And I remember both my embarrassment, for I felt that Mr. Stegner expected me to be at work, and how paltry, in light of his expectation, my excuses appeared in my own eyes. That was, maybe, my definitive encounter with the hardest truth that a writer—or any other worker—can learn: that, to all practical purposes, excuses are not available.

Mr. Stegner's teaching, then, as I have come to see it, was as important for what it did not do as for what it did. One thing he did not do was encourage us too much. If we wrote well, he said so. But he did not abet any suspicions we may have had that we were highly accomplished writers or that we were ever going to be highly accomplished. In this, I think, he respected the past, for he was better acquainted with the art of writing than we were and knew better than we did how much we had to learn. But he was being respectful of us too. He did not want to mislead us or help us to mislead ourselves. He did not say what he had not considered or did not mean. He did not deal in greetings at the beginnings of great careers.

He did not pontificate or indoctrinate or evangelize. We were not expected to become Stegnerians. None of us could have doubted that he wanted us to know and think and write as well as we possibly could. But no specific recipe or best way was recommended. The emphasis was on workmanship. What we were asked to be concerned with was the job of work at hand, what one or another of us had done or attempted to do. Our teacher was a writer, he too was at work on what he had chosen to do; he would help us if he could.

And so what I began by calling reticence—at some risk, for it is not a fashionable virtue now—finally declares itself as courtesy toward both past and future: courtesy toward the art of writing, which needs to be carefully learned and generously passed on; and courtesy toward us, who as young writers needed all the help we could get, but needed also to be left to our own ways.

In his conversations with Richard Etulain, there is a passage in which Mr. Stegner names several of his old students, speaks of their accomplishments, and then says, "I try not to take credit for any of that." In the mouths of some people, that statement would not be trustworthy; in the mouths of some it would contradict itself. Coming from Mr. Stegner, it is trustworthy, for in fact he has not been a taker of credit. The fellows *have* been left to their ways. They have come, benefited as they were able, and left free of obligation.

My bet, however, is that, by the fellows and other students, Wallace Stegner is given a great deal more credit than he would be comfortable in taking. The first fellowships were given in 1946. By 1971, when Mr. Stegner retired from teaching, there had been about a hundred of them, and the list of fellows contains a good many names that do it credit: Evan S. Connell, Dan Jacobson, Hannah Green, Eugene Burdick, Edward Abbey, Thomas McGuane, Ken Gangemi, Tillie Olsen, Robert Stone, Jim Houston, Larry McMurtry, Ernest Gaines, James Baker Hall, Gurney Norman, Ed McClanahan, Peter Beagle, Nancy Huddleston Packer, Max Apple, Blanche Boyd, Judith Rascoe, Max Crawford, Robin White, Merrill Joan Gerber,

Charlotte Painter, Al Young, Raymond Carver, William Wie-gand, William Kittredge.

My bet is that every one of those people owes something to Wallace Stegner. My further bet is that most of them, maybe years after their participation in the seminar, have been surprised by some recognition of what and how much they owe. And yet I know that a teacher is right in trying not to take credit for his students. The fellows, one assumes, must have learned something from attending the seminars given by Mr. Stegner, Richard Scowcraft, and other teachers. But they must have learned something too from coming to California and living there for a year. They must have learned something from one another, and from other people they came to know while they were there. They must have learned something from the books they read while they were there. How would you sort all that out, for the hundred fellows and the hundreds of others who sat in the Jones Room from 1946 to 1971, and make a precise estimate of the influence of one teacher? Obviously, you cannot. As thousands of faculty committees have found out, "teacher evaluation" is a hopeless business. There is, thank God, no teacher-meter, and there never is going to be one. A teacher's major contribution may pop out anonymously in the life of some ex-student's grandchild. A teacher, finally, has nothing to go on but faith, a student nothing to offer in return but testimony.

What I have to testify is that, although I do not think that Mr. Stegner thinks of himself as my teacher, my awareness of him as a teacher has grown over the years, and I think myself more than ever his student. And this has to do with the changes that have happened in me since my time at Stanford.

In those days I assumed, as my education had prepared me to assume, that I was going to follow a literary career that would lead me far from home. I assumed that I would teach (and write) in a university in a large city, and in fact I did so for a while. And then my family and I returned to my native county in Kentucky. That is, I became, in both habitat and subject, what is called a "regional writer."

There are, of course, problems in being a regional writer. If one is regional only in subject, then there is a temptation —and an abundance of precedent—for becoming a sort of industrialist of letters, mining one's province for whatever can be got out of it in the way of "raw material" for stories and novels. I would argue that it has been possible for such writers to write so exploitively, condescendingly, and contemptuously of their regions and their people as virtually to prepare the way for worse exploitation by their colleagues in other industries: if it's a god-forsaken boondocks full of ignorant hillbillies, or a god-forsaken desert populated by a few culturally deprived ranchers, why *not* strip-mine it?

On the other hand, if one both lives and writes in one's region, one becomes aware of good reasons to be more watchful and more careful. It was not until I began the struggle to live and write in my region that I began to be aware of Wallace Stegner as a writer struggling to live and write in *his* region —something I would never have done, I think, had I chosen to live in San Francisco or New York. Or at least his struggle would not have meant so much, would not have been so instructive and reassuring to me, had I lived in one of those places. In *Wolf Willow* and then in *The Sound of Mountain Water*, I saw him dealing head-on with the problems of the history and the literary history of the West, the challenges and flaws and threats of those histories, and the possible meanings and uses of his own history as a westerner. And of course, once one sees this effort in some of his work, one begins to see it in the rest. One sees him becoming a new kind of writer: one who not only writes about his region but also does his best to protect it, by writing and in other ways, from its would-be exploiters and destroyers.

As a regional writer he seems to me exemplary. He has worked strenuously to know his region. He has been not just a student of its history, but one of its historians. There is an instructive humility in his studentship as a historian of the West. It is hard to imagine Hemingway researching and writing a history of Michigan or Africa; to him, as to many writers now, history was immediate experience. To Mr. Stegner, it is also memory. He has the care and the scrupulousness of one who understands remembering as a duty, and who therefore understands historical insight and honesty as duties. He has endeavored to understand the differences of his region from other regions and also from its own pipe dreams and fantasies of itself. He has never condescended to his region—an impossibility, since he has so profoundly understood himself as a part of it. He has not dealt in the quaint, the fantastical, or the picturesque. And, above all, he has written well. He is a highly accomplished and an extremely versatile writer. One of the pleasures in reading him is to see how many kinds of writing he has done well. He does not allow suppositions about the homeliness or provinciality of his subject to qualify or limit his own powers. He writes as intelligently about cowboys as about historians and literary critics.

He is a re-readable writer. One reads out of interest or curiosity to see how the story or the argument or the explanation will play itself out. One can re-read, I think, only to be surprised. If re-reading does not yield surprises, one does not continue. Not long ago, when I re-read "Genesis," the long story in *Wolf Willow*, I was surprised by the excellence of the artistry, its clarity and crispness, the cleanness of movement. It is a story by a man who learned about work and hardship, weather and country, from experience, but who has also thought well about them, and has read well. More recently, re-reading some of the essays, I was surprised by the way they spoke, amplifying themselves, to the time and experience added to me since I read them last.

He is, then, a regional writer who has escaped the evils of regional-ism. And he has escaped, as well, the evils of idiosyncrasy. He has been a writer whose work has not exhausted literary possibility, but has made it, opened it up. Younger writers of the West, I would think, would find it possible to build on his work directly. And younger writers of other regions, I know, will find it illustrative of possibilities, a measure of excellence, and a source of comfort.

I have been using the adjective "regional" as if I am not aware of the antipathy attached to it in some literary circles. I am using it that way because I feel strongly the need for the sort of regional identification and commitment that is exemplified in Mr. Stegner's work. But I am, of course, inescapably aware of the scorn with which that identification and commitment are frequently met. Mr. Stegner himself cites a typical instance in his conversations with Richard Etulain: the *New York Times*, he says, called him "the Dean of Western Writers"—and got his name wrong. The adjective "western," as all regional writers will understand, would have been dismissive even if the name had been correctly given. This is the regionalism of New York, which will use the West, indeed depend on it, but not care for it. And this regionalism is opposed and corrected by writing that is authentically and faithfully regional.

In the face of this metropolitan regionalism which has so far been under no constraint to see itself as such, and which has condescended to and exploited all other regions, Mr. Stegner has taken his stand as a westerner and has produced work that seems to me not only excellent, but indispensable as well. And what I most respect in him is that in all the years of his effort he has not stooped to the self-promotion endemic to the literary regionalism of New York. He has represented himself solely by his writings and his acts of citizenship. In a self-exploiting, world-exploiting age, this is a high and admirable accomplishment in itself.

Writer and Region

I FIRST READ *Huckleberry Finn* when I was a young boy. My great-grandmother's copy was in the bookcase in my grandparents' living room in Port Royal, Kentucky. It was the Webster edition, with E. W. Kemble's illustrations. My mother may have told me that it was a classic, but I did not *know* that it was, for I had no understanding of that category, and I did not read books because they were classics. I don't remember starting to read *Huckleberry Finn*, or how many times I read it; I can only testify that it is a book that is, to me, literally familiar: involved in my family life.

I can say too that I "got a lot out of it." From early in my childhood I was not what was known as a good boy. My badness was that I was headstrong and did not respond positively to institutions. School and Sunday school and church were prisons to me. I loved being out of them, and I did not behave well in them. *Huckleberry Finn* gave me a comforting sense of precedent, and it refined my awareness of the open, outdoor world that my "badness" tended toward.

That is to say that *Huckleberry Finn* made my boyhood imaginable to me in a way that it otherwise would not have been. And later, it helped to make my grandfather's boyhood in Port Royal imaginable to me. Still later, when I had come to some knowledge of literature and history, I saw that that old green book had, fairly early, made imaginable to me my family's life as inhabitants of the great river system to which we, like Mark Twain, belonged. The world my grandfather had grown up in, in the eighties and nineties, was not greatly changed from the world of Mark Twain's boyhood in the thirties and forties. And the vestiges of that world had not entirely passed away by the time of my own boyhood in the thirties and forties of the next century.

My point is that *Huckleberry Finn* is about a world I know, or knew, which it both taught me about and taught me to imagine. That it did this before I could have known that it was doing so, and certainly before anybody told me to expect it to do so, suggests its greatness to me more forcibly than any critical assessment I have ever read. It is called a great American book; I think of it, because I have so experienced it, as a trans-figuring regional book.

As a boy resentful of enclosures, I think I felt immediately the great beauty, the great liberation, at first so fearful to him, of the passage in Chapter I when Huck, in a movement that happens over and over in his book, escapes the strictures of the evangelical Miss Watson and, before he even leaves the house, comes into the presence of the country:

> By-and-by they fetched the niggers in and had prayers, and then everybody was off to bed. I went up to my room with a piece of candle and put it on the table. Then I set down in a chair by the window and tried to think of something cheerful, but it warn't no use. I felt so lonesome I most wished I was dead. The stars was shining, and the leaves rustled in the woods ever so mournful; and I heard an owl, away off, who-whooing about somebody that was dead, and a whippoorwill and a dog crying about somebody that was going to die; and the wind was trying to whisper something to me and I couldn't make out what it was, and so it made the cold shivers run over me.

It is a fearful liberation because the country, so recently settled by white people, is already both haunted and threatened. But the liberation is nevertheless authentic, both for Huck and for the place and the people he speaks for. In the building and summoning rhythm of his catalog of the night sounds, in the sudden realization (his and ours) of the equality of his voice to his subject, we feel a young intelligence breaking the confines of convention and expectation to confront the world itself: the night, the woods, and eventually the river and all it would lead to.

By now we can see the kinship, in this respect, between Huck's voice and earlier ones to the east. We feel the same sort of outbreak as we read:

> When I wrote the following pages, or rather the bulk of them, I lived alone, in the woods, a mile from any neighbor, in a house which I had built myself, on the shore of Walden Pond, in Concord, Massachusetts, and earned my living by the work of my hands only.

That was thirty years before *Huckleberry Finn*. The voice is
certainly more cultivated, more adult, more reticent, but the
compulsion to get *out* is the same.

And a year after that we hear:

> I loaf and invite my soul,
> I lean and loaf at ease . . . observing a spear of summer
> grass.

And we literally *see* the outbreak here as Whitman's line grows
long and prehensile to include the objects and acts of a country's life that had not been included in verse before.

But Huck's voice is both fresher and historically more improbable than those. There is something miraculous about it.
It is not Mark Twain's voice. It is the voice, we can only say,
of a great genius named Huckleberry Finn, who inhabited a
somewhat lesser genius named Mark Twain, who inhabited
a frustrated businessman named Samuel Clemens. And Huck
speaks of and for and as his place, the gathering place of the
continent's inland waters. His is a voice governed always by the
need to flow, to move outward.

It seems miraculous also that this voice should have risen
suddenly out of the practice of "comic journalism," a genre
amusing enough sometimes, but extremely limited, now hard
to read and impossible to need. It was this way of writing that
gave us what I understand as region*alism*: work that is ostentatiously provincial, condescending, and exploitive. That *Huckleberry Finn* starts from there is evident from its first paragraph.
The wonder is that within three pages the genius of the book is
fully revealed, and it is a regional genius that for 220 pages (in
the Library of America edition) remains untainted by regionalism. The voice is sublimely confident of its own adequacy to
its own necessities, its eloquence. Throughout those pages the
book never condescends to its characters or its subject; it never
glances over its shoulder at literary opinion; it never fears for
its reputation in any "center of culture." It reposes, like Eliot's
Chinese jar, moving and still, at the center of its own occasion.

I should add too that the outbreak or upwelling of this
voice, impulsive and freedom-bent as it is, is not disorderly.
The freeing of Huck's voice is not a feat of power. The voice is
enabled by an economy and a sense of pace that are infallible,
and innately formal.

*

That the book fails toward the end (in the sixty-seven pages, to be exact, that follow the reappearance of Tom Sawyer) is pretty generally acknowledged. It does not fail exactly into the vice that is called regionalism, though its failure may have influenced or licensed the regionalism that followed; rather, it fails into a curious frivolity. It has been all along a story of escape. A runaway slave is an escaper, and Huck is deeply implicated, finally by his deliberate choice, in Jim's escape; but he is making his own escape as well, from Miss Watson's indoor piety. After Tom reenters the story, these authentic escapes culminate in a bogus one: the freeing of a slave who, as Tom knows, has already been freed. It is as though Mark Twain has recovered authorship of the book from Huck Finn at this point—only to discover that he does not know how to write it.

Then occurs the wounding and recovery of Tom and the surprising entrance of his Aunt Polly who, true to her character, clears things up in no time—a delightful scene; there is wonderful writing in the book right through to the end. But Mark Twain is not yet done with his theme of escape. The book ends with Huck's determination to "light out for the Territory" to escape being adopted and "sivilized" by Tom's Aunt Sally. And here, I think, we are left face-to-face with a flaw in Mark Twain's character that is also a flaw in our national character, a flaw in our history, and a flaw in much of our literature.

As I have said, Huck's point about Miss Watson is well taken and well made. There is an extremity, an enclosure, of conventional piety and propriety that needs to be escaped, and a part of the business of young people is to escape it. But this point, having been made once, does not need to be made again. In the last sentence, Huck is made to suggest a virtual identity between Miss Watson and Aunt Sally. But the two women are not at all alike. Aunt Sally is a sweet, motherly, entirely affectionate woman, from whom there is little need to escape because she has no aptitude for confinement. The only time she succeeds in confining Huck, she does so by *trusting* him. And so when the book says, "Aunt Sally she's going to adopt me and sivilize me and I can't stand it. I been there before," one can only conclude that it is not Huck talking about Aunt Sally, but Mark Twain talking, still, about the oppressive female piety of Miss Watson.

Something is badly awry here. At the end of this great book
we are asked to believe—or to believe that Huck believes—that
there are no choices between the "civilization" represented by
pious slave owners like Miss Watson or lethal "gentlemen"
like Colonel Sherburn and lighting out for the Territory. This
hopeless polarity marks the exit of Mark Twain's highest imagination from his work. Afterwards we get *Pudd'nhead Wilson*,
a fine book, but inferior to *Huckleberry Finn*, and then the
inconsolable grief, bitterness, and despair of the last years.

It is arguable, I think, that our country's culture is still sus-
pended as if at the end of *Huckleberry Finn*, assuming that its
only choices are either a deadly "civilization" of piety and violence or an escape into some "Territory" where we may remain
free of adulthood and community obligation. We want to be
free; we want to have rights; we want to have power; we do not
yet want much to do with responsibility. We have imagined the
great and estimable freedom of boyhood, of which Huck Finn
remains the finest spokesman. We have imagined the bache-
lorhoods of nature and genius and power: the contemplative,
the artist, the hunter, the cowboy, the general, the president
—lives dedicated and solitary in the Territory of individuality.
But boyhood and bachelorhood have remained our norms of
"liberation," for women as well as men. We have hardly begun
to imagine the coming to responsibility that is the meaning,
and the liberation, of growing up. We have hardly begun to
imagine community life, and the tragedy that is at the heart of
community life.

Mark Twain's avowed preference for boyhood, as the time
of truthfulness, is well known. Beyond boyhood, he glimpsed
the possibility of bachelorhood, an escape to "the Territory,"
where individual freedom and integrity might be maintained
—and so, perhaps, he imagined Pudd'nhead Wilson, a solitary
genius devoted to truth and justice, standing apart in the preserve of cynical honesty.

He also imagined Aunt Polly and Aunt Sally. They, I think,
are the true grown-ups of the Mississippi novels. They have
their faults, of course, which are the faults of their time and
place, but mainly they are decent people, responsible members of the community, faithful to duties, capable of love,

trust, and long-suffering, willing to care for orphan children. The characters of both women are affectionately drawn; Mark Twain evidently was moved by them. And yet he made no acknowledgment of their worth. He insists on regarding them as dampeners of youthful high spirits, and in the end he refuses to distinguish them at all from the objectionable Miss Watson.

There is, then, something stunted in *Huckleberry Finn*. I have hated to think so—for a long time I tried consciously *not* to think so—but it is so. What is stunted is the growth of Huck's character. When Mark Twain replaces Huck as author, he does so apparently to make sure that Huck remains a boy. Huck's growing up, which through the crisis of his fidelity to Jim ("All right, then, I'll *go* to hell") has been central to the drama of the book, is suddenly thwarted first by the Tomfoolery of Jim's "evasion" and then by Huck's planned escape to the "Territory." The real "evasion" of the last chapters is Huck's, or Mark Twain's, evasion of the community responsibility that would have been a natural and expectable next step after his declaration of loyalty to his friend. Mark Twain's failure or inability to imagine this possibility was a disaster for his finest character, Huck, whom we next see not as a grown man, but as partner in another boyish evasion, a fantastical balloon excursion to the Pyramids.

I am supposing, then, that *Huckleberry Finn* fails in failing to imagine a responsible, adult community life. And I am supposing further that this is the failure of Mark Twain's life, and of our life, so far, as a society.

Community life, as I suggested earlier, is tragic, and it is so because it involves unremittingly the need to survive mortality, partiality, and evil. Because Huck Finn and Mark Twain so clung to boyhood, and to the boy's vision of free bachelorhood, neither could enter community life as I am attempting to understand it. A boy can experience grief and horror, but he cannot experience that fulfillment and catharsis of grief, fear, and pity that we call tragedy and still remain a boy. Nor can he experience tragedy in solitude or as a stranger, for tragedy is experienceable only in the context of a beloved community. The fulfillment and catharsis that Aristotle described as the communal result of tragic drama is an artificial enactment of the way a mature community survives tragedy in fact. The community wisdom of tragic drama is in the implicit understanding that no community can survive that cannot survive the worst. Tragic drama attests to the community's need to survive the worst that it knows, or imagines, can happen.

In his own life, Mark Twain experienced deep grief over the deaths of loved ones, and also severe financial losses. But these experiences seem to have had the effect of isolating him, rather than binding him to a community. Great personal loss, moreover, is not much dealt with in those Mississippi books that are most native to his imagination: *Tom Sawyer*, *Huckleberry Finn*, and *Life on the Mississippi*. The only such event that I remember in those books is the story, in *Life on the Mississippi*, of his brother Henry's death after an explosion on the steamboat *Pennsylvania*. Twain's account of this is extremely moving, but it is peculiar in that he represents himself—though his mother, a brother, and a sister still lived—as Henry's *only* mourner. No other family member is mentioned.

What is wanting, apparently, is the tragic imagination that, through communal form or ceremony, permits great loss to be recognized, suffered, and borne, and that makes possible some sort of consolation and renewal. What is wanting is the return to the beloved community, or to the possibility of one. That would return us to a renewed and corrected awareness of our partiality and mortality, but also to healing and to joy in a renewed awareness of our love and hope for one another. Without that return we may know innocence and horror and grief, but not tragedy and joy, not consolation or forgiveness or redemption. There is grief and horror in Mark Twain's life and work, but not the tragic imagination or the imagined tragedy that finally delivers from grief and horror.

He seems instead to have gone deeper and deeper into grief and horror as his losses accumulated, and deeper into outrage as he continued to meditate on the injustices and cruelties of history. At the same time he withdrew further and further from community and the imagining of community, until at last his Hadleyburg—such a village as he had earlier written about critically enough, but with sympathy and good humor too—is used merely as a target. It receives an anonymous and indiscriminate retribution for its greed and self-righteousness —evils that community life has always had to oppose, correct, ignore, indulge, or forgive in order to survive. All observers of communities have been aware of such evils, Huck Finn having been one of the acutest of them, but now it is as if Huck has been replaced by Colonel Sherburn. "The Man that Corrupted Hadleyburg" is based on the devastating assumption that people are no better than their faults. In old age, Mark Twain had become obsessed with "the damned human race" and the malevolence of God—ideas that were severely isolating and, ultimately, self-indulgent. He was finally incapable of that magnanimity that is the most difficult and the most necessary: forgiveness of human nature and human circumstance. Given human nature and human circumstance, our only relief is in this forgiveness, which then restores us to community and its ancient cycle of loss and grief, hope and joy.

And so it seems to me that Mark Twain's example remains crucial for us, for both its virtues and its faults. He taught American writers to be writers by teaching them to be *regional* writers. The great gift of *Huckleberry Finn*, in itself and to us, is its ability to be regional without being provincial. The provincial is always self-conscious. It is the conscious sentimentalization of or condescension to or apology for a province, what I earlier called regionalism. At its most acute, it is the fear of provinciality. There is, as I said, none of that in the first thirty-two chapters of *Huckleberry Finn*. (In the final eleven chapters it is there in the person of Tom Sawyer, who is a self-made provincial.) Mark Twain apparently knew, or he had the grace to trust Huck to know, that *every* writer is a regional writer, even those who write about a fashionable region such as New York City. The value of this insight, embodied as it is in a great voice and a great tale, is simply unreckonable. If he had done nothing else, that would have made him indispensable to us.

But his faults are our own, just as much as his virtues. There are two chief faults and they are related: the yen to escape to the Territory, and retribution against the life that one has escaped or wishes to escape. Mark Twain was new, for his place, in his virtue. In his faults he was old, a spokesman for tendencies already long established in our history.

That these tendencies remain well established among us ought to be clear enough. Wallace Stegner had them in mind when he wrote in *The Sound of Mountain Water*:

For many, the whole process of intellectual and literary growth is a movement, not through or beyond, but away from the people and society they know best, the faiths they still at bottom accept, the little raw provincial world for which they keep an apologetic affection.

Mr. Stegner's "away from" indicates, of course, an escape to the Territory—and there are many kinds of Territory to escape to. The Territory that hinterland writers have escaped to has almost always been first of all that of some metropolis or "center of culture." This is not inevitably dangerous; great cities are probably necessary to the life of the arts, and all of us who have gone to them have benefited. But once one has reached the city, other Territories open up, and some of these *are* dangerous. There is, first, the Territory of retribution against one's origins. In our country, this is not just a Territory, but virtually a literary genre. From the sophisticated, cosmopolitan city, one's old home begins to look like a "little raw provincial world." One begins to deplore "small town gossip" and "the suffocating proprieties of small town life"—forgetting that gossip occurs only among people who know one another and that propriety is a dead issue only among strangers. The danger is not just in the falsification, the false generalization, that necessarily attends a *distant* scorn or anger, but also in the loss of the subject or the vision of community life and in the very questionable exemption that scorners and avengers customarily issue to themselves.

And so there is the Territory of self-righteousness. It is easy to assume that we do not participate in what we are not in the presence of. But if we are members of a society, we participate, willy-nilly, in its evils. Not to know this is obviously to be in error, but it is also to neglect some of the most necessary and the most interesting work. How do we reduce our dependency on what is wrong? The answer to that question will necessarily be practical; the wrong will be correctable by practice and by practical standards. Another name for self-righteousness is economic and political unconsciousness.

There is also the Territory of historical self-righteousness: if *we* had lived south of the Ohio in 1830, *we* would not have owned slaves; if *we* had lived on the frontier, *we* would have killed no Indians, violated no treaties, stolen no land. The probability is overwhelming that if we had belonged to the generations we

deplore, we too would have behaved deplorably. The probability is overwhelming that we *belong* to a generation that will
be found by its successors to have behaved deplorably. Not to
know that is, again, to be in error and to neglect essential work,
and some of this work, as before, is work of the imagination.
How can we imagine our situation or our history if we think
we are superior to it?

Then there is the Territory of despair, where it is assumed
that what is objectionable is "inevitable," and so again the essential work is neglected. How can we have something better
if we do not imagine it? How can we imagine it if we do not
hope for it? How can we hope for it if we do not attempt to
realize it?

There is the Territory of the national or the global point
of view, in which one does not pay attention to anything in
particular.

Akin to that is the Territory of abstraction, a regionalism of
the mind. This Territory originally belonged to philosophers,
mathematicians, economists, tank thinkers, and the like, but
now some claims are being staked out in it for literature. At a
meeting in honor of *The Southern Review*, held in the fall of
1985 at Baton Rouge, one of the needs identified, according to
an article in the *New York Times Book Review*, was "to redefine
Southernness without resort to geography." If the participants
all agreed on any one thing, the article concluded,

> it is perhaps that accepted definitions of regionalism have
> been unnecessarily self-limiting up to now. The gradual
> disappearance of the traditional, material South does not
> mean that Southernness is disappearing, any more than
> blackness is threatened by integration, or sacredness by
> secularization. If anything, these metaregions . . . ,
> based as they are upon values, achieve distinction in di-
> rect proportion to the homogenization of the physical
> world. By coming to terms with a concept of regionalism
> that is no longer based on geographical or material con-
> siderations, *The Southern Review* is sidestepping those
> forces that would organize the world around an unnatu-
> ral consensus.

Parts of that statement are not comprehensible. Blackness, I would think, *would* be threatened by integration, and

sacredness by secularization. Dilution, at least, is certainly implied in both instances. We might as well say that fire is a state of mind and thus is not threatened by water. And how might blackness and sacredness, which have never been regions, be "metaregions"? And is the natural world subject to limitless homogenization? There are, after all, southern species of plants and animals that will not thrive in the north, and vice versa.

This "metaregion," this region "without resort to geography," is a map without a territory, which is to say a map impossible to correct, a map subject to become fantastical and silly like that Southern chivalry-of-the-mind that Mark Twain so properly condemned. How this "metaregion" could resist homogenization and "unnatural consensus" is not clear. At any rate, it abandons the real region to the homogenizers: You just homogenize all you want to, and we will sit here being Southern in our minds.

Similar to the Territory of abstraction is the Territory of artistic primacy or autonomy, in which it is assumed that no value is inherent in subjects but that value is conferred upon subjects by the art and the attention of the artist. The subjects of the world are only "raw material." As William Matthews writes in a recent article, "A poet beginning to make something needs raw material, something to transform." For Marianne Moore, he says,

> subject matter is not in itself important, except that it gives her the opportunity to speak about something that engages her passions. What is important instead is what she can discover to say.

And he concludes:

> It is not, of course, the subject that is or isn't dull, but the quality of attention we do or do not pay to it, and the strength of our will to transform. Dull subjects are those we have failed.

This assumes that for the animals and humans who are not fine artists, who have discovered nothing to say, the world is dull, which is not true. It assumes also that attention is of interest in itself, which is not true either. In fact, attention is of value only insofar as it is paid in the proper discharge of an obligation. To pay attention is to come into the presence of a subject. In one of its root senses, it is to "stretch toward" a subject, in a kind of aspiration. We speak of "paying attention" because of a correct perception that attention is *owed*—that without our attention and our attending, our subjects, including ourselves, are endangered.

Mr. Matthews's trivializing of subjects in the interest of poetry industrializes the art. He is talking about an art oriented exclusively to production, like coal mining. Like an industrial entrepreneur, he regards the places and creatures and experiences of the world as "raw material," valueless until exploited.

The test of imagination, ultimately, is not the territory of art or the territory of the mind, but the territory underfoot. That is not to say that there is no territory of art or of the mind, only that it is not a separate territory. It is not exempt either from the principles above it or from the country below it. It is a territory, then, that is subject to correction—by, among other things, paying attention. To remove it from the possibility of correction is finally to destroy art and thought, and the territory underfoot as well.

Memory, for instance, must be a pattern upon the actual country, not a cluster of relics in a museum or a written history. What Barry Lopez speaks of as a sort of invisible landscape of communal association and usage must serve the visible as a guide and as a protector; the visible landscape must verify and correct the invisible. Alone, the invisible landscape becomes false, sentimental, and useless, just as the visible landscape, alone, becomes a strange land, threatening to humans and vulnerable to human abuse.

To assume that the context of literature is "the literary world" is, I believe, simply wrong. That its real habitat is the household and the community—that it can and does affect, even in practical ways, the life of a place—may not be recognized by most theorists and critics for a while yet. But they will finally come to it, because finally they will have to. And when they do, they will renew the study of literature and restore it to importance.

*

Emerson in "The American Scholar," worrying about the increasing specialization of human enterprises, thought that the individual, to be whole, "must sometimes return from his own labor to embrace all the other laborers"—a solution that he acknowledged to be impossible. The result, he saw, was that "man is thus metamorphosed into a thing, into many things." The solution that he apparently did think possible was a return out of specialization and separateness to the human definition, so that a thinker or scholar would not be a "mere thinker," a thinking specialist, but "Man Thinking." But this return is not meant to be a retreat into abstraction, for Emerson understood "Man Thinking" as a thinker committed to action: "Action is with the scholar subordinate, but it is essential. Without it, he is not yet man."

And action, of course, implies place and community. There can be disembodied thought, but not disembodied action. Action—embodied thought—requires local and communal reference. To act, in short, is to live. Living "is a total act. Thinking is a partial act." And one does not live alone. Living is a communal act, whether or not its communality is acknowledged. And so Emerson writes:

> I grasp the hands of those next me, and take my place in the ring to suffer and to work, taught by an instinct, that so shall the dumb abyss be vocal with speech.

Emerson's spiritual heroism can sometimes be questionable or tiresome, but he can also write splendidly accurate, exacting sentences, and that is one of them. We see how it legislates against what we now call "groupiness." Neighborhood is a given condition, not a contrived one; he is not talking about a "planned community" or a "network," but about the necessary interdependence of those who are "next" each other. We see how it invokes dance, acting in concert, as a metaphor of almost limitless reference. We see how the phrase "to suffer and to work" refuses sentimentalization. We see how common work, common suffering, and a common willingness to join and belong are understood as the conditions that make speech possible in "the dumb abyss" in which we are divided.

This leads us, probably, to as good a definition of the beloved community as we can hope for: common experience and common effort on a common ground to which one willingly belongs. The life of such a community has been very little regarded in American literature. Our writers have been much more concerned with the individual who is misunderstood or mistreated by a community that is in no sense beloved, as in *The Scarlet Letter*. From Thoreau to Hemingway and his successors, a great deal of sympathy and interest has been given to the individual as pariah or gadfly or exile. In Faulkner, a community is the subject, but it is a community disintegrating, as it was doomed to do by the original sins of land greed, violent honor, and slavery. There are in Faulkner some characters who keep alive the hope of community, or at least the fundamental decencies on which community depends, and in Faulkner, as in Mark Twain, these are chiefly women: Dilsey, Lena Grove, the properly outraged Mrs. Littlejohn.

The one American book I know that is about a beloved community— a settled, established white American community with a sustaining common culture, and mostly beneficent toward both its members and its place—is Sarah Orne Jewett's *The Country of the Pointed Firs*. The community that the book describes, the coastal village of Dunnet, Maine, and the neighboring islands and back country, is an endangered species on the book's own evidence: many of its characters are old and childless, without heirs or successors—and with the Twentieth Century ahead of it, it could not last. But though we see it in its last days, we see it whole.

We see it whole, I think, because we see it both in its time and in its timelessness. The centerpiece of the book, the Bowden family reunion, is described in the particularity of a present act, but it is perceived also—as such an event must be—as a reenactment; to see is to remember:

> There was a wide path mowed for us across the field, and, as we moved along, the birds flew up out of the thick second crop of clover, and the bees hummed as if it still were June. There was a flashing of white gulls over the water where the fleet of boats rode the low waves together in the cove, swaying their small masts as if they kept time to our steps. The plash of the water could be heard faintly, yet still be heard; we might have been a company of ancient Greeks.

Thus, though it precisely renders its place and time, the book never subsides into the flimsy contemporaneity of "local color." The narrator of the book is one who departs and returns, and her returns are homecomings—to herself as well as to the place:

> The first salt wind from the east, the first sight of a light-house set boldly on its outer rock, the flash of a gull, the waiting procession of seaward-bound firs on an island, made me feel solid and definite again, instead of a poor incoherent being. Life was resumed, and anxious living blew away as if it had not been. I could not breathe deep enough or long enough. It was a return to happiness.

Anyone acquainted with the sentimentalities of American regionalism will look on that word "happiness" with suspicion. But here it is not sentimental, for the work and suffering of the community are fully faced and acknowledged. The narrator's return is not to an idyll of the boondocks; it is a re-entrance into Emerson's "ring." The community is happy in that it has survived its remembered tragedies, has re-shaped itself coherently around its known losses, has included kindly its eccentrics, invalids, oddities, and even its one would-be exile. The wonderful heroine of the book, and its emblem, Mrs. Elmira Todd, a childless widow, who in her youth "had loved one who was far above her," is a healer—a grower, gatherer, and dispenser of medicinal herbs.

She is also a dispenser of intelligent talk about her kinfolk and neighbors. More than any book I know, this one makes its way by conversation, engrossing exchanges of talk in which Mrs. Todd and many others reveal to the narrator their life and history and geography. And perhaps the great cultural insight of the book is stated by Mrs. Todd:

> Conversation's got to have some root in the past, or else you've got to explain every remark you make, an' it wears a person out.

The conversation wells up out of memory, and in a sense *is* the community, the presence of its past and its hope, speaking in the dumb abyss.

An Argument for Diversity

Elegant solutions will be predicated upon the uniqueness of place.   JOHN TODD

I LIVE IN and have known all my life the northern corner of Henry County, Kentucky. The country here is narrowly creased and folded; it is a varied landscape whose main features are these:

1. A rolling upland of which some of the soil is good and some, because of abuse, is less so. This upland is well suited to mixed farming, which was, in fact, traditional to it, but which is less diversified now than it was a generation ago. Some row-cropping is possible here, but even the best-lying ridges are vulnerable to erosion and probably not more than ten percent should be broken in any year. It is a kind of land that needs grass and grazing animals, and it is excellent for this use.

2. Wooded bluffs where the upland breaks over into the valleys of the creeks and the Kentucky River. Along with virtually all of this region, most of these bluffs have been cleared and cropped at one time or another. They should never have been cropped, and because of their extreme vulnerability to erosion they should be logged only with the greatest skill and care. These bluffs are now generally forested, though not many old-growth stands remain.

3. Slopes of gentler declivity below the bluffs and elsewhere. Some of these slopes are grassed and, with close care, are maintainable as pasture. Until World War II they were periodically cropped, in a version of slash-and-burn agriculture that resulted in serious damage by erosion. Now much of this land is covered with trees thirty or forty years old.

4. Finally, there are the creek and river bottoms, some of which are subject to flooding. Much of this land is suitable for intensive row-cropping, which, under the regime of industrial agriculture, has sometimes been too intensive.

Within these four general divisions this country is extremely diverse. To familiarity and experience, the landscape divides into many small facets or aspects differentiated by the kind or quality of soil and by slope, exposure, drainage, rockiness, and so on. In the two centuries during which European races have occupied this part of the country, the best of the land has sometimes been well used, under the influence of good times and good intentions. But virtually none of it has escaped ill use under the influence of bad times or ignorance, need or greed. Some of it—the steeper, more marginal areas—never has been well used. Of virtually all this land it may be said that the national economy has prescribed ways of use but not ways of care. It is now impossible to imagine any immediate way that most of this land might receive excellent care. The economy, as it now is, prescribes plunder of the landowners and abuse of the land.

The connection of the American economy to this place—in comparison, say, to the connection of the American economy to just about any university—has been unregarding and ungenerous. Indeed, the connection has been almost entirely exploitive—and it has never been more exploitive than it is now. Increasingly, from the beginning, most of the money made on the products of this place has been made in other places. Increasingly the ablest young people of this place have gone away to receive a college education, which has given them a "professional status" too often understood as a license to become the predators of such places as this one that they came from. The destruction of the human community, the local economy, and the natural health of such a place as this is now looked upon not as a "trade-off," a possibly regrettable "price of progress," but as a good, virtually a national goal.

Recently I heard, on an early-morning radio program, a university economist explaining the benefits of off-farm work for farm women: that these women are increasingly employed off the farm, she said, has made them "full partners" in the farm's economy. Never mind that this is a symptom of economic desperation and great unhappiness on the farm. And never mind the value, which was more than economic, of these women's previous contribution *on* the farm to the farm family's life and economy—in what was, many of them would have said, a full partnership. *Now* they are "earning forty-five percent of total family income"; *now* they are playing "a major role." The forty-five percent and the "major role" are allowed to defray all other costs. That the farm family now furnishes labor and (by its increased consumption) income to the economy that is destroying it is seen simply as an improvement. Thus the abstract and extremely tentative value of money is thoughtlessly allowed to replace the particular and fundamental values of the lives of household and community. Obviously, we need to stop thinking about the economic functions of individuals for a while, and try to learn to think of the economic functions of communities and households. We need to try to understand the long-term economies of places—places, that is, that are considered as dwelling places for humans and their fellow creatures, not as exploitable resources.

What happens when farm people take up "off-farm work"? The immediate result is that they must be replaced by chemicals and machines and other purchases from an economy adverse and antipathetic to farming, which means that the remaining farmers are put under yet greater economic pressure to abuse their land. If under the pressure of an adverse economy the soil erodes, soil and water and air are poisoned, the woodlands are wastefully logged, and everything not producing an immediate economic return is neglected, that is apparently understood by most of the society as merely the normal cost of production.

This means, among other things, that the land and its human communities are not being thought about in places of study and leadership, and this failure to think is causing damage. But if one lives in a country place, and if one loves it, one must think about it. Under present circumstances, it is not easy to imagine what might be a proper human economy for the country I have just described. And yet, if one loves it, one must make the attempt; if one loves it, the attempt is irresistible.

Two facts are immediately apparent. One is that the present local economy, based like the economies of most rural places exclusively on the export of raw materials, is ruinous. Another is that the influence of a complex, aggressive national economy upon a simple, passive local economy will also be ruinous. In a varied and versatile countryside, fragile in its composition and extremely susceptible to abuse, requiring close human care and elaborate human skills, able to produce and needing to produce a great variety of products from its soils, what is needed, obviously, is a highly diversified local economy.

We should be producing the fullest variety of foods to be consumed locally, in the countryside itself and in nearby towns and cities: meats, grains, table vegetables, fruits and nuts, dairy products, poultry and eggs. We should be harvesting a sustainable yield of fish from our ponds and streams. Our woodlands, managed for continuous yields, selectively and carefully logged, should be yielding a variety of timber for a variety of purposes: firewood, fence posts, lumber for building, fine woods for furniture makers.

And we should be adding value locally to these local products. What is needed is not the large factory so dear to the hearts of government "developers." To set our whole population to making computers or automobiles would be as gross an error as to use the whole countryside for growing corn or Christmas trees or pulpwood; it would discount everything we have to offer as a community and a place; it would despise our talents and capacities as individuals.

We need, instead, a system of decentralized, small-scale industries to transform the products of our fields and woodlands and streams: small creameries, cheese factories, canneries, grain mills, saw mills, furniture factories, and the like. By "small" I mean simply a size that would not be destructive of the appearance, the health, and the quiet of the countryside. If a factory began to "grow" or to be noisy at night or on Sunday, that would mean that another such factory was needed somewhere else. If waste should occur at any point, that would indicate the need for an enterprise of some other sort. If poison or pollution resulted from any enterprise, that would be understood as an indication that something was absolutely wrong, and a correction would be made. Small scale, of course, makes such changes and corrections more thinkable and more possible than does large scale.

I realize that, by now, my argument has crossed a boundary line of which everyone in our "realistic" society is keenly aware. I will be perceived to have crossed over into "utopianism" or fantasy. Unless I take measures to prevent it, I am going to hear somebody say, "All that would be very nice, if it were possible. Can't you be realistic?"

Well, let me take measures to prevent it. I am not, I admit, optimistic about the success of this kind of thought. Otherwise, my intention, above all, is to be realistic; I wish to be practical. The question here is simply that of convention. Do I want to be realistic according to the conventions of the industrial economy and the military state, or according to what I know of reality? To me, an economy that sees the life of a community or a place as expendable, and reckons its value only in terms of money, is not acceptable because it is *not* realistic. I am thinking as I believe we must think if we wish to discuss the *best* uses of people, places, and things, and if we wish to give affection some standing in our thoughts.

If we wish to make the best use of people, places, and things, then we are going to have to deal with a law that reads about like this: as the quality of use increases, the scale of use (that is, the size of operations) will decline, the tools will become simpler, and the methods and the skills will become more complex. That is a difficult law for us to believe, because we have assumed otherwise for a long time, and yet our experience overwhelmingly suggests that it *is* a law, and that the penalties for disobeying it are severe.

I am making a plea for diversity not only because diversity exists and is pleasant, but also because it is necessary and we need more of it. For an example, let me return to the countryside I described at the beginning of this essay. From birth, I have been familiar with this place and have heard it talked about and thought about. For the last twenty-five years I have been increasingly involved in the use and improvement of a little part of it. As a result of some failures and some successes, I have learned some things about it. I am certain, however, that I do not know the best way to use this land. Nor do I believe that anyone else does. I no longer expect to live to see it come to its best use. But I am beginning to see what is needed, and everywhere the need is for diversity. This is the need of every American rural landscape that I am acquainted with. We need a greater range of species and varieties of plants and animals, of human skills and methods, so that the use may be fitted ever more sensitively and elegantly to the place. Our places, in short, are asking us questions, some of them urgent questions, and we do not have the answers.

The answers, if they are to come and if they are to work, must be developed in the presence of the user and the land; they must be developed to some degree *by* the user *on* the land. The present practice of handing down from on high policies and technologies developed without consideration of the nature and the needs of the land and the people has not worked, and it cannot work. Good agriculture and forestry cannot be "invented" by self-styled smart people in the offices and laboratories of a centralized economy and then sold at the highest possible profit to the supposedly dumb country people. That is not the way good land use comes about. And it does not matter how the methodologies so developed and handed down are labeled; whether "industrial" or "conventional" or "organic" or "sustainable," the professional or professorial condescension that is blind to the primacy of the union between individual people and individual places is ruinous. The challenge to the would-be scientists of an ecologically sane agriculture, as David Ehrenfeld has written, is "to provide unique and particular answers to questions about a farmer's unique and particular land." The proper goal, he adds, is *not* merely to "substitute the cult of the benevolent ecologist for the cult of the benevolent sales representative."

The question of what a beloved country is to be used for quickly becomes inseparable from the questions of who is to use it or who is to prescribe its uses, and what will be the ways of using it. If we speak simply of the use of "a country," then only the first question is asked, and it is asked only by its would-be users. It is not until we speak of "a beloved country" —a particular country, particularly loved—that the question about ways of use will arise. It arises because, loving our country, we see where we are, and we see that present ways of use are not adequate. They are not adequate because such local cultures and economies as we once had have been stunted or destroyed. As a nation, we have attempted to substitute the *concepts* of "land use," "agribusiness," "development," and the like for the *culture* of stewardship and husbandry. And this change is not a result merely of economic pressures and adverse social values; it comes also from the state of affairs in our educational system, especially in our universities.

It is readily evident, once affection is allowed into the

discussion of "land use," that the life of the mind, as presently constituted in the universities, is of no help. The sciences are of no help, indeed are destructive, because they work, by principle, outside the demands, checks, and corrections of affection. The problem with this "scientific objectivity" becomes immediately clear when science undertakes to "apply" itself to land use. The problem simply is that land users are using people, places, and things that cannot be well used without affection. To be well used, creatures and places must be used sympathetically, just as they must be known sympathetically to be well known. The economist to whom it is of no concern whether or not a family loves its farm will almost inevitably aid and abet the destruction of family farming. The "animal scientist" to whom it is of no concern whether or not animals suffer will almost inevitably aid and abet the destruction of the decent old ideal of animal husbandry and, as a consequence, increase the suffering of animals. I hope that my country may be delivered from the remote, cold abstractions of university science.

But "the humanities," as presently constituted in the universities, are of no help either, and indeed, with respect to the use of a beloved country, they too have been destructive. (The closer I have come to using the term "humanities," the less satisfactory it has seemed to me; by it I mean everything that is not a "science," another unsatisfactory term.) The humanities have been destructive not because they have been misapplied, but because they have been so frequently understood by their academic stewards as not applicable. The scientific ideals of objectivity and specialization have now crept into the humanities and made themselves at home. This has happened, I think, because the humanities have come to be infected with a suspicion of their uselessness or worthlessness in the face of the provability or workability or profitability of the applied sciences. The conviction is now widespread, for instance, that "a work of art" has no purpose but to be itself. Or if it is allowed that a poem, for instance, has a meaning, then it is a meaning peculiar to its author, its time, or its convention. A poem, in short, is a relic as soon as it is composed; it can be taught, but it cannot teach. The issue of its truth and pertinence is not raised because literary study is conducted with about the same anxiety for "control" as is scientific study. The context of a poem is its text, or the context of its history and criticism as a text. I have not, of course, read all the books or sat in all the classrooms, but my impression is that not much importance is attached to the question of the truth of poems. My impression is that "Comus," for example, is not often taught as an argument with a history and a sequel, with the gravest importance for us in our dilemma now. My impression is that the great works are taught less and less as Ananda Coomaraswamy said they should be: with the recognition "that nothing will have been accomplished unless men's lives are affected and their values changed by what we have to show." My impression is that in the humanities as in the sciences the world is increasingly disallowed as a context. I hope that my country may be delivered from the objectivity of the humanities.

Without a beloved country as context, the arts and the sciences become oriented to the careers of their practitioners, and the intellectual life to intellectual (and bureaucratic) procedures. And so in the universities we see forming an intellectual elite more and more exclusively accomplished in procedures such as promotion, technological innovation, publication, and grant-getting. The context of a beloved country, moreover, implies an academic standard that is not inflatable or deflatable. The standard—the physical, intellectual, political, ecological, economic, and spiritual health of the country—cannot be too high; it will be as high, simply, as we have the love, the vision, and the courage to make it.

I would like my country to be seen and known with an attentiveness that is schooled and skilled. I would like it to be loved with a minutely particular affection and loyalty. I would like the work in it to be practical and loving and respectful and forbearing. In order for these things to happen, the sciences and the humanities are going to have to come together again in the presence of the practical problems of individual places, and of local knowledge and local love in individual people— people able to see, know, think, feel, and act coherently and well without the modern instinct of deference to the "outside expert."

What should the sciences have to say to a citizen in search of the criteria by which to determine the best use of a beloved place or countryside, or of the technical or moral means by

which to limit that use to its best use? What should the humanities have to say to a scientist—or, for that matter, a citizen—in
search of the cultural instructions that might effectively govern
the use of a beloved place? These questions or such questions
could reunite the sciences and the humanities. That a scientist
and an artist can speak and work together in response to such
questions I know from my own experience. All that is necessary
is a mutuality of concern and a mutual willingness to speak
common English. When friends speak across these divisions
or out of their "departments," in mutual concern for a beloved country, then it is clear that these diverse disciplines are
not "competing interests," as the university structure and academic folklore suggest, but interests with legitimate claims on
all minds. It is only when the country becomes an abstraction,
a prize of conquest, that these interests compete—though, of
course, when that has happened *all* interests compete.

But in order to assure that a beloved country might be lovingly used, the sciences and the humanities will have to do
more than mend their divorce at "the university level"; they
will also have to mend their divorce from the common culture,
by which I do not mean the "popular culture," but rather the
low and local wisdom that is now either relegated to the com-
partments of anthropology or folklore or "oral history," or not
attended to at all.

Some time ago, after I had given a lecture at a college in
Ohio, a gentleman came up and introduced himself to me as a
fellow Kentuckian.

"Where in Kentucky are you from?" I asked.

"Oh, a little place you probably never heard of—North
Middletown."

"I *have* heard of North Middletown," I said. "It was the
home of my father's great friend John W. Jones."

"Well, John W. Jones was my uncle."

I told him then of my father's and my own respect for Mr.
Jones.

"I want to tell you a story about Uncle John," he said. And
he told me this:

When his Uncle John was president of the bank in North
Middletown, his policy was to give a loan to any graduate
of the North Middletown high school who wanted to go to

college and needed the money. This practice caused great con-
sternation to the bank examiners who came and found those
unsecured loans on the books and no justification for them
except Mr. Jones's conviction that it was right to make them.

As it turned out, it was right in more than principle, for
in the many years that Mr. Jones was president of the bank,
making those "unsound loans," *all* of the loans were repaid; he
never lost a dime on a one of them.

I do not mean to raise here the question of the invariable
goodness of a college education, which I doubt. My point in
telling this story is that Mr. Jones was acting from a kind of
knowledge, inestimably valuable and probably indispensable,
that comes out of common culture and that cannot be taught
as a part of the formal curriculum of a school. The students
whose education he enabled were not taught this knowledge at
the colleges they attended. What he knew—and this involved
his knowledge of himself, his tradition, his community, and
everybody in it—was that trust, in the circumstances then present, could beget trustworthiness. This is the kind of knowledge,
obviously, that is fundamental to the possibility of community
life and to certain good possibilities in the characters of people.
Though I don't believe that it can be taught and learned in a
university, I think that it should be known about and respected
in a university, and I don't know where, in the sciences and the
humanities as presently constituted, students would be led to
suspect, much less to honor, its existence. It is certainly no part
of banking or of economics as now taught and practiced. It is
a part of community life, which most scientists ignore in their
professional pursuits, and which most people in the humanities
seem to regard as belonging to a past now useless or lost or
dispensed with.

Let me give another, more fundamental example. My
brother, who is a lawyer, recently had as a client an elderly man
named Bennie Yeary who had farmed for many years a farm of
about three hundred acres of hilly and partly forested land. His
farm and the road to his house had been damaged by a power
company.

Seeking to determine the value of the land, my brother asked
him if he had ever logged his woodlands. Mr. Yeary answered,
"Yes, sir, since 1944. . . . I have never robbed [the land]. I

have always just cut a little out where I thought it needed it. I have got as much timber right now, I am satisfied, . . . as I had when I started mill runs here in '44."

That we should not rob the land is a principle to be found readily enough in the literary culture. That it came into literature out of the common culture is suggested by the fact that it is commonly phrased in this way by people who have not inherited the literary culture. That we should not rob the land, anyhow, is a principle that can be learned from books. But the ways of living on the land so as not to rob it probably cannot be learned from books, and this is made clear by a further exchange between my brother and Mr. Yeary.

They came to the question of what was involved in the damage to the road, and the old farmer said that the power company had destroyed thirteen or fourteen water breaks. A water break is a low mound of rock and earth built to divert the water out of a hilly road. It is a means of preventing erosion both of the roadbed and of the land alongside it, one of the ways of living on the land without robbing it.

"How long . . . had it been since you had those water breaks constructed in there?"

"I had been working on them . . . off and on, for about twelve years, putting them water breaks in. I hauled rocks out of my fields . . . and I would dig out, bury these rocks down, and take the sledgehammer and beat rock in here and make this water break."

The way to make a farm road that will not rob the land cannot be learned from books, then, because the long use of such a road is a part of the proper way of making it, and because the use and improvement of the road are intimately involved with the use and improvement of the place. It is of the utmost importance that the rocks to make the water breaks were hauled from the fields. Mr. Yeary's solution did not, like the typical industrial solution, involve the making of a problem, or a series of problems, elsewhere. It involved the making of a solution elsewhere: the same work that improved the road improved the fields. Such work requires not only correct principles, skill, and industry, but a knowledge of local particulars and many years; it involves slow, small adjustments in response to questions asked by a particular place. And this is true in general of

the patterns and structures of a proper human use of a beloved
country, as examination of the traditional landscapes of the
Old World will readily show: they were made by use as much
as by skill.

This implication of use in the making of essential artifacts
and the maintenance of the landscape—which are to so large
an extent the making and the maintenance of culture—brings
us to the inescapable final step in an argument for diversity:
the realization that without a diversity of people we cannot
maintain a diversity of anything else. By a diversity of people I
do not mean a diversity of specialists, but a diversity of people
elegantly suited to live in their places and to bring them to
their best use, whether the use is that of uselessness, as in a
place left wild, or that of the highest sustainable productivity.
The most abundant diversity of creatures and ways cannot be
maintained in preserves, zoos, museums, and the like, but only
in the occupations and the pleasures of an appropriately diversified human economy.

The proper ways of using a beloved country are "humanities," I think, and are as complex, difficult, interesting, and
worthy as any of the rest. But they defy the present intellectual and academic categories. They are *both* science and art,
knowing and doing. Indispensable as these ways are to the
success of human life, they have no place and no standing in
the present structures of our intellectual life. The purpose, indeed, of the present structures of our intellectual life has been
to educate them out of existence. I think I know where in any
university my brother's client, Mr. Yeary, would be laughed at
or ignored or tape-recorded or classified. I don't know where
he would be appropriately honored. The scientific disciplines
certainly do not honor him, and the "humane" ones almost
as certainly do not. We would have to go some distance back
in the literary tradition—back to Thomas Hardy at least, and
before Hardy to Wordsworth—to find the due respect paid to
such a person. He *has* been educated almost out of existence,
and yet an understanding of his importance and worth would
renew the life of the mind in this country, in the university
and out.

The Pleasures of Eating

M ANY TIMES, after I have finished a lecture on the decline of American farming and rural life, someone in the audience has asked, "What can city people do?"

"Eat responsibly," I have usually answered. Of course, I have tried to explain what I meant by that, but afterwards I have invariably felt that there was more to be said than I had been able to say. Now I would like to attempt a better explanation.

I begin with the proposition that eating is an agricultural act. Eating ends the annual drama of the food economy that begins with planting and birth. Most eaters, however, are no longer aware that this is true. They think of food as an agricultural product, perhaps, but they do not think of themselves as participants in agriculture. They think of themselves as "consumers." If they think beyond that, they recognize that they are passive consumers. They buy what they want—or what they have been persuaded to want—within the limits of what they can get. They pay, mostly without protest, what they are charged. And they mostly ignore certain critical questions about the quality and the cost of what they are sold: How fresh is it? How pure or clean is it, how free of dangerous chemicals? How far was it transported, and what did transportation add to the cost? How much did manufacturing or packaging or advertising add to the cost? When the food product has been manufactured or "processed" or "precooked," how has that affected its quality or price or nutritional value?

Most urban shoppers would tell you that food is produced on farms. But most of them do not know what farms, or what kinds of farms, or where the farms are, or what knowledge or skills are involved in farming. They apparently have little doubt that farms will continue to produce, but they do not know how or over what obstacles. For them, then, food is pretty much an abstract idea—something they do not know or imagine—until it appears on the grocery shelf or on the table.

The specialization of production induces specialization of consumption. Patrons of the entertainment industry, for example, entertain themselves less and less and have become more and more passively dependent on commercial suppliers. This is certainly true also of patrons of the food industry, who have tended more and more to be *mere* consumers—passive, uncritical, and dependent. Indeed, this sort of consumption may be said to be one of the chief goals of industrial production. The food industrialists have by now persuaded millions of consumers to prefer food that is already prepared. They will grow, deliver, and cook your food for you and (just like your mother) beg you to eat it. That they do not yet offer to insert it, prechewed, into your mouth is only because they have found no profitable way to do so. We may rest assured that they would be glad to find such a way. The ideal industrial food consumer would be strapped to a table with a tube running from the food factory directly into his or her stomach.

Perhaps I exaggerate, but not by much. The industrial eater is, in fact, one who does not know that eating is an agricultural act, who no longer knows or imagines the connections between eating and the land, and who is therefore necessarily passive and uncritical—in short, a victim. When food, in the minds of eaters, is no longer associated with farming and with the land, then the eaters are suffering a kind of cultural amnesia that is misleading and dangerous. The current version of the "dream home" of the future involves "effortless" shopping from a list of available goods on a television monitor and heating precooked food by remote control. Of course, this implies and depends on a perfect ignorance of the history of the food that is consumed. It requires that the citizenry should give up their hereditary and sensible aversion to buying a pig in a poke. It wishes to make the selling of pigs in pokes an honorable and glamorous activity. The dreamer in this dream home will perforce know nothing about the kind or quality of this food, or where it came from, or how it was produced and prepared, or what ingredients, additives, and residues it contains—unless, that is, the dreamer undertakes a close and constant study of the food industry, in which case he or she might as well wake up and play an active and responsible part in the economy of food.

There is, then, a politics of food that, like any politics, involves our freedom. We still (sometimes) remember that we cannot be free if our minds and voices are controlled by someone else. But we have neglected to understand that we cannot be free if our food and its sources are controlled by someone else. The condition of the passive consumer of food is not a democratic condition. One reason to eat responsibly is to live free.

But if there is a food politics, there are also a food esthetics and a food ethics, neither of which is dissociated from politics. Like industrial sex, industrial eating has become a degraded, poor, and paltry thing. Our kitchens and other eating places more and more resemble filling stations, as our homes more and more resemble motels. "Life is not very interesting," we seem to have decided. "Let its satisfactions be minimal, perfunctory, and fast." We hurry through our meals to go to work and hurry through our work in order to "recreate" ourselves in the evenings and on weekends and vacations. And then we hurry, with the greatest possible speed and noise and violence, through our recreation—for what? To eat the billionth hamburger at some fast-food joint hellbent on increasing the "quality" of our life? And all this is carried out in a remarkable obliviousness to the causes and effects, the possibilities and the purposes, of the life of the body in this world.

One will find this obliviousness represented in virgin purity in the advertisements of the food industry, in which food wears as much makeup as the actors. If one gained one's whole knowledge of food from these advertisements (as some presumably do), one would not know that the various edibles were ever living creatures, or that they all come from the soil, or that they were produced by work. The passive American consumer, sitting down to a meal of pre-prepared or fast food, confronts a platter covered with inert, anonymous substances that have been processed, dyed, breaded, sauced, gravied, ground, pulped, strained, blended, prettified, and sanitized beyond resemblance to any part of any creature that ever lived. The products of nature and agriculture have been made, to all appearances, the products of industry. Both eater and eaten are thus in exile from biological reality. And the result is a kind of solitude, unprecedented in human experience, in which the eater may think of eating as, first, a purely commercial transaction between him and a supplier and then as a purely appetitive transaction between him and his food.

And this peculiar specialization of the act of eating is, again, of obvious benefit to the food industry, which has good reasons to obscure the connection between food and farming. It would not do for the consumer to know that the hamburger she is eating came from a steer who spent much of his life standing deep in his own excrement in a feedlot, helping to pollute the local streams, or that the calf that yielded the veal cutlet on her plate spent its life in a box in which it did not have room to turn around. And, though her sympathy for the slaw might be less tender, she should not be encouraged to meditate on the hygienic and biological implications of mile-square fields of cabbage, for vegetables grown in huge monocultures are dependent on toxic chemicals—just as animals in close confinement are dependent on antibiotics and other drugs.

The consumer, that is to say, must be kept from discovering that, in the food industry—as in any other industry—the overriding concerns are not quality and health, but volume and price. For decades now the entire industrial food economy, from the large farms and feedlots to the chains of supermarkets and fast-food restaurants, has been obsessed with volume. It has relentlessly increased scale in order to increase volume in order (presumably) to reduce costs. But as scale increases, diversity declines; as diversity declines, so does health; as health declines, the dependence on drugs and chemicals necessarily increases. As capital replaces labor, it does so by substituting machines, drugs, and chemicals for human workers and for the natural health and fertility of the soil. The food is produced by any means or any shortcut that will increase profits. And the business of the cosmeticians of advertising is to persuade the consumer that food so produced is good, tasty, healthful, and a guarantee of marital fidelity and long life.

It is possible, then, to be liberated from the husbandry and wifery of the old household food economy. But one can be thus liberated only by entering a trap (unless one sees ignorance and helplessness as the signs of privilege, as many people apparently do). The trap is the ideal of industrialism: a walled city surrounded by valves that let merchandise in but no consciousness out. How does one escape this trap? Only voluntarily, the same way that one went in: by restoring one's consciousness of what is involved in eating; by reclaiming responsibility for one's own

part in the food economy. One might begin with the illuminating principle of Sir Albert Howard's *The Soil and Health*,
that we should understand "the whole problem of health in
soil, plant, animal, and man as one great subject." Eaters, that
is, must understand that eating takes place inescapably in the
world, that it is inescapably an agricultural act, and that how
we eat determines, to a considerable extent, how the world is
used. This is a simple way of describing a relationship that is
inexpressibly complex. To eat responsibly is to understand and
enact, so far as one can, this complex relationship. What can
one do? Here is a list, probably not definitive:

1. Participate in food production to the extent that you can.
If you have a yard or even just a porch box or a pot in a sunny
window, grow something to eat in it. Make a little compost of
your kitchen scraps and use it for fertilizer. Only by growing
some food for yourself can you become acquainted with the
beautiful energy cycle that revolves from soil to seed to flower
to fruit to food to offal to decay, and around again. You will be
fully responsible for any food that you grow for yourself, and
you will know all about it. You will appreciate it fully, having
known it all its life.

2. Prepare your own food. This means reviving in your own
mind and life the arts of kitchen and household. This should
enable you to eat more cheaply, and it will give you a measure
of "quality control": you will have some reliable knowledge of
what has been added to the food you eat.

3. Learn the origins of the food you buy, and buy the food
that is produced closest to your home. The idea that every
locality should be, as much as possible, the source of its own
food makes several kinds of sense. The locally produced food
supply is the most secure, the freshest, and the easiest for local
consumers to know about and to influence.

4. Whenever possible, deal directly with a local farmer, gardener, or orchardist. All the reasons listed for the previous suggestion apply here. In addition, by such dealing you eliminate
the whole pack of merchants, transporters, processors, packag-
ers, and advertisers who thrive at the expense of both producers and consumers.

5. Learn, in self-defense, as much as you can of the economy and technology of industrial food production. What is

added to food that is not food, and what do you pay for these additions?

6. Learn what is involved in the *best* farming and gardening.

7. Learn as much as you can, by direct observation and experience if possible, of the life histories of the food species.

The last suggestion seems particularly important to me. Many people are now as much estranged from the lives of domestic plants and animals (except for flowers and dogs and cats) as they are from the lives of the wild ones. This is regrettable, for these domestic creatures are in diverse ways attractive; there is much pleasure in knowing them. And farming, animal husbandry, horticulture, and gardening, at their best, are complex and comely arts; there is much pleasure in knowing them, too.

It follows that there is great *dis*pleasure in knowing about a food economy that degrades and abuses those arts and those plants and animals and the soil from which they come. For anyone who does know something of the modern history of food, eating away from home can be a chore. My own inclination is to eat seafood instead of red meat or poultry when I am traveling. Though I am by no means a vegetarian, I dislike the thought that some animal has been made miserable in order to feed me. If I am going to eat meat, I want it to be from an animal that has lived a pleasant, uncrowded life outdoors, on bountiful pasture, with good water nearby and trees for shade. And I am getting almost as fussy about food plants. I like to eat vegetables and fruits that I know have lived happily and healthily in good soil, not the products of the huge, bechemicaled factory-fields that I have seen, for example, in the Central Valley of California. The industrial farm is said to have been patterned on the factory production line. In practice, it looks more like a concentration camp.

The pleasure of eating should be an *extensive* pleasure, not that of the mere gourmet. People who know the garden in which their vegetables have grown and know that the garden is healthy will remember the beauty of the growing plants, perhaps in the dewy first light of morning when gardens are at their best. Such a memory involves itself with the food and is one of the pleasures of eating. The knowledge of the good health of the garden relieves and frees and comforts the eater. The same goes for eating meat. The thought of the good pasture and of the calf contentedly grazing flavors the steak. Some, I know, will think it bloodthirsty or worse to eat a fellow creature you have known all its life. On the contrary, I think it means that you eat with understanding and with gratitude. A significant part of the pleasure of eating is in one's accurate consciousness of the lives and the world from which food comes. The pleasure of eating, then, may be the best available standard of our health. And this pleasure, I think, is pretty fully available to the urban consumer who will make the necessary effort.

I mentioned earlier the politics, esthetics, and ethics of food. But to speak of the pleasure of eating is to go beyond those categories. Eating with the fullest pleasure—pleasure, that is, that does not depend on ignorance—is perhaps the profoundest enactment of our connection with the world. In this pleasure we experience and celebrate our dependence and our gratitude, for we are living from mystery, from creatures we did not make and powers we cannot comprehend. When I think of the meaning of food, I always remember these lines by the poet William Carlos Williams, which seem to me merely honest:

> There is nothing to eat,
>   seek it where you will,
>   but of the body of the Lord.
> The blessed plants
>   and the sea, yield it
>   to the imagination
> intact.

The Work of Local Culture

FOR MANY YEARS, my walks have taken me down an old fencerow in a wooded hollow on what was once my grandfather's farm. A battered galvanized bucket is hanging on a fence post near the head of the hollow, and I never go by it without stopping to look inside. For what is going on in that bucket is the most momentous thing I know, the greatest miracle that I have ever heard of: it is making earth. The old bucket has hung there through many autumns, and the leaves have fallen around it and some have fallen into it. Rain and snow have fallen into it, and the fallen leaves have held the moisture and so have rotted. Nuts have fallen into it, or been carried into it by squirrels; mice and squirrels have eaten the meat of the nuts and left the shells; they and other animals have left their droppings; insects have flown into the bucket and died and decayed; birds have scratched in it and left their droppings or perhaps a feather or two. This slow work of growth and death, gravity and decay, which is the chief work of the world, has by now produced in the bottom of the bucket several inches of black humus. I look into that bucket with fascination because I am a farmer of sorts and an artist of sorts, and I recognize there an artistry and a farming far superior to mine, or to that of any human. I have seen the same process at work on the tops of boulders in a forest, and it has been at work immemorially over most of the land surface of the world. All creatures die into it, and they live by it.

The old bucket started out a far better one than you can buy now. I think it has been hanging on that post for something like fifty years. I think so because I remember hearing, when I was just a small boy, a story about a bucket that must have been this one. Several of my grandfather's black hired hands went out on an early spring day to burn a tobacco plant bed, and they took along some eggs to boil to eat with their dinner. When dinner time came and they looked around for something to boil the eggs in, they could find only an old bucket that at one time had been filled with tar. The boiling water softened the residue of tar, and one of the eggs came out of the water black. The hands made much sport of seeing who would have to eat the black egg, welcoming their laughter in the midst of their day's work. The man who had to eat the black egg was Floyd Scott, whom I remember well. Dry scales of tar still adhere to the inside of the bucket.

However small a landmark the old bucket is, it is not trivial. It is one of the signs by which I know my country and myself. And to me it is irresistibly suggestive in the way it collects leaves and other woodland sheddings as they fall through time. It collects stories, too, as they fall through time. It is irresistibly metaphorical. It is doing in a passive way what a human community must do actively and thoughtfully. A human community, too, must collect leaves and stories, and turn them to account. It must build soil, and build that memory of itself —in lore and story and song—that will be its culture. These two kinds of accumulation, of local soil and local culture, are intimately related.

In the woods, the bucket is no metaphor; it simply reveals what is always happening in the woods, if the woods is let alone. Of course, in most places in my part of the country, the human community did not leave the woods alone. It felled the trees and replaced them with pastures and crops. But this did not revoke the law of the woods, which is that the ground must be protected by a cover of vegetation and that the growth of the years must return—or be returned—to the ground to rot and build soil. A good local culture, in one of its most important functions, is a collection of the memories, ways, and skills necessary for the observance, within the bounds of domesticity, of this natural law. If the local culture cannot preserve and improve the local soil, then, as both reason and history inform us, the local community will decay and perish, and the work of soil building will be resumed by nature.

A human community, then, if it is to last long, must exert a sort of centripetal force, holding local soil and local memory in place. Practically speaking, human society has no work more important than this. Once we have acknowledged this principle, we can only be alarmed at the extent to which it has been ignored. For although our present society does generate a centripetal force of great power, this is not a local force, but one centered almost exclusively in our great commercial and industrial cities, which have drawn irresistibly into themselves both the products of the countryside and the people and talents of the country communities.

There is, as one assumes there must be, a countervailing or centrifugal force that also operates in our society, but this returns to the countryside not the residue of the land's growth to re-fertilize the fields, not the learning and experience of the greater world ready to go to work locally, and not—or not often—even a just monetary compensation. What are returned, instead, are overpriced manufactured goods, pollution in various forms, and garbage. A landfill on the edge of my own rural county in Kentucky, for example, daily receives about eighty truckloads of garbage. Fifty to sixty of these loads come from cities in New York, New Jersey, and Pennsylvania. Thus, the end result of the phenomenal modern productivity of the countryside is a debased countryside, which becomes daily less pleasant, and which will inevitably become less productive.

The cities, which have imposed this inversion of forces on the country, have been unable to preserve themselves from it. The typical modern city is surrounded by a circle of affluent suburbs, eating its way outward, like ringworm, leaving the so-called inner city desolate, filthy, ugly, and dangerous.

My walks in the hills and hollows around my home have inevitably produced in my mind the awareness that I live in a diminished country. The country has been and is being reduced by the great centralizing process that is our national economy. As I walk, I am always reminded of the slow, patient building of soil in the woods. And I am reminded of the events and companions of my life—for my walks, after so long, are cultural events. But under the trees and in the fields I see also the gullies and scars, healed or healing or fresh, left by careless logging and bad farming. I see the crumbling stone walls and the wire fences that have been rusting out ever since the 1930s. In the returning woods growth of the hollows, I see the sagging and the fallen barns, the empty and ruining houses, the houseless chimneys and foundations. As I look at this evidence of human life poorly founded, played out, and gone, I try to recover some understanding, some vision, of what this country was at the beginning: the great oaks and beeches and hickories, walnuts and maples, lindens and ashes, tulip poplars, standing in beauty and dignity now unimaginable, the black soil of their making, also no longer imaginable, lying deep at their feet—an incalculable birthright sold for money, most of which we did not receive. Most of the money made on the products of this place has gone to fill the pockets of people in distant cities who did not produce the products.

If my walks take me along the roads and streams, I see also the trash and the junk, carelessly manufactured and carelessly thrown away, the glass and the broken glass and the plastic and the aluminum that will lie here longer than the lifetime of trees—longer than the lifetime of our species, perhaps. And I know that this also is what we have to show for our participation in the American economy, for most of the money made on these things too has been made elsewhere.

It would be somewhat more pleasant for country people if they could blame all this on city people. But the old opposition of country versus city—though still true, and truer than ever economically, for the country is more than ever the colony of the city—is far too simple to explain our problem. For country people more and more live like city people, and so connive in their own ruin. More and more country people, like city people, allow their economic and social standards to be set by television and salesmen and outside experts. Our garbage mingles with New Jersey garbage in our local landfill, and it would be hard to tell which is which.

As local community decays along with local economy, a vast amnesia settles over the countryside. As the exposed and disregarded soil departs with the rains, so local knowledge and local memory move away to the cities or are forgotten under the influence of homogenized salestalk, entertainment, and education. This loss of local knowledge and local memory—that is, of local culture—has been ignored, or written off as one of the cheaper "prices of progress," or made the business of folklorists. Nevertheless, local culture has a value, and part of its value is economic. This can be demonstrated readily enough.

For example, when a community loses its memory, its members no longer know one another. How can they know one another if they have forgotten or have never learned one another's stories? If they do not know one another's stories, how can they know whether or not to trust one another? People who do not trust one another do not help one another, and moreover they fear one another. And this is our predicament now. Because of a general distrust and suspicion, we not only lose one another's help and companionship, but we are all now living in jeopardy of being sued.

We don't trust our "public servants" because we know that they don't respect us. They don't respect us, as we understand, because they don't know us; they don't know our stories. They expect us to sue them if they make mistakes, and so they must insure themselves, at great expense to them and to us. Doctors in a country community must send their patients to specialists in the city, not necessarily because they believe that they are wrong in their diagnoses, but because they know that they are not infallible and they must protect themselves against lawsuits, at great expense to us.

The government of my home county, which has a population of about ten thousand people, pays an annual liability insurance premium of about $34,000. Add to this the liability premiums that are paid by every professional person who is "at risk" in the county, and you get some idea of the load we are carrying. Several decent family livelihoods are annually paid out of the county to insurance companies for a service that is only negative and provisional.

All of this money is lost to us by the failure of community. A good community, as we know, insures itself by trust, by good faith and good will, by mutual help. A good community, in other words, is a good local economy. It depends on itself for many of its essential needs and is thus shaped, so to speak, from the inside—unlike most modern populations that depend on distant purchases for almost everything and are thus shaped from the outside by the purposes and the influence of salesmen.

I was walking one Sunday afternoon several years ago with an older friend. We went by the ruining log house that had belonged to his grandparents and great-grandparents. The house stirred my friend's memory, and he told how the oldtime people used to visit each other in the evenings, especially in the long evenings of winter. There used to be a sort of institution in our part of the country known as "sitting till bedtime." After supper, when they weren't too tired, neighbors would walk across the fields to visit each other. They popped corn, my friend said, and ate apples and talked. They told each other stories. They told each other stories, as I knew myself, that they all had heard before. Sometimes they told stories about each other, about themselves, living again in their own memories and thus keeping their memories alive. Among the hearers of these stories were always the children. When bedtime came, the visitors lit their lanterns and went home. My friend talked about this, and thought about it, and then he said, "They had everything but money."

They were poor, as country people have often been, but they had each other, they had their local economy in which they helped each other, they had each other's comfort when they needed it, and they had their stories, their history together in that place. To have everything but money is to have much. And most people of the present can only marvel to think of neighbors entertaining themselves for a whole evening without a single imported pleasure and without listening to a single minute of sales talk.

Most of the descendants of those people have now moved away, partly because of the cultural and economic failures that I mentioned earlier, and most of them no longer sit in the evenings and talk to anyone. Most of them now sit until bedtime watching TV, submitting every few minutes to a sales talk. The message of both the TV programs and the sales talks is that the watchers should spend whatever is necessary to be like everybody else.

By television and other public means, we are encouraged to believe that we are far advanced beyond sitting till bedtime with the neighbors on a Kentucky ridgetop, and indeed beyond anything we ever were before. But if, for example, there should occur a forty-eight-hour power failure, we would find ourselves in much more backward circumstances than our ancestors. What, for starters, would we do for entertainment? Tell each other stories? But most of us no longer talk with each other, much less tell each other stories. We tell our stories now mostly to doctors or lawyers or psychiatrists or insurance adjusters or the police, not to our neighbors for their (and our) entertainment. The stories that now entertain us are made up for us in New York or Los Angeles or other centers of such commerce.

But a forty-eight-hour power failure would involve almost unimaginable deprivations. It would be difficult to travel, especially in cities. Most of the essential work could not be done. Our windowless modern schools and other such buildings that depend on air conditioning could not be used. Refrigeration would be impossible; food would spoil. It would be difficult or impossible to prepare meals. If it was winter, heating systems would fail. At the end of forty-eight hours many of us would be hungry.

Such a calamity (and it is a modest one among those that our time has made possible) would thus reveal how far most of us are now living from our cultural and economic sources, and how extensively we have destroyed the foundations of local life. It would show us how far we have strayed from the locally centered life of such neighborhoods as the one my friend described—a life based to a considerable extent on what we now call solar energy, which is decentralized, democratic, clean, and free. If we note that much of the difference we are talking about can be accounted for as an increasing dependence on energy sources that are centralized, undemocratic, filthy, and expensive, we will have completed a sort of historical parable.

How has this happened? There are many reasons for it. One of the chief reasons is that everywhere in our country the local succession of the generations has been broken. We can trace this change through a series of stories that we may think of as cultural landmarks.

Throughout most of our literature, the normal thing was for the generations to succeed one another in place. The memorable stories occurred when this succession failed or became difficult or was somehow threatened. The norm is given in Psalm 128, in which this succession is seen as one of the rewards of righteousness: "Thou shalt see thy children's children, and peace upon Israel."

The longing for this result seems to have been universal. It presides also over *The Odyssey*, in which Odysseus's desire to return home is certainly regarded as normal. And this story is also much concerned with the psychology of family succession. Telemachus, Odysseus's son, comes of age in preparing for the return of his long-absent father; and it seems almost that Odysseus is enabled to return home by his son's achievement of enough manhood to go in search of him. Long after the return of both father and son, Odysseus's life will complete itself, as we know from Teiresias's prophecy in Book XI, much in the spirit of Psalm 128:

> A seaborne death
> soft as this hand of mist will come upon you
> when you are wearied out with sick old age,
> your country folk in blessed peace around you.

The Bible makes much of what it sees as the normal succession, in such stories as those of Abraham, Isaac, and Jacob, or of David and Solomon, in which the son completes the work or the destiny of the father. The parable of the prodigal son is prepared for by such Old Testament stories as that of Jacob, who errs, wanders, returns, is forgiven, and takes his place in the family lineage.

Shakespeare was concerned throughout his working life with the theme of the separation and rejoining of parents and children. It is there at the beginning in *The Comedy of Errors*, and he is still thinking about it when he gets to *King Lear* and *Pericles* and *The Tempest*. When Lear walks onstage with Cordelia dead in his arms, the theme of return is fulfilled, only this time in the way of tragedy.

Wordsworth's poem "Michael," written in 1800, is in the same line of descent. It is the story of a prodigal son, and return is still understood as the norm; before the boy's departure, he and his father make a "covenant" that he will return home and carry on his father's life as a shepherd on their ancestral pastures. But the ancient theme here has two significant differences: the son leaves home for an economic reason, and he does not return. Old Michael, the father, was long ago "bound / In surety for his brother's son." This nephew has failed in his business, and Michael is "summoned to discharge the forfeiture." Rather than do this by selling a portion of their patrimony, the aged parents decide that they must send their son to work for another kinsman in the city in order to earn the necessary money. The country people all are poor; there is no money to be earned at home. When the son has cleared the debt from the land, he will return to it to "possess it, free as the wind / That passes over it." But the son goes to the city, is corrupted by it, eventually commits a crime, and is forced "to seek a hiding place beyond the seas."

"Michael" is a sort of cultural watershed. It carries on the theme of return that goes back to the beginnings of Western culture, but that return now is only a desire and a memory; in the poem it fails to happen. Because of that failure, we see in "Michael" not just a local story of the Lake District of England, which it is, but the story of rural families in the industrial nations from Wordsworth's time until today. The children go to the cities, for reasons imposed by the external economy, and they do not return; eventually the parents die and the family land, like Michael's, is sold to a stranger. By now it has happened millions of times.

And by now the transformation of the ancient story is nearly complete. Our society, on the whole, has forgotten or repudiated the theme of return. Young people still grow up in rural families and go off to the cities, not to return. But now it is felt that this is what they *should* do. Now the norm is to leave and not return. And this applies as much to urban families as to rural ones. In the present urban economy the parent-child succession is possible only among the economically privileged. The children of industrial underlings are not likely to succeed their parents at work, and there is no reason for them to wish to do so. We are not going to have an industrial "Michael" in which it is perceived as tragic that a son fails to succeed his father on an assembly line.

According to the new norm, the child's destiny is not to succeed the parents, but to outmode them; succession has given way to supersession. And this norm is institutionalized not in great communal stories, but in the education system. The schools are no longer oriented to a cultural inheritance that it is their duty to pass on unimpaired, but to the career, which is to say the future, of the child. The orientation is thus necessarily theoretical, speculative, and mercenary. The child is not educated to return home and be of use to the place and community; he or she is educated to *leave* home and earn money in a provisional future that has nothing to do with place or community. And parents with children in school are likely to find themselves immediately separated from their children, and made useless to them, by the intervention of new educational techniques, technologies, methods, and languages. School systems innovate as compulsively and as eagerly as factories. It is no wonder that, under these circumstances, "educators" tend to look upon the parents as a bad influence and wish to take the children away from home as early as possible. And many parents, in truth, are now finding their children an encumbrance at home, where there is no useful work for them to do, and are glad enough to turn them over to the state for the use of the future. The extent to which this order of things is now dominant is suggested by a recent magazine article on the discovery of what purports to be a new idea:

> The idea that a parent can be a teacher at home has caught the attention of educators. . . . Parents don't have to be graduates of Harvard or Yale to help their kids learn and achieve.

Thus the home as a place where a child can learn becomes an *idea* of the professional "educator," who retains control of the idea. The home, as the article makes clear, is not to be a place where children may learn on their own, but a place where they are taught by parents according to the instructions of professional "educators." In fact, the Home and School Institute, Inc., of Washington, D.C. (known, of course, as "the HSI") has been "founded to show . . . how to involve families in their kids' educations."

In such ways as this, the nuclei of home and community have been invaded by the organizations, just as have the nuclei of cells and atoms. And we must be careful to see that the old cultural centers of home and community were made vulnerable to this invasion by their failure as economies. If there is no household or community economy, then family members and neighbors are no longer useful to one another. When people are no longer useful to one another, then the centripetal force of family and community fails, and people fall into dependence on exterior economies and organizations. The hegemony of professionals and professionalism erects itself on local failure, and from then on the locality exists merely as a market for consumer goods and as a source of "raw material," human and natural. The local schools no longer serve the local community; they serve the government's economy and the economy's government. Unlike the local community, the government and the economy cannot be served with affection, but only with professional zeal or professional boredom. Professionalism means more interest in salaries and less interest in what used to be known as disciplines. And so we arrive at the idea, endlessly reiterated in the news media, that education can be improved by bigger salaries for teachers—which may be true, but education cannot be improved, as the proponents too often imply, by bigger salaries alone. There must also be love of learning and of the cultural tradition and of excellence—and this love cannot exist, because it makes no sense, apart from the love of a place and a community. Without this love, education is only the importation into a local community of centrally prescribed "career preparation" designed to facilitate the export of young careerists.

Our children are educated, then, to leave home, not to stay home, and the costs of this education have been far too little acknowledged. One of the costs is psychological, and the other is at once cultural and ecological.

The natural or normal course of human growing up must begin with some sort of rebellion against one's parents, for it is clearly impossible to grow up if one remains a child. But the child, in the process of rebellion and of achieving the emotional and economic independence that rebellion ought to lead to, finally comes to understand the parents as fellow humans and fellow sufferers, and in some manner returns to them as their friend, forgiven and forgiving the inevitable wrongs of family life. That is the old norm.

The new norm, according to which the child leaves home as a student and never lives at home again, interrupts the old course of coming of age at the point of rebellion, so that the child is apt to remain stalled in adolescence, never achieving any kind of reconciliation or friendship with the parents. Of course, such a return and reconciliation cannot be achieved without the recognition of mutual practical need. In the present economy, however, where individual dependences are so much exterior to both household and community, family members often have no practical need or use for one another. Hence the frequent futility of attempts at a purely psychological or emotional reconciliation.

And this interposition of rebellion and then of geographical and occupational distance between parents and children may account for the peculiar emotional intensity that our society attaches to innovation. We appear to hate whatever went before, very much as an adolescent hates parental rule, and to look on its obsolescence as a kind of vengeance. Thus we may explain industry's obsessive emphasis on "this year's model," or the preoccupation of the professional "educators" with theoretical and methodological innovation. Similarly, in modern literature we have had for many years an emphasis on "originality" and "the anxiety of influence" (an adolescent critical theory), as opposed, say, to Spenser's filial admiration for Chaucer, or Dante's for Virgil.

But if the new norm interrupts the development of the relation between children and parents, that same interruption, ramifying through a community, destroys the continuity and so the integrity of local life. As the children depart, generation after generation, the place loses its memory of itself, which is its history and its culture. And the local history, if it survives at all, loses its place. It does no good for historians, folklorists, and anthropologists to collect the songs and the stories and the lore that make up local culture and store them in books and archives. They cannot collect and store—because they cannot know—the pattern of reminding that can survive only in the living human community in its place. It is this pattern that is the life of local culture and that brings it usefully or pleasurably to mind. Apart from its local landmarks and occasions, the local culture may be the subject of curiosity or of study, but it is also dead.

The loss of local culture is, in part, a practical loss and an economic one. For one thing, such a culture contains, and conveys to succeeding generations, the history of the use of the place and the knowledge of how the place may be lived in and used. For another, the pattern of reminding implies affection for the place and respect for it, and so, finally, the local culture will carry the knowledge of how the place may be well and lovingly used, and also the implicit command to use it *only* well and lovingly. The only true and effective "operator's manual for spaceship earth" is not a book that any human will ever write; it is hundreds of thousands of local cultures.

Lacking an authentic local culture, a place is open to exploitation, and ultimately destruction, from the center. Recently, for example, I heard the dean of a prominent college of agriculture interviewed on the radio. What have we learned, he was asked, from last summer's drouth? And he replied that "we" need to breed more drouth resistance into plants, and that "we" need a government "safety net" for farmers. He might have said that farmers need to re-examine their farms and their circumstances in light of the drouth, and to think again on such subjects as diversification, scale, and the mutual helpfulness of neighbors. But he did not say that. To him, the drouth was merely an opportunity for agribusiness corporations and the government, by which the farmers and rural communities could only become more dependent on the economy that is destroying them. This is as good an example as any of the centralized thinking of a centralized economy—to which the only effective answer that I know is a strong local community with a strong local economy and a strong local culture.

For a long time now, the prevailing assumption has been that if the nation is all right, then all the localities within it will be all right also. I see little reason to believe that this is true. At present, in fact, both the nation and the national economy are living at the expense of localities and local communities—as all small-town and country people have reason to know. In rural America, which is in many ways a colony of what the government and the corporations think of as the nation, most of us have experienced the losses that I have been talking about: the departure of young people, of soil and other so-called natural resources, and of local memory. We feel ourselves crowded more and more into a dimensionless present, in which the past is forgotten and the future, even in our most optimistic "projections," is forbidding and fearful. Who can desire a future that is determined entirely by the purposes of the most wealthy and the most powerful, and by the capacities of machines?

Two questions, then, remain: Is a change for the better possible? And who has the power to make such a change? I still believe that a change for the better is possible, but I confess that my belief is partly hope and partly faith. No one who hopes for improvement should fail to see and respect the signs that we may be approaching some sort of historical waterfall, past which we will not, by changing our minds, be able to change anything else. We know that at any time an ecological or a technological or a political event that we will have allowed may remove from us the power to make change and leave us with the mere necessity to submit to it. Beyond that, the two questions are one: the possibility of change depends on the existence of people who have the power to change.

Does this power reside at present in the national government? That seems to me extremely doubtful. To anyone who has read the papers during the recent presidential campaign, it must be clear that at the highest level of government there is, properly speaking, no political discussion. Are the corporations likely to help us? We know, from long experience, that the corporations will assume no responsibility that is not forcibly imposed upon them by government. The record of the corporations is written too plainly in verifiable damage to permit us to expect much from them. May we look for help to the universities? Well, the universities are more and more the servants of government and the corporations.

Most urban people evidently assume that all is well. They live too far from the exploited and endangered sources of their economy to need to assume otherwise. Some urban people are becoming disturbed about the contamination of air, water, and food, and that is promising, but there are not enough of them yet to make much difference. There is enough trouble in the "inner cities" to make them likely places of change, and evidently change is in them, but it is desperate and destructive change. As if to perfect their exploitation by other people, the people of the "inner cities" are destroying both themselves and their places.

My feeling is that if improvement is going to begin anywhere, it will have to begin out in the country and in the country towns. This is not because of any intrinsic virtue that can be ascribed to rural people, but because of their circumstances. Rural people are living, and have lived for a long time, at the site of the trouble. They see all around them, every day, the marks and scars of an exploitive national economy. They have much reason, by now, to know how little real help is to be expected from somewhere else. They still have, moreover, the remnants of local memory and local community. And in rural communities there are still farms and small businesses that can be changed according to the will and the desire of individual people.

In this difficult time of failed public expectations, when thoughtful people wonder where to look for hope, I keep returning in my own mind to the thought of the renewal of the rural communities. I know that one revived rural community would be more convincing and more encouraging than all the government and university programs of the last fifty years, and I think that it could be the beginning of the renewal of our country, for the renewal of rural communities ultimately implies the renewal of urban ones. But to be authentic, a true encouragement and a true beginning, this would have to be a revival accomplished mainly by the community itself. It would have to be done not from the outside by the instruction of visiting experts, but from the inside by the ancient rule of neighborliness, by the love of precious things, and by the wish to be at home.

Why I Am Not Going to Buy a Computer

L IKE ALMOST EVERYBODY ELSE, I am hooked to the energy corporations, which I do not admire. I hope to become less hooked to them. In my work, I try to be as little hooked to them as possible. As a farmer, I do almost all of my work with horses. As a writer, I work with a pencil or a pen and a piece of paper.

My wife types my work on a Royal standard typewriter bought new in 1956 and as good now as it was then. As she types, she sees things that are wrong and marks them with small checks in the margins. She is my best critic because she is the one most familiar with my habitual errors and weaknesses. She also understands, sometimes better than I do, what *ought* to be said. We have, I think, a literary cottage industry that works well and pleasantly. I do not see anything wrong with it.

A number of people, by now, have told me that I could greatly improve things by buying a computer. My answer is that I am not going to do it. I have several reasons, and they are good ones.

The first is the one I mentioned at the beginning. I would hate to think that my work as a writer could not be done without a direct dependence on strip-mined coal. How could I write conscientiously against the rape of nature if I were, in the act of writing, implicated in the rape? For the same reason, it matters to me that my writing is done in the daytime, without electric light.

I do not admire the computer manufacturers a great deal more than I admire the energy industries. I have seen their advertisements, attempting to seduce struggling or failing farmers into the belief that they can solve their problems by buying yet another piece of expensive equipment. I am familiar with their propaganda campaigns that have put computers into public schools in need of books. That computers are expected to become as common as TV sets in "the future" does not impress me or matter to me. I do not own a TV set. I do not see that computers are bringing us one step nearer to anything that does matter to me: peace, economic justice, ecological health, political honesty, family and community stability, good work.

What would a computer cost me? More money, for one thing, than I can afford, and more than I wish to pay to people whom I do not admire. But the cost would not be just monetary. It is well understood that technological innovation always requires the discarding of the "old model"—the "old model" in this case being not just our old Royal standard, but my wife, my critic, my closest reader, my fellow worker. Thus (and I think this is typical of present-day technological innovation), what would be superseded would be not only something, but somebody. In order to be technologically up-to-date as a writer, I would have to sacrifice an association that I am dependent upon and that I treasure.

My final and perhaps my best reason for not owning a computer is that I do not wish to fool myself. I disbelieve, and therefore strongly resent, the assertion that I or anybody else could write better or more easily with a computer than with a pencil. I do not see why I should not be as scientific about this as the next fellow: when somebody has used a computer to write work that is demonstrably better than Dante's, and when this better is demonstrably attributable to the use of a computer, then I will speak of computers with a more respectful tone of voice, though I still will not buy one.

To make myself as plain as I can, I should give my standards for technological innovation in my own work. They are as follows:

1. The new tool should be cheaper than the one it replaces.

2. It should be at least as small in scale as the one it replaces.

3. It should do work that is clearly and demonstrably better than the one it replaces.

4. It should use less energy than the one it replaces.

5. If possible, it should use some form of solar energy, such as that of the body.

6. It should be repairable by a person of ordinary intelligence, provided that he or she has the necessary tools.

7. It should be purchasable and repairable as near to home as possible.

8. It should come from a small, privately owned shop or store that will take it back for maintenance and repair.

9. It should not replace or disrupt anything good that already exists, and this includes family and community relationships.

1987

After the foregoing essay, first published in the *New England Review and Bread Loaf Quarterly*, was reprinted in *Harper's*, the *Harper's* editors published the following letters in response and permitted me a reply.   W. B.

LETTERS

Wendell Berry provides writers enslaved by the computer with a handy alternative: Wife—a low-tech energy-saving device. Drop a pile of handwritten notes on Wife and you get back a finished manuscript, edited while it was typed. What computer can do that? Wife meets all of Berry's uncompromising standards for technological innovation: she's cheap, repairable near home, and good for the family structure. Best of all, Wife is politically correct because she breaks a writer's "direct dependence on strip-mined coal."

History teaches us that Wife can also be used to beat rugs and wash clothes by hand, thus eliminating the need for the vacuum cleaner and washing machine, two more nasty machines that threaten the act of writing.

Gordon Inkeles
Miranda, Calif.

I have no quarrel with Berry because he prefers to write with pencil and paper; that is his choice. But he implies that I and others are somehow impure because we choose to write on a computer. I do not admire the energy corporations, either. Their shortcoming is not that they produce electricity but how they go about it. They are poorly managed because they are blind to long-term consequences. To solve this problem, wouldn't it make more sense to correct the precise error they are making rather than simply ignore their product? I would be happy to join Berry in a protest against strip mining, but

I intend to keep plugging this computer into the wall
with a clear conscience.

James Rhoads
Battle Creek, Mich.

I enjoyed reading Berry's declaration of intent never to
buy a personal computer in the same way that I enjoy
reading about the belief systems of unfamiliar tribal cultures. I tried to imagine a tool that would meet Berry's
criteria for superiority to his old manual typewriter. The
clear winner is the quill pen. It is cheaper, smaller, more
energy-efficient, human-powered, easily repaired, and
non-disruptive of existing relationships.

Berry also requires that this tool must be "clearly and
demonstrably better" than the one it replaces. But surely
we all recognize by now that "better" is in the mind of
the beholder. To the quill pen aficionado, the benefits
obtained from elegant calligraphy might well outweigh
all others.

I have no particular desire to see Berry use a word processor; if he doesn't like computers, that's fine with me.
However, I do object to his portrayal of this reluctance
as a moral virtue. Many of us have found that computers
can be an invaluable tool in the fight to protect our environment. In addition to helping me write, my personal
computer gives me access to up-to-the-minute reports
on the workings of the EPA and the nuclear industry. I
participate in electronic bulletin boards on which environmental activists discuss strategy and warn each other
about urgent legislative issues. Perhaps Berry feels that
the Sierra Club should eschew modern printing technology, which is highly wasteful of energy, in favor of having
its members hand-copy the club's magazines and other
mailings each month?

Nathaniel S. Borenstein
Pittsburgh, Pa.

The value of a computer to a writer is that it is a tool not
for generating ideas but for typing and editing words. It
is cheaper than a secretary (or a wife!) and arguably more

fuel-efficient. And it enables spouses who are not inclined to provide free labor more time to concentrate on *their* own work.

We should support alternatives both to coal-generated electricity and to IBM-style technocracy. But I am reluctant to entertain alternatives that presuppose the traditional subservience of one class to another. Let the PCs come and the wives and servants go seek more meaningful work.

Toby Koosman
Knoxville, Tenn.

Berry asks how he could write conscientiously against the rape of nature if in the act of writing on a computer he was implicated in the rape. I find it ironic that a writer who sees the underlying connectedness of things would allow his diatribe against computers to be published in a magazine that carries ads for the National Rural Electric Cooperative Association, Marlboro, Phillips Petroleum, McDonnell Douglas, and yes, even Smith-Corona. If Berry rests comfortably at night, he must be using sleeping pills.

Bradley C. Johnson
Grand Forks, N.D.

WENDELL BERRY REPLIES:

The foregoing letters surprised me with the intensity of the feelings they expressed. According to the writers' testimony, there is nothing wrong with their computers; they are utterly satisfied with them and all that they stand for. My correspondents are certain that I am wrong and that I am, moreover, on the losing side, a side already relegated to the dustbin of history. And yet they grow huffy and condescending over my tiny dissent. What are they so anxious about?

I can only conclude that I have scratched the skin of a technological fundamentalism that, like other fundamentalisms, wishes to monopolize a whole society and, therefore, cannot tolerate the smallest difference of opinion. At the slightest hint of a threat to their complacency, they repeat, like a chorus of toads, the notes sounded by their leaders in industry. The past was gloomy, drudgery-ridden, servile, meaningless, and slow. The present, thanks only to purchasable products, is meaningful, bright, lively, centralized, and fast. The future, thanks only to more purchasable products, is going to be even better. Thus consumers become salesmen, and the world is made safer for corporations.

I am also surprised by the meanness with which two of these writers refer to my wife. In order to imply that I am a tyrant, they suggest by both direct statement and innuendo that she is subservient, characterless, and stupid—a mere "device" easily forced to provide meaningless "free labor." I understand that it is impossible to make an adequate public defense of one's private life, and so I will only point out that there are a number of kinder possibilities that my critics have disdained to imagine: that my wife may do this work because she wants to and likes to; that she may find some use and some meaning in it; that she may not work for nothing. These gentlemen obviously think themselves feminists of the most correct and principled sort, and yet they do not hesitate to stereotype and insult, on the basis of one fact, a woman they do not know. They are audacious and irresponsible gossips.

In his letter, Bradley C. Johnson rushes past the possibility of sense in what I said in my essay by implying that I am or ought to be a fanatic. That I am a person of this century and am implicated in many practices that I regret is fully acknowledged at the beginning of my essay. I did not say that I proposed to end forthwith all my involvement in harmful technology, for I do not know how to do that. I said merely that I want to limit such involvement, and to a certain extent I do know how to do that. If some technology does damage to the world—as two of the above letters seem to agree that it does—then why is it not reasonable, and indeed moral, to try to limit one's use of that technology? *Of course*, I think that I am right to do this.

I would not think so, obviously, if I agreed with Nathaniel S. Borenstein that "'better' is in the mind of
the beholder." But if he truly believes this, I do not see
why he bothers with his personal computer's "up-to-
the-minute reports on the workings of the EPA and the
nuclear industry" or why he wishes to be warned about
"urgent legislative issues." According to his system, the
"better" in a bureaucratic, industrial, or legislative mind
is as good as the "better" in his. His mind apparently is
being subverted by an objective standard of some sort,
and he had better look out.

Borenstein does not say what he does after his computer
has drummed him awake. I assume from his letter that he
must send donations to conservation organizations and
letters to officials. Like James Rhoads, at any rate, he has
a clear conscience. But this is what is wrong with the
conservation movement. It has a clear conscience. The
guilty are always other people, and the wrong is always
somewhere else. That is why Borenstein finds his "electronic bulletin board" so handy. To the conservation
movement, it is only production that causes environmental degradation; the consumption that supports the production is rarely acknowledged to be at fault. The ideal of
the run-of-the-mill conservationist is to impose restraints
upon production without limiting consumption or bur-
dening the consciences of consumers.

But virtually all of our consumption now is extravagant, and virtually all of it consumes the world. It is not
beside the point that most electrical power comes from
strip-mined coal. The history of the exploitation of the
Appalachian coal fields is long, and it is available to readers. I do not see how anyone can read it and plug in any
appliance with a clear conscience. If Rhoads can do so,
that does not mean that his conscience is clear; it means
that his conscience is not working.

To the extent that we consume, in our present circumstances, we are guilty. To the extent that we guilty
consumers are conservationists, we are absurd. But what
can we do? Must we go on writing letters to politicians
and donating to conservation organizations until the

majority of our fellow citizens agree with us? Or can we do something directly to solve our share of the problem?

I am a conservationist. I believe wholeheartedly in putting pressure on the politicians and in maintaining the conservation organizations. But I wrote my little essay partly in distrust of centralization. I don't think that the government and the conservation organizations alone will ever make us a conserving society. Why do I need a centralized computer system to alert me to environmental crises? That I live every hour of every day in an environmental crisis I know from all my senses. Why then is not my first duty to reduce, so far as I can, my own consumption?

Finally, it seems to me that none of my correspondents recognizes the innovativeness of my essay. If the use of a computer is a new idea, then a newer idea is not to use one.

Feminism, the Body, and the Machine

SOME TIME AGO *Harper's* reprinted a short essay of mine in which I gave some of my reasons for refusing to buy a computer. Until that time, the vast numbers of people who disagree with my writings had mostly ignored them. An unusual number of people, however, neglected to ignore my insensitivity to the wonders of computer enhancement. Some of us, it seems, would be better off if we would just realize that this is already the best of all possible worlds, and is going to get even better if we will just buy the right equipment.

Harper's published only five of the letters the editors received in response to my essay, and they published only negative letters. But of the twenty letters received by the *Harper's* editors, who forwarded copies to me, three were favorable. This I look upon as extremely gratifying. If these letters may be taken as a fair sample, then one in seven of *Harper's* readers agreed with me. If I had guessed beforehand, I would have guessed that my supporters would have been fewer than one in a thousand. And so I suppose, after further reflection, that my surprise at the intensity of the attacks on me is mistaken. There are more of us than I thought. Maybe there is even a "significant number" of us.

Only one of the negative letters seemed to me to have much intelligence in it. That one was from R. N. Neff of Arlington, Virginia, who scored a direct hit: "Not to be obtuse, but being willing to bare my illiterate soul for all to see, is there indeed a 'work demonstrably better than Dante's' . . . which was written on a Royal standard typewriter?" I like this retort so well that I am tempted to count it a favorable response, raising the total to four. The rest of the negative replies, like the five published ones, were more feeling than intelligent. Some of them, indeed, might be fairly described as exclamatory.

One of the letter writers described me as "a fool" and "doubly a fool," but fortunately misspelled my name, leaving me a speck of hope that I am not the "Wendell Barry" he was talking about. Two others accused me of self-righteousness, by which they seem to have meant that they think they are righter

than I think I am. And another accused me of being more concerned about my own moral purity than with "any ecological
effect," thereby making the sort of razor-sharp philosophical
distinction that could cause a person to be elected president.

But most of my attackers deal in feelings either feminist or
technological, or both. The feelings expressed seem to be representative of what the state of public feeling currently permits
to be felt, and of what public rhetoric currently permits to be
said. The feelings, that is, are similar enough, from one letter
to another, to be thought representative, and as representative
letters they have an interest greater than the quarrel that occa-
sioned them.

Without exception, the feminist letters accuse me of exploiting
my wife, and they do not scruple to allow the most insulting
implications of their indictment to fall upon my wife. They fail
entirely to see that my essay does not give any support to their
accusation—or if they see it, they do not care. My essay, in fact,
does not characterize my wife beyond saying that she types
my manuscripts and tells me what she thinks about them. It
does not say what her motives are, how much work she does,
or whether or how she is paid. Aside from saying that she is
my wife and that I value the help she gives me with my work,
it says nothing about our marriage. It says nothing about our
economy.

There is no way, then, to escape the conclusion that my wife
and I are subjected in these letters to a condemnation by category. My offense is that I am a man who receives some help
from his wife; my wife's offense is that she is a woman who
does some work for her husband—which work, according to
her critics and mine, makes her a drudge, exploited by a conventional subservience. And my detractors have, as I say, no
evidence to support any of this. Their accusation rests on a
syllogism of the flimsiest sort: my wife helps me in my work,
some wives who have helped their husbands in their work have
been exploited, therefore my wife is exploited.

This, of course, outrages justice to about the same extent
that it insults intelligence. Any respectable system of justice
exists in part as a protection against such accusations. In a
just society nobody is expected to plead guilty to a general

indictment, because in a just society nobody can be convicted on a general indictment. What is required for a just conviction is a particular accusation that can be *proved*. My accusers have made no such accusation against me.

That feminists or any other advocates of human liberty and dignity should resort to insult and injustice is regrettable. It is equally regrettable that all of the feminist attacks on my essay implicitly deny the validity of two decent and probably necessary possibilities: marriage as a state of mutual help, and the household as an economy.

Marriage, in what is evidently its most popular version, is now on the one hand an intimate "relationship" involving (ideally) two successful careerists in the same bed, and on the other hand a sort of private political system in which rights and interests must be constantly asserted and defended. Marriage, in other words, has now taken the form of divorce: a prolonged and impassioned negotiation as to how things shall be divided. During their understandably temporary association, the "married" couple will typically consume a large quantity of merchandise and a large portion of each other.

The modern household is the place where the consumptive couple do their consuming. Nothing productive is done there. Such work as is done there is done at the expense of the resident couple or family, and to the profit of suppliers of energy and household technology. For entertainment, the inmates consume television or purchase other consumable diversion elsewhere.

There are, however, still some married couples who understand themselves as belonging to their marriage, to each other, and to their children. What they have they have in common, and so, to them, helping each other does not seem merely to damage their ability to compete against each other. To them, "mine" is not so powerful or necessary a pronoun as "ours."

This sort of marriage usually has at its heart a household that is to some extent productive. The couple, that is, makes around itself a household economy that involves the work of both wife and husband, that gives them a measure of economic independence and self-protection, a measure of self-employment, a measure of freedom, as well as a common ground and a common satisfaction. Such a household economy may employ the disciplines and skills of housewifery, of carpentry and other trades of building and maintenance, of gardening and other branches of subsistence agriculture, and even of woodlot management and woodcutting. It may also involve a "cottage industry" of some kind, such as a small literary enterprise.

It is obvious how much skill and industry either partner may put into such a household and what a good economic result such work may have, and yet it is a kind of work now frequently held in contempt. Men in general were the first to hold it in contempt as they departed from it for the sake of the professional salary or the hourly wage, and now it is held in contempt by such feminists as those who attacked my essay. Thus farm wives who help to run the kind of household economy that I have described are apt to be asked by feminists, and with great condescension, "But what do you *do*?" By this they invariably mean that there is something better to do than to make one's marriage and household, and by better they invariably mean "employment outside the home."

I know that I am in dangerous territory, and so I had better be plain: what I have to say about marriage and household I mean to apply to men as much as to women. I do not believe that there is anything better to do than to make one's marriage and household, whether one is a man or a woman. I do not believe that "employment outside the home" is as valuable or important or satisfying as employment at home, for either men or women. It is clear to me from my experience as a teacher, for example, that children need an ordinary daily association with *both* parents. They need to see their parents at work; they need, at first, to play at the work they see their parents doing, and then they need to work with their parents. It does not matter so much that this working together should be what is called "quality time," but it matters a great deal that the work done should have the dignity of economic value.

I should say too that I understand how fortunate I have been in being able to do an appreciable part of my work at home. I know that in many marriages both husband and wife are now finding it necessary to work away from home. This issue, of course, is troubled by the question of what is meant by "necessary," but it is true that a family living that not so long ago was ordinarily supplied by one job now routinely requires two or more. My interest is not to quarrel with individuals, men or women, who work away from home, but rather to ask why we should consider this general working away from home to be a desirable state of things, either for people or for marriage, for our society or for our country.

If I had written in my essay that my wife worked as a typist and editor for a publisher, doing the same work that she does for me, no feminists, I daresay, would have written to *Harper's* to attack me for exploiting her—even though, for all they knew, I might have forced her to do such work in order to keep me in gambling money. It would have been assumed as a matter of course that if she had a job away from home she was a "liberated woman," possessed of a dignity that no home could confer upon her.

As I have said before, I understand that one cannot construct an adequate public defense of a private life. Anything that I might say here about my marriage would be immediately (and rightly) suspect on the ground that it would be only *my* testimony. But for the sake of argument, let us suppose that whatever work my wife does, as a member of our marriage and household, she does both as a full economic partner and as her own boss, and let us suppose that the economy we have is adequate to our needs. Why, granting that supposition, should anyone assume that my wife would increase her freedom or dignity or satisfaction by becoming the employee of a boss, who would be in turn also a corporate underling and in no sense a partner?

Why would any woman who would refuse, properly, to take the marital vow of obedience (on the ground, presumably, that subservience to a mere human being is beneath human dignity) then regard as "liberating" a job that puts her under the authority of a boss (man or woman) whose authority specifically requires and expects obedience? It is easy enough to see why women came to object to the role of Blondie, a mostly decorative custodian of a degraded, consumptive modern household, preoccupied with clothes, shopping, gossip, and outwitting her husband. But are we to assume that one may fittingly cease to be Blondie by becoming Dagwood? Is the life of a corporate

underling—even acknowledging that corporate underlings are
well paid—an acceptable end to our quest for human dignity
and worth? It is clear enough by now that one does not cease
to be an underling by reaching "the top." Corporate life is
composed only of lower underlings and higher underlings.
Bosses are everywhere, and all the bosses are underlings. This is
invariably revealed when the time comes for accepting responsibility for something unpleasant, such as the Exxon fiasco in
Prince William Sound, for which certain lower underlings are
blamed but no higher underling is responsible. The underlings
at the top, like telephone operators, have authority and power,
but no responsibility.

And the oppressiveness of some of this office work defies
belief. Edward Mendelson (in the *New Republic*, February 22,
1988) speaks of "the office worker whose computer keystrokes
are monitored by the central computer in the personnel office, and who will be fired if the keystrokes-per-minute figure
doesn't match the corporate quota." (Mr. Mendelson does not
say what form of drudgery this worker is being saved from.)
And what are we to say of the diversely skilled country housewife who now bores the same six holes day after day on an
assembly line? What higher form of womanhood or humanity
is she evolving toward?

How, I am asking, can women improve themselves by submitting to the same specialization, degradation, trivialization,
and tyrannization of work that men have submitted to? And
that question is made legitimate by another: How have men
improved themselves by submitting to it? The answer is that
men have not, and women cannot, improve themselves by submitting to it.

Women have complained, justly, about the behavior of
"macho" men. But despite their he-man pretensions and
their captivation by masculine heroes of sports, war, and the
Old West, most men are now entirely accustomed to obey-
ing and currying the favor of their bosses. Because of this,
of course, they hate their jobs—they mutter, "Thank God
it's Friday" and "Pretty good for Monday"—but they do as
they are told. They are more compliant than most housewives have been. Their characters combine feudal submissive-
ness with modern helplessness. They have accepted almost

without protest, and often with relief, their dispossession of any usable property and, with that, their loss of economic independence and their consequent subordination to bosses. They have submitted to the destruction of the household economy and thus of the household, to the loss of home employment and self-employment, to the disintegration of their families and communities, to the desecration and pillage of their country, and they have continued abjectly to believe, obey, and vote for the people who have most eagerly abetted this ruin and who have most profited from it. These men, moreover, are helpless to do anything for themselves or anyone else without money, and so for money they do whatever they are told. They know that their ability to be useful is precisely defined by their willingness to be somebody else's tool. Is it any wonder that they talk tough and worship athletes and cowboys? Is it any wonder that some of them are violent?

It is clear that women cannot justly be excluded from the daily fracas by which the industrial economy divides the spoils of society and nature, but their inclusion is a poor justice and no reason for applause. The enterprise is as devastating with women in it as it was before. There is no sign that women are exerting a "civilizing influence" upon it. To have an equal part in our juggernaut of national vandalism is to be a vandal. To call this vandalism "liberation" is to prolong, and even ratify, a dangerous confusion that was once principally masculine.

A broader, deeper criticism is necessary. The problem is not just the exploitation of women by men. A greater problem is that women and men alike are consenting to an economy that exploits women and men and everything else.

Another decent possibility my critics implicitly deny is that of work as a gift. Not one of them supposed that my wife may be a consulting engineer who helps me in her spare time out of the goodness of her heart; instead they suppose that she is "a household drudge." But what appears to infuriate them the most is their supposition that she works for nothing. They assume—and this is the orthodox assumption of the industrial economy—that the only help worth giving is not given at all, but sold. Love, friendship, neighborliness, compassion, duty— what are they? We are realists. We will be most happy to receive your check.

*

The various reductions I have been describing are fairly directly the results of the ongoing revolution of applied science known as "technological progress." This revolution has provided the means by which both the productive and the consumptive capacities of people could be detached from household and community and made to serve other people's purely economic ends. It has provided as well a glamor of newness, ease, and affluence that made it seductive even to those who suffered most from it. In its more recent history especially, this revolution has been successful in putting unheard-of quantities of consumer goods and services within the reach of ordinary people. But the technical means of this popular "affluence" has at the same time made possible the gathering of the real property and the real power of the country into fewer and fewer hands.

Some people would like to think that this long sequence of industrial innovations has changed human life and even human nature in fundamental ways. Perhaps it has—but, arguably, almost always for the worse. I know that "technological progress" can be defended, but I observe that the defenses are invariably quantitative—catalogs of statistics on the ownership of automobiles and television sets, for example, or on the increase of life expectancy—and I see that these statistics are always kept carefully apart from the related statistics of soil loss, pollution, social disintegration, and so forth. That is to say, there is never an effort to determine the *net* result of this progress. The voice of its defenders is not that of the responsible bookkeeper, but that of the propagandist or salesman, who says that the net gain is more than 100 percent—that the thing we have bought has perfectly replaced everything it has cost, and added a great deal more: "You just can't lose!" We thus have got rich by spending, just as the advertisers have told us we would, and the best of all possible worlds is getting better every day.

The statistics of life expectancy are favorites of the industrial apologists, because they are perhaps the hardest to argue with. Nevertheless, this emphasis on longevity is an excellent example of the way the isolated aims of the industrial mind reduce and distort human life, and also the way statistics corrupt the truth. A long life has indeed always been thought desirable; everything that is alive apparently wishes to continue to live.

But until our own time, that sentence would have been qualified: long life is desirable and everything wishes to live *up to a point*. Past a certain point, and in certain conditions, death becomes preferable to life. Moreover, it was generally agreed that a good life was preferable to one that was merely long, and that the goodness of a life could not be determined by its length. The statisticians of longevity ignore good in both its senses; they do not ask if the prolonged life is virtuous, or if it is satisfactory. If the life is that of a vicious criminal, or if it is inched out in a veritable hell of captivity within the medical industry, no matter—both become statistics to "prove" the good luck of living in our time.

But in general, apart from its own highly specialized standards of quantity and efficiency, "technological progress" has produced a social and ecological decline. Industrial war, except by the most fanatically narrow standards, is worse than war used to be. Industrial agriculture, except by the standards of quantity and mechanical efficiency, diminishes everything it affects. Industrial workmanship is certainly worse than traditional workmanship, and is getting shoddier every day. After forty-odd years, the evidence is everywhere that television, far from proving a great tool of education, is a tool of stupefaction and disintegration. Industrial education has abandoned the old duty of passing on the cultural and intellectual inheritance in favor of baby-sitting and career preparation.

After several generations of "technological progress," in fact, we have become a people who *cannot* think about anything important. How far down in the natural order do we have to go to find creatures who raise their young as indifferently as industrial humans now do? Even the English sparrows do not let loose into the streets young sparrows who have no notion of their identity or their adult responsibilities. When else in history would you find "educated" people who know more about sports than about the history of their country, or uneducated people who do not know the stories of their families and communities?

To ask a still more obvious question, what is the purpose of this technological progress? What higher aim do we think it is serving? Surely the aim cannot be the integrity or happiness of our families, which we have made subordinate to the education system, the television industry, and the consumer economy. Surely it cannot be the integrity or health of our communities, which we esteem even less than we esteem our families. Surely it cannot be love of our country, for we are far more concerned about the desecration of the flag than we are about the desecration of our land. Surely it cannot be the love of God, which counts for at least as little in the daily order of business as the love of family, community, and country.

The higher aims of "technological progress" are money and ease. And this exalted greed for money and ease is disguised and justified by an obscure, cultish faith in "the future." We do as we do, we say, "for the sake of the future" or "to make a better future for our children." How we can hope to make a good future by doing badly in the present, we do not say. We cannot think about the future, of course, for the future does not exist: the existence of the future is an article of faith. We can be assured only that, if there is to be a future, the good of it is already implicit in the good things of the present. We do not need to plan or devise a "world of the future"; if we take care of the world of the present, the future will have received full justice from us. A good future is implicit in the soils, forests, grasslands, marshes, deserts, mountains, rivers, lakes, and oceans that we have now, and in the good things of human culture that we have now; the only valid "futurology" available to us is to take care of those things. We have no need to contrive and dabble at "the future of the human race"; we have the same pressing need that we have always had—to love, care for, and teach our children.

And so the question of the desirability of adopting any technological innovation is a question with two possible answers —not one, as has been commonly assumed. If one's motives are money, ease, and haste to arrive in a technologically determined future, then the answer is foregone, and there is, in fact, no question, and no thought. If one's motive is the love of family, community, country, and God, then one will have to think, and one may have to decide that the proposed innovation is undesirable.

The question of how to end or reduce dependence on some of the technological innovations already adopted is a baffling one. At least, it baffles me. I have not been able to see, for example, how people living in the country, where there is no public transportation, can give up their automobiles without becoming less useful to each other. And this is because, owing largely to the influence of the automobile, we live too far from each other, and from the things we need, to be able to get about by any other means. Of course, you *could* do without an automobile, but to do so you would have to disconnect yourself from many obligations. Nothing I have so far been able to think about this problem has satisfied me.

But if we have paid attention to the influence of the automobile on country communities, we know that the desirability of technological innovation is an issue that requires thinking about, and we should have acquired some ability to think about it. Thus if I am partly a writer, and I am offered an expensive machine to help me write, I ought to ask whether or not such a machine is desirable.

I should ask, in the first place, whether or not I wish to purchase a solution to a problem that I do not have. I acknowledge that, as a writer, I need a lot of help. And I have received an abundance of the best of help from my wife, from other members of my family, from friends, from teachers, from editors, and sometimes from readers. These people have helped me out of love or friendship, and perhaps in exchange for some help that I have given them. I suppose I should leave open the possibility that I need more help than I am getting, but I would certainly be ungrateful and greedy to think so.

But a computer, I am told, offers a kind of help that you can't get from other humans; a computer will help you to write faster, easier, and more. For a while, it seemed to me that every university professor I met told me this. Do I, then, want to write faster, easier, and more? No. My standards are not speed, ease, and quantity. I have already left behind too much evidence that, writing with a pencil, I have written too fast, too easily, and too much. I would like to be a *better* writer, and for that I need help from other humans, not a machine.

The professors who recommended speed, ease, and quantity to me were, of course, quoting the standards of their universities. The chief concern of the industrial system, which is to say the present university system, is to cheapen work by increasing volume. But implicit in the professors' recommendation was the idea that one needs to be up with the times. The pace-setting academic intellectuals have lately had a great hankering to be up with the times. They don't worry about keeping up with the Joneses: as intellectuals, they know that they are supposed to be Nonconformists and Independent Thinkers living at the Cutting Edge of Human Thought. And so they are all a-dither to keep up with the times—which means adopting the latest technological innovations as soon as the Joneses do.

Do I wish to keep up with the times? No.

My wish simply is to live my life as fully as I can. In both our work and our leisure, I think, we should be so employed. And in our time this means that we must save ourselves from the products that we are asked to buy in order, ultimately, to re-place ourselves.

The danger most immediately to be feared in "technological progress" is the degradation and obsolescence of the body. Implicit in the technological revolution from the beginning has been a new version of an old dualism, one always destructive, and now more destructive than ever. For many centuries there have been people who looked upon the body, as upon the natural world, as an encumbrance of the soul, and so have hated the body, as they have hated the natural world, and longed to be free of it. They have seen the body as intolerably imperfect by spiritual standards. More recently, since the beginning of the technological revolution, more and more people have looked upon the body, along with the rest of the natural creation, as intolerably imperfect by mechanical standards. They see the body as an encumbrance of the mind—the mind, that is, as reduced to a set of mechanical ideas that can be implemented in machines—and so they hate it and long to be free of it. The body has limits that the machine does not have; there-fore, remove the body from the machine so that the machine can continue as an unlimited idea.

It is odd that simply because of its "sexual freedom" our time should be considered extraordinarily physical. In fact, our "sexual revolution" is mostly an industrial phenomenon, in which the body is used as an idea of pleasure or a pleasure machine with the aim of "freeing" natural pleasure from natural consequence. Like any other industrial enterprise, industrial sexuality seeks to conquer nature by exploiting it and ignoring the consequences, by denying any connection between nature and spirit or body and soul, and by evading social responsibility. The spiritual, physical, and economic costs of this "freedom" are immense, and are characteristically belittled or ignored. The diseases of sexual irresponsibility are regarded as a technological problem and an affront to liberty. Industrial sex, characteristically, establishes its freeness and goodness by an industrial accounting, dutifully toting up numbers of "sexual partners," orgasms, and so on, with the inevitable industrial implication that the body is somehow a limit on the idea of sex, which will be a great deal more abundant as soon as it can be done by robots.

This hatred of the body and of the body's life in the natural world, always inherent in the technological revolution (and sometimes explicitly and vengefully so), is of concern to an artist because art, like sexual love, is of the body. Like sexual love, art is of the mind and spirit also, but it is made with the body and it appeals to the senses. To reduce or shortcut the intimacy of the body's involvement in the making of a work of art (that is, of any artifice, anything made by art) inevitably risks reducing the work of art and the art itself. In addition to the reasons I gave previously, which I still believe are good reasons, I am not going to use a computer because I don't want to diminish or distort my bodily involvement in my work. I don't want to deny myself the *pleasure* of bodily involvement in my work, for that pleasure seems to me to be the sign of an indispensable integrity.

At first glance, writing may seem not nearly so much an art of the body as, say, dancing or gardening or carpentry. And yet language is the most intimately physical of all the artistic means. We have it palpably in our mouths; it is our *langue*, our tongue. Writing it, we shape it with our hands. Reading aloud what we have written—as we must do, if we are writing carefully—our language passes in at the eyes, out at the mouth, in at the ears; the words are immersed and steeped in the senses of the body before they make sense in the mind. They *cannot* make sense in the mind until they have made sense in the body.

Does shaping one's words with one's own hand impart character and quality to them, as does speaking them with one's own tongue to the satisfaction of one's own ear? There is no way to prove that it does. On the other hand, there is no way to prove that it does not, and I believe that it does.

The act of writing language down is not so insistently tangible an act as the act of building a house or playing the violin. But to the extent that it is tangible, I love the tangibility of it. The computer apologists, it seems to me, have greatly underrated the value of the handwritten manuscript as an artifact. I don't mean that a writer should be a fine calligrapher and write for exhibition, but rather that handwriting has a valuable influence on the work so written. I am certainly no calligrapher, but my handwritten pages have a homemade, handmade look to them that both pleases me in itself and suggests the possibility of ready correction. It looks hospitable to improvement. As the longhand is transformed into typescript and then into galley proofs and the printed page, it seems increasingly to resist improvement. More and more spunk is required to mar the clean, final-looking lines of type. I have the notion—again not provable—that the longer I keep a piece of work in longhand, the better it will be.

To me, also, there is a significant difference between ready correction and easy correction. Much is made of the ease of correction in computer work, owing to the insubstantiality of the light-image on the screen; one presses a button and the old version disappears, to be replaced by the new. But because of the substantiality of paper and the consequent difficulty involved, one does not handwrite or typewrite a new page every time a correction is made. A handwritten or typewritten page therefore is usually to some degree a palimpsest; it contains parts and relics of its own history—erasures, passages crossed out, interlineations—suggesting that there is something to go back to as well as something to go forward to. The light-text on the computer screen, by contrast, is an artifact typical of what can only be called the industrial present, a present absolute. A computer destroys the sense of historical succession, just as do other forms of mechanization. The well-crafted table or cabinet embodies the memory of (because it embodies respect for) the tree it was made of and the forest in which the tree stood. The work of certain potters embodies the memory that the clay was dug from the earth. Certain farms contain hospitably the remnants and reminders of the forest or prairie that preceded them. It is possible even for towns and cities to remember farms and forests or prairies. All good human work remembers its history. The best writing, even when printed, is full of intimations that it is the present version of earlier versions of itself, and that its maker inherited the work and the ways of earlier makers. It thus keeps, even in print, a suggestion of the quality of the handwritten page; it is a palimpsest.

Something of this undoubtedly carries over into industrial products. The plastic Clorox jug has a shape and a loop for the forefinger that recalls the stoneware jug that went before it. But something vital is missing. It embodies no memory of its source or sources in the earth or of any human hand involved in its shaping. Or look at a large factory or a power plant or an airport, and see if you can imagine—even if you know—what was there before. In such things the materials of the world have entered a kind of orphanhood.

It would be uncharitable and foolish of me to suggest that nothing good will ever be written on a computer. Some of my best friends have computers. I have only said that a computer cannot help you to write *better*, and I stand by that. (In fact, I know a publisher who says that under the influence of computers—or of the immaculate copy that computers produce—many writers are now writing worse.) But I do say that in using computers writers are flirting with a radical separation of mind and body, the elimination of the work of the body from the work of the mind. The text on the computer screen, and the computer printout too, has a sterile, untouched, factory-made look, like that of a plastic whistle or a new car. The body does not do work like that. The body *characterizes* everything it touches. What it makes it traces over with the marks of its pulses and breathings, its excitements, hesitations, flaws, and mistakes. On its good work, it leaves the marks of skill, care, and love persisting through hesitations, flaws, and mistakes. And to those of us who love and honor the life of the body in this world, these marks are precious things, necessities of life.

But writing is of the body in yet another way. It is preeminently a walker's art. It can be done on foot and at large. The beauty of its traditional equipment is simplicity. And cheapness. Going off to the woods, I take a pencil and some paper (*any* paper—a small notebook, an old envelope, a piece of a feed sack), and I am as well equipped for my work as the president of IBM. I am also free, for the time being at least, of everything that IBM is hooked to. My thoughts will not be coming to me from the power structure or the power grid, but from another direction and way entirely. My mind is free to go with my feet.

I know that there are some people, perhaps many, to whom you cannot appeal on behalf of the body. To them, disembodiment is a goal, and they long for the realm of pure mind— or pure machine; the difference is negligible. Their departure from their bodies, obviously, is much to be desired, but the rest of us had better be warned: they are going to cause a lot of dangerous commotion on their way out.

Some of my critics were happy to say that my refusal to use a computer would not do any good. I have argued, and am convinced, that it will at least do *me* some good, and that it may involve me in the preservation of some cultural goods. But what they meant was real, practical, public good. They meant that the materials and energy I save by not buying a computer will not be "significant." They meant that no individual's restraint in the use of technology or energy will be "significant." That is true.

But each one of us, by "insignificant" individual abuse of the world, contributes to a general abuse that is devastating. And if I were one of thousands or millions of people who could afford a piece of equipment, even one for which they had a conceivable "need," and yet did not buy it, *that* would be "significant." Why, then, should I hesitate for even a moment to be one, even the first one, of that "significant" number? Thoreau gave the definitive reply to the folly of "significant numbers" a long time ago: Why should anybody wait to do what is right until everybody does it? It is not "significant" to love your own children or to eat your own dinner, either. But normal humans will not wait to love or eat until it is mandated by an act of Congress.

One of my correspondents asked where one is to draw the line. That question returns me to the bewilderment I mentioned earlier: I am unsure where the line ought to be drawn, or how to draw it. But it is an intelligent question, worth losing some sleep over.

I know how to draw the line only where it is easy to draw. It is easy—it is even a luxury—to deny oneself the use of a television set, and I zealously practice that form of self-denial. Every time I see television (at other people's houses), I am more inclined to congratulate myself on my deprivation. I have no doubt, as I have said, that I am better off without a computer. I joyfully deny myself a motorboat, a camping van, an off-road vehicle, and every other kind of recreational machinery. I have, and want, no "second home." I suffer very comfortably the lack of colas, TV dinners, and other counterfeit foods and beverages.

I am, however, still in bondage to the automobile industry and the energy companies, which have nothing to recommend them except our dependence on them. I still fly on airplanes, which have nothing to recommend them but speed; they are inconvenient, uncomfortable, undependable, ugly, stinky, and scary. I still cut my wood with a chainsaw, which has nothing to recommend it but speed, and has all the faults of an airplane, except it does not fly.

It is plain to me that the line ought to be drawn without fail wherever it can be drawn easily. And it ought to be easy (though many do not find it so) to refuse to buy what one does not need. If you are already solving your problem with the equipment you have—a pencil, say—why solve it with something more expensive and more damaging? If you don't have a problem, why pay for a solution? If you love the freedom and elegance of simple tools, why encumber yourself with something complicated?

And yet, if we are ever again to have a world fit and pleasant for little children, we are surely going to have to draw the line where it is *not* easily drawn. We are going to have to learn to give up things that we have learned (in only a few years, after all) to "need." I am not an optimist; I am afraid that I won't live long enough to escape my bondage to the machines. Nevertheless, on every day left to me I will search my mind and circumstances for the means of escape. And I am not without hope. I knew a man who, in the age of chainsaws, went right on cutting his wood with a handsaw and an axe. He was a healthier and a saner man than I am. I shall let his memory trouble my thoughts.

Word and Flesh

Toward the end of *As You Like It*, Orlando says: "I can live no longer by thinking." He is ready to marry Rosalind. It is time for incarnation. Having thought too much, he is at one of the limits of human experience, or of human sanity. If his love does put on flesh, we know he must sooner or later arrive at the opposite limit, at which he will say, "I can live no longer without thinking." Thought—even consciousness—seems to live between these limits: the abstract and the particular, the word and the flesh.

All public movements of thought quickly produce a language that works as a code, useless to the extent that it is abstract. It is readily evident, for example, that you can't conduct a relationship with another person in terms of the rhetoric of the civil rights movement or the women's movement—as useful as those rhetorics may initially have been to personal relationships.

The same is true of the environment movement. The favorite adjective of this movement now seems to be "planetary." This word is used, properly enough, to refer to the interdependence of places, and to the recognition, which is desirable and growing, that no place on the earth can be completely healthy until all places are.

But the word "planetary" also refers to an abstract anxiety or an abstract passion that is desperate and useless exactly to the extent that it is abstract. How, after all, can anybody—any particular body—do anything to heal a planet? The suggestion that anybody could do so is preposterous. The heroes of abstraction keep galloping in on their white horses to save the planet—and they keep falling off in front of the grandstand.

What we need, obviously, is a more intelligent—which is to say, a more accurate—description of the problem. The description of a problem as planetary arouses a motivation for which, of necessity, there is no employment. The adjective "planetary" describes a problem in such a way that it cannot be solved. In fact, though we now have serious problems nearly everywhere on the planet, we have no problem that can accurately be described as planetary. And, short of the total annihilation of the human race, there is no planetary solution.

There are also no national, state, or county problems, and no national, state, or county solutions. That will-o'-the-wisp, the large-scale solution to the large-scale problem, which is so dear to governments, universities, and corporations, serves mostly to distract people from the small, private problems that they may, in fact, have the power to solve.

The problems, if we describe them accurately, are all private and small. Or they are so initially.

The problems are our lives. In the "developed" countries, at least, the large problems occur because all of us are living either partly wrong or almost entirely wrong. It was not just the greed of corporate shareholders and the hubris of corporate executives that put the fate of Prince William Sound into one ship; it was also our demand that energy be cheap and plentiful.

The economies of our communities and households are wrong. The answers to the human problems of ecology are to be found in economy. And the answers to the problems of economy are to be found in culture and in character. To fail to see this is to go on dividing the world falsely between guilty producers and innocent consumers.

The planetary versions—the heroic versions—of our problems have attracted great intelligence. But these problems, as they are caused and suffered in our lives, our households, and our communities, have attracted very little intelligence.

There are some notable exceptions. A few people have learned to do a few things better. But it is discouraging to reflect that, though we have been talking about most of our problems for decades, we are still mainly *talking* about them. The civil rights movement has not given us better communities. The women's movement has not given us better marriages or better households. The environment movement has not changed our parasitic relationship to nature.

We have failed to produce new examples of good home and community economies, and we have nearly completed the destruction of the examples we once had. Without examples, we are left with theory and the bureaucracy and meddling that come with theory. We change our principles, our thoughts, and our words, but these are changes made in the air. Our lives go on unchanged.

For the most part, the subcultures, the countercultures, the dissenters, and the opponents continue mindlessly—or perhaps just helplessly—to follow the pattern of the dominant society in its extravagance, its wastefulness, its dependencies, and its addictions. The old problem remains: How do you get intelligence *out* of an institution or an organization?

My small community in Kentucky has lived and dwindled for at least a century under the influence of four kinds of organizations: governments, corporations, schools, and churches—all of which are distant (either actually or in interest), centralized, and consequently abstract in their concerns.

Governments and corporations (except for employees) have no presence in our community at all, which is perhaps fortunate for us, but we nevertheless feel the indifference or the contempt of governments and corporations for communities such as ours.

We have had no school of our own for nearly thirty years. The school system takes our young people, prepares them for "the world of tomorrow"—which it does not expect to take place in any rural area—and gives back "expert" (that is, extremely generalized) ideas.

The church is present in the town. We have two churches. But both have been used by their denominations, for almost a century, to provide training and income for student ministers, who do not stay long enough even to become disillusioned.

For a long time, then, the minds that have most influenced our town have not been *of* the town and so have not tried even to perceive, much less to honor, the good possibilities that are there. They have not wondered on what terms a good and conserving life might be lived there. In this my community is not unique but is like almost every other neighborhood in our country and in the "developed" world.

The question that *must* be addressed, therefore, is not how to care for the planet, but how to care for each of the planet's millions of human and natural neighborhoods, each of its millions of small pieces and parcels of land, each one of which is in some precious way different from all the others. Our understandable wish to preserve the planet must somehow be reduced to the scale of our competence—that is, to the wish to preserve all of its humble households and neighborhoods.

What can accomplish this reduction? I will say again, without overweening hope but with certainty nonetheless, that only love can do it. Only love can bring intelligence out of the institutions and organizations, where it aggrandizes itself, into the presence of the work that must be done.

Love is never abstract. It does not adhere to the universe or the planet or the nation or the institution or the profession, but to the singular sparrows of the street, the lilies of the field, "the least of these my brethren." Love is not, by its own desire, heroic. It is heroic only when compelled to be. It exists by its willingness to be anonymous, humble, and unrewarded.

The older love becomes, the more clearly it understands its involvement in partiality, imperfection, suffering, and mortality. Even so, it longs for incarnation. It can live no longer by thinking.

And yet to put on flesh and do the flesh's work, it must think.

In his essay on Kipling, George Orwell wrote: "All left-wing parties in the highly industrialized countries are at bottom a sham, because they make it their business to fight against something which they do not really wish to destroy. They have internationalist aims, and at the same time they struggle to keep up a standard of life with which those aims are incompatible. We all live by robbing Asiatic coolies, and those of us who are 'enlightened' all maintain that those coolies ought to be set free; but our standard of living, and hence our 'enlightenment,' demands that the robbery shall continue."

This statement of Orwell's is clearly applicable to our situation now; all we need to do is change a few nouns. The religion and the environmentalism of the highly industrialized countries are at bottom a sham, because they make it their business to fight against something that they do not really wish to destroy. We all live by robbing nature, but our standard of living demands that the robbery shall continue.

We must achieve the character and acquire the skills to live much poorer than we do. We must waste less. We must do more for ourselves and each other. It is either that or continue merely to think and talk about changes that we are inviting catastrophe to make.

The great obstacle is simply this: the conviction that we cannot change because we are dependent on what is wrong. But that is the addict's excuse, and we know that it will not do.

How dependent, in fact, are we? How dependent are our neighborhoods and communities? How might our dependences be reduced? To answer these questions will require better thoughts and better deeds than we have been capable of so far.

We must have the sense and the courage, for example, to see that the ability to transport food for hundreds or thousands of miles does not necessarily mean that we are well off. It means that the food supply is more vulnerable and more costly than a local food supply would be. It means that consumers do not control or influence the healthfulness of their food supply and that they are at the mercy of the people who have the control and influence. It means that, in eating, people are using large quantities of petroleum that other people in another time are almost certain to need.

Our most serious problem, perhaps, is that we have become a nation of fantasists. We believe, apparently, in the infinite availability of finite resources. We persist in land-use methods that reduce the potentially infinite power of soil fertility to a finite quantity, which we then proceed to waste as if it were an infinite quantity. We have an economy that depends not on the quality and quantity of necessary goods and services but on the moods of a few stockbrokers. We believe that democratic freedom can be preserved by people ignorant of the history of democracy and indifferent to the responsibilities of freedom.

Our leaders have been for many years as oblivious to the realities and dangers of their time as were George III and Lord North. They believe that the difference between war and peace is still the overriding political difference—when, in fact, the difference has diminished to the point of insignificance. How would you describe the difference between modern war and modern industry—between, say, bombing and strip mining, or between chemical warfare and chemical manufacturing? The difference seems to be only that in war the victimization of humans is directly intentional and in industry it is "accepted" as a "trade-off."

Were the catastrophes of Love Canal, Bhopal, Chernobyl, and the *Exxon Valdez* episodes of war or of peace? They were, in fact, peacetime acts of aggression, intentional to the extent that the risks were known and ignored.

We are involved unremittingly in a war not against "foreign enemies," but against the world, against our freedom, and indeed against our existence. Our so-called industrial accidents should be looked upon as revenges of Nature. We forget that Nature is necessarily party to all our enterprises and that she imposes conditions of her own.

Now she is plainly saying to us: "If you put the fates of whole communities or cities or regions or ecosystems at risk in single ships or factories or power plants, then I will furnish the drunk or the fool or the imbecile who will make the necessary small mistake."

Nature as Measure

I LIVE IN a part of the country that at one time a good farmer could take some pleasure in looking at. When I first became aware of it, in the 1940s, the better land, at least, was generally well farmed. The farms were mostly small and were highly diversified, producing cattle, sheep, and hogs, tobacco, corn, and the small grains; nearly all the farmers milked a few cows for home use and to market milk or cream. Nearly every farm household maintained a garden, kept a flock of poultry, and fattened its own meat hogs. There was also an extensive "support system" for agriculture: every community had its blacksmith shop, shops that repaired harness and machinery, and stores that dealt in farm equipment and supplies.

Now the country is not well farmed, and driving through it has become a depressing experience. Some good small farmers remain, and their farms stand out in the landscape like jewels. But they are few and far between, and they are getting fewer every year. The buildings and other improvements of the old farming are everywhere in decay or have vanished altogether. The produce of the country is increasingly specialized. The small dairies are gone. Most of the sheep flocks are gone, and so are most of the enterprises of the old household economy. There is less livestock and more cash-grain farming. When cash-grain farming comes in, the fences go, the livestock goes, erosion increases, and the fields become weedy.

Like the farm land, the farm communities are declining and eroding. The farmers who are still farming do not farm with as much skill as they did forty years ago, and there are not nearly so many farmers farming as there were forty years ago. As the old have died, they have not been replaced; as the young come of age, they leave farming or leave the community. And as the land and the people deteriorate, so necessarily must the support system. None of the small rural towns is thriving as it did forty years ago. The proprietors of small businesses give up or die and are not replaced. As the farm trade declines, farm equipment franchises are revoked. The remaining farmers must drive longer and longer distances for machines and parts and repairs.

Looking at the country now, one cannot escape the conclusion that there are no longer enough people on the land to farm it well and to take proper care of it. A further and more ominous conclusion is that there is no longer a considerable number of people knowledgeable enough to look at the country and see that it is not properly cared for—though the face of the country is now everywhere marked by the agony of our enterprise of self-destruction.

And suddenly in this wasting countryside there is talk of raising production quotas on Burley tobacco by twenty-four percent, and tobacco growers are coming under pressure from the manufacturers to decrease their use of chemicals. Everyone I have talked to is doubtful that we have enough people left in farming to meet the increased demand for either quantity or quality, and doubtful that we still have the barnroom to house the increased acreage. In other words, the demand going up has met the culture coming down. No one can be optimistic about the results.

Tobacco, I know, is not a food, but it comes from the same resources of land and people that food comes from, and this emerging dilemma in the production of tobacco can only foreshadow a similar dilemma in the production of food. At every point in our food economy, present conditions remaining, we must expect to come to a time when demand (for quantity or quality) going up will meet the culture coming down. The fact is that we have nearly destroyed American farming, and in the process have nearly destroyed our country.

How has this happened? It has happened because of the application to farming of far too simple a standard. For many years, as a nation, we have asked our land only to produce, and we have asked our farmers only to produce. We have believed that this single economic standard not only guaranteed good performance but also preserved the ultimate truth and rightness of our aims. We have bought unconditionally the economists' line that competition and innovation would solve all problems, and that we would finally accomplish a technological end-run around biological reality and the human condition.

Competition and innovation have indeed solved, for the time being, the problem of production. But the solution has been extravagant, thoughtless, and far too expensive. We have been winning, to our inestimable loss, a competition against our own land and our own people. At present, what we have to show for this "victory" is a surplus of food. But this is a surplus achieved by the ruin of its sources, and it has been used, by apologists for our present economy, to disguise the damage by which it was produced. Food, clearly, is the most important economic product—except when there is a surplus. When there is a surplus, according to our present economic assumptions, food is the *least* important product. The surplus becomes famous as evidence to consumers that they have nothing to worry about, that there is no problem, that present economic assumptions are correct.

But our present economic assumptions are failing in agriculture, and to those having eyes to see the evidence is everywhere, in the cities as well as in the countryside. The singular demand for production has been unable to acknowledge the importance of the sources of production in nature and in human culture. Of course agriculture must be productive; that is a requirement as urgent as it is obvious. But urgent as it is, it is not the *first* requirement; there are two more requirements equally important and equally urgent. One is that if agriculture is to remain productive, it must preserve the land, and the fertility and ecological health of the land; the land, that is, must be used *well*. A further requirement, therefore, is that if the land is to be used well, the people who use it must know it well, must be highly motivated to use it well, must know how to use it well, must have time to use it well, and must be able to afford to use it well. Nothing that has happened in the agricultural revolution of the last fifty years has disproved or invalidated these requirements, though everything that has happened has ignored or defied them.

In light of the necessity that the farm land and the farm people should thrive while producing, we can see that the single standard of productivity has failed.

Now we must learn to replace that standard by one that is more comprehensive: the standard of nature. The effort to do this is not new. It was begun early in this century by Liberty Hyde Bailey of the Cornell University College of Agriculture, by F. H. King of the University of Wisconsin College of Agriculture and the United States Department of Agriculture, by J. Russell Smith, professor of economic geography at Columbia University, by the British agricultural scientist Sir Albert Howard, and by others; and it has continued into our own time in the work of such scientists as John Todd, Wes Jackson, and others. The standard of nature is not so simple or so easy a standard as the standard of productivity. The term "nature" is not so definite or stable a concept as the weights and measures of productivity. But we know what we mean when we say that the first settlers in any American place recognized that place's agricultural potential "by its nature"—that is, by the depth and quality of its soil, the kind and quality of its native vegetation, and so on. And we know what we mean when we say that all too often we have proceeded to ignore the nature of our places in farming them. By returning to "the nature of the place" as standard, we acknowledge the necessary limits of our own intentions. Farming cannot take place except in nature; therefore, if nature does not thrive, farming cannot thrive. But we know too that nature includes us. It is not a place into which we reach from some safe standpoint outside it. We are in it and are a part of it while we use it. If it does not thrive, we cannot thrive. The appropriate measure of farming then is the world's health and our health, and this is inescapably *one* measure.

But the oneness of this measure is far different from the singularity of the standard of productivity that we have been using; it is far more complex. One of its concerns, one of the inevitable natural measures, is productivity; but it is also concerned for the health of all the creatures belonging to a given place, from the creatures of the soil and water to the humans and other creatures of the land surface to the birds of the air. The use of nature as measure proposes an atonement between ourselves and our world, between economy and ecology, between the domestic and the wild. Or it proposes a conscious and careful recognition of the interdependence between ourselves and nature that in fact has always existed and, if we are to live, must always exist.

Industrial agriculture, built according to the single standard of productivity, has dealt with nature, including human nature, in the manner of a monologist or an orator. It has not asked for anything, or waited to hear any response. It has told nature what it wanted, and in various clever ways has taken what it wanted. And since it proposed no limit on its wants, exhaustion has been its inevitable and foreseeable result. This, clearly, is a dictatorial or totalitarian form of behavior, and it is as totalitarian in its use of people as it is in its use of nature. Its connections to the world and to humans and the other creatures become more and more abstract, as its economy, its authority, and its power become more and more centralized.

On the other hand, an agriculture using nature, including human nature, as its measure would approach the world in the manner of a conversationalist. It would not impose its vision and its demands upon a world that it conceives of as a stockpile of raw material, inert and indifferent to any use that may be made of it. It would not proceed directly or soon to some supposedly ideal state of things. It *would* proceed directly and soon to serious thought about our condition and our predicament. On all farms, farmers would undertake to know responsibly where they are and to "consult the genius of the place." They would ask what nature would be doing there if no one were farming there. They would ask what nature would permit them to do there, and what they could do there with the least harm to the place and to their natural and human neighbors. And they would ask what nature would *help* them to do there. And after each asking, knowing that nature will respond, they would attend carefully to her response. The use of the place would necessarily change, and the response of the place to that use would necessarily change the user. The conversation itself would thus assume a kind of creaturely life, binding the place and its inhabitants together, changing and growing to no end, no final accomplishment, that can be conceived or foreseen.

Farming in this way, though it certainly would proceed by desire, is not visionary in the political or utopian sense. In a conversation, you always expect a reply. And if you honor the other party to the conversation, if you honor the *otherness* of the other party, you understand that you must not expect always to receive a reply that you foresee or a reply that you will like. A conversation is immitigably two-sided and always to some degree mysterious; it requires faith.

For a long time now we have understood ourselves as traveling toward some sort of industrial paradise, some new Eden conceived and constructed entirely by human ingenuity. And we have thought ourselves free to use and abuse nature in any way that might further this enterprise. Now we face overwhelming evidence that we are not smart enough to recover Eden by assault, and that nature does not tolerate or excuse our abuses. If, in spite of the evidence against us, we are finding it hard to relinquish our old ambition, we are also seeing more clearly every day how that ambition has reduced and enslaved us. We see how everything—the whole world—is belittled by the idea that all creation is moving or ought to move toward an end that some body, some human body, has thought up. To be free of that end and that ambition would be a delightful and precious thing. Once free of it, we might again go about our work and our lives with a seriousness and pleasure denied to us when we merely submit to a fate already determined by gigantic politics, economics, and technology.

Such freedom is implicit in the adoption of nature as the measure of economic life. The reunion of nature and economy proposes a necessary democracy, for neither economy nor nature can be abstract in practice. When we adopt nature as measure, we require practice that is locally knowledgeable. The particular farm, that is, must not be treated as any farm. And the particular knowledge of particular places is beyond the competence of any centralized power or authority. Farming by the measure of nature, which is to say the nature of the particular place, means that farmers must tend farms that they know and love, farms small enough to know and love, using tools and methods that they know and love, in the company of neighbors that they know and love.

In recent years, our society has been required to think again of the issues of the use and abuse of human beings. We understand, for instance, that the inability to distinguish between a particular woman and any woman is a condition predisposing to abuse. It is time that we learn to apply the same understanding to our country. The inability to distinguish between a farm and any farm is a condition predisposing to abuse, and abuse has been the result. Rape, indeed, has been the result, and we have seen that we are not exempt from the damage we have inflicted. Now we must think of marriage.

Chronology

1934   Born Wendell Erdman Berry on August 5 in Henry County,
        Kentucky, the first child of John Marshall and Virginia Erd-
        man Perry Berry. Father, born November 8, 1900, near
        Lacie, Kentucky, earned a bachelor's degree at Georgetown
        College in central Kentucky in 1922, and then worked as
        secretary to Congressman Virgil Chapman while earning
        a law degree from The George Washington University in
        Washington, D.C. He graduated in 1927 and later that year
        opened a private legal practice in Henry County. Mother
        born November 6, 1907, in Port Royal, in Henry County.
        She attended Randolph-Macon Woman's College in Lynch-
        burg, Virginia, a Methodist women's college tied to the
        all-male Randolph-Macon College in Ashland, Virginia,
        graduating with a degree in English in 1930. Both families
        can trace roots back to early years of Henry County's his-
        tory; both families also owned slaves before the Civil War
        (at time of Berry's birth, there are many children of both
        slaves and slaveholders in the community). Parents married
        on July 14, 1933.

1935   Brother John Marshall Jr. born October 13.

1936   Family moves to New Castle, Kentucky, where father prac-
        tices law and helps with paternal grandfather's farm, the
        "Home Place."

1938   Sister Mary Jo born March 18.

1939   Sister Martha Frances born July 22.

1940   Enters New Castle Elementary School in Henry County.
        Over the next few years, mother introduces Berry to books
        about King Arthur's knights and Robin Hood, as well as
        The Swiss Family Robinson, *Treasure Island*, and *The Year-
        ling*. Later will read Mary O'Hara's *My Friend Flicka*,
        Thunderhead, and *Green Grass of Wyoming*.

1941   The Burley Tobacco Growers Cooperative Association,
        representing burley tobacco farmers in Kentucky, Mis-
        souri, Indiana, Ohio, and West Virginia with an average
        farm size of 150 acres, which had been founded in 1921,

is revived under the New Deal, establishing parity prices and production control as protections for farmers. Father is one of the principal authors of the program and serves as counsel and vice-president and later president.

1942   Uncle Morgan Perry serves in the navy medical corps during World War II, and because of his knowledge of agriculture, is placed in charge of the garden of the naval hospital at Pearl Harbor.

1944   On July 3, uncle Wendell Holmes Berry, with whom Berry was extremely close, is shot at a defunct lead mine property, where he had been working with a crew salvaging lumber and roofing, after a quarrel with Floyd Martin, one of the owners of the property, and dies the same day.

1948   Finishes the eighth grade at the New Castle School. Enters Millersburg Military Institute in Millersburg, Kentucky, at the same time as his brother, John, Jr. Berry will remember, "It was very confining and military, of course. I never liked the military part of it. I did have some good teachers there, and I began there to read more seriously than I had before. I began to have a favorite subject: literature." Father speaks before Congress on behalf of the Tobacco Program. In high school, begins tentatively to write poems.

1952   Graduates from Millersburg Military Institute. In the fall, enters University of Kentucky, in Lexington, Kentucky. Co-edits freshman magazine, *Green Pen*.

1953   Takes "Introduction to Literature" from Thomas Stroup. Begins bringing Stroup poems for criticism. Later takes classes from Hollis Summers and Robert Hazel. Reads T. S. Eliot and Ezra Pound. In May, publishes first essay, "The Wings of the Future," in *Green Pen*, about the survival of society in the light of two world wars and the entrance of the U.S. into the Korean conflict; he writes that it is the "American debt to tomorrow" to ensure that a democratic form of government succeeds over Soviet-style communism. During a composition class, the instructor Robert D. Jacobs introduces Berry to the term "agrarian" and to the 1930s manifesto of the twelve Southern Agrarian writers (who include Allen Tate, John Crowe Ransom, and Robert Penn Warren), *I'll Take My Stand*, a work that will be significant (both in influence and in reaction) to the development of Berry's thought.

1954   Publishes first poem, "Spring," and first short story, "Summer Crop," in the spring issue of *Stylus*, the university literary magazine. Becomes editor of *Stylus*. At a writers' conference at Morehead State College in Kentucky meets writer James Still, whose work will become an influence (Berry will later write, "His short stories, I think, are as near to perfect as any writing I know"). In fall, meets James Baker Hall, who will become a close friend, in a creative writing class taught by Hollis Summers, for which Berry writes story "The Brothers" (later part of *Nathan Coulter*).

1955   Wins *Stylus*'s Dantzler Award for short story "The Brothers." In fall, meets Tanya Amyx (born April 30, 1936, in Berkeley, California), a French and music major and daughter of University of Kentucky art professor Clifford Amyx and textile artist Dee Amyx. "The first time I ever saw Tanya," Berry would later remark, "she was standing by [a] wooden newel post in Miller Hall at the University of Kentucky. Years later, they started to remodel the place. I went over and said, 'Look. When you tear that post out, I want it.'" (The post is now in the Berrys' home.) With fellow student Edward M. Coffman, travels to Kenyon College to meet John Crowe Ransom.

1956   Wins *Stylus*'s Dantzler Award for short story "The Chestnut Stud." Meets fellow student Gurney Norman. On May 16, submits term paper on *I'll Take My Stand*: "The Regional Context: A Consideration of the Southern Agrarians as Southerners." Graduates from the University of Kentucky with a bachelor's degree in English. In summer, Berry and James Baker Hall join literature seminars at Indiana University School of Letters; Berry takes course in Joyce and Yeats from Richard Ellman, and course in modern poetry from Karl Shapiro. "The Brothers" wins first prize in the *Carolina Quarterly*'s 6th Annual Fiction Contest and is published in its summer issue. In the fall, enters the master's program in English at the University of Kentucky.

1957   In February, poem "Rain Crow" is published in *Poetry* magazine; in the next five years, his work will be published in *Poetry* five times. Meets writer Ed McClanahan during a graduate course. Completes MA in English. "Elegy" wins the Farquhar Award for Poetry from the University of Kentucky. Two short stories, "Whippoorwills" and "Apples," are published in *Coraddi: Arts*

Forum 1957. Marries Tanya Amyx, May 29. In summer, they live in a cabin built by Berry's great-uncle Curran Mathews, called "the Camp," on the Kentucky River near Port Royal, which was used as a family retreat from the 1920s. In the fall, they move to an apartment in Georgetown, Kentucky; Berry begins teaching freshman composition and sophomore literature at Georgetown College, father's alma mater, while Tanya continues to study at the University of Kentucky.

1958   Daughter Mary Dee born May 10 in Lexington. Awarded a Wallace Stegner Fellowship at Stanford University; in August, family moves to Mill Valley, California, to live while he studies creative writing at Stanford University under Stegner (about whom Berry will later write, "My debt to him is probably greater than I know") and Richard Scowcroft; fellow members of the fiction seminars include Ernest J. Gaines (also a Stegner fellow), Ken Kesey, and Nancy Packer. Lives with family in small house owned by Tanya's aunt and uncle, Ann and Dick O'Hanlon, which would later become part of the O'Hanlon Center. Works on novel *Nathan Coulter*: "I had about half of it written when I went out there, and I just continued to work on it." Introduced by Holman Hamilton, one of Berry's history professors at the University of Kentucky, to Craig Wylie of Houghton Mifflin, to whom Berry submits the manuscript of *Nathan Coulter*. Houghton Mifflin purchases an option on the book for $250.

1959   Accepts appointment as Edward H. Jones Lecturer in Creative Writing at Stanford University for one year.

1960   In January, begins writing second novel, *A Place on Earth*. First novel, *Nathan Coulter*, published in April by Houghton Mifflin. In spring, family moves back to Kentucky, living on the "Home Place" in Henry County, and also farming with Berry's friend and neighbor Owen Flood.

1961   Receives a Guggenheim Foundation Fellowship. Will later write of Henry County that summer: "I began to understand that so long as I did not know the place fully, or even adequately, I belonged to it only partially. That summer I began to see, however dimly, that one of my ambitions, perhaps my governing ambition, was to belong fully to this place, to belong as the thrushes and the herons and the muskrats belonged, to be altogether at home here." In August, leaves with family for a year in Tuscany and southern France to study and write: "the land in Tuscany has had about 2,000 years of good care. And it looked like it when I was there. . . . The sight of that changed my mind about what was possible in land use."

1962   Meets critic and translator Wallace Fowlie when both stay at La Napoule Arts Foundation near Cannes. Awarded Vachel Lindsay Prize by *Poetry* magazine. Returns to Kentucky from Europe in July. Son Pryor Clifford (Den) born August 19 in Lexington. In the fall, moves with family to New York to take up post as assistant professor of English and director of freshman writing at New York University's University Heights campus in the Bronx. Rents apartment in New Rochelle, New York. Meets Bobbie Ann Mason.

1963   Family moves to loft at 277 Greenwich Street, New York City, downstairs from poet Denise Levertov (whom Berry had met the previous fall, although they had corresponded since 1958, when Levertov wrote Berry about a poem he had published in *Poetry*) and writer Mitchell Goodman. Tanya arranges playdates for daughter Mary and a fellow student at St. Luke's School, the daughter of James Parks Morton, later dean of St. John the Divine. Spends summer in Kentucky. Dismantles the Camp, which has flooded several times over the years, and rebuilds it higher on the bank of the river. Works on novel *A Place on Earth*. Reads Harry Caudill's *Night Comes to the Cumberlands*, an important influence ("It showed me what it might mean to be a responsible Kentucky writer living in Kentucky, and it affected me deeply"). Returns to New York for fall semester. Meets poet Donald Hall, who becomes friend and correspondent, at a cocktail party on Riverside Drive in Manhattan. Offered and accepts position as professor of English at the University of Kentucky to begin in the fall of 1964, where he will teach creative writing and other courses.

1964   In May, visits Levertov and Goodman's Maine farm, where he reads in typescript Hayden Carruth's poetry collection *North Winter* (Berry will later begin a correspondence with Carruth that will last until Carruth's death in 2008). Leaves New York for Kentucky soon after. A poem about Kennedy's assassination, *November Twenty Six Nineteen Hundred Sixty Three* (with drawings by Ben Shahn) published by George Braziller in May. Begins commuting weekly from Lexington (where family lives) to the Camp to write. Meets Kentucky artist Harlan Hubbard and his wife Anna by chance while on a canoe trip on the Ohio River; later writes that their homestead "differed from Thoreau's economy radically in some respects, and also advanced and improved upon it. The main differences were that, whereas Thoreau's was a bachelor's economy, the Hubbards' was that of a married couple." First poetry collection *The Broken Ground* published by Harcourt, Brace & World in September (a *New York Times* review will begin: "The quiet but sure and melodious voice of Wendell Berry makes 'The Broken Ground' an immediate pleasure"). In November, purchases the Lanes Landing property as a weekend place, a twelve-acre hillside farm adjacent to the Camp and bordering his maternal grandfather's farm on the west side of the Kentucky River. Berry's editor at Harcourt, Dan Wickenden, introduces him to the work of organic farming pioneer Sir Albert Howard.

1965   Awarded a Rockefeller Foundation Fellowship. On July 4, moves with family to Lanes Landing Farm, where they will continue to live and farm year round. "I remember a moment—in 1965, or a little after—when I realized that I didn't have to be a writer; there were other kinds of work also that required artistry and offered satisfaction. From here, looking back, I can see what a defining moment that was. I had, in effect, decided not to be a 'professional' writer, but instead, in the literal sense, an amateur: I would work for love." Over time, expands the farm until it consists of about 117 acres in two tracts; at first, raises only food for the family. In July, sees first strip mine above Hardburly, Kentucky, while visiting writer Gurney Norman; Norman introduces him to Kentucky lawyer, writer, and environmentalist Harry Caudill and Caudill's wife Anne, who will become lifelong friends, at a meeting of the Appalachian Group to Save the Land and People. Will one day write, "Gurney has been my Virgil on the upper end (the mountain end) of the Kentucky River watershed, where he is at home. I'm at home on the lower end."

1967   In summer, makes first notes for a work that will become *The Unsettling of America* after reading about the report of President Johnson's "special commission on federal food and fiber policies" that defined the nation's greatest agriculture problem as a surplus of farmers. Novel *A Place on Earth* published by Harcourt, Brace & World in September. Receives Bess Hoken Prize from *Poetry* magazine.

Becomes president of the Cumberland Chapter of the Sierra Club in Kentucky. Begins work with farmers and landowners to oppose a dam across the Red River Gorge that would destroy the gorge's unique topography and ecosystem. In December, visits Trappist monk Thomas Merton at Abbey of Our Lady of Gethsemani near Bardstown, Kentucky, with Ralph Eugene Meatyard and his wife Madelyn, Tanya, and Denise Levertov.

1968　On February 10, delivers speech, "A Statement Against the War in Vietnam," during the Kentucky Conference on War and the Draft in Lexington, Kentucky: "I have come to the realization that I can no longer imagine a war that I would believe to be either useful or necessary. I would be against any war." In fall, takes leave from University of Kentucky to become visiting professor of creative writing at Stanford University during fall 1968 and winter 1969 quarters; lives with family in Menlo Park. Colleagues at Stanford include Wallace Stegner, Ed McClanahan, and Ken Fields. Meets writer John Haines when he gives a reading at Stanford. During the Christmas holiday, writes *The Hidden Wound*, an extended meditation on race and memories of two older black people who were his friends during his childhood. Poetry collection *Openings* published by Harcourt, Brace & World in October; *The New York Times* reviewer calls it a book "to win the respect of anyone who cares about contemporary verse."

1969　Returns with family to Kentucky. Essay collection *The Long-Legged House*, of which the title essay is about the history of the Camp, published by Harcourt, Brace & World in April; other essays concern place and belonging, agriculture, community, small farming, the Vietnam War, and consumerism. The same month, poetry collection *Findings* published by The Prairie Press. Wins first prize in the Borestone Mountain Poetry Awards. Receives grant from the National Endowment for the Arts.

1970　On March 7, speaks at the march at the state capitol in Frankfort, Kentucky, to protest the war in Vietnam; his speech notes government disdain for the will of the people and the huge expense of the war: "With the very earth dying under our feet, with the air full of pollution; we are spending billions of dollars and thousands of lives to assure the success of tyranny in Vietnam." At the first Earth Day

celebration at the University of Kentucky, delivers speech
"Think Little." Publishes essay "The Regional Motive" in
the autumn issue of *The Southern Review*, a response to
the Southern Agrarians; in it, he is critical of those Agrarians who had moved to northern universities, saying that
it "invalidated their thinking, and reduced their effort to
the level of an academic exercise." (Receives letter from
Allen Tate in response, and when the essay is included in
A Continuous Harmony in 1972, Berry appends an apologetic footnote; he later writes that "whatever the amount
of truth in that statement, and there is some, it is also a
piece of smartassery.") Poetry collection *Farming: A Hand
Book* published by Harcourt Brace Jovanovich in September. Meets writer and farmer Gene Logsdon, who becomes
a friend and ally, after Logsdon reads *Farming: A Hand
Book*. *The Hidden Wound* published by Houghton Mifflin
in September.

1971   *The Unforeseen Wilderness: An Essay on Kentucky's Red
          River Gorge* published by University Press of Kentucky in
          April, with photographs by Ralph Eugene Meatyard; it is
          a response to the plan to dam the Red River, which Berry
          had opposed since 1967. (Congress will withdraw funds for
          the project in 1976. In 1993 the river is designated by Con-
          gress in the National Wild and Scenic Rivers System Act.)
          Elected Distinguished Professor of the Year by the Uni-
          versity of Kentucky. Receives the Arts and Letters Literary
          Award from the American Academy of Arts and Letters.

1972   With his brother John Berry Jr. and the Save Our Land
          Committee, helps in the successful opposition to the Jef-
          ferson County Air Board's plan for an international jet-
          port in Henry and Shelby Counties. In the fall, takes a
          one-year sabbatical from the University of Kentucky; his
          courses are covered by Ed McClanahan. Essay collection
          A Continuous Harmony: Essays Cultural and Agricultural
          published by Harcourt Brace Jovanovich in September.
          Meets bookseller and future publisher Jack Shoemaker
          when Shoemaker visits Port Royal after reading *The Long-
          Legged House*.

1973   *The Country of Marriage* published by Harcourt Brace
          Jovanovich in February, his fifth poetry collection, with
          thirty-five poems, including the title poem, "Prayer Af-
          ter Eating," and "Manifesto: The Mad Farmer Liberation

Front." Buys neighboring forty-acre farm that had been sold to a developer, bulldozed, and graveled; begins process of repair and restoration. Invites writer and photographer James Baker Hall to take photographs during a tobacco harvest (they will be published with an essay by Berry in 2004). Begins correspondence with poet and essayist Gary Snyder, and after a few months Snyder travels to Port Royal to visit Lanes Landing. In November, brother elected to the Kentucky senate, in which he will serve until 1981.

1974   Serves as Elliston Poetry Lecturer at the University of Cincinnati for winter 1974. Novel *The Memory of Old Jack* published by Harcourt Brace Jovanovich in February (*The New York Times Book Review* calls it "a slab of rich Americana" and *Library Journal* says the novel is "worthy of a place among the best pieces of prose written by American writers of this century"). Meets Maurice Telleen at a draft horse sale in Waverly, Iowa, whose *Draft Horse Journal* will publish many of Berry's stories. Gives speech on July 1 at "Agriculture for a Small Planet Symposium" in Spokane, Washington, which will form the basis for the first chapters of *The Unsettling of America*: "I was asked to talk about 'Labor Intensive Micro-Systems Agriculture.' That's not my language, and it's not the sort of language I wish to use because it's the way people speak when they don't want to be understood by most people. I'm not sure what to make of these particular phrases, but they seem to suggest a very methodological or technological approach to agriculture. Part of my purpose here is to suggest that any such approach will necessarily be too simple." Mentions critically Secretary of Agriculture Earl Butz's "adapt or die" policy. Speech helps encourage the Tilth movement, a regional network of organic farmers in the Northwest (still active today). Poetry collection *An Eastward Look* published by Sand Dollar Press in the fall, Berry's first publication with editor Jack Shoemaker. Wins the Emily Clark Balch Prize from *The Virginia Quarterly Review*.

1975   *The Memory of Old Jack* wins the Friends of American Writers Award. A poetry pamphlet, *Horses*, published by Larkspur Press in Kentucky in April; another, *To What Listens*, published by Best Cellar Press. Poetry collection *Sayings and Doings* published by Gnomon Press in December.

1977   Serves as writer-in-residence for the winter quarter at Centre College in Danville, Kentucky. Poetry collection *Clearing* published by Harcourt Brace Jovanovich in March. *The Unsettling of America: Culture and Agriculture* published by Sierra Club Books in August and dedicated to Maurice Telleen, the editor of *Draft Horse Journal*. Donald Hall reviews both volumes for *The New York Times*, writing, "Berry is a prophet of our healing, a utopian poet-legislator like William Blake." Resigns from the University of Kentucky to become contributing editor at Rodale Press, including its magazines *Organic Gardening* and *The New Farm*.

1978   With Tanya, buys first flock (six ewes and one buck) of Border Cheviot sheep, a Scottish breed, which they will raise for the next four decades. In November, debates Secretary of Agriculture Earl Butz on the agricultural crisis at Manchester University, North Manchester, Indiana: "As I see it, the farmer standing in his field is not simply a component of a production machine. He stands where lots of cultural lines cross. The traditional farmer, that is the farmer who first fed himself off his farm and then fed other people, who farmed with his family, who passed the land on down to people who knew it and had the best reasons to take care of it—that farmer stood at the convergence of traditional values, our values: independence, thrift, stewardship, private property, political liberties, family, marriage, parenthood, neighborhood—values that decline as that farmer is replaced by a technologist whose only standard is efficiency."

1979   On June 3, opposes and with eighty-nine others is arrested during nonviolent protests against the construction of the Marble Hill nuclear power plant on the Ohio River near Madison, Indiana. The company eventually abandons the effort to finish it. Poetry collection *The Gift of Gravity* published by Deerfield Press.

1980   Fired from Rodale Press: "I think this was because I was more for small farmers than I was for organic farmers. Also, I don't think I've ever been a very good employee." Ernest J. Gaines visits the Berrys in Kentucky. Begins working with editor Jack Shoemaker, who had left bookselling to cofound North Point Press in Berkeley, California, in 1979, with William Turnbull. (Most of Berry's books will henceforward be published with Shoemaker.) Poetry collection

A Part published in October. Joins with brother and other residents of Henry County along with the group Kentuckians for the Commonwealth in a successful effort to oppose the building of a hazardous chemical waste incinerator in the county. Publishes poetry pamphlet *The Salad*. On November 11, writes to Wes Jackson, founder of The Land Institute, after reading Jackson's book, *New Roots for Agriculture*. Visits Jackson in December to write an article about The Land Institute for *The New Farm* magazine. The two begin a long correspondence and friendship.

1981   *Recollected Essays: 1965–1980* published in August. Essay collection *The Gift of Good Land: Further Essays Cultural and Agricultural*, dedicated to Gene Logsdon, published in November. Both volumes are reviewed in *The Washington Post* by Larry Woiwode, who calls them "reference works of the body and soul." Granddaughter Katie Jean Smith born, December 16.

1982   Poetry collection *The Wheel* published in October.

1983   Speaking at Oberlin College in February, meets David and Elsie Kline, an Amish couple from Holmes County who had read *The Unsettling of America* and came to hear him speak. Kline (with Meatyard, Telleen, and Jackson) is one of the only four agrarian friends Berry has from outside of Henry County for the next decade. Essay collection *Standing by Words* published in October, which *The Christian Science Monitor* review calls "nothing short of splendid." Publishes substantially revised and shortened version of 1967 novel *A Place on Earth* in March; in a new introduction, writes that the original book was "clumsy, overwritten, wasteful."

1985   Co-edits with Wes Jackson and Bruce Coleman *Meeting the Expectations of the Land: Essays in Sustainable Agriculture and Stewardship*, published by North Point in February; Berry's contribution is the essay "Whose Head is the Farmer Using? Whose Head is Using the Farmer?" Granddaughter Virginia Dee Smith born, March 6. *Collected Poems 1957–1982* published in May; *The New York Times* reviewer writes that Berry "can be said to have returned American poetry to a Wordsworthian clarity of purpose." Publishes substantially revised version of *Nathan Coulter* in May. Helps found Community Farm Alliance, a group of Kentucky farmers, bankers, businessmen, and clergy

members organized to help small farmers shift from tobacco to other products. Begins correspondence with
writer John Haines.

1986   *The Wild Birds: Six Stories of the Port William Membership*
published in March. Receives honorary doctorate from
the University of Kentucky. In November, delivers lecture
"Preserving Wildness" at the first Temenos Academy Conference in Devon, England.

1987   Returns to the English Department of the University of Ken-
tucky, now teaching courses for "future teachers and farmers
or anyone preparing to work in practical and comprehensive
ways with young minds or with nature"; courses include
"Composition for Teachers," aimed at public school English
teachers; "Readings in Agriculture," which assigns Spenser,
Milton, Shakespeare, Pope, and Wordsworth, as well as Sir
Albert Howard, J. Russell Smith, Wes Jackson, and Gene
Logsdon; and "The Pastoral." Publishes *Home Economics:
Fourteen Essays* in June; the *Christian Science Monitor*, in reviewing it, calls Berry "*the* prophetic American voice of our
day." Serves as writer-in-residence for the interim term at
Bucknell University. Receives the American Academy of Arts
and Letters Jean Stein Award and the Kentucky Governor's
Milner Award in the Arts. In the fall, "Why I Am Not Going
to Buy a Computer" published in the *New England Review
and Bread Loaf Quarterly* and reprinted in *Harper's*; the
essay inspires many critical responses. *Sabbaths: Poems* published in September, the first of what will become a sustained
project of poems on the themes of work and rest, fields and
woods. *The Landscape of Harmony: Two Essays on Wildness
and Community*, including the lecture "Preserving Wildness," published by Five Seasons Press, United Kingdom, in
September. Poetry collection *Some Differences* published by
Confluence Press in December.

1988   Novel *Remembering* published in October and reviewed
favorably in the *Los Angeles Times*.

1989   Receives Lannan Foundation Award for nonfiction. Poetry
collection *Traveling at Home* published in November. Delivers the Blazer Lecture on the life and work of Kentucky
artist and author Harlan Hubbard at the University of
Kentucky.

1990   Granddaughter Tanya Christine Smith born February 19.

Essay collection *What Are People For?*, dedicated to Gurney Norman, published in March; it includes the oft-repeated line, "Eating is an agricultural act." (Berry later calls this quote, as repeated without context, "an oversimplification he is now damned sorry to have written.") Bill McKibben publishes a major profile of Berry in *The New York Review of Books*, writing that "wherever we live, however we do so, we desperately need a prophet of responsibility; and although the days of the prophets seem past to many of us, Berry may be the closest to one we have. But, fortunately, he is also a poet of responsibility. He makes one believe that the good life may not only be harder than we're used to but sweeter as well." Inducted into the Fellowship of Southern Writers. Berry's Blazer lecture, *Harlan Hubbard: Life and Work*, published by University Press of Kentucky in November.

1991   *Standing on Earth: Selected Essays* published in the United Kingdom by Brian Keeble's Golgonooza Press in April. Publishes poetry collection *Sabbaths 1987* with Larkspur Press in October. Father dies October 31. North Point Press closes, and Berry follows Shoemaker to Pantheon.

1992   Receives the Victory of Spirit Ethics Award from the University of Louisville and the Louisville Community Foundation. From late August to late September, teaches a course at Schumacher College in South Devon, England, on "Nature as Teacher: The Lineage of Writings Which Link Culture and Agriculture"; writers assigned include Shakespeare, Sir Albert Howard, and Wes Jackson. On September 8, meets Charles, Prince of Wales. "It was a much easier meeting than I expected, for he is intelligent, considerate, and talks and listens well. He is an ally, is deeply concerned about what is happening to rural life and to civilization," Berry wrote in a letter to Wes Jackson. *Fidelity: Five Stories* published in October to favorable reviews, including in *The New York Times* and *The New Criterion*.

1993   Granddaughter Emily Rose Berry born, May 2. *Sex, Economy, Freedom and Community: Eight Essays* published in October; *The New York Times* writes that the book "in its eight essays distills the author's radically conservative views (in the good senses of both words)." Receives The Orion Society's John Hay Award for nature writing. Quits his job at the University of Kentucky.

1994   Receives T. S. Eliot Award for creative writing from the

Ingersoll Foundation. Poetry collection *Entries* published in May. *Watch With Me and Six Other Stories of the Yet-Remembered Ptolemy Proudfoot and His Wife, Miss Minnie, Née Quinch* published in August. Berry follows his editor to Counterpoint Press, which Shoemaker cofounds in Washington, D.C., with Frank H. Pearl.

1995   Grandson Marshall Amyx Berry born June 22. In October, long poem *The Farm* (one of the *Sabbath* series) published by Larkspur Press and the essay collection *Another Turn of the Crank* published by Counterpoint.

1996   Co-authors *Three on Community* with Gary Snyder and Carole Koda, published by Limberlost Press in April. Receives the Harry M. Caudill Conservationist Award from the Cumberland Chapter of the Sierra Club. Novel *A World Lost* published in October.

1997   Mother dies, January 3. Father-in-law Clifford Amyx dies, July 30. *Two More Stories of the Port William Membership* published by Gnomon Press. Receives the Lyndhurst Prize for individuals making a significant contribution to the arts. Speaks at a benefit for The Garden Project in San Francisco, November 10.

1998   Gives keynote address, "In Distrust of Movements," at the Northeast Organic Farming Association conference. *A Timbered Choir: The Sabbath Poems 1979–1997* published in April. *The Selected Poems of Wendell Berry* published in October.

1999   Receives the Thomas Merton Award from the Thomas Merton Center for Peace and Social Justice in Pittsburgh. On November 10, joins Gary Snyder and Jack Shoemaker in reading and conversation at a Lannan Foundation event in Santa Fe, New Mexico.

2000   Wins Poets' Prize for *The Selected Poems of Wendell Berry*. In May publishes *Life Is a Miracle: An Essay Against Modern Superstition*; Bill McKibben, reviewing it for *The Washington Monthly*, writes, "it's hard to imagine that the millennium has seen a more important book than this slim volume from our finest essayist." Novel *Jayber Crow* published in September; *The New York Times* reviewer writes that "by the end this melancholy barber has won both our attention and our hearts." In October, teaches one-week course at Schumacher College in England on "Community, Sustainability and Globalisation."

2001   In response to the events following 9/11, writes "Thoughts in the Presence of Fear," which is published on October 30 by *The Land Report*; in December it is joined by two other essays, "The Idea of a Local Economy," and "In Distrust of Movements," and published by The Orion Society in paperback as *In the Presence of Fear: Three Essays for a Changed World*. Publishes *Sonata at Payne Hollow: A Play* with Larkspur Press.

2002   *The Art of the Commonplace: The Agrarian Essays of Wendell Berry*, edited and introduced by Norman Wirzba, published in April. Berry follows his editor to Shoemaker & Hoard, which Shoemaker cofounds with Trish Hoard.

2003   On February 9, essay "A Citizen's Response to the National Security Strategy of the United States," critical of the Bush administration's post-9/11 strategy, published as a full-page advertisement in *The New York Times*. Essay collection *Citizenship Papers* published in August. *Citizen's Dissent: Security, Morality, and Leadership in an Age of Terror: Essays*, co-authored with David James Duncan, published by The Orion Society. Receives Lifetime Achievement Award from the Cathedral Heritage Foundation in Louisville, Kentucky.

2004   *That Distant Land: The Collected Stories* published in February. Receives the Eli M. Oboler Memorial Award in Orlando in June, shared by David James Duncan, awarded by the American Library Association Intellectual Freedom Round Table, for *Citizen's Dissent*. Mother-in-law Dee Rice Amyx dies, July 3. Awarded the Charity Randall Citation from the International Poetry Forum, as well as the Soil and Water Conservation Society Honor Award from the Kentucky Bluegrass Chapter for dedicated efforts toward the preservation of unique rural and cultural resources. Writes essay for *Tobacco Harvest: An Elegy*, to accompany photographs by James Baker Hall; published by University Press of Kentucky in September. Novel *Hannah Coulter* published in September. Poetry collection *Sabbaths 2002* published by Larkspur Press.

2005   Receives O. Henry Prize for short story "The Hurt Man." *Given: Poems* published in May. Inducted into the University of Kentucky Arts and Sciences Hall of Fame. *Blessed Are the Peacemakers: Christ's Teachings about Love, Compassion & Forgiveness*, selections of the gospels compiled and introduced by Berry, published in October. *The Way of Ignorance and Other Essays* published in October. Essay "Not a Vision of our Future, But of Ourselves," about the earth's destruction due to coal mining, included in *Missing Mountains: We Went to the Mountaintop but It Wasn't There*, published by Wind Publications in October. Receives Lifetime Achievement Award from the Conference on Christianity and Literature, December 29.

2006　In August, speaks at "One Thing to Do About Food: A Forum" with Alice Waters, Eric Schlosser, Marion Nestle, Peter Singer, and others. In October, gives the keynote address for the thirtieth anniversary of The Land Institute in Salina, Kansas. Novel *Andy Catlett: Early Travels* published in November.

2007　Visits Ernest J. Gaines in Louisiana. Shoemaker and Charlie Winton acquire both Counterpoint Press and Soft Skull Press; merge both with Shoemaker & Hoard to re-form Counterpoint.

2008　On February 14, at the Kentuckians for the Commonwealth's annual I Love Mountains Day, delivers speech at the Kentucky state capitol in Frankfort calling for nonviolent resistance and civil disobedience to protest mountaintop removal in Kentucky. In May, awarded honorary doctorate by Duke University. *The Mad Farmer Poems* published by Counterpoint in November. Children's book *Whitefoot: A Story from the Center of the World* published in December. Receives the Cynthia Pratt Laughlin Medal from the Garden Club of America.

2009　On January 4, publishes an op-ed in *The New York Times* with Wes Jackson, "A 50-Year Farm Bill," calling for legislation that would address soil loss, pollution, fossil fuels, and the preservation of rural communities. On January 23, releases public statement against the death penalty through the Kentucky Coalition to Abolish the Death Penalty: "As I am made deeply uncomfortable by the taking of a human life before birth, I am also made deeply uncomfortable by the taking of a human life after birth." On March 2, joins protests in Washington, D.C., against mountaintop removal and burning of fossil fuels. On April 3, awarded the Fellowship of Southern Writers' Cleanth Brooks Medal for Excellence in Southern Letters. Essay collection *Bringing It to the Table: On Farming and Food* with an introduction by Michael Pollan, published in August. Poetry collection *Leavings* published in October. In November, joins public letter signed by forty Kentucky writers to Kentucky governor and attorney general asking for a moratorium on the death penalty. On December 20, removes personal papers on loan to the University of Kentucky Archives as a result of the University's too close relationship with the coal industry. (In August 2012, those papers will be donated to the Kentucky Historical Society in Frankfort.)

2010   Receives O. Henry Prize for short story "Stand By Me." Essay collections *Imagination in Place* (including reflections on Wallace Stegner, Gurney Norman, Hayden Carruth, Donald Hall, Jane Kenyon, John Haines, James Still, Gary Snyder, and Kathleen Raine) published in January, and *What Matters? Economics for a Renewed Commonwealth*, with a foreword by Herman E. Daly, published in May.

2011   In mid-February, joins nineteen other activists protesting mountaintop removal in eastern Kentucky by occupying the office of Governor Steve Beshear in Frankfort for a weekend. The governor agrees only to tour coal communities to see the damage. *The Poetry of William Carlos Williams of Rutherford*, a personal tribute to the poet, published in February and is reviewed in *The New York Review of Books*. On March 2, awarded National Humanities Medal by President Barack Obama at the White House. On May 4, speaks at the Future of Food conference at Georgetown University: "the future of food is not distinguishable from the future of the land, which is indistinguishable, in turn, from the future of human care." Daughter Mary Berry establishes The Berry Center in New Castle, Kentucky, to continue the work of Berry and his father and brother toward healthy and sustainable agriculture.

2012   Great-granddaughter Charlcye Anne Johnson born February 7. Receives O. Henry Prize for short story "Nothing Living Lives Alone." *New Collected Poems* published in March; *Christian Science Monitor* hails it as "a welcome complement to the rest of his work." Receives the Steward of God's Creation Award at the National Cathedral in Washington, D.C., April 22. Delivers the 41st Jefferson Lecture, the highest honor the federal government confers

for distinguished intellectual achievement in the humanities, at the John F. Kennedy Center for the Performing
Arts on April 23: "We cannot know the whole truth,
which belongs to God alone, but our task nevertheless is
to seek to know what is true. And if we offend gravely
enough against what we know to be true, as by failing
badly enough to deal affectionately and responsibly with
our land and our neighbors, truth will retaliate with ugliness, poverty, and disease." *It All Turns on Affection: The
Jefferson Lecture and Other Essays* published in September.
*A Place in Time: Twenty Stories of the Port William Mem-
bership* published in October. On October 17, receives the
James Beard Foundation Leadership Award for efforts toward a healthier, safer, and more sustainable food system,
in New York City. On October 30, receives the inaugural
Green Cross Award, given by the Bishop of California, for
increasing cultural engagement in environmental issues.

2013   In April, elected into the American Academy of Arts and
Sciences. The Berry Center and St. Catharine College in
Springfield, Kentucky, establish The Berry Farming Program, an undergraduate, multidisciplinary degree inspired
by Wes Jackson and Berry's philosophy and work, and designed for students from generational farm families. The
program accepts its first students in the fall (St. Catharine
College will close in 2016, and the Berry Farming Program
will move to partnership with Sterling College in Vermont). *This Day: Sabbath Poems Collected & New 1979–2013*
published in October. On October 17, awarded the Freedom Medal from the Roosevelt Institute's Four Freedoms
Awards in New York City. Daughter Mary marries Steve
Smith, October 26.

2014   *Distant Neighbors: The Selected Letters of Wendell Berry
and Gary Snyder*, edited by Chad Wriglesworth, published
in June with reviews in the *Los Angeles Review of Books*,
Commonweal, and other publications. *Terrapin and Other
Poems*, with illustrations by Tom Pohrt, published in November. Poetry collection *Roots to the Earth*, with wood-
cuts by Wesley Bates, published by Larkspur Press.

2015   On January 28, inducted into the Kentucky Writers Hall
of Fame, the first living writer to be so honored. *Our Only
World: Ten Essays* published in February. Poetry collection
Sabbaths 2013 published by Larkspur Press.

2016   Receives O. Henry Prize for short story "Dismember-
        ment." On March 17, receives the Ivan Sandrof Lifetime
        Achievement Award from the National Book Critics Circle
        in New York City, presented by Nick Offerman. *A Small
        Porch: Sabbath Poems 2014 and 2015* together with *"The Pres-
        ence of Nature in the Natural World: A Long Conversation"*
        published in June. On September 25, delivers the Strachan
        Donnelley Lecture on Conservation and Restoration at
        the 2016 Prairie Festival in Salina, Kansas, celebrating the
        fortieth anniversary of The Land Institute and the ten-
        ure and retirement of Wes Jackson as president. Brother
        John M. Berry Jr. dies, October 27. Speaks in conversa-
        tion with Wes Jackson, and moderated by Mary Berry,
        at the 36th Annual Schumacher Lectures, October 22.
        Great-granddaughter Wendy Jean Johnson born, October
        29. In December, speaks with Eric Schlosser at the twenti-
        eth anniversary of the Center for a Livable Future at Johns
        Hopkins University.

2017   *The World Ending Fire: The Essential Wendell Berry*, se-
        lected by Paul Kingsnorth, published by Allen Lane in
        the United Kingdom in January. Contributes to *Letters to
        a Young Farmer: On Food, Farming, and Our Future* by
        Stone Barns Center for Food and Agriculture, published
        by Princeton Architectural Press. *The Art of Loading Brush*
        published by Counterpoint Press in October.

Note on the Texts

This volume—the first of a two-volume set, *What I Stand On: The Collected Essays of Wendell Berry 1969–2017*—presents a selection of Berry's nonfiction from the first half of his career. It contains the whole of the author's long argumentative book *The Unsettling of America* (1977) and thirty-two essays culled from eight other books by Berry published from 1969 to 1990: *The Long-Legged House* (1969), *The Hidden Wound* (1970), *A Continuous Harmony* (1972), *Recollected Essays: 1965–1980* (1981), *The Gift of Good Land* (1981), *Standing by Words* (1983), *Home Economics* (1987), and *What Are People For?* (1990). The contents were chosen by the author and his longtime editor and publisher, Jack Shoemaker, and are arranged in chronological order by first book publication. In accord with Shoemaker and Berry's wishes, and as described below, this volume prints the texts of the essays from the most recent Counterpoint Press printings, which reflect corrections made by the author. The companion volume presents selections from books published from 1993 to 2017.

After the publication of *The Unsettling of America* (1977) by Sierra Club Books, Berry began to look for a new publisher. In 1980 he embarked on his long collaboration with Jack Shoemaker, then a thirty-four-year-old editor who, with William Turnbull, had recently cofounded North Point Press in Berkeley, California. From that point forward most of Berry's books would be edited by Shoemaker. Berry followed his editor first to Pantheon when North Point closed in 1991; then to Counterpoint Press, which Shoemaker cofounded with Frank H. Pearl in Washington, D.C., in 1994; then to Shoemaker & Hoard in 2002, which Shoemaker cofounded with Trish Hoard; and finally back to Counterpoint, which merged with Shoemaker & Hoard when Shoemaker and Charlie Winton purchased both Counterpoint Press and Soft Skull Press in 2007. None of the books from which selections have been made for this volume appeared in British editions.

The Long-Legged House was published by Harcourt, Brace & World on April 9, 1969; the most recent Counterpoint Press paperback printing, the source for the texts printed here, was published in 2012. The first publication of the essays selected from *The Long-Legged House* is given below.

The Rise. *The Rise* (Lexington: The University of Kentucky Library Press, 1968).

The Long-Legged House. *The Long-Legged House* (New York: Harcourt, Brace & World, 1969).

A Native Hill. *The Hudson Review*, Winter 1968–1969.

The Hidden Wound was published by Houghton Mifflin Co. on September 14, 1970; the most recent Counterpoint Press paperback printing, the source for the text of the excerpt printed here (Chapters 4 through 8), was published in 2010. These chapters first appeared in the 1970 Houghton Mifflin edition of *The Hidden Wound*. They were later published, in a slightly different version, as "Nick and Aunt Georgie," in *Recollected Essays: 1965–1980* (1981).

A Continuous Harmony was published by Harcourt Brace Jovanovich on September 20, 1972; the most recent Counterpoint Press paperback printing, the source for the texts printed here, was published in 2012. The first publication of the essays selected from *A Continuous Harmony* is given below.

Think Little. *Whole Earth Catalog*, September 1970.

Discipline and Hope. *A Continuous Harmony* (New York: Harcourt Brace Jovanovich, 1972). (A portion of the essay was delivered at the University of Kentucky as the Distinguished Professor Lecture on November 17, 1971.)

In Defense of Literacy. *A Continuous Harmony* (New York: Harcourt Brace Jovanovich, 1972).

The Unsettling of America: Culture and Agriculture was published by Sierra Club Books in San Francisco on August 27, 1977. Parts of *The Unsettling of America* appeared previously in slightly different form in *The Nation* and *CoEvolution Quarterly*. Sierra Club Books issued a revised second edition in paperback on August 12, 1986, and ten years later, on March 1, 1996, a revised third edition. The text of *The Unsettling of America* used in this volume is that of the most recent Counterpoint Press paperback printing, published September 15, 2015, which includes the author's Preface to the Second Edition (1986) and his Afterword to the Third Edition (1995).

Recollected Essays: 1965–1980 was published by North Point Press in August 1981. While "The Making of a Marginal Farm" was selected from *Recollected Essays: 1965–1980*, the text of the essay used in this volume is taken from *The World-Ending Fire: The Essential Wendell Berry*, published by Counterpoint Press in 2018. "The Making of a Marginal Farm" was first published in *The Smithsonian*, August 1980.

The Gift of Good Land was published by North Point Press on November 30, 1981; the most recent Counterpoint Press paperback printing, the source for the texts printed here, was published in 2012. The first publication of the essays selected from *The Gift of Good Land* is given below.

Horse-Drawn Tools and the Doctrine of Labor Saving. Lane de Moll and Gigi Coe, eds., *Stepping Stones: Appropriate Technology and Beyond* (New York: Schocken Books, 1978).

Solving for Pattern. *The New Farm*, January 1981.

Family Work. Roger B. Yepsen, Jr., ed., *Home Food Systems* (Emmaus, PA: Rodale Press, 1981).

A Few Words for Motherhood. *The New Farm*, May/June 1981.

A Talent for Necessity. *The New Farm*, February 1981.

Seven Amish Farms. *The Gift of Good Land* (San Francisco: North Point Press, 1981).

The Gift of Good Land. *Sierra Club Bulletin*, November/December 1979.

Standing by Words was published by North Point Press on October 30, 1983; the most recent Counterpoint Press paperback printing, the source for the texts printed here, was published in 2011. The first publication of the essays selected from *Standing by Words* is given below.

Standing by Words. *The Hudson Review*, Winter 1980/1981. (An earlier version of this essay was delivered as a speech at the annual meeting of the Lindisfarne Fellows, Summer 1978.)

Poetry and Marriage: The Use of Old Forms. *CoEvolutionary Quarterly*, Winter 1982.

Home Economics was published by North Point Press on June 24, 1987; the most recent Counterpoint Press paperback printing, the source for the texts printed here, was published in 2009. The first publication of the essays selected from *Home Economics* is given below.

Getting Along with Nature. *The Land Report*, Summer 1982.

Two Economies. *Review and Expositor*, May 1984.

The Loss of the University. Gerald Graff and Reginald Gibbons, eds., *Criticism in the University* (Evanston, Ill: Northwestern University Press, 1985).

Preserving Wildness. Speech delivered at the Temenos Conference at Dartington Hall, in November 1986, first published in *Wilderness*, Spring 1987.

A Good Farmer of the Old School. *Draft Horse Journal*, Spring 1986.

What Are People For? was published by North Point Press on April 1, 1990; the most recent Counterpoint Press paperback printing, the source for the texts printed here, was published in 2010. The first publication of the essays selected from *What Are People For?* is given below.

Damage. *The American Poetry Review*, May–June 1975.

Wallace Stegner and the Great Community. *South Dakota Review*, Winter 1985.

Writer and Region. *The Hudson Review*, Spring 1987.

An Argument for Diversity. *The Hudson Review*, Winter 1980.

The Pleasures of Eating. *What Are People For?* (San Francisco: North Point Press, 1990).

The Work of Local Culture. An Iowa Humanities Lecture delivered at Harlan Community High School, Harlan, Iowa, November 13, 1988, and published after the occasion as a pamphlet by the Iowa Humanities Board, 1988.

Why I Am Not Going to Buy a Computer. *New England Review and Bread Loaf Quarterly*, Autumn 1987, reprinted in *Harper's* in May 1987 with the letters and response.

Feminism, the Body, and the Machine. *What Are People For?* (San Francisco: North Point Press, 1990).

Word and Flesh. *Whole Earth Review*, Spring 1990.

Nature as Measure. *What Are People For?* (San Francisco: North Point Press, 1990).

This volume presents the texts of the printings chosen for inclusion here, but it does not attempt to reproduce features of their typographic design. Original citations have been altered to conform to Library of America practices; however, original footnotes that take the form of personal observation or argument have been preserved verbatim. In a few instances quotations have been corrected with the author's approval. The texts are presented without change, except for the correction of typographical errors and minor changes necessitated by the formatting of notes. Spelling, punctuation, and capitalization are often expressive features, and they are not altered, even when inconsistent or irregular. The following is a list of typographical errors corrected, cited by page and line number: 15.9, kidnaped; 16.12, W.O.A.; 25.16, and end; 26.27, simple; 29.10, back to the; 37.24, And how; 38.20, 'tis; 40.11, now the; 44.20, camp; 45.33, content; 53.12, though; 63.15, camp; 71.20, Ever; 73.39, of Kentucky; 77.21, to more; 100.34, father; 161.2 and 5, Castenada; 210.14, "destiny,"; 213.11, furnance; 291.21, [California}; 314.18, clams.; 380.8, THE LAND GRANT-COLLEGES; 384.2, Allis-Chalmer; 408.12, suceeded; 409.16, *a*; 446.28, WD Allis-Chalmer; 463.15, with.; 475.2, principle; 500.24, farm, It; 538.15, best"'; 548.40, Commissioners; 580.23, occuring; 625.35, specialists) seem; 653.3, principle; 699.28, asked; 704.26, on, a; 718.19, sociey,.

Notes

In the notes below, the reference numbers denote page and line of this volume (the line count includes headings). No note is made for material included in standard desk-reference books. Quotations from Shakespeare are keyed to *The Riverside Shakespeare*, ed. G. Blakemore Evans (Boston: Houghton Mifflin, 1974). Biblical quotations are keyed to the King James Version. Original citations have been altered to conform to Library of America practices; however, original footnotes that take the form of personal observation or argument have been preserved verbatim. For references to other studies and further information than is included in the Chronology, see *Wendell Berry: Life and Work*, ed. Jason Peters (Lexington: The University Press of Kentucky, 2007); and *Conversations with Wendell Berry*, ed. Morris Allen Grubbs (Jackson: University Press of Mississippi, 2007).

From THE LONG-LEGGED HOUSE

8.26 George Caleb Bingham's painting of trappers] American painter George Caleb Bingham (1811–1879) portrayed frontier life along the Missouri and Mississippi Rivers. His *Fur Traders Descending the Missouri* (1845)—originally entitled *French Trader and His Half-Breed Son*—was followed by the similar but often seen as less skillful painting *The Trappers' Return* (1851).

36.29–30 Thoreau's idea of a hypaethral book] "Hypaethral" means roofless, open to the sky. Writing in his *Journal* on June 29, 1851, Henry David Thoreau (1817–1862) characterized *A Week on the Concord and Merrimack Rivers* (1849) as "a hypaethral or unroofed book, lying open under the ether and permeated by it, open to all weathers, not easy to be kept on a shelf."

36.33–34 a long poem . . . titled "Diagon,"] Published in the February 1959 issue of *Poetry*, when Berry was twenty-four.

36.40 "Upon Appleton House, to my Lord *Fairfax*."] A country-house poem written by Marvel in 1651 for his patron Thomas, Lord Fairfax, while the poet was a tutor at Appleton House, the Fairfax Yorkshire estate.

53.40 *A Place on Earth*] Berry's second novel, published in 1967 (revised 1983).

64.30–32 "It requires . . . a day."] Thoreau, "Life Without Principle," published in the *Atlantic Monthly Magazine* (October 1863).

65.17–19 "For an inheritance . . . not be seen."] René Char (1907–1988), *Hypnos Waking: Poetry and Prose by René Char*, selected and translated by Jackson Matthews (1956).

73.21–22 the autobiography of Rev. Jacob Young] One of many itinerant Methodist preachers in the early American republic, Jacob Young (1776–1859) traveled Kentucky's Salt River Circuit. His *Autobiography of a Pioneer: Or, The Nativity, Experience, Travels, and Ministerial Labors of Rev. Jacob Young, with Incidents, Observations, and Reflections* was published in 1857.

86.9–10 the music of the spheres] The idea that the celestial bodies—the sun, the moon, the planets—resonated with harmonious sound dates back to antiquity. According to Aristotle in *On the Heavens*, the Pythagoreans supposed the music of the spheres could not be heard because, having become habituated to it, we could not distinguish it from silence.

92.29 Solomon in all his glory . . . one of these.] Matthew 6:29.

97.38–39 like Valéry's sycamore] An allusion to the poem "Au Platane" [The Plane Tree] by French poet Paul Valéry (1871–1945), published in *Charmes* (1922). The American plane tree is also known as the sycamore.

From THE HIDDEN WOUND

110.25 "Get along home . . . along home!"] "Cindy," a popular Southern folk song, possibly African American in origin, dating back to the nineteenth century.

113.12–13 Witch of Endor] In 1 Samuel 28:3–25, the witch of Endor summons the prophet Samuel from the dead, at King Saul's request.

113.24 the Back to Africa movement.] Originating in the nineteenth century, the Back to Africa movement gained new life and momentum in the 1920s through the efforts of the Jamaican-born social activist Marcus Garvey (1887–1940).

117.9–10 Dr. Bell's Pine Tar Honey] A patented elixir manufactured by E. E. Sutherland Medicine Company of Paducah, Kentucky, from 1894 to 1921. It advertised itself as a remedy for coughs, colds, croup, whooping cough, and lung soreness: "Like the sun's rays through a cloud comes Dr. Bell's Pine-Tar-Honey to the weak and weary cough-worn lungs."

122.33 Joe Louis] World heavyweight boxing champion from 1937 to 1949, nicknamed the "Brown Bomber" (1914–1981).

From A CONTINUOUS HARMONY

136.13 My Lai] My Lai Massacre of approximately 500 Vietnamese civilians by U.S. troops during the Vietnam War, on March 16, 1968.

137.14   Frankfort] Frankfort, Kentucky, home to the state government of the Commonwealth of Kentucky.

138.31–34   "chief way . . . consumers be few. . . ."] See Ezra Pound's translation of *The Great Digest* (1928) in his anthology *Confucius* (1951).

141.3–4   The lotus-eaters] In Homer's *Odyssey*, Book IX, a people living in stupefied indolence.

142.16   who turns up the air conditioner] President Nixon, who liked to read by the fireside in the White House, sometimes ran the air conditioner in summer months while burning a fire in the fireplace.

143.1–2   Louis Bromfield] American novelist and conservationist (1896–1956); Bromfield and his family spent more than a decade living in France.

143.5–9   F. H. King . . . cultivated land."] Franklin Hiram King (1848–1911), American agricultural scientist and inventor of the cylindrical-tower storage silo. See the Introduction to his book *Farmers of Forty Centuries: Or Permanent Agriculture in China, Korea and Japan* (1911), drawn from observations made during his extensive travels to Asia.

144.27–30   "It is the story . . . green things. . . ."] See Chapter 1 of John G. Neihardt's *Black Elk Speaks* (1932), based on the author's conversations with the Oglala Lakota medicine man.

144.31–35   "I saw . . . was holy."] *Black Elk Speaks*, Chapter 3.

149.13–16   "We are fatalists . . . other way."] Edward Dahlberg, "Thoreau and *Walden*," collected in *Do These Bones Live* (1941), later retitled *Can These Bones Live* (1960).

152.27–28   Sir Albert Howard . . . *An Agricultural Testament*] Albert Howard (1873–1947), British botanist and agriculturist. Howard's publications drew on extensive practical research in India, where, working as an agricultural advisor, he came to see the wisdom of traditional Indian farming practices. *An Agricultural Testament* (1940) addresses the proper management of soil fertility. The quotation concerning "the wheel of life" comes from Chapter II.

154.1–2   F. H. King's *Farmers of Forty Centuries*] See note 143.5–9.

155.39   John Collier] Advocate of indigenous rights and cultural pluralism (1884–1968) whose appointment as commissioner for the Bureau of Indian Affairs by Franklin D. Roosevelt in 1933 would alter federal assimilationist policies toward Native Americans. His book *The Indians of the Americas* appeared in 1947.

156.4–5   Chief Rekayi . . . refusing to leave his ancestral home] Chief Rekayi Tangwena (1910–1984) and the people of Kaerezi, on Zimbabwe's eastern border (then Rhodesia), resisted the colonial government's evictions from their ancestral lands in the late 1960s and early 1970s.

157.10–12   "are tied . . . lasting bonds."] Thomas Jefferson, letter to John Jay, August 23, 1785.

157.12–13   ". . . legislators cannot invent . . . subdividing property. . . ."] Thomas Jefferson, letter to James Madison, October 28, 1785.

157.14–16   ". . . it is not . . . a state."] Thomas Jefferson, letter to James Madison, October 28, 1785.

157.18–19   "natural aristocracy . . . government of society,"] Thomas Jefferson, letter to John Adams, October 28, 1813.

157.38–40   ". . . that the producers . . . the consumers temperate. . . ."] See Ezra Pound's translation of *The Great Digest* (1928) in his anthology *Confucius*.

160.16   Paul Radin] American anthropologist and ethnographer (1883–1959).

160.36   the Native American Church] A pan-Native American religion, also known as Peyotism, combining traditional Native American beliefs and Christianity. Meetings involve the ingestion of peyote, a hallucinogen containing mescaline.

161.2–3   *The Teachings of Don Juan: A Yaqui Way of Knowledge*] Published in 1968, the first in a series of popular books by anthropologist Carlos Castaneda (1925–1998) based on the teaching of Yaqui shaman don Juan Matus. Critics of Castaneda's work have questioned whether or not the shaman don Juan existed.

166.33   "the police of meaningless labor."] Thoreau, "Life Without Principle" (1863).

167.31–37   Thoreau in his *Journal* . . . music and poetry."] November 20, 1851.

168.11–12   the Twelve Southerners] The Twelve Southerners, also known as the Southern Agrarians, were a circle of Southern writers and academics formed in the 1920s and united in the cause of preserving an agrarian way of life against the forces of industrialism. Among the group's most prominent members were John Crowe Ransom, Robert Penn Warren, Donald Davidson, and Allen Tate. Each of the Twelve Southerners contributed an essay to *I'll Take My Stand: The South and Agrarian Tradition* (1930), sometimes described as an "agrarian manifesto."

172.23   Sir Albert Howard] See note 152.27–28.

172.27–31   "I think I have told you . . . to see."] John G. Neihardt, *Black Elk Speaks* (1932), Chapter 18. See also note 144.27–30.

172.32–33   "No ideas but in things."] William Carlos Williams (1883–1963), *Paterson*, Book I (1946).

174.23–25   "by the alacrity . . . its way."] Thoreau, "Civil Disobedience" (1849), published initially under the title "Resistance to Civil Government" in Elizabeth Peabody's *Aesthetic Papers*.

176.2–11   The *I Ching* says . . . the good."] See hexagram 43 of the *I Ching*, one of the *Wujing* (Five Classics) of Confucianism. Berry quotes from the English translation by Cary F. Baynes (1950), based on the German translation by Richard Wilhelm.

176.11–12   "if not obtained . . . do not last."] The Confucian *Analects*, 4:5, translation by Ezra Pound (1950), in his anthology *Confucius*.

176.13–15   "Ye have heard . . . not evil . . ."] Matthew 5:38–39.

176.16–17   "As soon as . . . based on."] Ken Kesey in conversation with Berry.

176.17–22   In 1931, Judge Lusk . . . non-existent."] After a jury had found three Communist Party members in Chattanooga guilty of violating the Sedition Statute in March 1931, Lusk granted the defendants a new trial.

176.23   the University of Emily's Run] Emily's Run, a creek in Henry County, Kentucky.

178.19–20   J. Russell Smith's . . . *Tree Crops*] *Tree Crops: A Permanent Agriculture* (1929) describes the traditional use of tree crops and how trees such as the carob, honey locust, mulberry, chestnut, and black walnut can be an important source for food, soil conservation, and sustainable agriculture.

181.16–17   "Take therefore . . . of itself."] Matthew 6:34.

182.4–16   "Everything the Power . . . our children."] *Black Elk Speaks*, Chapter 17.

182.23–25   "the sacred hoop . . . as starlight. . . ."] *Black Elk Speaks*, Chapter 3.

183.11–14   "The Six Grandfathers . . . make happy."] *Black Elk Speaks*, Chapter 16.

183.35–37   "medicine seems . . . more disease."] Leon R. Kass, *Toward a More Natural Science: Biology and Human Affairs* (1985), Chapter 1.

184.3–6   "Our father . . . for them."] Paul Radin, *The Road of Life and Death: A Ritual of the American Indian* (1945), Prologue.

186.5–10   Justus von Liebig wrote . . . of Africa."] See Letter XII in *Letters on Modern Agriculture* (1859) by the German agricultural chemist Justus Freiherr von Liebig (1803–1873).

187.6–7   "Art," A. R. Ammons says . . . unconscious event. . . ."] In his long poem "Essay on Poetics" (1970), reprinted in his *Collected Poems: 1951–1971* (1972).

187.18–20 "We have constructed . . . returning curve. . . ."] Thoreau, *Walden* (1854), Chapter 4.

192.3–18 ". . . though I recognize . . . American citizen."] Robert E. Lee, letter to Anne R. Marshall, April 20, 1861.

193.3–4 ". . . all things whatsoever . . . to them. . . ."] Matthew 7:12.

193.12–16 ". . . whosoever shall smite . . . persecute you. . . ."] Matthew 5:39–44.

193.19–20 "An eye . . . a tooth"] Matthew 5:38.

194.28–30 "a trumpet . . . the pitchers."] Judges 7:16.

201.17–18 "neither the day nor the hour. . . ."] Matthew 25:13.

201.25–27 "archer, when he misses . . . in himself."] *The Unwobbling Pivot*, one of the Four Books (*Sìshū*) of Confucianism. Berry quotes from Ezra Pound's translation (1947) in his anthology *Confucius*.

202.3–5 Thomas Merton . . . work well."] Merton, American Trappist monk of the Abbey of Our Lady of Gethsemani, in Nelson County, Kentucky (1915–1968), whose numerous books include *The Seven Storey Mountain* (1948). He had a deep and abiding interest in Shakers and their work. Not far from Gethsemani is the Pleasant Hill Shaker Village in Harrodsburg, Kentucky, which Merton visited at least twice. Merton's comment about Shakers and craftsmanship was made in conversation with Berry.

205.26 "Read not the Times. Read the Eternities,"] H. D. Thoreau, "Life without Principle" (1863).

205.27 "literature is news that STAYS news."] Ezra Pound, *ABC of Reading* (1939), Chapter 2.

205.31–37 Men are . . . of the heart . . .] Edwin Muir, "The Island," first collected in *One Foot in Eden* (1956) and reprinted in *Collected Poems* (1965).

From RECOLLECTED ESSAYS

212.12 a Gravely "walking tractor"] A multipurpose two-wheeled walking tractor, also known as a "walk behind," manufactured by Gravely Tractor, Inc.

212.12–13 an old Farmall A] A compact row-crop tractor with a four-cylinder engine, manufactured by International Harvester from 1939 to 1947. In 1947 Harvester introduced the Farmall Super A, manufactured until 1954.

THE UNSETTLING OF AMERICA

220.1 *For Maurice Telleen*] American farmer, writer, founder of *The Draft Horse Journal*, and one of Berry's close personal friends (1928–2011). See Berry's tribute to Telleen in *It All Turns on Affection* (2012).

223.13   former Secretary of Agriculture Earl L. Butz] Butz (1909–2008) was Secretary of Agriculture from 1971 to 1976 under Richard Nixon and Gerald Ford.

223.29–30   an article in the *Louisville Courier-Journal*] "Surplus of People Called Biggest Farm Problem," the Louisville *Courier-Journal* (July 24, 1967).

229.1–2   Who so hath . . . hath taken.] Montaigne, "Of Coaches," *The Essayes of Michael Lord of Montaigne*, John Florio translation (1603).

230.1–7   *So many goodly citties . . . base conquest.*] Montaigne, "Of Coaches," *The Essayes of Michael Lord of Montaigne*, John Florio translation (1603).

232.13   El Dorado] Spanish: literally "the golden one," referring initially to a (mythical) king of the ancient indigenous Muisca culture of Colombia, and later synonymous with a lost kingdom of gold. The legend of El Dorado led many Spanish conquistadors (and other Europeans) to South America in search of riches.

233.25–40   "affected every aspect . . . the red man."] Bernard DeVoto, *The Course of Empire* (1952).

234.24–29   "More than four-fifths . . . adverse interest."] DeVoto, *The Course of Empire*.

234.31–33   "The interest . . . northern Mexico . . ."] DeVoto, *The Course of Empire*.

235.9   Tenochtitlán] Capital city of the Aztec Empire, located on an island on Lake Texcoco (present-day Mexico City).

236.39–40   our recent secretary of agriculture remarked . . . "Food is a weapon."] Earl L. Butz, see note 221.13. Butz's comment appeared in "What to Do: Costly Choices," *Time* magazine (November 11, 1974): "Food is a weapon. It is now one of the principal tools in our negotiating kit." See also Butz's speech to the Advertising Council, Washington, D.C., June 24, 1974: ". . . food was . . . a major weapon in achieving an honorable peace in Vietnam."

237.1–2   a secretary of defense . . . "palatable" levels of devastation.] James R. Schlesinger (1929–2014), Secretary of Defense from 1973 to 1975 under Richard Nixon and Gerald Ford, advocated making plans for limited nuclear war a part of U.S. military strategy.

237.24–32   Men are made . . . the rose.] Edwin Muir, "The Island," collected in *One Foot in Eden* (1956).

239.2   "Snirt"] Dirty snow.

239.16–17   "a farmer . . . about $320,000"] Roy Reed, "'Paper Rich'—The Farmer's Plaint," the Louisville *Courier-Journal* (February 18, 1976).

240.5–6   "If a man . . . about him. . . ."] Ezra Pound, Canto XIII, *The Cantos*.

242.11 Homestead Act of 1862] Legislative action signed into law by Lincoln that granted land to private citizens (up to 160 acres), encouraging settlement of the West.

243.5–12 Only a man . . . Dynasties pass.] Thomas Hardy, "In a Time of 'The Breaking of Nations,'" first published in the *Saturday Review* (January 1916).

244.1–3 . . . *wanting good government . . . disciplined themselves . . .*] See Ezra Pound's translation of *The Unwobbling Pivot* (1947) in his anthology *Confucius*.

245.3–4 William Rood in the *Los Angeles Times*] William Rood, "Environment Groups Invest in Polluters," the *Los Angeles Times* (July 29, 1975).

245.8 The Sierra Club] National conservation organization founded in 1872 by Scottish American naturalist John Muir (1838–1914).

247.3–4 methods of transportation . . . bridge."] Ivan Illich, *Tools for Conviviality* (1973). Illich (1926–2002) was an Austrian philosopher and Roman Catholic priest who was a fierce critic of modern institutions, including public schooling and medicine.

257.1–2 David Budbill] American poet and playwright (1940–2016).

259.22–25 Sir Albert Howard suggests . . . balance] See Howard's Introduction to *An Agricultural Testament* (1940).

261.34–36 "feeds himself . . . food production"] Richard E. Bell (1934–2015), Assistant Secretary of Agriculture, International Affairs and Commodity Programs from 1975 to 1977, "Meeting World Food Needs," speech to the Fertilizer Institute Annual Meeting, Chicago, February 3, 1976.

261.39–40 a recent assistant secretary of agriculture] Richard E. Bell.

262.2–5 former Secretary of Agriculture Butz could say . . . farm goods,"] In "Agriculture—200 Years After," a speech to the National Council of Farmer Cooperatives Annual Meeting, Washington, D.C., January 15, 1976. See also note 223.13.

263.20–21 one of these speeches . . . by former Assistant Secretary Richard E. Bell] Bell's comments were made at a speech before the Fall Convention of the National Farm Broadcasters Association, Kansas City, Kansas, November 15, 1975.

270.21–22 what Maurice Telleen calls . . . hedonism."] In a conversation with Berry.

271.39–272.3 Dr. John Nicolai . . . needed to be eliminated.] James R. Russell, "Dairy farming in Kentucky registers sharp decline," the Louisville *Courier-Journal* (April 22, 1974).

276.22–24 "that dreary precept . . . after truth."] Howard, *The Soil and Health* (1947), Chapter 6.

276.25–29   "The natural universe . . . group of experts."] *The Soil and Health*, Chapter 6.

276.30–31   "I could not . . . other people."] *The Soil and Health*, Introduction.

276.32–33   "wide chasm . . . in the field."] *The Soil and Health*, Introduction.

276.37–38   "for treating . . . one great subject."] *The Soil and Health*, Introduction.

277.5–6   "Death supersedes . . . dead and decayed."] *The Soil and Health*, Chapter 2.

283.39–284.15   Carl Sauer wrote . . . European nations."] Carl Sauer, *Northern Mists* (1968). An influential American geographer, Sauer (1889–1975) explored in his many writings the relationship between human cultures and physical environments. *Northern Mists* examines the challenges faced by pre-Columbian explorers in North America.

291.19–20   former Agriculture Secretary Clifford Hardin] Hardin (1915–2010) was Secretary of Agriculture from 1969 to 1971 under Richard Nixon.

296.5   Freedom City] Experimental 400-acre commune near Greenville, Mississippi, intended to aid African Americans who had lost their farm jobs as a result of mechanization and the introduction of chemical herbicides. Freedom City foundered in the mid-1970s.

309.4   F. M. Esfandiary] Fereidoun M. Esfandiary (1930–2000), American futurist, transhumanist, and author of several books including *Are You Transhuman?: Monitoring and Stimulating Your Personal Rate of Growth in a Rapidly Changing World* (1989), who would change his name to FM-2030, signifying in part his belief that by 2030 science and technology would overcome death. Esfandiary was vitrified after his death.

309.6–7   an article entitled "Homo Sapiens, the Manna Maker,"] F. M. Esfandiary, "Homo Sapiens, the Manna Maker," *The New York Times* (August 9, 1975).

311.13–14   "All these things . . . worship me."] Matthew 4:9.

312.3   "Energy . . . Eternal Delight."] William Blake, *The Marriage of Heaven and Hell* (1790–93), plate 4.

312.9–10   "Energy is the only life . . ."] *The Marriage of Heaven and Hell*, plate 4.

312.28–29   F. H. King's *Farmers of Forty Centuries*] See note 143.5–9.

312.31–32   Sir Albert Howard . . . the "Wheel of Life."] See note 152.27–28. In *The Soil and Health* (1947), Howard writes: "[Stability in nature] dominates by means of an ever-recurring cycle, a cycle which, repeating itself silently and ceaselessly, ensures the continuation of living matter. This cycle is constituted of the successive and repeated processes of birth, growth, maturity, death, and decay. An eastern religion calls this cycle the Wheel of Life and no better name could be given to it. The revolutions of this Wheel never falter and are perfect. Death supersedes life and life rises again from what is dead and decayed."

320.10–17 According to Lauren Soth . . . Indiana."] Lauren Soth, "Politics and Agribuz: The Operations of Dr. Butz," the *Nation* (October 26, 1974).

323.38 The digging stick] An agricultural tool for digging, a sharpened stick.

327.3–5 Can we . . . from its use"?] Ivan Illich, *Energy and Equity* (1974). See also note 247.3–4.

328.1–3 *But just stop . . . own food.*] James E. Bostic, Jr., "Rural America: Where the Action Is," speech to the Farmers Home Administration State Conference, Raleigh, North Carolina, April 16, 1976.

328.6–7 *Find the shortest . . . the mouth.*] Richard Deats, "A Conversation with Lanza del Vasto," *Fellowship* (September 1975).

330.11–18 The crows . . . for sight.] *King Lear*, IV.vi.13–20.

330.24–25 a Bedlamite] An insane person, literally a resident or ex-patient of Bedlam, as London's Bethlem Royal Hospital came to be known.

331.8–10 You ever-gentle gods . . . you please.] *King Lear*, IV.vi.217–19.

331.15–16 "'Twixt two extremes . . . and grief . . ."] *King Lear*, V.iii.199.

332.19 Lookout Mountain] Located at the border between Tennessee and Georgia, a popular tourist destination. From the rock formations known as "Rock City," it is possible, the tourism industry claims, to see seven states: Tennessee, Kentucky, Virginia, South Carolina, North Carolina, Georgia, and Alabama.

333.22–25 Out of . . . speechless caravan.] Hart Crane, "Proem: To Brooklyn Bridge," lines 17–18, *The Bridge* (1930).

335.22–23 "Man has no Body distinct from his Soul . . ."] Blake, *The Marriage of Heaven and Hell* (1790–93), plate 4.

342.20–30 I went by the field . . . armed man.] Proverbs 24:30–34.

356.26–30 Maybe the bride-bed . . . a single light . . .] W. B. Yeats, "Solomon and the Witch," collected in *Michael Robartes and the Dancer* (1921).

358.36–37 "revel and rest softly, side by side,"] *The Odyssey*, Book V, line 236, translation by Robert Fitzgerald (1961). All quotations of Homer are from the Fitzgerald translation, and citations are keyed to the 1998 edition.

359.4–11 If you could . . . grace and form?] *The Odyssey*, Book V, lines 215–222.

359.13–17 My quiet Penélopê . . . for home . . .] *The Odyssey*, Book V, lines 225–229.

361.20–24   grew like a pillar . . . a bedpost . . .] *The Odyssey*, Book XXIII, lines 217–223.

361.27–32   Now from his breast . . . went down . . .] *The Odyssey*, Book XXIII, lines 259–264.

362.23–24   "The government of the state . . . family order."] See Ezra Pound's translation of *The Great Digest* (1928) in his anthology *Confucius*.

363.6–10   Odysseus found . . . the brambles . . .] *The Odyssey*, Book XXIV, lines 251–255.

363.31–32   "beat their swords . . . learn war any more,"] Isaiah 2:4.

363.38   "in blessed peace."] *The Odyssey*, Book XI, line 151.

365.17–19   To learn to preserve . . . study the forest.] See the Introduction to Albert Howard's *An Agricultural Testament*: "The main characteristic of Nature's farming can therefore be summed up in a few words. Mother earth never attempts to farm without live stock; she always raises mixed crops; great pains are taken to preserve the soil and to prevent erosion; the mixed vegetable and animal wastes are converted into humus; there is no waste; the processes of growth and the processes of decay balance one another; ample provision is made to maintain large reserves of fertility; the greatest care is taken to store the rainfall; both plants and animals are left to protect themselves against disease."

367.8–10   "so that perhaps acres . . . that danger. . . ."] See "The Danger," by American poet and novelist Millen Brand (1906–1980), collected in *Local Lives* (1975), a commonplace history of the Pennsylvania Dutch.

367.37–368.1   a recent *National Geographic* article] Sabrina Michaud and Roland Michaud, "Trek to Lofty Hunza—and Beyond," *National Geographic* (November 1975).

375.20–27   ". . . the ploughman . . . oxen's work . . ."] See *The Horse in the Furrow* (1960), Chapter 2, by George Ewart Evans (1909–1988), an Anglo Welsh oral historian.

377.7–10   "that the mass . . . grace of God."] Thomas Jefferson, letter to Roger C. Weightman, June 24, 1826.

377.17–21   "Cultivators of the earth . . . lasting bonds."] Thomas Jefferson, letter to John Jay, August 23, 1785.

377.25–27   "I consider . . . generally overturned."] Thomas Jefferson, letter to John Jay, August 23, 1785.

378.2–7   ". . . I do most anxiously wish . . . distrustful superintendence."] Thomas Jefferson, letter to Mann Page, August 30, 1795.

378.10–16   ". . . the long succession . . . the Atlantic . . ."] Thomas Jefferson, letter to James Madison, February 17, 1826.

453.27–28 a letter to the editor of the *Draft Horse Journal*] *Draft Horse Journal* (Autumn 1976).

457.5–6 the New Alchemy Institute, the Farallones Institute, Rodale Press] The New Alchemy Institute (1969–91), located on a former dairy farm on Cape Cod, was a research center focused on decreasing human dependence on fossil fuels through ecological approaches such as organic agriculture, aquaculture, and bioshelter; the Farallones Institute, in Berkeley, California, founded in 1972, was a community of Northern California scientists, designers, and horticulturalists dedicated to the development of ecologically integrated living design; Rodale Press, in Emmaus, Pennsylvania, founded in 1930, was a family-owned independent publisher of health and wellness books and magazines, until its sale in 2018 to Hearst Communications.

463.15–18 "about 1.5 million chickens . . . 5 million eggs."] "Michigan Episode," *The Shepherd* (July 1976).

463.21–22 "may have been . . . the wrong lever."] "Michigan Episode," *The Shepherd*.

468.38 Marty Strange, Gene Logsdon, and Wes Jackson] Marty Strange (b. 1947), environmentalist, author of *Family Farming: A New Economic Vision* (1988), and cofounder of the Center for Rural Affairs in Nebraska, championing the sustainable agriculture movement; Gene Logsdon (1931–2016), American farmer and author of numerous farm-related books including *The Contrary Farmer* (1994) and *Letter to a Young Farmer* (2017), with a foreword by Wendell Berry; Wes Jackson (b. 1936), American plant geneticist and advocate for sustainable practices, author of several books including *New Roots for Agriculture* (1980) and *Becoming Native to This Place* (1994), and cofounder of The Land Institute in Salina, Kansas, dedicated to the development of perennial crops.

468.39–469.1 the Land Institute, the Center for Rural Affairs, the Land Stewardship Project, Tilth, and the E. F. Schumacher Society] The Land Institute, see note 468.38; the Center for Rural Affairs, see note 468.38; the Land Stewardship Project, a Minnesota-based organization dedicated to promoting an ethic of farmland stewardship; Tilth, the Tilth Alliance, a sustainable agricultural movement and organization in the Pacific Northwest inspired by a speech made by Berry at the "Agriculture for a Small Planet" symposium in Spokane, Washington, July 1, 1974; E. F. Schumacher Society, now the Schumacher Center for a New Economics, an organization based in Great Barrington, Massachusetts, working to build a just and sustainable global economy.

From THE GIFT OF GOOD LAND

473.6 a Gravely walking tractor] See note 212.12.

473.28 breaking plow . . . cultivator . . . grain drill.] A ground-breaking plow with a moldboard; cultivator, a secondary tillage plow often used for weeding; grain drill, an implement for planting small grains.

378.17 JUSTIN MORRILL] Justin Smith Morrill (1810–1898), a representative and senator from Vermont, best known for his sponsorship of the Land-Grant College Act of 1862 and the Agricultural College Act of 1890.

378.20–23 "an amount . . . in Congress . . ."] Morrill Act, U.S.C., Title 7, Section 301.

378.24–29 "to the endowment . . . professions in life."] Morrill Act, U.S.C., Title 7, Section 304.

378.32–34 "a sound . . . prosperity and security."] Hatch Act, U.S.C., Title 7, Section 361b.

378.35–38 "It is also . . . the economy."] Hatch Act, U.S.C., Title 7, Section 361b.

379.2–9 "It shall be . . . rural life . . ."] Hatch Act, U.S.C., Title 7, Section 361b.

379.11–14 "In order to aid . . . the same . . ."] Smith-Lever Act, U.S.C., Title 7, Section 341.

379.19–20 Morrill's statement . . . written "apparently in 1874."] William Belmont Parker, *The Life and Public Services of Justin Smith Morrill* (1924). Morrill's autobiographical memorandum, found among his papers after his death, enumerates five reasons for writing the Land-Grant College Act.

380.19–20 "natural aristocracy" of "virtue and talents"] Thomas Jefferson, letter to John Adams, October 28, 1813: "For I agree with you that there is a natural aristocracy among men. The grounds of this are virtue and talents."

380.34 "degrees of genius."] Thomas Jefferson, letter to John Page, August 30, 1795: "I do most anxiously wish to see the highest degrees of education given to the higher degrees of genius, and to all degrees of it, so much as may enable them to read & understand what is going on in the world, and to keep their part of it going on right: for nothing can keep it right but their own vigilant and distrustful superintendence."

383.34–35 Jim Hightower and Susan DeMarco give . . . the central argument] Jim Hightower and Susan DeMarco of the Agribusiness Accountability Project, statements before the Senate Subcommittee on Migratory Labor, June 19, 1972. Berry's source is Jim Hightower, *Hard Tomatoes, Hard Times: The Failure of the Land Grant College Complex* (1972).

384.40–385.1 An article in the *Louisville Courier-Journal*] Phil Norman, "UK Agriculture College touches many lives," the Louisville *Courier-Journal* (October 10, 1976).

385.40–386.3 "to assume . . . better citizenship."] House Committee on Agriculture, U.S. Congress, "Cooperative Agricultural Extension Work," *Report No. 110*, 63rd Congress, 2nd Session (1913).

386.22–23 "corporate control . . . agriculture production"] Sarah Shaver Hughes, "Agricultural Surpluses and American Foreign Policy 1952–60," an unpublished master's thesis, University of Wisconsin (1964). Quoted in Darryl McLeod, "Urban-Rural Food Alliances: A Perspective on Recent Community Food Organizing," *Radical Agriculture*, edited by Richard Merrill (1976).

386.24–25 then Secretary of Agriculture Ezra Taft Benson] Ezra Taft Benson (1899–1994) served as Secretary of Agriculture from 1953 to 1961, under Dwight D. Eisenhower.

391.27 Gresham's Law] Economic principle that "bad money drives out good."

393.25 a textbook of rhetoric] W. Ross Winterowd, *The Contemporary Writer* (1975).

396.32 F. H. King] See note 143.5–9.

401.27–30 "mechanized agriculture . . . systems are."] Sterling Wortman, "Food and Agriculture," *Scientific American* (September 1976).

407.6–7 Professor Philip M. Raup . . . testified] Philip M. Raup, "Needed Research into the Effects of Large Scale Farm and Business Firms on Rural America," testimony before the Subcommittee on Monopoly of the United States Senate Small Business Committee, March 1, 1972.

411.28–29 an unpublished paper by Professor Stephen B. Brush] "Andean Culture and Agriculture: Perspectives on Development," presented at the International Hill Land Symposium, West Virginia University, Morgantown, West Virginia, October 5, 1976.

416.15–16 "the regional specialization . . . individual specialization."] Maurice Telleen, personal correspondence with Berry.

424.19 ladino] A variety of white clover.

424.20 Korean lespedeza] A very good annual clover for pasture and hay.

424.24 harrow] An agricultural instrument consisting of spikes, teeth, or discs, used specifically for the purpose of breaking up and smoothing the surface of the soil.

425.29–426.2 "In Kentucky . . . horses or mules."] Thomas P. Cooper, comments in University of Kentucky College of Agriculture Circular No. 306 (November 1937).

427.4 Percheron horses] A breed of tall heavy draft horses, usually gray or black.

430.27–28 "Except a corn . . . much fruit."] John 12:24.

432.2 chisel plow] A plow that allows reduced tillage, loosening but not inverting soil.

432.37 Charolais cows] A breed of large white beef cattle, developed in France.

434.16 bu. per acre] Bushels per acre.

435.29–32 a report published in 1975 and entitled *A Comparison . . . Pesticides.*] William Lockeretz, Robert Klepper, Barry Commoner, Michael Gertler, Sarah Fast, Daniel O'Leary, and Roger Blobaum, "A Comparison of the Production, Economic Returns, and Energy Intensiveness of Corn Belt Farms That Do and Do Not Use Inorganic Fertilizers and Pesticides," Center for the Biology of Natural Systems at Washington University (1975).

435.32–33 Barry Commoner] American cellular biologist and pioneer in the modern environmental movement (1917–2012). Berry quotes from Chapter 7 of Commoner's *The Poverty of Power: Energy and the Economic Crisis* (1976).

437.25 BTU] British thermal unit, the amount of heat necessary to raise the temperature of one pound of water by one degree Fahrenheit.

438.5 vetch] Any of several legumes of the genus *Vicia*, grown for fodder or green manure.

439.34 in a speech by . . . Butz himself.] Earl L. Butz, speech to the American Chemical Society Meeting, New York, April 8, 1976. Subsequent quotations of Butz refer to this speech.

440.6–7 As the *Draft Horse Journal* noted in an editorial] *Draft Horse Journal* (Winter 1976).

441.9–10 Morrison's *Feeds and Feeding*] First published in 1898, a guide to the feeding, care, and management of livestock that underwent numerous revisions and editions through the mid-twentieth century.

446.5 H Farmall tractors] H Farmall tractor, a row-crop tractor manufactured by International Harvester from 1939 to 1953.

446.27–28 a 1946 WC Allis-Chalmers] A nimble row-crop tractor manufactured by Allis-Chalmers from 1948 to 1953.

446.30–31 Percheron stud] See note 427.4.

448.7 as Maurice Telleen points out] In a conversation with Berry. See note 220.1.

450.29 "among God . . . community."] "Agricultural Alternatives," *CBNS Notes*, Center for the Biology of Natural Systems, Washington University (March–April 1972).

451.29–33 "technological innovation . . . as a whole."] "Agricultural Alternatives," *CBNS Notes*.

452.12–13 "take their own . . . other people."] Albert Howard, *The Soil and Health: A Study of Organic Agriculture* (1947), Introduction.

473.36 hay conditioners and chisel plows] Hay conditioner, tool that crimps newly cut hay to promote faster drying; chisel plow, see note 432.2.

475.17–18 Harry Groom, as quoted in . . . George Ewart Evans's *The Horse in the Furrow*] *The Horse in the Furrow* (1960), Chapter 3. See also note 375.20–27.

479.28 Earl Butz] See note 223.13.

483.15 John Todd] A Canadian ecological designer (b. 1939) whose design innovations address challenges of food production, waste treatment, and environmental repair through ecosystem technologies that emulate nature's patterns and strategies. Todd was one of the founders of Cape Cod's New Alchemy Institute (1969–91), a research center dedicated to ecological design solutions.

483.20 *The New Farm*] A magazine devoted to organic farming, published by the Rodale Press from 1979 to 1994. Berry served as a contributing editor to Rodale in the late 1970s.

486.40–487.1 an insight from . . . Timothy Taylor] Timothy H. Taylor (1918–2010), a faculty member in the College of Agriculture at University of Kentucky whose expertise was forage production and grassland ecology, and a friend of Wendell Berry.

488.7–8 a good farm . . . "manures itself."] Albert Howard, *An Agricultural Testament* (1940), Introduction.

495.24–25 Thoreau . . . the first to assert that people should not belong to farm animals] In the first chapter of *Walden* (1854), Thoreau writes: "I see young men, my townsmen, whose misfortune it is to have inherited farms, houses, barns, cattle, and farming tools; for these are more easily acquired than got rid of. Better if they had been born in the open pasture and sucked by a wolf, that they might have seen with clearer eyes field they were called to labor in. Who made them serfs of the soil? Why should they eat their sixty acres, when man is condemned to eat only his peck of dirt? Why should they begin digging their graves as soon as they are born?"

498.2 the Southdown ram] Southdown sheep, a hornless breed of sheep prized for its fleece and meat and used for the improvement of other breeds.

498.3 Henry Besuden] Henry Carlisle Besuden (1904–1985), sheep breeder, conservationist, and the writer of the column "Sheep Sense" for *The Sheepman* magazine, 1945–47.

498.35–36 *The Farm Quarterly*] Published from 1946 to 1972, a large-format popular farm magazine that included photographs and human-interest stories.

499.37 Kentucky fescue 31] A low-maintenance fescue that tolerates heat and drought.

499.39 Korean lespedeza] See note 424.20.

507.36 Belgians or Percherons] Belgian, a breed of heavy draft horse; Percheron, see note 427.4.

508.37–38 all nubbins] Nubbin, a small and stunted ear of corn.

509.22 silage] Fermented fodder, chopped and stored, for cattle, sheep, and other cud-chewing farm animals.

510.33 Model D John Deere tractor] The first tractor built by John Deere, manufactured from 1923 until 1953.

512.9 a farrier] A craftsman who specializes in shoeing horses.

513.27 John A. Hostetler] American writer and scholar of Amish and Hutterite societies (1918–2001). His *Amish Society* (1963) saw an expanded fourth edition in 1993.

519.19–21 what Arthur O. Lovejoy . . . Biblical thought.] Arthur Oncken Lovejoy (1873–1962), American philosopher and historian of ideas. In *The Great Chain of Being: A Study of the History of an Idea* (1936, 1963), Lovejoy writes: "through the Middle Ages there were at least kept alive, in an age of which the official doctrine was predominantly otherworldly, certain roots of an essentially 'this-worldly' philosophy: the assumption that there is a true and intrinsic multiplicity in the divine nature, that is to say, in the world of Ideas . . . that the world of temporal and sensible experience is thus good, and the supreme manifestation of the divine."

519.29–30 Lynn White, Jr. . . . "The Historical Roots of Our Ecologic Crisis"] Delivered as an address at the annual meeting of the American Association for the Advancement of Science in Washington, D.C., in December 1966, prior to its publication in *Science* (March 10, 1967).

519.34 Genesis 1:28] "And God blessed them, and God said unto them, Be fruitful, and multiply, and replenish the earth, and subdue it: and have dominion over the fish of the sea, and over the fowl of the air, and over every living thing that moveth upon the earth."

522.17 "year of jubilee,"] Leviticus 25:10.

522.22–23 "my power . . . mine hand"] Deuteronomy 8:17.

524.25 "the fowls of the air" and "the lilies of the field."] Matthew 6:26, 6:28.

526.25–31 "otherworldly philosophy" . . . Nature."] Lovejoy, *The Great Chain of Being*, Chapter II.

528.20–21 As C. S. Lewis . . . material things] In *Mere Christianity* (1952), Book II, Chapter 5, Lewis writes: "There is no good trying to be more spiritual than God. God never meant man to be a purely spiritual creature. That is why He uses material things like bread and wine to put the new life into us. We may think this rather crude and unspiritual. God does not: He invented eating. He likes matter. He invented it."

From STANDING BY WORDS

538.26–27 Against Mr. Winterowd's . . . a definition by Gary Snyder] See Gary Snyder's "Poetry, Community, & Climax," *Field 20* (Spring 1979). Snyder (b. 1930), American poet of the Pacific Rim, essayist, and environmental activist, whose many works include the 1974 Pulitzer Prize–winning poetry collection *Turtle Island*. Berry and Snyder have been friends and correspondents since the 1970s. *Distant Neighbors* (2015), edited by Chad Wriglesworth, is a selection of their correspondence.

540.34–37 For if any . . . he was.] James 1:23–24.

541.3 "Honesty is the treasure of states."] See Ezra Pound's translation of *The Great Digest* (1928) in his anthology *Confucius*.

541.4–7 Pound's observation . . . exaggerations of dogma."] See Pound's translator's note ("Procedure") in his translation of the Confucian *Analects* (1950) in his anthology *Confucius*.

541.8–10 "Where shall . . . a standard man?"] Thoreau, *A Week on the Concord and Merrimack Rivers* (1849), the "Sunday" chapter.

541.26–27 "possibilities . . . inward and outward."] Gary Snyder, "Poetry, Community, & Climax," *Field 20* (Spring 1979).

541.33–34 "I talk . . . no language."] T. S. Eliot, *The Family Reunion*, Part I, Scene I, a verse play, first performed in 1939.

542.25–26 "create the object . . . they contemplate."] T. S. Eliot, "John Dryden" (1921), reprinted in *Selected Essays of T. S. Eliot* (1950).

542.30–33 Shelley's first wife . . . a poet.] *The Norton Anthology of English Literature* (1962), Volume 2, edited by M. H. Abrams et al.

544.24–25 "Men must endure . . . coming hither."] *King Lear*, V.ii.9–10.

544.36 Out of the world he must, who once comes in . . .] Robert Herrick, "None Free from Fault," collected in *Hesperides* (1648).

545.3–5 Old as I am . . . my Wit.] The opening lines of John Dryden's "Cymon and Iphigenia, from Boccace," collected in *Fables Ancient and Modern* (1700).

546.4–7 To be free . . . magnanimous and brave.] John Milton, *The Second Defense of the People of England*, published originally in Latin in 1654, in which the English poet offers legal justification for Parliament's execution of King Charles. It follows and continues Milton's first defense, published three years earlier.

546.25–26 technological romanticism of Buckminster Fuller] Richard Buckminster Fuller (1895–1983), American inventor, futurist, and self-described "design scientist," believed that through technological innovation humans would one day transform the planet, feeding and housing themselves in workless leisure.

547.7–9 transcribed conversations . . . of the Nuclear Regulatory Commission . . . Three Mile Island] Berry, in an original note, acknowledges his indebtedness, for a sampling and commentary on the Nuclear Regulatory Commission transcripts, to Paul Trachtman, "Phenomena, comment, and notes," *Smithsonian* (July 1979). Through mechanical failures and human errors, the Three Mile Island nuclear power station, near Harrisburg, Pennsylvania, had a partial meltdown of its nuclear core on March 28, 1979, releasing radioactive gases into the atmosphere.

549.5–13 And they . . . another's speech.] Genesis 11:4–6.

552.2–4 In an article . . . transformation of American agriculture] G. W. Salisbury and R. G. Hart, "The Evolution and Future of American Animal Culture," *Perspective in Biology and Medicine* (Spring 1979).

559.34–35 in Gary Snyder's phrase, "at one with each other."] Gary Snyder, "Poetry, Community, & Climax," *Field 20* (Spring 1979).

561.6–24 Conceptualizing realistically . . . coconut milk.] See Buckminster Fuller's comments in the anthology *Space Colonies* (1977), edited by Stewart Brand, which gathers in book form the debate on space colonization that took place in the 1970s in the pages of *CoEvolution Quarterly*.

562.30–40 There wanted . . . God Supreme . . .] Milton, *Paradise Lost*, VII.505–515.

563.19–33 First the humans . . . the Moon.] Buckminster Fuller, comments in *Space Colonies*.

563.37–564.9 He created . . . for bread.] William Faulkner, "The Bear," *Go Down, Moses* (1942).

565.27–30 A mind . . . still the same . . .] Milton, *Paradise Lost*, I.253–256.

567.19–32 "academy of projectors . . . miserably waste. . . .] Jonathan Swift, *Gulliver's Travels* (1726), Part III, Chapter IV. Among the many impractical experiments conducted by Lagado's Academy of Projectors are a project to extract sunshine from cucumbers, an operation to reduce human waste to its original food, and another to turn ice into gunpowder.

568.1–4 But past a certain scale, as C. S. Lewis wrote . . . *all* others.] See Lewis's *Abolition of Man* (1943): "Man's conquest of Nature, if the dreams of some scientific planners are realized, means the rule of a few hundreds of men over billions upon billions of men. . . . Each new power won *by* man is a power *over* man as well. . . . For the power of Man to make himself what he pleases means, as we have seen, the power of some men to make other men what *they* please."

569.18–22 the Taoist village-as-globe . . . roosters crow.] *Tao Te Ching*, section 80.

569.35–40 "Alert as a winter-farer . . . a valley . . ."] *Tao Te Ching*, section 15, Witter Bynner translation (1944).

573.34–37 through infusion sweete . . . rather meete.] Edmund Spenser, *The Faerie Queen*, IV, canto ii, stanza 34.

575.2–4 consigning all . . . "to cold oblivion," as Shelley wrongly believed marriage was supposed to do.] Percy Bysshe Shelley, "Epipsychidion" (1821), lines 149–154.

577.9–13 I simply want . . . Claverton . . .] T. S. Eliot, *The Elder Statesman*, Act II, a verse play, first performed in 1958.

577.17–21 to go abroad . . . both ways.] Eliot, *The Elder Statesman*, Act II.

578.7–9 the Confucian principles, dear to Pound . . . stood by).] See Pound's glossary to his translation of *The Great Digest* (1928) in his anthology *Confucius*.

581.36–38 The negro . . . string-piece . . .] Walt Whitman, "Song of Myself," section 13, first published as one of twelve untitled poems in the 1855 edition of *Leaves of Grass*.

582.16–17 Seasons pursuing . . . the ground . . .] Whitman, "Song of Myself," section 15.

From HOME ECONOMICS

588.21–28 as the poet Edmund Spenser put it . . . brother unto brother."] *The Faerie Queen*, VII, canto vii, stanza 14.

591.26 "In wildness . . . the world,"] Thoreau, "Walking," delivered at the Concord Lyceum on April 23, 1851, and printed in the June 1862 issue of the *Atlantic Monthly* shortly after the author's death.

591.34–35 Wes Jackson of the Land Institute] See note 468.38.

596.14–16 "The proper scale," a friend wrote to me . . . life and health."] Perhaps a reference to a postcard from Jane Kenyon, according to Berry.

601.16–22 The ancient Greeks . . . inevitable punishment."] Aubrey de Sélincourt, *The World of Herodotus* (1982).

601.30–31 "Mine own hand hath saved me."] Judges 7:2.

602.32–36 "Religion in every age . . . his wits."] De Sélincourt, *The World of Herodotus*.

603.6–10 "Therefore take . . . unto you."] Matthew 6:31, 6:33.

604.2–3 "take no thought for the morrow."] Matthew 6:34.

604.28–30 the aim of "Buddhist economics . . . minimum of consumption,"] E. F. Schumacher, *Small Is Beautiful* (1973), Part I, Chapter 4.

608.26–609.4   Henry Besuden . . . depth and quality."] From Henry Besuden's speech to the International Stockmen's School, San Antonio, Texas,
January 2–6, 1983.

610.2–4   "my engineer's mind . . . practical and possible,"] Gordon Millar,
"Agriculture: Is Small Really Beautiful?," *World Research INK* (January 1978).

612.5–6   "All flesh . . . unto dust."] Job 34:15.

612.25–26   "Our bread," Guy Davenport . . . our movies."] Personal remark made by Davenport to Berry. Guy Mattison Davenport (1927–2005),
American fiction writer, essayist, translator, teacher, and longtime resident of
Lexington, Kentucky. Davenport began teaching at the University of Kentucky in 1964, the same year Berry joined the faculty as a creative writing
instructor.

613.35   "Satanic wheels"] William Blake, *Jerusalem* (1804–20), plate 12, line
44, and plate 13, line 37.

613.35–36   "Satanic mills"] Blake, *Milton* (1810), Preface.

613.36–37   "wheel without wheel . . . each other."] *Jerusalem*, plate 15, lines
18–19.

614.4–5   "Wheel within wheel . . . harmony and peace."] *Jerusalem*, plate
15, line 20.

614.6   "wheel in the middle of a wheel"] Ezekiel 1:16.

614.30   "only with spring," as e. e. cummings said] In his poem "O sweet
spontaneous," first published in *The Dial* in 1920 and collected in *Tulips and
Chimneys* (1923).

617.5–6   "He who would . . . Minute Particulars."] *Jerusalem*, plate 55, line
60.

617.9–11   Labour well . . . teeming Earth.] *Jerusalem*, plate 55, lines 51–53.

617.31–36   When people . . . as loyal.] *Tao Te Ching*, section 18, Witter Bynner translation (1944).

617.37–39   warning . . . sixth chapter of Matthew.] Matthew 6:1: "Take
heed that ye do not your alms before men, to be seen of them: otherwise ye
have no reward of your Father which is in heaven."

619.28–34   "Skill in making . . . concern of his."] Eric Gill, "What Is Art?,"
A Holy Tradition of Working: Passages from the Writings of Eric Gill (1983), edited by Brian Keeble. Gill (1882–1940), an English sculptor and type designer.

624.18–23   Dr. Johnson . . . all the Branches."] W. Jackson Bate, *Samuel
Johnson* (1975), Chapter 4.

624.30–35   "Modern knowledge . . . in heaven."] H. J. Massingham, *The
Tree of Life* (1943).

627.12–14   "the head . . . toward jobs."] "Texas School Board Chief Wants Sixth-Graders to Pick Job 'Tracks,'" Louisville *Courier-Journal* (August 13, 1983).

632.32–633.6   Judge Jackson Kiser . . . that way."] James J. Kilpatrick, "Plan to Teach the Bible as Literature May Wind up in the Supreme Court," Louisville *Courier-Journal* (September 15, 1983).

634.24–28   Coleridge's statement . . . poetic faith."] Samuel Taylor Coleridge, *Biographia Literaria* (1817), Chapter 14.

635.3–4   A spring of love . . . them unaware . . .] Coleridge, "The Rime of the Ancient Mariner," Part IV, stanza 14, published in *Lyrical Ballads* (1798).

635.10–11   "one need . . . an abyss."] Harry Mason, in a letter to Berry.

638.26–27   According to a recent press release] Larry Green, "Family Farm Threatened as Technology Raises Productivity: Science Reshaping Agriculture's Future," *Los Angeles Times* (July 14, 1985).

642.26–27   "To know . . . whom to trust."] The opening lines of Pound's Canto LXXXV, *The Cantos*.

642.30–36   the trees rise . . . are voices . . .] Pound, Canto XC, *The Cantos*.

645.37–38   pointing out, as the Reagan administration has done] When James Watt (b. 1938) was appointed Secretary of the Interior in 1981 by Ronald Reagan, he pledged: "We will mine more, drill more, cut more timber to use our resources rather than simply keep them locked up."

646.8–10   I would agree with Edward Abbey that we need . . . "absolute wilderness,"] Edward Abbey, *Desert Solitaire* (1968), chapter entitled "Polemic: Industrial Tourism and the National Parks."

652.2–3   Maury Telleen] See note 220.1.

656.20–21   10-10-20] A fertilizer grade. The numbers refer to the primary nutrients needed by plant life: nitrogen, phosphorus, and potash. 10-10-20 fertilizer contains 10 percent nitrogen, 10 percent phosphorus, and 20 percent potash.

659.18   alsike clover] Typically used for hay, pasture, and soil improvement. Alsike produces white or pale pink flowers.

From WHAT ARE PEOPLE FOR?

665.13–14   "You never . . . enough."] One of the "Proverbs of Hell" in Blake's *The Marriage of Heaven and Hell* (1790–93), plate 7.

665.25   "No bird . . . own wings."] *The Marriage of Heaven and Hell*, plate 7. Another of the "Proverbs of Hell."

667.1   *Wallace Stegner*] Stegner (1909–1993) was an American novelist, short story writer, historian, environmentalist, and Berry's friend as well as his teacher. He won the Pulitzer Prize for his novel *Angle of Repose* (1971) and the National Book Award for the novel that followed, *The Spectator Bird* (1977).

669.11–12   the Harvard Five-Foot Shelf.] In 1909 P. F. Collier published its highly successful fifty-volume library of the world's classics, edited by retired Harvard University president Charles W. Eliot, who promoted the idea that a five-foot shelf "would hold books enough to afford a good substitute for a liberal education to anyone who would read them with devotion, even if he could spare but fifteen minutes a day for reading."

676.5   Webster edition, with E. W. Kemble's illustrations.] Samuel Clemens, who wrote under the pen name Mark Twain, started and personally financed the publishing firm Charles L. Webster & Co. *Adventures of Huckleberry Finn* (1884), illustrated by E. W. Kemble, was one of the firm's first publications. Despite notable successes, the firm went bankrupt in 1894, after only ten years of operation.

677.35–39   When I wrote . . . my hands only.] The opening sentence of Thoreau's *Walden* (1854).

678.5–7   I loaf . . . summer grass.] Walt Whitman, "Song of Myself," section 1.

678.34–35   like Eliot's Chinese jar] In "Burnt Norton," the first of his *Four Quartets* (1943), Eliot asserts, "Only by the form, the pattern, / Can words or music reach / The stillness, as a Chinese jar still / Moves perpetually in its stillness."

681.37–38   The fulfillment . . . Aristotle described] In his *Poetics*, Aristotle says the purpose of dramatic tragedy is to bring about catharsis in theater-goers, to arouse sensations such as pity and fear, thereby cleansing them of these passions.

685.22–23   according . . . *Times Book Review*] Mark K. Stengel, "Modernism on the Mississippi: The Southern Review 1935–85," *The New York Times Book Review* (November 24, 1985).

686.22–23   As William Matthews writes in a recent article] William Matthews, "Dull Subjects," *New England Review and Bread Loaf Quarterly* (Winter 1985).

691.2–3   *Elegant solutions . . . uniqueness of place.*] John Todd, "Tomorrow Is Our Permanent Address," published in *The Book of the New Alchemists*, edited by Nancy Jack Todd (1977). See note 483.15.

698.5   "Comus,"] A poem by English poet John Milton, whose actual title is "A Mask Presented at Ludlow Castle," performed on Michaelmas, 1634, before John Egerton, Earl of Bridgewater, at Ludlow Castle and published anonymously in 1637.

698.9–11 "that nothing . . . to show."] Ananda K. Coomaraswamy, "Why Exhibit Works of Art?," collected in his book of essays *Christian and Oriental Philosophy of Art* (1956).

705.17–18 billionth hamburger] On November 20, 1984, in a highly publicized event, McDonald's served its 50 billionth hamburger.

707.3–4 "the whole problem . . . one great subject."] Albert Howard, *The Soil and Health: A Study of Organic Agriculture*, Introduction, originally published under the title *Farming and Gardening for Health or Disease* (1945) before its 1947 reissue. See note 152.27–28.

709.21–27 There is nothing . . . intact.] William Carlos Williams, "The Host," collected in *The Desert Music and Other Poems* (1954).

717.7–8 in the spirit of Psalm 128] Psalms 128:3–6: "Thy wife shall be as a fruitful vine by the sides of thine house: thy children like olive plants round about thy table. . . . and thou shalt see the good of Jerusalem all the days of thy life. Yea, thou shalt see thy children's children, and peace upon Israel."

717.9–12 A seaborne death . . . around you.] *The Odyssey*, Book XI, lines 148–151, translation by Robert Fitzgerald.

717.35 "bound / In surety for his brother's son."] Wordsworth, "Michael," lines 210–211, first published in the 1800 edition of Wordsworth and Coleridge's *Lyrical Ballads*.

717.36–37 "summoned to discharge the forfeiture."] Wordsworth, "Michael," line 215.

718.3–4 "possess it . . . over it."] Wordsworth, "Michael," lines 246–247.

718.5–6 "to seek . . . the seas."] Wordsworth, "Michael," line 449.

719.14 a recent magazine article] Marianne Merrill Moates, "Learning . . . Every Day," *Creative Ideas for Living* (July/August 1988).

721.15 "the anxiety of influence"] A term employed by the American critic Harold Bloom (b. 1930) to characterize the poetic tradition: more particularly the anxiety aroused by predecessors as an individual poet attempts to make his or her own lasting contribution. This idea was first outlined at length in Bloom's book *The Anxiety of Influence: A Theory of Poetry* (1973).

737.39–40 cease to be Blondie . . . becoming Dagwood?] Blondie (Boopadoop) Bumstead and her husband Dagwood Bumstead, the principal characters in the still-running syndicated comic strip *Blondie* created by American cartoonist Chic Young (1901–1973).

738.8–9 the Exxon fiasco in Prince William Sound] On March 24, 1989, the oil tanker *Exxon Valdez*, bound for California, ran aground on Alaska's Bligh Reef, spilling nearly 11 million gallons of crude oil into Prince William Sound.

748.33–36 Thoreau . . . until everybody does it?] An allusion to H. D. Thoreau's essay "Civil Disobedience" (1849), in which he writes: "The only obligation which I have a right to assume is to do at any time what I think right."

751.2–3 "I can live no longer by thinking."] *As You Like It*, V.ii.50.

752.15 Prince William Sound] See note 738.8–9.

754.11 the singular sparrows of the street] Cf. Matthew 10:29.

754.11 the lilies of the field] Matthew 6:28.

754.12 "the least of these my brethren."] Matthew 25:40.

754.21–30 "All left-wing parties . . . shall continue."] George Orwell, "Rudyard Kipling" (1942), reprinted in *Collected Essays* (1961).

755.32–34 oblivious . . . as were George III and Lord North.] King George III of Great Britain and Ireland (1738–1820) and his prime minister Lord North (1732–1792) lost the American colonies.

756.4–5 Love Canal, Bhopal, Chernobyl, and the *Exxon Valdez*] Love Canal, near Niagara Falls, New York, the site of a chemical waste dump in the 1940s and 1950s that forced the evacuation and resettlement of more than a thousand families in the Love Canal neighborhood, after President Jimmy Carter (twice) declared a state of emergency in the area, in 1978 and 1981; Bhopal, the capital city in Madhya Pradesh, India, which gained international attention in 1984 when a pesticide-manufacturing plant leaked toxic gas, killing thousands; Chernobyl, an uninhabitable city in northern Ukraine, synonymous with the nuclear disaster there in 1986, when the Chernobyl nuclear power station emitted large amounts of radiation into the atmosphere; *Exxon Valdez*, see note 738.8–9.

758.12 Burley tobacco] A light air-cured tobacco used in cigarettes, once the "king" of crops in Kentucky.

760.1–8 Liberty Hyde Bailey . . . F. H. King . . . J. Russell Smith . . . Sir Albert Howard . . . John Todd, Wes Jackson] Liberty Hyde Bailey (1858–1954), American botanist, horticulturalist, and cofounder of the American Society for Horticultural Science; F. H. King, see note 143.5–9; Joseph Russell Smith (1874–1966), American economic geographer, conservationist, and writer whose books include *Tree Crops: A Permanent Agriculture* (1929); Sir Albert Howard, see note 152.27–28; John Todd, see note 483.15; Wes Jackson (b. 1936), see note 468.38.

Index

*This book is set in 10 point ITC Galliard, a face
designed for digital composition by Matthew Carter and based
on the sixteenth-century face Granjon. The paper is acid-free
lightweight opaque that will not turn yellow or brittle with age.
The binding is sewn, which allows the book to open easily and lie flat.
The binding board is covered in Brillianta, a woven rayon cloth
made by Van Heek–Scholco Textielfabrieken, Holland.
Composition by Dedicated Book Services.
Printing and binding by LSC Communications.
Designed by Bruce Campbell.*

THE LIBRARY OF AMERICA SERIES

Library of America fosters appreciation of America's literary heritage by publishing, and keeping permanently in print, authoritative editions of America's best and most significant writing. An independent nonprofit organization, it was founded in 1979 with seed funding from the National Endowment for the Humanities and the Ford Foundation.

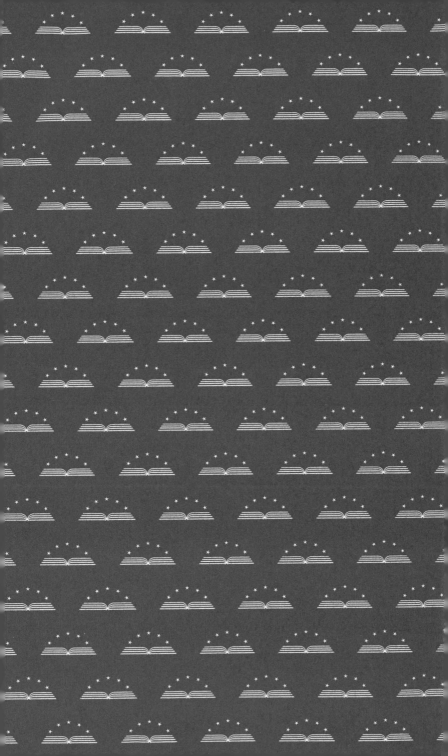